Top-Down Digital
VLSI Design

Top-Down Digital VLSI Design

Hubert Kaeslin
Microelectronics Design Center
Department of Information Technology and Electrical Engineering
ETH Zürich
Switzerland

AMSTERDAM • BOSTON • HEIDELBERG • LONDON
NEW YORK • OXFORD • PARIS • SAN DIEGO
SAN FRANCISCO • SINGAPORE • SYDNEY • TOKYO

Morgan Kaufmann is an imprint of Elsevier

Acquiring Editor: Stephen Merken
Editorial Project Manager: Nate McFadden
Project Manager: Poulouse Joseph
Cover Designer: Maria Inês Cruz

Morgan Kaufmann is an imprint of Elsevier
225 Wyman Street, Waltham, MA 02451, USA

Library of Congress Cataloging-in-Publication Data
Kaeslin, Hubert, author.
  Top-down digital VLSI design : from architectures to gate-level circuits and FPGAS / Hubert Kaeslin, Microelectronics Design Center, Dept. of Information Technology and Electrical Engineering, ETH Zurich, Switzerland.
    pages cm
  ISBN 978-0-12-800730-3
  1. Digital integrated circuits–Design and construction. 2. Integrated circuits–Very large scale integration–Design and construction. I. Title.
  TK7874.65.K336 2015
  621.39′5–dc23
                                                                                2014035133

British Library Cataloguing in Publication Data
A catalogue record for this book is available from the British Library

ISBN: 978-0-12-800730-3

For information on all MK publications
visit our website at www.mkp.com

Contents

Designing integrated electronics has become a multidisciplinary enterprise that involves solving problems from fields as disparate as

- Hardware architecture
- Software engineering
- Marketing and investment
- Semiconductor physics
- Systems engineering and verification
- Circuit design
- Discrete mathematics
- Layout design
- Electronic design automation
- Hardware test equipment and measurement techniques

Covering all these subjects is clearly beyond the scope of this text and also beyond the author's proficiency. Yet, I have made an attempt to collect material from the above fields that I have found to be relevant for making major design decisions and for carrying out the actual engineering work when developing Very Large Scale Integration (VLSI) circuits.

The present volume covers front-end design, that is all steps required to turn a software model into a gate-level netlist or, alternatively, into a bit stream for configuring field-programmable logic devices. A second volume on back-end design may follow at a later date.

The text has been written with two audiences in mind. As a textbook, it wants to introduce engineering students to the beauty and the challenges of digital VLSI design while preventing them from repeating mistakes that others have made before. Practising electronics engineers should find it appealing as a reference book because of the many tables, checklists, diagrams, and case studies intended to help them not to overlook important action items and alternative options when planning to develop their own circuits.

What sets this book apart from others in the field is its top-down approach. Beginning with hardware architectures, rather than with solid-state physics, naturally follows the normal VLSI design flow and makes the material more accessible to readers with a background in systems engineering, information technology, digital signal processing, or management.

- Top-down approach.
- Systematic overview on architecture optimization techniques.
- A chapter on field-programmable logic devices, their technologies and architectures.
- Key concepts behind both VHDL and SystemVerilog without too many syntactical details.
- A proven naming convention for signals and variables.
- Introduction to assertion-based verification.
- Concepts for re-usable simulation testbenches.
- Emphasis on synchronous design and HDL code portability.
- Comprehensive discussion of clocking disciplines.
- Largely self-contained (required previous knowledge summarized in two appendices).
- Emphasis on knowledge likely to remain useful in the years to come.
- Plenty of detailed illustrations.
- Checklists, hints and warnings for various situations.
- A concept proven in classroom teaching and actual design projects.

NOTES TO INSTRUCTORS

Over the past decade, the capabilities of Field-Programmable Gate Arrays (FPGA) have grown to a point where they compete with custom-fabricated ICs in many electronic designs, especially for products marketed by small and medium enterprises.

Beginning with the higher levels of abstraction enables instructors to focus on those topics that are equally relevant irrespective of whether a design eventually gets implemented as "mask-programmed" chip or from components that are configured electrically. That material is collected in chapters 1 to 6 of the book and best taught as part of a Bachelor program for maximum dissemination. No prior introduction to semiconductor physics or devices is required. For audiences with little exposure to digital logic and finite state machines, the material can always be complemented with appendices A and B.

Chapter 7 is the only one to have a clear orientation towards mask-programmed circuits as clocking and clock distribution are largely pre-defined in field-programmable logic devices. As opposed to this, the material on synchronization in chapter 8 is equally important to FPGA users and to persons specializing in full- or semi-custom design.

Probably the best way of preparing for an engineering career in the electronics and microelectronics industry is to complete a design project where circuits are not just being modeled and simulated on a computer but actually fabricated and tested. At ETH Zürich, students are given this opportunity as part of a three-semester course given by the author and his colleagues, see figure below. The 6th term covers front-end design. Building a circuit of modest size with an FPGA is practiced in a series of exercises. Provided they come up with a meaningful proposal, students then get accepted for a much more substantial project that runs in parallel with their regular lectures and exercises during the 7th term.

Typically working in teams of two, students are expected to devote at least half of their working time to that project. Following tapeout at the end of the term, chip fabrication via an external multi-project wafer service roughly takes three months. Circuit samples then get systematically tested by their very developers in the 8th and final term. Needless to say that students accepting this offer feel very motivated and that industry highly values the practical experience of graduates formed in this way.

Most chapters in this book come with student problems. Some of them expose ideas left aside in the main text for the sake of conciseness. Problems are graded as a function of the effort required to solve them.

* A few thoughts lead to a brief answer.
** Details need to be worked out, count between 20 min and 90 min.
*** A small engineering project, multiple solutions may exist. Access to EDA and other computer tools may help.

Solutions and presentation slides are available to instructors who register with the publisher from the book's companion **website** http://booksite.elsevier.com/9780128007303.

Syllabus of ETH Zurich in
Digital VLSI Design and Test

Microelectronics Design Center
Prof. Hubert Kaeslin

Integrated Systems Laboratory
Dr. Norbert Felber

Degree

Bachelor — Master

Calendar

| Feb | Mar | Apr | May | Jun | Jul | Aug | Sep | Oct | Nov | Dec || Jan | Feb | Mar | Apr | May |

6th term | 7th term | 8th term

Specials

student projects ☆ | low pwr research ☆ | industry ☆

Lectures

VLSI I:
Architectures of VLSI Circuits

VLSI II:
Design of VLSI Circuits

VLSI III: **Fabrication and
Testing of VLSI Circuits**

Key topics

top ——————————————— down

- ◆ Architecture design
- ◆ HW Description Languages
- ◆ Functional verification

- ◆ Clocking & synchronization
- ◆ Back-end design
- ◆ Parasitic effects
- ◆ VLSI economics

- ◆ VLSI testing
- ◆ CMOS fabrication
- ◆ Technology outlook

Exercises

From VHDL to FPGA | IC design through to final layout | Testing of fabricated ICs

Student project (optional)

Speci-fication | Modeling | Circuit design | Fabrication on MPWs | Testing

Milestones

accepted proposal | software model | overall architecture | synthesis model | verified netlist | verified layout, project report | test vectors | test report

Acknowledgments

This text collects the insight and the experience that many persons have accumulated over more than twenty years. While I was fortunate enough to author the book, I am indebted to all those who have been willing to share their expertise with me.

My thanks thus go to many past and present colleagues of mine including Christoph Balmer, David Bellasi, Prof. Andreas Burg, Dr. Felix Bürgin, Dr. Norbert Felber, Prof. em. Wolfgang Fichtner, Michael Gautschi, Dr. Pierre Greisen, Dr. Frank Gürkaynak, Christoph Keller, Prof. Mathieu Luisier, Dr. Patrick Mächler, Beat Muheim, Michael Mühlberghuber, Michael Schaffner, Prof. Christoph Studer, Prof. Jürgen Wassner, Dr. Markus Wenk, Prof. Paul Zbinden, and Dr. Reto Zimmermann. As long-time VHDL users, our staff and me are grateful to Dr. Christoph Sühnel who made us become fluent in SystemVerilog with as few detours and misunderstandings as possible. Most of these experts have contributed examples, have reviewed parts of my manuscript, or have otherwise helped improve its quality. Still, the only person to blame for all errors and other shortcomings that have remained in the text is me.

Next, I would like to extend my sincere thanks to all students who have followed our courses on Digital VLSI Design and Test. Not only their comments and questions, but also results and data from many of their projects have found their way into this text. Sebastian Axmann deserves special credit for helping with the solutions on a voluntary basis.

Giving students the opportunity to design microchips, to have them fabricated, and to test physical samples is a rather onerous undertaking that would clearly have been impossible without the continuous funding by ETH Zürich. Let me express our gratitude for that on behalf of all our graduates.

In cooperation with Christoph Wicki and his IT support team, the staff of the local Microelectronics Design Center does a superb job in setting up and maintaining the EDA infrastructure and the services indispensable for VLSI design in spite of the frequent landslides caused by rapid technological evolution and by unforeseeable business changes. I am particularly grateful to them for occasionally filling all sorts of gaps in my technical knowledge without making me feel too badly about it.

I am further indebted to Todd Green, Nate McFadden, Poulouse Joseph, and many more members of the staff at Morgan Kaufmann Publishers for their support with turning my LaTeX manuscript into a printed book. Finally, I would like to thank all persons and organizations who have taken their time to answer my reprint requests and who have granted me the right to reproduce illustrations of theirs.

1

INTRODUCTION TO MICROELECTRONICS

1.1 ECONOMIC IMPACT

Let us begin by relating the worldwide sales of semiconductor products to the world's gross domestic product (GDP)[1]. In 2012, this proportion was 300 GUSD out of 72.0 TUSD (0.42%).

Assessing the significance of semiconductors on the basis of sales volume grossly underestimates its impact on world economy, however. This is because microelectronics is acting as a technology driver that enables or expedites a range of other industrial, commercial, and service activities. Just consider

- Computer and software industry,
- Telecommunications and media industry,
- Commerce, logistics, and transportation,
- Natural science and medicine,
- Power generation and distribution, and — last but not least —
- Finance and administration.

Microelectronics thus has an enormous economic leverage as any progress there spurs many, if not most, innovations in "downstream" industries and services.

A popular example ...

After a rapid growth during the last four decades, the electric and electronic content of passenger cars nowadays makes up more than 20% of the total value in simpler cars and close to 40% in well-equipped vehicles.

What's more, microelectronics is responsible for the vast majority of improvements that we have witnessed. Just consider electronic ignition and injection that have subsequently merged and evolved to become electronic engine management. Add to that anti-lock brakes and anti-skid stability programs, trigger circuits for airbags, anti-theft equipment, automatic air conditioning, navigation aids, multiplexed busses, electronically controlled drive train and suspension, audio/video information and entertainment, LED illumination and headlights, night vision and collision avoidance systems, hybrid propulsion, and regenerative braking.

---

[1] The GDP indicates the value of all goods and services sold during some specified year.

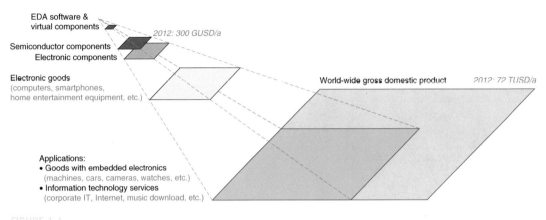

FIGURE 1.1

Economic leverage of microelectronics on "downstream" industries and services.

Forthcoming innovations include an 48 V on-board supply network, electronically driven valve trains, brake by wire, drive by wire, advanced driver assistance systems, and, possibly, entirely autonomous vehicles. And any future transition to propulsion by other forms of energy is bound to intensify the importance of semiconductors in the automotive industry even further.

... and its less evident face

Perhaps less obvious but as important are the many contributions of electronics to the processes of development, manufacturing and servicing. Innovations behind the scene of the automotive industry include computer-aided design (CAD) and finite element analysis, virtual crash tests, computational fluid dynamics, computer numeric controlled (CNC) machine tools, welding and assembly robots, computer-integrated manufacturing (CIM), quality control and process monitoring, order processing, supply chain management, and diagnostic procedures.

□

This almost total penetration has been made possible by a long-running drop of **cost per function** at a rate of 25 to 29% per year. While computing, telecommunication, and entertainment products existed before the advent of microelectronics, today's anywhere, anytime information and telecommunication society would not have been possible without, just compare the electronic devices in fig.1.2.

Observation 1.1. *Microelectronics is <u>the</u> enabler of information technology.*

On the positive side, microelectronics and information technology improve speed, efficiency, safety, comfort, and pollution control of industrial products and commercial processes thereby bringing competitive advantages to those companies that take advantage of them.

On the negative side, the rapid progress, most of which is ultimately fueled by advances in semiconductor manufacturing technology, also implies a rapid obsoletion of hardware and software products, services, know-how, and organizations. A highly cyclic economy is another unfortunate trait of the semiconductor industry [1].

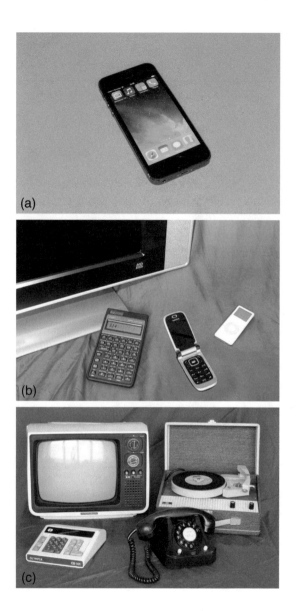

FIGURE 1.2

The impact of microelectronics on consumer goods. A smartphone that takes advantage of advanced application-specific integrated circuits to combine a TV player, a jukebox, a calculator, a mobile phone, and much more in one handheld device (a). The same four functions just a couple of years earlier (ca. 2005) (b). Similar products from the 1970s that operate with vacuum tubes, discrete solid-state devices, and other components but include no large-scale integrated circuits (c) (photos courtesy of Alain Kaeslin).

1.2 MICROELECTRONICS VIEWED FROM DIFFERENT PERSPECTIVES

An **integrated circuit** (IC) is an electronic component that incorporates and interconnects a multitude of miniature electronic devices, mostly transistors, on a single piece of semiconductor material, typically **silicon**.[2] Many such circuits are jointly manufactured on a thin semiconductor wafer with a diameter of typically 300 mm before they get cut apart to become (naked) **dies**. The sizes of typical dies range between a pinhead and a large poststamp. The vast majority of ICs or **(micro)chips**, as they are colloquially referred to, gets individually encapsulated in a hermetic package before being soldered onto **printed circuit boards** (PCB).

The rapid progress of semiconductor technology in conjunction with marketing activities of many competing companies — notably trade mark registration and eye catching — has led to a plethora of terms and acronyms, the meaning of which is not consistently understood by all members of the microelectronics community. This section introduces the most important terms, clarifies what they mean, and so prepares the ground for more in-depth discussions.

Depending on perspective, microchips are classified according to different criteria.

1.2.1 THE GUINNESS BOOK OF RECORDS POINT OF VIEW

In a world obsessed with records, a foremost question asks "How large is that circuit?".

Die size is a poor metric for design complexity because the geometric dimensions of a circuit greatly vary as a function of technology generation, fabrication depth, and design style.

Transistor count is a much better indication. Still, comparing across logic families is problematic as the number of devices necessary to implement some given function varies.[3]

Gate equivalents attempt to capture a design's hardware complexity independently from its actual circuit style and fabrication technology. One gate equivalent (GE) stands for a two-input NAND gate and corresponds to four MOSFETs in static CMOS; a flip-flop takes roughly 7 GEs. Memory circuits are rated according to storage capacity in bits. Gate equivalents and memory capacities are at the basis of the naming convention below.

---

[2] This is a note to non-Angloamerican readers made necessary by a tricky translation of the term silicon.

English	German	French	Italian	meaning
silicon	Silizium	silicium	silicio	Si, the chemical element with atomic number 14
silicone	Silikon	silicone	silicone	a broad family of polymers of Si with hydrocarbon groups that comprises viscous liquids, greases and rubber-like solids

---

[3] Consistent with our top-down approach, there is no need to know the technicalities of CMOS, TTL and other logic families at this point. Interested readers will find a minimum of information in appendix 1.5.

circuit complexity	GEs of logic + bits of memory
small-scale integration (SSI)	1 ... 10
medium-scale integration (MSI)	10 ... 100
large-scale integration (LSI)	100 ... 10 000
very-large-scale integration (VLSI)	10 000 ... 1 000 000
ultra-large-scale integration (ULSI)	1 000 000 ...

Clearly, this type of classification is a very arbitrary one in that it attempts to impose boundaries where there are none. Also, it equates one storage bit to one gate equivalent. While this is approximately correct when talking of static RAM (SRAM) with its four- or six-transistor cells, the single-transistor cells found in dynamic RAMs (DRAM) and in ROMs cannot be likened to a two-input NAND gate. A better idea is to state storage capacities separately from logic complexity and along with the memory type concerned, e.g. 75 000 GE of logic + 32 Kibit SRAM + 512 bit flash $\approx$ 108 000 GE overall complexity.[4]

One should not forget that circuit complexity per se is of no merit. Rather than coming up with inflated designs, engineers are challenged to find the most simple and elegant solutions that satisfy the specifications given in an efficient and dependable way.

1.2.2 THE MARKETING POINT OF VIEW

In this section, let us adopt a market-oriented perspective and ask
"How do functionality and target markets relate to each other?"

General-purpose ICs

The function of a general-purpose IC is either so simple or so generic that the component is being used in a multitude of applications and typically sold in huge quantities. Examples include gates, flip-flops, counters, adders, and other components of the various 7400 families but also RAMs, ROMs, microcomputers, and many digital signal processors (DSP).

Application-specific integrated circuit

Application-specific integrated circuits (ASIC) are being specified and designed with a particular purpose, equipment, or processing algorithm in mind. Initially, the term had been closely associated with **glue logic**, that is to all those bus drivers, decoders, multiplexers, registers, interfaces, etc. found in many system assembled from integrated circuits. ASICs have evolved from substituting a single package for such ancillary functions that were originally dispersed over several SSI/MSI circuits.

---

[4] The binary prefixes kibi- (Ki), mebi- (Mi), gibi- (Gi) and tebi- (Ti) are recommended by various standard bodies for 2^{10}, 2^{20}, 2^{30} and 2^{40} respectively because the more common decimal SI prefixes kilo- (k), mega- (M), giga- (G) and tera- (T) give rise to ambiguity since $2^{10} \neq 10^3$. As an example, 1 Mibyte = 8 Mibit = $8 \cdot 2^{20}$ bit with Mibyte being pronounced as "me bi byte", a contraction of "**me**ga **bi**nary **byte**".

Today's highly-integrated ASICs are much more complex and include powerful systems or subsystems that implement highly specialized tasks in data and/or signal processing. The term **system-on-a-chip** (SoC) has been coined to reflect this development. Overall manufacturing costs, performance, miniaturization, and energy efficiency are key reasons for opting for ASICs.

Still from a marketing point of view, ASICs are subdivided further into application-specific standard products and user-specific ICs.

Application-specific standard product (ASSP). While designed and optimized for a highly specific task, an application-specific standard product circuit is being sold to various customers for incorporation into their own products. Examples include graphics accelerators, multimedia chips, data compression circuits, forward error correction devices, ciphering/deciphering circuits, smart card chips, chip sets for cellular radio, PCIe and Ethernet interfaces, wireless LAN chips, and driver circuits for power semiconductor devices, just to name a few.[5]

User-specific integrated circuit (USIC). As opposed to ASSPs, user-specific ICs are being designed and produced for a single company that seeks a competitive advantage for their products, they are not intended to be marketed as such. Control of innovation and protection of proprietary know-how are high-ranking motivations for designing circuits of this category. Popular USICs include the Apple A4 SoC introduced with the iPad in 2010 and its successors A5, A6, A7, etc. Various audio processor chips for hearing aids have been developed for similar reasons.

User-specific parts are sometimes fabricated in relatively modest quantities. As an example, consider the USIC ($14 \cdot 10^6$ GE, 90 nm CMOS, 60 W) that forms the heart of the first oscilloscope family by Rhode & Schwarz.

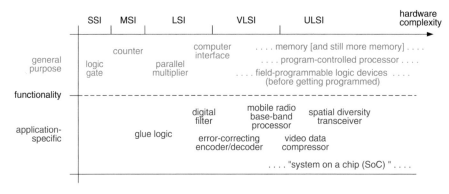

FIGURE 1.3

ICs classified as a function of functionality and hardware complexity.

---

[5] Microprocessors that have their instruction sets, input/output capabilities, memory configurations, timers, and other auxiliary features tailored to meet specific needs also belong to the ASSP category.

1.2.3 THE FABRICATION POINT OF VIEW

Another natural question is
"To what extent is a circuit manufactured according to user specifications?".

Full-custom ICs

Integrated circuits are manufactured by patterning multiple layers of semiconductor materials, metals and dielectrics. In a full-custom IC, all such layers are patterned according to user specifications. Fabricating a particular design requires wafers to go through all processing steps under control of a full set of lithographic **photomasks** all of which are made to order for this very design, see fig.1.4. This is relevant from an economic point of view because mask manufacturing is a dominant contribution to non-recurring VLSI fabrication costs. A very basic CMOS process featuring two layers of metal requires some 10 to 12 fabrication masks, any additional metal layer asks for two more masks. At the 65 nm generation, an advanced CMOS process comprises twelve layers of metal and involves some 45 lithography cycles.

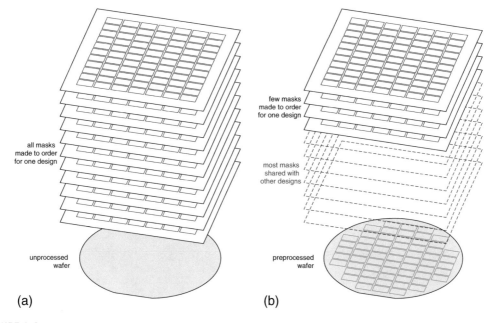

FIGURE 1.4

Full-custom (a) and semi-custom (b) mask sets compared.

Semi-custom ICs

Only a small subset of fabrication layers is unique to each design. Customization starts from preprocessed wafers that include large quantities of prefabricated but largely uncommitted primitive items such as transistors or logic gates. These so-called **master wafers** then undergo a few more processing steps during which those primitives get interconnected such as to complete the electrical and logic circuitry required for a particular design. As an example, fig.1.5 shows how a logic gate is being manufactured from a few pre-existing MOSFETs by etching open contact holes followed by deposition and patterning of one metal layer.

In order to accommodate designs of different complexities, vendors make masters available in various sizes ranging from a couple of thousands to millions of usable gate equivalents. Organization and customization of semi-custom ICs have evolved over the years.

Gate array, aka channeled gate array. Originally, sites of a few uncommitted transistors each were arranged in long rows that extended across most of the die's width. Metal lines were then used to connect the prefabricated transistors into gates, and the gates into circuits. The number of custom photomasks was twice that of metal layers made to order. As long as no more than two layers of metal were available, special routing channels had to be set aside in between to accommodate the necessary intercell wiring, see fig.1.6a.

Sea-of-gates. When more metals became available in the early 1990s, those early components got displaced by channelless sea-of-gate circuits because of their superior layout density. The availability of higher-level metals allowed for routing over the gates customized on the layers underneath, dispensing with the waste of routing channels, see fig.1.6b. Thanks to a trick called separation gate, sea-of-gates also did away with the periodic gaps in the layout that had grouped MOSFETs into sites. All this together afforded the flexibility to accommodate highly repetitive structures such as RAMs and ROMs.

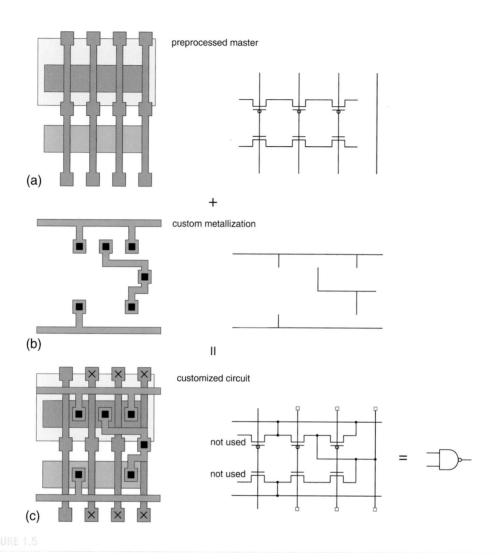

preprocessed master

(a)

+

custom metallization

(b)

||

customized circuit

not used

not used

(c)

FIGURE 1.5

Customization of a gate array site (simplified). A six-pack of prefabricated MOS transistors (a), metal pattern with contact openings (b), and finished 2-input NAND gate (c).

Structured ASIC. A decade later, the number of metal layers had grown to a point where it became uneconomical to customize them all. Instead, transistors are preconnected into small generic subcircuits such as NANDs, MUXes, AOI gates, full-adders, lookup tables (LUT), and bistables on the lower layers of metal. Customization is essentially confined to interconnecting those subcircuits on the top two to four metal layers. The design process is also accelerated as power and clock distribution networks are largely prefabricated.

Fabric. Exploding mask costs and the limitations of sub-wavelength lithography currently work against many custom-made photomasks. The idea behind fabrics is to standardize the metal layers as much as possible. A subset of them are patterned into fixed segments of predetermined lengths which get pieced together by short metal straps, aka jumpers, on the next metal layer below or above to obtain the desired wiring. Customization is via the vertical contact plugs, called vias, that connect between two adjacent layers.

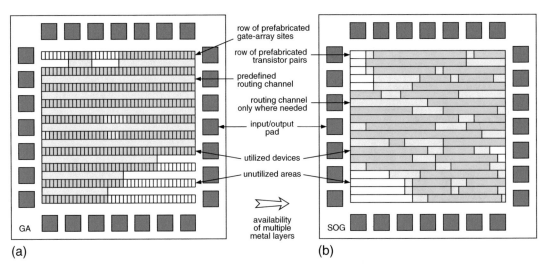

(a)   (b)

FIGURE 1.6

Floorplan of channeled gate-array (a) versus channelless semi-custom circuits (b).

Due to the small number of design-specific photomasks and processing steps, semi-custom manufacturing significantly reduces the non-recurring costs as well as the turnaround time.[6] Conversely, prefabrication necessarily results in non-optimal layouts. Note the unused transistor pair in fig.1.5, for instance, or think of the extra parasitic capacitances and resistances caused by standardized wiring. Prefabrication also implies a self-restraint to fixed transistor geometries thereby further limiting circuit density, speed, and energy efficiency. Lastly, not all semi-custom masters accommodate on-chip memories equally well.

The concept of metal customization is also applied to analog and mixed-signal circuits. Prefabricated masters then essentially consist of uncommitted transistors (MOSFETs and/or BJTs) and of passive devices.[7]

---

[6] **Turnaround time** denotes the lapse of time from coming up with a finalized set of design data until physical samples become available for testing.

[7] Microdul MD300 and MD500 are just two examples.

Field-programmable logic

Rather than manufacturing dedicated layout structures, a generic part is made to assume a user-defined circuit configuration by purely electrical means. No custom photomasks are involved. Field-programmable logic (FPL) devices are best viewed as "soft hardware".[8] Unlike semi- or full-custom ASICs, FPL offers turnaround times in the order of seconds. Most of today's product families allow for in-system configuration (ISC).

The key to obtaining various gate-level networks from the same hardware resources are a multitude of electrical links that can be done — and in many cases also undone — long after a device has left the factory. For the moment being, you can think of a programmable link as some kind of fuse.

Commercial parts further differ in how the on-chip hardware resources are organized. **Field-programmable gate arrays** (FPGA), for instance, resemble mask-programmed gate arrays (MPGA) in that they are organized into a multitude of logic sites and interconnect channels.

In this text, we will be using the term **field-programmable logic** (FPL) as a collective term for any kind of electrically configurable IC regardless of its capabilities, organization, and configuration technology.[9]

Their important market share affords FPL devices a more detailed discussion in chapter 2.

Standard parts

By standard part, aka commercial off-the-shelf (COTS) component, we mean a catalog part with no customization of the circuit hardware whatsoever.

1.2.4 THE DESIGN ENGINEER'S POINT OF VIEW

Hardware designers will want to know
"Which levels of detail are being addressed during a part's design process?".

Hand layout

In this design style, an IC or some subblock thereof gets entered into the CAD database by delineating individual transistors, wires, and other circuit elements at the layout level. To that end, designers use a **layout editor**, essentially a color graphics editing tool, to draw the desired geometric shapes to scale, much as in the illustration of fig.1.5c. Any design so established must conform with the layout rules imposed by the target process. Porting it to some other process requires the layout to be redesigned unless the new set of rules is obtained from the previous one by simple scaling operations. Editing **geometric layout** is slow, cumbersome, and prone to errors. Productivity is estimated to lie somewhere between 5 and 10 transistors drawn per day, including the indispensable verification, correction, and documentation steps which makes this approach prohibitively expensive.

---

[8] As opposed to firmware which can be understood as frozen instruction sequences or "hard software".

[9] "Field-configurable" would better reflect what actually happens and also avoid confusion with program-controlled processors. Yet, the term "programmable" has gained so much acceptance in acronyms such as PLA, PAL, CPLD, FPGA, etc. that we stay with it. Just make sure to understand there is no instruction sequence to execute.

Conversely, manual editing gives designers full control over their layouts when in search of maximum density, performance, and/or electrical matching. Geometric layout, which in the early days had been the only avenue to IC design, continues to play a dominant role in memory and analog circuit design. In digital design, it is considered archaic, although a fully handcrafted circuit may outperform a synthesis-based equivalent by a factor of three or more.

Cell libraries and schematic entry

Design capture here occurs by drawing circuit diagrams where subfunctions are instantiated and interconnected by wires as illustrated in fig.1.9c. All the details of those elementary subcircuits, aka cells, have been established before, collected in **cell libraries**, and made available to VLSI designers. For the sake of economy, cell libraries are shared among numerous designs. A **schematic editor** differs from a standard drawing tool in several ways:

- Circuit connectivity is maintained when components are being relocated.
- A schematic editor is capable of reading and writing both circuit diagrams and netlists.[10]
- It supports circuit concepts such as connectors, busses, node names, and instance identifiers.

The resulting circuits and netlists are then verified by simulation and other means. Compared to manual layout entry, **cell-based** design represented a marked step towards abstracting from process-dependent details. Physical design does not go beyond **place and route** (P&R) where each cell is assigned a geometric location and connected to other cells by way of metal lines. As this is done by automatic tools, the resulting layouts are almost always correct by construction and **design productivity** is much better than for manual layout. Another advantage is that any electronics engineer can start to develop cell-based ASICs with little extra training.

---

[10] The difference between a netlist and a circuit diagram, aka schematic (drawing), is explained in section 1.6.

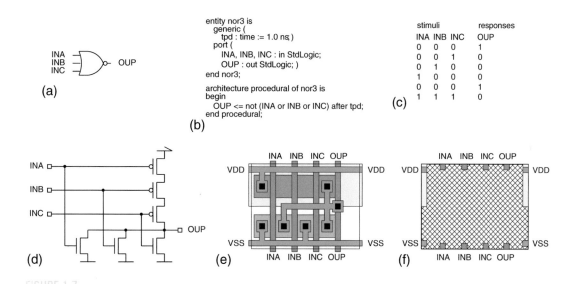

FIGURE 1.7

Views of a library cell or of any other subcircuit shown for a 3-input NOR gate. Icon (a), simulation model (b), test vector set (c), transistor-level schematic (d), detailed layout (e), and cell abstract (f) (simplified).

Library elements are differentiated into standard cells, macrocells, and megacells.

Standard cells are small but universal building blocks such as logic gates, latches, flip-flops, multiplexers, adder slices, and the like with preestablished layout and defined electrical characteristics.[11] They are the preferred means for implementing random logic as there is virtually no restriction on the functionality that can be assembled from them. Commercial libraries include between 300 and 500 standard cells with logic complexities ranging from 1/2 to some 60 gate equivalents; the pertaining collection of datasheets typically occupies between 400 and more than 1000 pages.

On the semiconductor die, standard cells get arranged in adjoining parallel rows with the interconnecting wires running over the top of them. This so-called **over-the-cell routing** style is being practiced ever since three and more layers of metal have become available.[12]

Megacells also come with a ready-to-use layout. What sets them apart from standard cells is their larger size and complexity. Typical examples include microprocessor cores and peripherals such as direct memory access controllers, various serial and parallel communication interfaces, timers, A/D and D/A converters, and the like. Megacells are ideal for piecing together a microcomputer or an ASIC with comparatively very little effort. Typical application areas are in telecommunications equipment, automotive equipment, instrumentation, and control systems.

---

[11] Standard cells are also termed "books" (within IBM) and **macros** (in the context of semi-custom ICs).

[12] Older processes did not afford that much routing resources and the wires had to be inserted between the rows such as to form well-defined **routing channels**. The resulting separation between adjacent cell rows obviously made a poor usage of silicon. In fact, it was not uncommon that routing channels occupied twice or even three times as much area as the active cells themselves.

Macrocells, in contrast, have their layout assembled on a per case basis according to designer specifications. The software tool that does so is called a **macrocell generator** and is also in charge of providing a simulation model, an icon, a datasheet, and other views of the macrocell. For reasons of area and design efficiency, this approach is essentially limited to a few common building blocks of medium complexity such as RAMs and ROMs. This is because all such structures show fairly regular geometries that lend themselves well to being put together from a limited collection of layout tiles. Those tiles are manually designed, optimized and verified before being stored as part of the generator package.

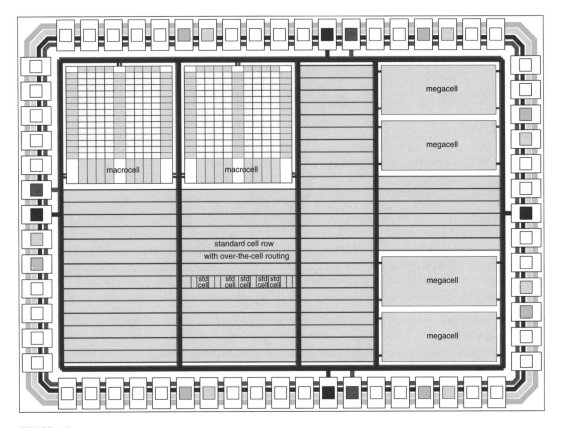

FIGURE 1.8

Typical cell mix in a full-custom IC.

As standard cells, macrocells, megacells, and hand layout all have their specific merits and drawbacks, they are often combined in the design of full-custom ICs.[13] The resulting mix of cells is illustrated in fig.1.8. While design productivity in terms of transistors instantiated per day is clearly superior for megacells and macrocells than for standard cells, expect an average of some 15 to 20 GEs per day from cell-based design. Schematic entry was important at a time. Today, it is confined to functions that are neither available as library items nor amenable to automatic synthesis. Schematic entry further continues to be essential in analog circuit design.

Automatic circuit synthesis

The entry level here is a formal description of an entire chip or of a major subblock therein. Most such **synthesis models** are established using a text editor and look like software code. Yet, they are typically written in a **hardware description language** (HDL) such as VHDL or SystemVerilog, see fig.1.9b. The output from the automatic synthesis procedure is a gate-level netlist. That netlist then forms the starting point for place and route (P&R) or for preparing a bit stream that will eventually serve to configure an FPL device.

Logic synthesis implies the generation of combinational networks and — as an extension — of fairly simple finite state machines (FSM). A synthesis tool accepts logic equations built from operators such as NOT, AND, OR, XOR, etc., truth tables, state graphs, and the like. Automatic tools for logic synthesis and optimization have been in routine use for a long time, they have been completely absorbed in today's EDA flows.

Register transfer level (RTL) **synthesis** goes one step further in that an entire circuit is viewed as a network made up of storage elements — registers and possibly also RAMs — that are held together by combinational building blocks, see fig.1.9a. Also, behavioral specifications are no longer limited to simple logic operations but are allowed to include arithmetic functions (e.g. comparison, addition, subtraction, multiplication), string operations (e.g. concatenation), arrays, enumerated types, and other more powerful constructs.

The synthesis process essentially begins with the registers that store the circuit's state. Next, the combinational networks required to process data words while they are moving back and forth between those registers are generated and optimized. Command on a circuit's structure is otherwise left to the designer who decides on the number of registers, on the concurrency of operations, on the necessary computational resources, etc.

RTL synthesis dispenses with the need of manually assembling a given functionality from logic gates and, therefore, greatly facilitates design parametrization and maintenance. Synthesis further enables engineers to render their work portable, that is to capture all relevant characteristics of a circuit design in a form that is virtually technology-independent. It so becomes possible to defer the commitment to a specific silicon foundry, to a particular cell library, or to subordinate idiosyncrasies of some FPL family until late in the design process. As fabrication processes are frequently being upgraded, making designs portable and reusable is extremely valuable.

---

[13] In a microcomputer, for instance, the datapath might be implemented in hand layout, data RAM and program ROM generated as macrocells, the controller as a network of standard cells obtained from automatic synthesis, while a serial interface from an earlier design might get reused as a megacell.

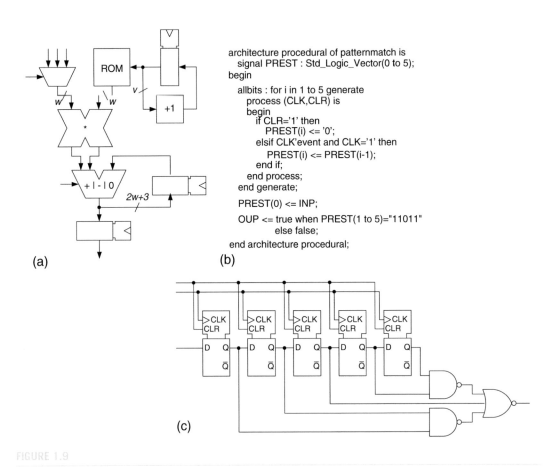

architecture procedural of patternmatch is
  signal PREST : Std_Logic_Vector(0 to 5);
begin

  allbits : for i in 1 to 5 generate
    process (CLK,CLR) is
    begin
      if CLR='1' then
        PREST(i) <= '0';
      elsif CLK'event and CLK='1' then
        PREST(i) <= PREST(i-1);
      end if;
    end process;
  end generate;

  PREST(0) <= INP;

  OUP <= true when PREST(1 to 5)="11011"
      else false;
end architecture procedural;

(a)   (b)

(c)

FIGURE 1.9

RTL diagram (a), RTL synthesis model (b), and gate-level schematic (c) (simplified, note that (a) and (b) refer to different circuits).

Architecture synthesis, which is also referred to as high-level synthesis in VLSI circles, starts from an algorithmic description such as a C++ program or a MATLAB model, for instance. As opposed to an RTL model, the source code is purely behavioral and includes no explicit indications for how to marshal data processing operations and the necessary hardware resources. Rather, these elements must be obtained in an automatic process that essentially works in five major phases:

1. Identify the computational and storage requirements of the algorithm.
2. From a virtual library of common hardware building blocks, select a suitable item for each kind of processing and storage operation.
3. Establish a cycle-based schedule for carrying out the algorithm with those resources. Where there is a choice, indicate which building block is to process what data item.

4. Decide on a hardware organization able to execute the resulting work plan. Specify the architecture in terms of combinational logic blocks, data registers, on- and off-chip memories, busses, switches, signals, and finite state machines.

5. Keeping track of data moves and operations for each clock cycle, translate all this into the necessary instructions for synthesis at the RTL level.

Generating a close-to-optimum architecture under performance, power, cost and further constraints represents a formidable optimization problem, especially if a tool is expected to work for arbitrary applications. To get an idea, consult the lists of issues in section 1.3.2. Architecture synthesis allows for rapid exploration of the design space and performs well in a couple of specialized areas (e.g. digital filters and ASIPs). Apart from that, however, high-level synthesis does not — up to now — produce results comparable to those of inspired and experienced engineers. Nonetheless, architecture synthesis continues to be an active field of research as VLSI design can no longer afford to deal with low-level details.

Even an experienced RTL code writer cannot be expected to complete much more than 40 lines of code per day. Estimates say that design productivity ranges from 20 to 400 GE per working day.[14] Albeit quite impressive, these figures are actually insufficient to keep pace with the rapid advances of fabrication technology.

Design with virtual components

In the late 1990s, synthesis technology together with HDL standardization has opened up the door for an entirely new approach to designing digital VLSI circuits. A **virtual component** (VC)[15] essentially is a HDL synthesis package that is made available to others on a commercial basis for incorporation into their own ICs. VLSI design teams across the electronics industry are so put in a position to purchase hardware designs for major subfunctions on the merchant market, dispensing with the need to write too much HDL source code on their own. The licensees just remain in charge of synthesis, place and route (P&R), and overall verification.

Though of highly specific nature, most VCs implement fairly common subfunctions; some degree of parametrization is sought to cover more potential applications. Examples include, but are not limited to, micro- and signal processor cores, all sorts of filters, audio and/or video en/decoders, cipher functions, error correction en/decoders, USB, FireWire, and many other interfaces.

---

[14] Be warned that design productivity is extremely dependent on circumstances:
- The effort per transistor is not the same for memories, logic, and mixed-signal designs.
- The more circuit blocks can be reused that have been validated before, the better.
- Skilled engineering teams not only work faster but also manage with less design iterations.
- Powerful EDA tools can work out many minor circuit and layout details automatically.
- The existence of an established and proven design flow benefits the design process.
- Tight timing, power and layout density budgets ask for more human attention.
- Unstable specifications and rapidly changing teams are detrimental to productivity.

[15] Virtual components are better known as **intellectual property modules** or IP modules for short. We prefer the term "virtual component" because IP does not point to electronics in any way and because the acronym might easily be misunderstood as "Internet protocol". Other synonyms include "core" and "core ware".

While hard modules such as standard cells, macrocells and megacells had freed most IC designers from addressing transistor-level issues and detailed layout by the mid 1980s, the soft VCs have extended these benefits to higher levels of abstraction in a natural way. New business opportunities have opened up and companies that specialize in marketing synthesis models have emerged.

A classification scheme depicted in table 1.1 nicely complements the one of fig.1.3.

Electronic system-level (ESL) design automation

More recently, competitive pressure has incited the industry to look at design productivity from a wider perspective. ESL design automation is a collective term for efforts that take inspirations from numerous ideas.

- Enforce a correct-by-construction methodology by supporting progressive refinement starting with a virtual prototype of the system to be.
- Recur to architecture synthesis to explore the solution space more systematically and more rapidly than with conventional, e.g. RTL synthesis, methods.

Table 1.1   IC families as a function of fabrication depth and design abstraction level.

Fabrication depth	Electrical configuration	Semi-custom fabrication	Full-custom fabrication	
Design level	Cell-based as obtained from ∘ synthesis with VCs in HDL form, ∘ synthesis from captive HDL code, ∘ schematic entry, or a mix of these			Hand layout
Product name	Field-programmable logic device (FPGA, CPLD)	Gate-array, sea-of-gates, or structured ASIC	Std. cell IC (with or w/o macrocells and megacells)	Full-custom IC

- Support hardware-software co-design by making it possible to start software development before hardware design is completed.
- Improve the coverage and efficiency of functional verification by dealing with system-level transactions and by taking advantage of formal verification techniques where possible.

1.2.5 THE BUSINESS POINT OF VIEW

Our final question is concerned with the sharing of industrial activities among partners.
"What is the business model of a given semiconductor company?"

Integrated device manufacturer (IDM) is the name for a company that not only designs and markets microchips but that also does the wafer processing in-house in their own semiconductor fabrication plant or **fab** for short.
Examples: Intel, Samsung, Toshiba, ST-Microelectronics, Texas Instruments, Cypress Semiconductor, AMS.

Silicon foundry, albeit technically incorrect, designates a company that operates a complete wafer processing line and that offers its manufacturing services to others.
Examples: TSMC, UMC, SMIC, etc.

Fabless vendor. A company that develops and markets proprietary semiconductor components but has their manufacturing commissioned to an independent silicon foundry rather than operating any wafer processing facilities of its own.
Examples: Altera (FPL), Actel (FPL), Broadcom (networking components), Cirrus Logic-Crystal (audio and video chips), Lattice Semiconductor (FPL), Nvidia (graphics accelerators), PMC-Sierra (networking components), Qualcomm (chipsets for wireless telecommunication), and Xilinx (FPL).

Fab-lite vendor is the name for a company that outsources standard wafer processing steps but retains the limited and highly specialized manufacturing capabilities required to integrate sensors, actuators, microelectromechanical systems, RF components (such as high-Q inductors and RF switches), photonic devices, or the like, in a silicon substrate along with electronic circuitry. This approach typically implies that fully or partially processed CMOS wafers undergo a series of final processing steps at the vendor's own facilities.
Examples: Luxtera, Sensirion.

Intellectual property (IP) **vendor.** A fabless company that makes it a business to develop hardware subfunctions and to license them to others for incorporation into their ICs. Intellectual property here refers to any kind of predeveloped electronic subfunction such as standard cells, macrocells, megacells, or virtual components.
Examples: ARM, Faraday, Sci-worx, Synopsys.

Originally, all IC business had been confined to vertically integrated semiconductor companies that designed and manufactured standard parts for the markets they perceived. Opening VLSI to other players was essential to instilling new and highly successful fabless business models. Three factors came together in the 1980s to make this possible.

- Generous integration densities at low costs.
- Proliferation of high-performance engineering workstations and EDA software.
- Availability of know-how in VLSI design outside IC manufacturing companies.

This text wants to contribute to the third item with a focus on synthesis-based design.

1.3 THE VLSI DESIGN FLOW

1.3.1 THE Y-CHART, A MAP OF DIGITAL ELECTRONIC SYSTEMS

The Y-chart by Daniel Gajski is convenient for situating the various stages of digital design and the numerous attempts to automate them. Three axes stand for three different ways to look at a digital system and concentric circles represent various levels of abstraction, see fig.1.10.

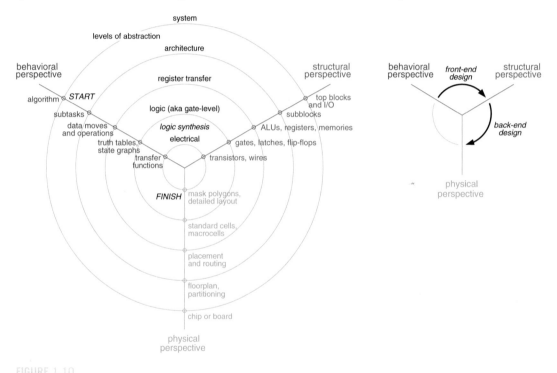

FIGURE 1.10

The Y-chart of digital electronic systems.

From a **behavioral** perspective we are only interested in what a circuit or system does, not in how it is actually built. Put differently, the design is viewed as a black box that processes information by producing some output symbols in response to some input symbols. What matters most is the dependency of the output from past and present inputs, but timing relationships between input data, output data, and clock signals may be relevant too.

A **structural** way of looking at electronic circuits is concerned with connectivity, that is with the building blocks from which a circuit is composed and with how they connect to each other. Given some behavioral specification, it is always possible to come up with more than one network for implementing it. Structural alternatives typically differ in terms of circuit complexity, performance, energy efficiency, and in other characteristics of practical interest.

What counts from a **physical** point of view is how the various hardware components and wires are arranged in the space available on a semiconductor chip or on a printed circuit board. Again, there is a one-to-many relationship between structural description and physical arrangement.

Illustrations of circuits viewed at different levels of abstraction and from all three perspectives have been given in figs.1.7 and 1.9 while fig.1.11 adds some new ones. In addition, table 1.2 lists the objects that are of interest for the individual views. It is interesting to note that different time units are used depending on the abstraction level on which behavior is described.

Table 1.2   Views and levels of abstraction in digital design.

level of abstraction	view			concept of time
	behavioral	structural	physical	
system	input/output relationship	system with input/output	chip, board, or cabinet	sequence, throughput
architecture	bus functional model (BFM)	organization into subsystems	partitioning, floorplan	partial ordering relationships
register transfer	data transfers and operations	ALUs, muxes, and registers	placement and routing	clock cycles (cycle-true)
logic	truth tables, state graphs	gates, latches, and flip-flops	standard cells or components	events, delays, timing params[a]
electrical	transfer functions	transistors, wires, R, L, C	detailed layout, mask polygons	continuous

[a] Such as t_{pd}, t_{su}, t_{ho}, and the like. Glitches are also accounted for at this level of abstraction.

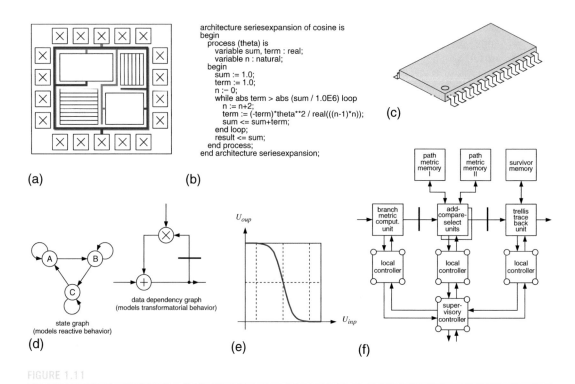

```
architecture seriesexpansion of cosine is
begin
   process (theta) is
      variable sum, term : real;
      variable n : natural;
   begin
      sum := 1.0;
      term := 1.0;
      n :- 0;
      while abs term > abs (sum / 1.0E6) loop
         n := n+2;
         term := (-term)*theta**2 / real(((n-1)*n));
         sum <= sum+term;
      end loop;
      result <= sum;
   end process;
end architecture seriesexpansion;
```

(a)

(b)

(c)

(d)

state graph
(models reactive behavior)

data dependency graph
(models transformatorial behavior)

U_{oup}

U_{inp}

(e)

(f)

path
metric
memory
I

path
metric
memory
II

survivor
memory

branch
metric
comput.
unit

add-
compare-
select
units

trellis
trace
back
unit

local
controller

local
controller

local
controller

super-
visory
controller

FIGURE 1.11

Floorplan of a VLSI chip (a), software model (b), encapsulated chip (c), graphical formalisms (d), transfer
characteristic of an inverter (e), and block diagram (f) (simplified).

1.3.2 MAJOR STAGES IN DIGITAL VLSI DESIGN

The development cycle of VLSI circuits comprises a multitude of steps, illustrated by way of two
drawings that partially overlap. Fig.1.12 focuses on system-level issues and reduces all activities that
are related to actual IC design to their most simple expression while fig.1.13 does the opposite. Again,
figs.1.7, 1.9 and 1.11 help to clarify what is meant.

The remainder of this chapter explains how everything fits together before important choices that have
to be made as part of front-end design are being discussed in the upcoming chapters. Depending on
your preferences, you may skim over sections 1.3.2 to 1.3.4 for a first reading and come back later
when having developed a better understanding of the details.

System-level design. The decisions taken during this stage are most important as they determine the
final outcome more than anything else.

- Specify the functionality, operating conditions, and desired characteristics
  (in terms of performance, power, form factor, costs, etc.) of the system to be.
- Partition the system's functionality into subtasks.

- Explore alternative hardware and software tradeoffs.
- Decide on make or buy for all major building blocks.
- Decide on interfaces and protocols for data exchange.
- Decide on data formats, operating modes, exception handling procedures, etc.
- Define, model, evaluate and refine the various subtasks from a behavioral perspective.

It is a characteristic trait of this stage that acceptance criteria, design procedures, design expertise, and the software tools that are being put to service vary greatly with the nature of the overall application and of the subsystem currently being considered.

Examples

	packet router	audio compressor	microprocessor
mode of operation	reactive	transformatorial	must fit both
data manipulation	shallow	deep	must fit both
real-time processing	yes	typically yes	maybe, maybe not
numerical precision	known a priori	to be determined	must fit both
design focus	architecture	algorithm	instruction set
background	queuing theory	human perception	code analysis
evaluation tool	traffic simulation	algorithmic simulation	benchmark programs

☐

Fig.1.12 exposes another difficulty of system-level design that has its roots in the highly heterogeneous nature of electronic systems. At various points, some fairly abstract design description must be propagated from one software tool to the next. Yet, there are no mathematical formalisms and agreed-on computer languages of adequate scope to capture a sufficient portion of a system, let alone a system as a whole. Some relief comes from the system-level modeling language SystemC that supports piecing together partial models.

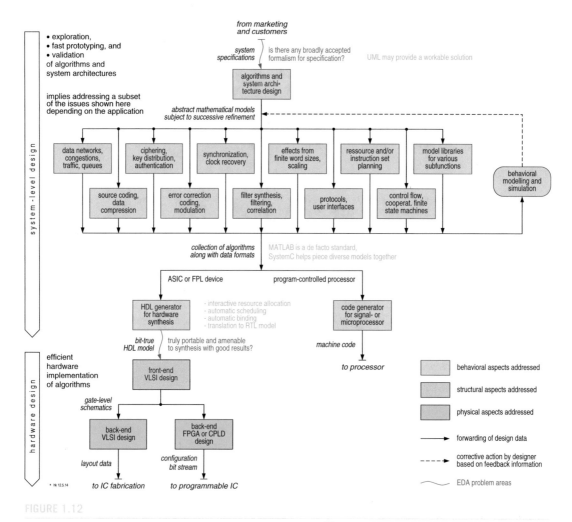

FIGURE 1.12

Design flow from a system-level perspective (greatly simplified).[16]

Algorithm design.   The central theme is to meet the data and/or signal processing requirements defined before with a series of computations that are streamlined in view of their implementation in hardware. The subsequent assignments make part of algorithm design.

- Coming up with a collection of suitable algorithms or computational paradigms.[17]
- Cut down computational burden and memory requirements.
- Find acceptable compromises between computational complexity and accuracy.

---

[16] This text focuses on hardware; implementing the software components in a system is beyond its scope.

[17] The term "computational paradigm" has been chosen to include finite state machines, cellular automata, neural networks, fuzzy logic, and other computational schemes that are not necessarily covered by the word "algorithm" as it is normally understood in the context of software engineering.

- Analyze and contain effects of finite word-length computation.
- Decide on number representation schemes.
- Evaluate alternatives and select the one best suited for the situation at hand.
- Quantify the minimum required computational resources (in terms of memory, word widths, arithmetic and logic operations, and their frequencies of occurrence).

Algorithm design culminates in a bit-true software model which is indispensable for checking figures of merit relevant for the application at hand, e.g. signal-to-noise ratio, coding gain, data compression factor, error rate, and the like against specifications.

Architecture design. VLSI architects essentially decide on the necessary hardware resources and organize their interplay such as to implement a known computational algorithm under the performance, cost, power and other constraints imposed by the target application. The hardware arrangement they have to come up with must capture the essential structural characteristics of the future circuit but, at the same time, abstracts from implementation details. Still, architecture design also implies selecting a target technology and taking into account its possibilities and limitations.[18]

Architecture design starts from fairly abstract notions of a circuit's functionality and gradually proceeds to more detailed representations. The process is understood to happen in two substages, namely high-level architecture design and register transfer level design. The former involves.

- Partition a computational task in view of a hardware realization.
- Organize the interplay of the various subtasks.
- Decide on the hardware resources to allocate to each subtask. (**allocation**)
- Define datapaths and controllers.[19]
- Decide between off-chip RAMs, on-chip RAMs, and registers.
- Decide on communication topologies and protocols (parallel, serial).
- Define how much parallelism to provide in hardware.
- Decide where to opt for pipelining and to what degree.
- Decide on a circuit style, fabrication technology, and manufacturing process.
- What abstraction level to design at and what cell libraries to use, if any?
- Get a first estimate of the circuit's size and cost.
- etc.

The result is captured in a high-level block diagram that includes datapaths, controllers, memories, interfaces, and key signals. A preliminary floorplan is also being established. Verification of an architecture typically occurs by way of simulations, where each major building block is represented by a behavioral model of its own.

---

[18] Take this as an analogy from everyday life. Assume you were given the recipe for a phantastic cake by your grandmother and you were now to make a business out of it by setting up a bakery to mass-produce the cake. The recipe corresponds to the algorithm or software model that specifies how the various ingredients must be processed in order to obtain the final product. Architecture design can then be likened to deciding on the mixers, kneaders, ovens and other machines for processing the ingredients, and to planning the material flow in an industrial bakery. Observe that you will arrive at different factory layouts depending on the quantity of cakes that you intend to produce and depending on the availability and costs of labor and equipment.

[19] These and other circuit-related terms are explained in section 1.6.

The work is then carried down to the more detailed **register transfer level** (RTL) where the circuit gets modeled as a collection of storage elements interconnected by purely combinational subcircuits. Relevant issues at this stage include

- How to implement arithmetic and logic units
  (e.g. ripple-carry, carry-lookahead, or carry-select adder)?
- Whether to use hardwired logic or microcode to implement a controller?
- When to use a ROM rather than random logic?
- What operations to perform during which clock cycle? (**scheduling**)
- What operations to carry out on which processing unit? (**binding**)
- Where to insert pipelining and shimming registers?
- How to balance combinational depth between registers?
- What clocking discipline to adopt?
- What time interval to use as the basic clock period?
- Where to prefer a bidirectional bus over a unidirectional one?
- How to control the access to a bus with multiple drivers?
- By what test strategy to ensure testability?
- How to initialize the circuit?
- etc.

A hardware description language (HDL) is used to hold the outcome of this step. **Simulations** augmented with assertion-based techniques are instrumental in debugging the RTL code. The floorplan is refined on the basis of the more detailed data that are now available and compared against the die size and cost targets for the final product. This is also the point to decide on the most appropriate design level — synthesis, schematic entry, hand layout — for each circuit block.

Logic design. The translation into a gate-level netlist and its Boolean optimization are largely automatic (**HDL synthesis**). The design is now definitively being committed to

- A fabrication depth (e.g. full-custom vs. semi-custom vs. FPL),
- One or more cell libraries (e.g. by AMS vs. Faraday vs. Xilinx),
- A circuit style (e.g. static vs. dynamic CMOS logic), and
- A manufacturing process (e.g. 28HPM by UMC vs. 28HP by TSMC).

The delays and energy dissipation figures associated with the various computational and storage operations are being calculated. Subcircuits that are found to limit performance during pre-layout analysis are identified and redesigned or reoptimized where possible. The result is a complete set of gate-level schematics and/or netlists validated by **electrical rule check** (ERC), gate-level simulation, **timing verification**, and power estimation.

Testability improvement. A malfunctioning IC is the result of design flaws, fabrication defects, or both. Special provisions are necessary to ascertain the correct operation of millions of transistors enclosed in a package with a couple of hundred pins at most. **Design for test** (DFT) implies improving the controllability and observability of inner circuit nodes by adding auxiliary circuitry on top of the payload logic.[20]

In addition, a **test vector set** is generated for telling faulty circuits from correct ones. Such a vector set typically includes thousands or millions of stimuli and expected responses. In a procedure referred to as **fault grading**, testability is rated by relating the number of fabrication defects that can in fact be detected with a test vector set under consideration to the total number of conceivable faults. Both the test circuitry and the test patterns are iteratively refined until a satisfactory fault coverage is obtained.

---

[20] Standard techniques include block isolation, scan testing, and BIST. Block isolation makes major circuit blocks accessible from outside a chip with the aid of extra multiplexers so that stimuli can be applied and responses evaluated via package pins while in test mode. Scan testing is to be outlined in section 7.2.2. The idea behind built-in self test (BIST) is to move stimuli generation and response checking onto the chip itself, and to essentially output a "go/no go" result [2]. BIST and block isolation are popular for testing on-chip memories. As DFT, test vector preparation, and automated test equipment (ATE) are not part of this text, the reader is referred to the specialized literature such as [3] [4], for instance.

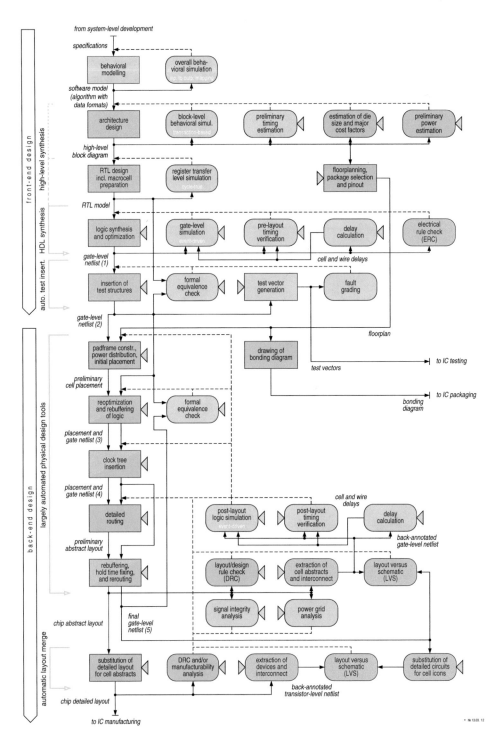

FIGURE 1.13

Digital VLSI design flow (simplified). See fig.1.14 for an explanation of symbols.

Physical design. Physical design addresses all issues of arranging the multitude of subcircuits and
devices along with their interconnections on a piece of semiconductor material. **Floorplanning**
is concerned with organizing the major circuit blocks into a rectangular area as small as possible
while, at the same time, limiting the effects of interconnect delays on the chip's performance.[21]
Power and clock distribution are also to be dealt with. A padframe must be generated to hold
the bond pads and the top-level layout blocks. During the subsequent **place and route** (P&R)
steps, each cell gets assigned a specific location on the die before the courses of myriads of metal
wires that are to carry electrical signals between those cells get defined. The final phase where
the global wires running between padframe and core get routed is also known as **chip assembly**.

As the inner layout details of the cells do not really matter for floorplanning, place, and route,
cells are typically abstracted to their outlines up to this point. To prepare for IC manufacturing,
detailed layout data must be filled in for those abstract views. The outcome is a huge set of
polygons that involves all mask layers. Prior to fabrication, the complete layout data need to
be checked carefully to protect against fatal mishaps. **Physical design verification** relies on a
number of software tools.

- Layout rule check — better known as **design rule check** (DRC) — examines
conformity of layout with geometric rules imposed by the target process.
- **Manufacturability analysis** searches for layout patterns likely to be detrimental to
fabrication yield.
- **Layout extraction** (re-)obtains the actual circuit netlist in preparation of
- **Layout versus schematic** (LVS) where it gets compared against the desired one.
- Post-layout timing verification.
- Post-layout simulation.

Sign-off. By accepting a design for prototype fabrication, an IC vendor commits himself to delivering
circuits that behave exactly like its post-layout simulation model. As no customer is willing to pay
for fabricated parts that do not conform with this requirement, the vendor wants to make sure the
design is consistent with good engineering practice and with company-specific guidelines before
doing so. DRC, manufacturability, ERC, LVS, post-layout simulation, and fault coverage are
routinely examined. Inspection often extends to timing verification, power and clock distribution,
test structures, and more.

A couple of comments are due after this rather general overview.

- In reality, the separation into individual subtasks is not as nice and clear as in fig.1.13. Various
side effects of nanometer-size technologies and the quest for optimum results make it
necessary for most software tools to work across several levels of abstraction. As an example,
it is no longer possible to place and route a gate-level netlist without adapting the circuit logic
as a function of the resulting layout parasitics and interconnect delays. In the drawing, this gets
reflected by the joint refinement of layout data and netlists.

---

[21] Floorplanning makes part of physical design much as layout design does. What is the difference then? As an analogy,
floorplanning is concerned with the partitioning of a flat into rooms and hallways whereas layout design deals with tiny
geometric patterns on a carpet.

- Only ideally does design occur as a linear sequence of steps. Some back and forth between the various subtasks is inevitable to obtain a truly satisfactory result. Also, not all design stages are explicitly covered in every IC development project. Depending on the circuit's nature, fabrication depth, and design level, some of the design stages are skipped or outsourced, i.e. delegated to specialists at third-party companies.[22]

- Note the presence of two kinds of boxes in fig.1.13. While angular boxes refer to construction activities, the rounded ones stand for analysis and verification steps. A backward arrow implies that any problem uncovered during such an analysis triggers corrective action. That results from construction steps are subject to immediate verification is typical for VLSI.
  The reason is that correcting a mistake becomes more and more onerous the further the design process has progressed. Correcting a minor functional bug after layout design, for instance, would require redoing several design stages and would waste many hours of labor and computer time. Also, a functional bug can be uncovered more effectively from a behavioral or RTL model than from a post-layout transistor-level netlist because simulation speed is orders of magnitudes higher and because automatic response checking is much easier to implement for logic and numeric data types than for analog waveforms.

- A critical point is reached when first silicon is going to be produced. While it is possible to cut and add wires using advanced and expensive equipment such as focused ion-beam (FIB) technology to patch a malfunctioning prototype [5], there is virtually no way to fix bugs in volume production. Depending on the circuit's size, fabrication depth, process, and manufacturer, expenses up to 1 MUSD are involved with preparation of photomasks, tooling, wafer processing, preparation of probe cards and evaluation of pre-production samples. Any design flaw found after prototype fabrication thus implies the waste of important sums of money.
  To make things worse, with turnaround times ranging between three weeks and four months, a product's arrival on the market is delayed so much that the chip is likely to miss its window of opportunity.

Observation 1.2. *Redesigns are so devastating for the business that the entire semiconductor industry has committed itself to "first-time-right" design as a guiding principle. To avoid them, VLSI engineers typically spend much more time verifying a circuit than actually designing it.*

- Fig.1.13 also includes a number of forward arrows that bypass one or two construction steps. They suggest how electronic design automation, cell libraries, and purchased know-how help speed up the design process. Keeping pace with the breathtaking progress of fabrication technology is in fact one of the major challenges for today's VLSI designers.

---

[22] The design of a simple glue logic chip, for instance, begins at the logic level as there are no algorithmic or architectural questions to deal with.

1.3.3 CELL LIBRARIES

Library development occurs quite separately from actual IC design.[23] Cell libraries are typically licensed to IC developers by specialized library vendors since silicon vendors have largely withdrawn from this business.

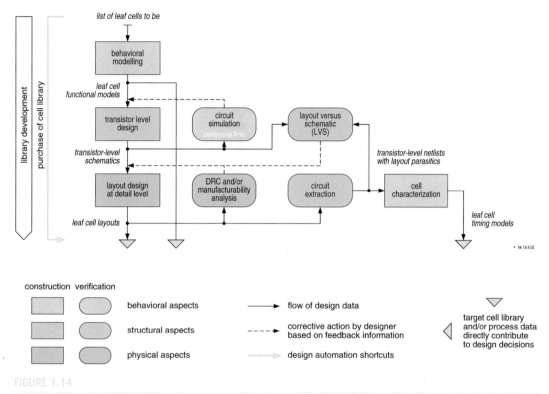

Library design flow (simplified).

Once the set of prospective library cells has been defined functionally, library development proceeds in three major phases, see fig.1.14. **Electrical design** deals with implementing logic functions as transistor-level networks and with sizing the individual devices such as to find an optimum trade-off between performance, circuit complexity, and energy efficiency.

---

[23] Semi-custom ICs and FPL rely on prefabricated primitives anyway.

During the subsequent **layout design**, the locations and geometric shapes of individual devices are defined along with the shapes of the connecting wires. The goal is to obtain leaf cells that are compact, fast, energy-efficient, suitable for automatic place and route (P&R), and that can be manufactured with maximum yield.

Verification includes the customary ERC, DRC, manufacturability analysis, extraction, and LVS procedures. Next the electrical and timing parameters that are to be included into data sheets and simulation models of the cells are determined. This **library characterization** step typically relies on repeated continuous-time continuous-value simulations under varying load, input ramp, and operating conditions.

Designing, characterizing, documenting, and maintaining a cell library is a considerable effort as multiple **design views** must be prepared for each cell, including

- A **datasheet** with functional, electrical, and timing specifications.
- A graphical **icon** or symbol for inclusion into schematic drawings.
- An accurate behavioral models for simulation and timing analysis.
- A set of simulation and test vectors.
- A transistor-level netlist or schematic.
- A detailed layout.
- A simplified layout view showing cell outline and connector locations for the purpose of place and route known as **cell abstract**, floorplanning abstract, or phantom cell.

Please refer back to fig.1.7 for illustrations.

In order to protect their investments, most library vendors consider their library cells to be proprietary and are not willing to disclose how they are constructed internally. They supply datasheets, icons, simulation models, and abstracts, but no transistor-level schematics and no layouts. Under this scheme, detailed layouts are to be substituted for all cell abstracts by the vendor before mask preparation can begin. Note this extra step is reflected in fig.1.13.

1.3.4 ELECTRONIC DESIGN AUTOMATION SOFTWARE

The VLSI industry has long become entirely dependent on electronic design automation (EDA) software. There is not one single step that could possibly be brought to an end without the assistance of sophisticated computer programs. The sheer quantity of data necessary to describe a multi-million transistor chip makes this impossible. The design flow outlined in the previous section gives a rough idea on the variety of CAE/CAD programs that are required to pave the way for VLSI and FPL design. Almost each box in fig.1.13 stands for yet another tool.

While a few vendors can take pride in offering a range of products that covers all stages from system-level decision making down to physical layout, their efforts tend to focus on relatively small portions of the overall flow for reasons of market penetration and profitability. Frequent mergers and acquisitions are another characteristic trait of the EDA industry. Truly integrated design environments and seamless design flows are hardly available out of the box.

Also, the idea of integrating numerous EDA tools over a common design database and with a consistent user interface, once promoted as front-to-back environments, aka frameworks, has lost momentum in the marketplace in favor of point tools and the "best in class" approach. Design flows are typically pieced together from software components of various origins.[24] The prevalence of software tools, design kits, and cell libraries from multiple sources in conjunction with the absence of industry-wide standards adds to the complexity of maintaining coherent design environments. Many of the practical difficulties with setting up efficient design flows are left to EDA customers and can sometimes become a real nightmare. Hopefully this trend will be reversed one day when customers are willing to pay more attention to design productivity than to layout density and circuit performance.

---

[24] Fascinating accounts of the evolution of the EDA industry are given in [6] [7].

1. ∗ Various examples of design views have been given in figs.1.7, 1.9, and 1.11. Locate them in the Y-chart of fig.1.10.

2. ∗ Think of some industrial product family of your own liking (computer, record/music player, smartphone, camera, TV set/video recorder, car, locomotive, airplane, photocopier, light source, building control equipment, etc.). Discuss what microelectronics has contributed towards making these products possible in their present form. How has the microelectronic content evolved over the years? Where do you see challenges for improving these products and their microelectronic content?

1.5 APPENDIX I: A BRIEF GLOSSARY OF LOGIC FAMILIES

A logic family is a collection of digital subfunctions that

- support the construction of arbitrary logic, arithmetic and storage functions,
- are compatible among themselves electrically, and that
- share a common fabrication technology.

A logic family must be available either as physical parts (SSI/MSI/LSI components for board design) or in virtual form as a set of library cells to be instantiated and manufactured together on a semiconductor chip (for IC design).

Table 1.3   Major semiconductor technologies and logic families with their acronyms.

Acronym	Meaning
MOS	Metal Oxide Semiconductor.
FET	Field Effect Transistor (either of n- or p-channel type).
BJT	Bipolar Junction Transistor (either of npn or pnp type).
NMOS	n-MOS (transistor, circuit style, or fabrication technology).
PMOS	p-MOS (transistor, circuit style, or fabrication technology).
CMOS	Complementary MOS (circuit style or fabrication technology) where pairs of n- and p-channel MOSFETs cooperate in each logic gate; features zero quiescent power dissipation, or almost so; supply voltages have evolved from up to 15 V down to below 1 V.
static CMOS	Circuit style that supports suspending all switching activities indefinitely and in any state with no loss of state or data.
dynamic CMOS	Circuit style where data and/or state are kept as electrical charges that need to be refreshed or computed anew at regular intervals as data and/or state are otherwise lost.
TTL	Transistor Transistor Logic, made up of BJTs and passive devices; first logic family to gain widespread acceptance as SSI/MSI parts, has evolved over many generations all of which share a 5 V supply.
ECL	Emitter-Coupled Logic, non-saturating current switching circuits built on the basis of BJTs, provides complementary outputs with a mere 0.5 V swing; exhibits prohibitive static power dissipation.
ESCL	Enhancement Source-Coupled Logic, similar but built from MOSFETs.
BiCMOS	CMOS subcircuits combined with bipolar devices on a single chip.

Originally a low-power but slow alternative to TTL, CMOS has become the technology that almost totally dominates VLSI today. This is essentially because layout density, operating speed, energy efficiency, and manufacturing costs per function benefit from the geometric down-scaling that comes with every process generation. In addition, the simplicity and comparatively low power dissipation of CMOS circuits have allowed for integration densities not possible on the basis of BJTs.

Fig.1.15 opposes samples from various logic families. The focus of this text is on static circuits in CMOS technology. However, as designing digital VLSI systems and developing with PLDs only loosely depend on technology, there is no need to worry about further details at this stage.

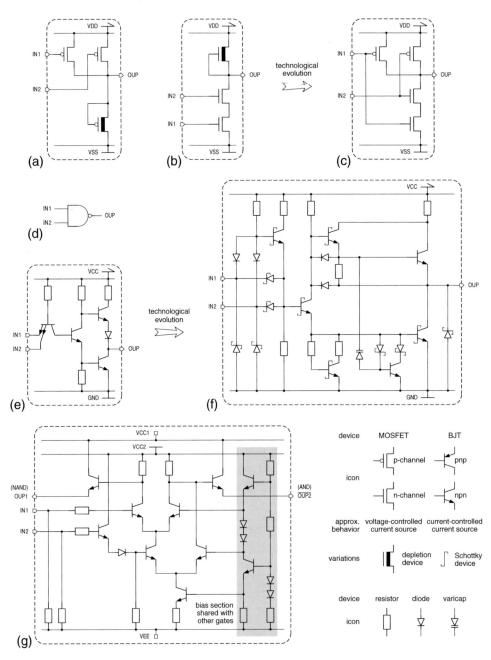

FIGURE 1.15

Major logic families exemplified by way of a 2-input NAND gate. Icon (d), PMOS (a), NMOS (b), static CMOS (c), TTL (e,f), and ECL circuits (g). (e) shows the original multi-emitter structure that gave TTL its name whereas (f) refers to a more recent F generation part that includes many auxiliary devices for clamping and speedup. Observe how disparate one gate equivalent (GE) can be.

1.6 APPENDIX II: AN ILLUSTRATED GLOSSARY OF CIRCUIT-RELATED TERMS

Table 1.4 lists important terms from digital circuits, microelectronics, and electronic design automation (EDA). Two illustrations follow. Fig.1.16 identifies most of the underlying concepts by way of a circuit diagram while fig.1.17 shows how they reflect in a hardware description language (HDL) model. Although those concepts are applied throughout the EDA community, the terms being used and their meanings vary from one company to the next.

Note the difference between a **schematic** and a **netlist**. Either one unambiguously specifies a circuit as a collection of interconnected components. On top of this, schematic data include information that indicate where and how to draw icons, wires, busses, etc. on a computer screen or on a piece of paper. While totally irrelevant from an electrical or functional point of view, the graphical arrangement matters when humans want to grasp a circuit's organization and understand its operation. A netlist is easily derived from a schematic, but the converse is not obvious. Except for trivial examples, circuit diagrams obtained from netlists by automatic means lack the clarity and expressiveness of human-made schematics.

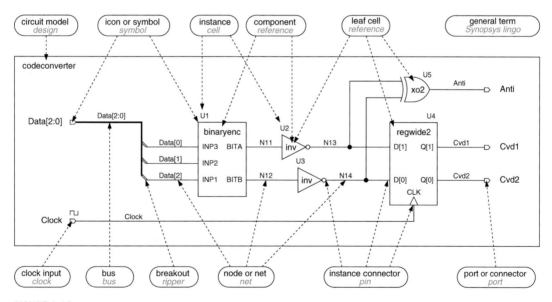

FIGURE 1.16

Circuit-related terms illustrated by way of a schematic drawing. Make sure you understand why U2 and U3 relate to the same component but to distinct instances. Also note that inv and xo2 are leaf cells whereas binaryenc and regwide2 are not.

Relating to information-processing hardware, **datapath** is a generic term for all those subcircuits that manipulate payload data, see fig.1.18. That is, a datapath is not confined to arithmetic/logic units (ALU) that carry out operations on data words, but also includes short-term data storage (accumulators, registers, FIFOs) plus the necessary data routing facilities (busses and switches). Datapaths tend to be highly regular as similar functions are carried out on multiple bits at a time.

Datapath operation is governed by a control section that also coordinates activities with surrounding circuits. The **controller** does so by interpreting various status signals and by piloting datapath operation via control signals in response. A controller is either implemented as a hardwired finite state machine (FSM), as a stored program (program counter plus microcoded instruction sequence), or as a combination of both. In a computer-type architecture, all facilities dedicated to the sole purpose of address processing must be considered part of the controller, not of the datapath, even if they are ALUs or registers by nature.

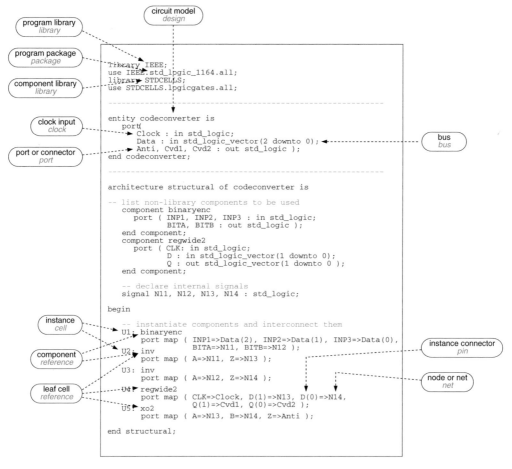

FIGURE 1.17

Circuit- and software-related terms in a structural VHDL model.

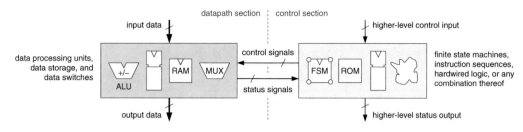

FIGURE 1.18

Interplay of datapath and controller in a typical information-processing circuit.

Table 1.4   A Glossary of terms from electronic design.

General term	Synopsys lingo	Meaning
Circuit elements		
circuit model	design	a description of an electronic circuit or subcircuit
component	reference	a self-contained subcircuit of well-defined functionality
component library	library	a named collection of components
(leaf) cell	reference	an atomic component typically available from a library that can not be decomposed into smaller components
instance	cell	one specific copy of a subcircuit that is being used as part of a larger circuit
Interconnect		
node aka net	net	an electrical node or — which is the same — a wire that runs between two or more (instance) connectors
port aka terminal aka connector	port	a node that can be electrically contacted from the next higher level of circuit hierarchy
instance connector	pin	a connector of an instance
clock input	clock	a connector explicitly defined as clock source
bus	bus	a named set of nodes with cardinality > 1
special net		a net not shown but tacitly implied in schematics, examples: ground and power
Circuit drawings		
icon aka symbol	symbol	a graphical symbol for a component or a connector
schematic diagram	schematic	a drawing of a (sub)circuit that is made up of icons and of wires where the latter are graphically shown as lines
netlist	netlist	a data structure that captures what instances make up for a (sub)circuit and how they are interconnected
breakout	ripper	a special icon that indicates where a net or a subbus leaves or enters the graphical representation for a bus

Continued

Table 1.4   A Glossary of terms from electronic design—cont'd

General term	Synopsys lingo	Meaning
Integrated circuits		
die aka chip		a fully processed but unencapsulated IC
package		the encapsulation around a die
(package) pin		a connector on the outside of an IC package
pad	pad	a connector on a die that is intended to be wired or otherwise electrically connected to a package pin, the term is often meant to include interface circuitry
HDL software		
program package	package	a named collection of data types, subprograms, etc.
program library	library	a named repository for compiled program packages
Functional verification		
model under test	design ...	a circuit model subject to simulation
circuit under test		a physical circuit, e.g. a chip, subject to testing
testbench		HDL code written for driving the simulation of a model under test; not meant to be turned into a physical circuit
Layout items		
layout		a 2D drawing that captures a component's detailed geometry layer by layer and that guides IC fabrication
(cell) row		many standard cells arranged in a row such as to share common ground lines, power lines, and wells
well		a volume that accommodates MOSFETs of identical polarity; doping is opposite to the source and drain islands embedded
row end cell		a special cell void of functionality to be instantiated at either end of a cell row to properly end the wells
fillcap cell aka decap		a special cell void of functionality to be instantiated between two regular cells to add decoupling capacitance typically where dense wiring asks for extra room anyway
tie-off cell		a special cell void of functionality to be instantiated where a regular net must connect to ground or power
cell outline aka abstract		a simplified view where a cell's layout is reduced to the outline and the locations of all of its connectors
routing channel		space set aside between adjacent cell rows for wiring, no longer needed with today's multi-metal processes
contact		a galvanic connection between a metal and a silicon layer
via		a galvanic connection between two superimposed metal layers
bonding area		a square opening in the protective overglass exposing a die's top-level metal for connecting to a package pin

FIELD-PROGRAMMABLE LOGIC

2

What makes field-programmable logic (FPL) attractive in many applications are their low up-front costs and short turnaround times. That it should be possible to turn a finished piece of silicon into an application-specific circuit by purely electrical means — i.e. with no bespoke photomasks or wafer processing steps — may seem quite surprising at first, and section 2.1 demonstrates the basic approach. Sections from 2.2 onwards then give technical details and explain the major differences that set apart the various product families from each other, before the particularities of the FPL design flow get discussed in section 2.6.

2.1 GENERAL IDEA

The term "programmable" in field-programmable logic is a misnomer as there is no program, no instruction sequence to execute. Instead, pre-manufactured subcircuits get configured into the target circuit via electrically programmable links that can be done — and in many cases also undone — as dictated by so called **configuration bits**. This is nicely illustrated in figs.2.1 to 2.3 from [8] (copyright Wiley-VCH Verlag GmbH & Co. KG, reprinted with permission).

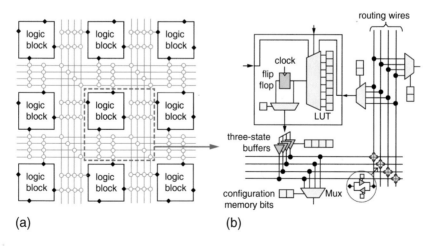

(a)   (b)

FIGURE 2.1

FPGA hardware resources before configuration. Top-level organization (a) and one configurable logic cell (b). In panel (b), each small square represents one configuration bit.

41

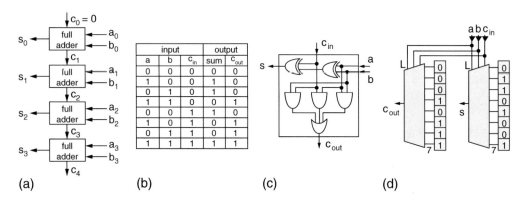

FIGURE 2.2

Target functionality ripple carry adder. General diagram (a), truth table of one full adder slice (b), gate-level circuit (c), and implementation as a lookup table (LUT) (d).

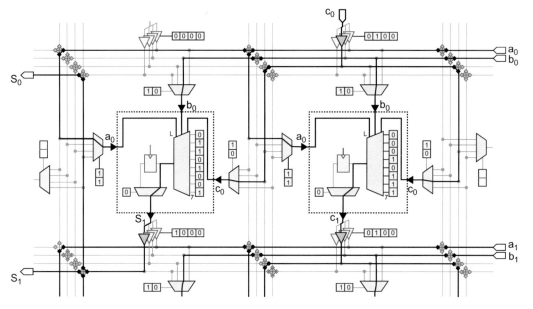

FIGURE 2.3

FPGA after configuration. One full adder slice implemented using the pre-fabricated resources. Highlighted lines show the wires activated by configuration bits.

Key properties of any FPL device depend on decisions taken by its developers along two dimensions. A first choice refers to how the device is actually being configured and how its configuration is stored electrically, while a second choice is concerned with the overall organization of the hardware resources made available to customers.[1]

---

[1] Customers, in this case, are design engineers who want to create their circuits using an FPL device (rather than as a semi or full-custom IC).

2.2 CONFIGURATION TECHNOLOGIES

Three configuration technologies coexist today, they all have their roots in memory technology.

2.2.1 STATIC MEMORY

The key element here is an electronic switch — such as a transmission gate, a pass transistor, or a three-state buffer — that gets turned "on" or "off" under control of a configuration bit. Unlimited reprogrammability is obtained from storing the configuration data in static memory (SRAM) cells or in similar on-chip subcircuits built from two cross-coupled inverters, see fig.2.4a.

Reconfigurability is very helpful for debugging. It permits one to probe inner nodes, to alternate between normal operation and various diagnostic modes, and to patch a design once a flaw has been located. Many RAM-based FPL devices further allow for reconfiguring their inner logic during operation, a capability known as **in-system configuration** (ISC) that opens a door towards reconfigurable computing.

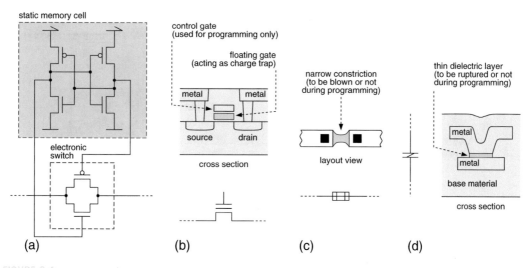

FIGURE 2.4

FPL configuration technologies (simplified, programming circuitry not shown). Switch steered by static memory cell (a), MOSFET controlled by a charge trapped on a floating gate (b), fuse (c), and antifuse (d).

As a major drawback of SRAM-based storage, an FPL device must (re-)obtain the entire configuration — the settings of all its programmable links — from outside whenever it is being powered up. The problem is solved in one of three possible ways, namely

(a) by reading from a dedicated bit-serial or bit-parallel off-chip ROM,
(b) by downloading a bit stream from a host computer, or
(c) by long-term battery backup.

2.2.2 FLASH MEMORY

Flash memories rely on special MOSFETs where a second gate electrode is sandwiched between the transistor's bulk material underneath and a control gate above, see fig.2.4b. The name **floating gate** captures the fact that this gate is entirely surrounded by dielectric material. An electrical charge trapped there determines whether the MOSFET, and hence the programmable link too, is "on"or "off".[2]

Charging occurs by way of hot electron injection from the channel. That is, a strong lateral field applied between source and drain accelerates electrons to the point where they get injected through the thin dielectric layer into the floating gate, see fig.2.5a. The necessary programming voltage on the order of 5 to 20 V is typically generated internally by an on-chip charge pump.

Erasure occurs by allowing the electrons trapped on the floating gate to tunnel through the oxide layer underneath the floating gate. The secret is a quantum-mechanical effect known as Fowler-Nordheim tunneling that comes into play when a strong vertical field (8 ... 10 MV/cm or so) is applied across the gate oxide.

Flash FPL devices are non-volatile and immediately live at power up, thereby doing away with the need for any kind of configuration-backup apparatus. The fact that erasure must occur in chunks, that is to say many bits at a time, is perfectly adequate in the context of FPL. Data retention times vary between 10 and 40 years. Endurance of flash FPL is typically specified with 100 to 1000 configure-erase cycles, which is much less than for flash memory chips.

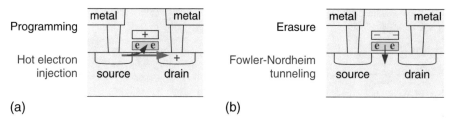

(a)   (b)

FIGURE 2.5

Flash memory operation during programming (a) and erasure (b).

2.2.3 ANTIFUSES

Fuses, which were used in earlier bipolar PROMs and SPLDs, are narrow bridges of conducting material that blow in a controlled fashion when a programming current is forced through. Antifuses, such as those employed in today's FPGAs, are thin dielectrics separating two conducting layers that are made to rupture upon applying a programming voltage, thereby establishing a conductive path of low impedance.

---

[2] More precisely, the presence or absence of an electrical charge modifies the MOSFET's threshold voltage and so determines whether the transistor will conduct or not when a voltage gets applied to its control gate during memory readout operations.

In either case, programming is permanent. Whether this is desirable or not depends on the application. Full factory testing prior to programming of one-time programmable links is impossible for obvious reasons. Special circuitry is incorporated to test the logic devices and routing tracks at the manufacturer before the unprogrammed devices are being shipped. On the other hand, antifuses are only about the size of a contact or via and, therefore, allow for higher densities than reprogrammable links, see fig.2.4c and d. Antifuse-based FPL is also less sensitive to radiation effects, offers superior protection against unauthorized cloning, and does not need to be configured following power-up.

Table 2.1   FPL configuration technologies compared.

Configuration technology	Non vola- tile	Live at power up	Reconfi- gurable	Unlimit. endu- rance	Radiation tolerance of config.	Area occupation per link	Extra fabr. steps
SRAM	no	no	in circuit	yes	poor	large	0
Flash memory	yes	yes	in circuit	no	good	small	>5
Antifuse PROM	yes	yes	no	n.a.	best	smallest	3

2.3 ORGANIZATION OF HARDWARE RESOURCES

2.3.1 SIMPLE PROGRAMMABLE LOGIC DEVICES (SPLD)

Historically, FPL has evolved from purely combinational devices with just one or two programmable levels of logic such as ROMs, PALs and PLAs. Flip-flops and local feedback paths were added later to allow for the construction of finite state machines, see fig.2.6a and b. Products of this kind continue to be commercially available for glue logic applications. Classic SPLD examples include the 18P8 (combinational) and the 22V10 (sequential).

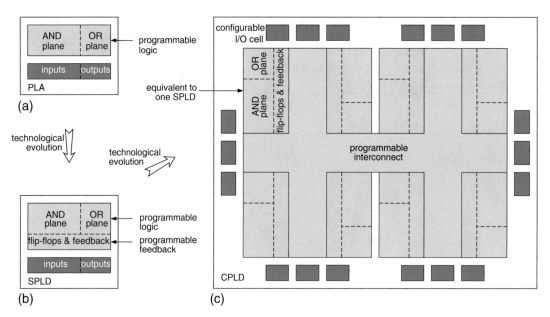

FIGURE 2.6

General architecture of CPLDs (c) along with precursors (a,b).

The rigid two-level-logic-plus-register architecture and the scanty resources (number of inputs, outputs, product terms, flip-flops) naturally restrict SPLDs to small applications. More powerful architectures had thus to be sought, and the spectacular progress of VLSI technology has made their implementation economically feasible from the late 1980's onwards.

2.3.2 COMPLEX PROGRAMMABLE LOGIC DEVICES (CPLD)

CPLDs simply followed the motto "more of the same", see fig.2.6c. Many identical subcircuits, each of which conforms to a classic SPLD, are combined on a single chip together with a large programmable interconnect matrix or network. A difficulty with this type of organization is that a partitioning into a bunch of cooperating SPLDs has to be imposed artificially on any given application which neither benefits hardware nor design efficiency.

2.3.3 FIELD-PROGRAMMABLE GATE ARRAYS (FPGA)

FPGAs have their overall organization patterned after that of gate-arrays. Many **configurable logic cells** are arranged in a two-dimensional array with bundles of parallel wires in between. A switchbox is present wherever two wiring channels intersect, see fig.2.7.[3] Depending on the product, each logic cell can be configured such as to carry out some not-too-complex combinational operation, to store a bit or two, or both. As opposed to traditional gate arrays, it is the state of programmable links rather than fabrication masks that decide on logic functions and signal routing. The number of configurable logic cells greatly varies between products, with typical figures ranging between a few dozens and hundred thousands.

Owing to their more scalable and flexible organization, FPGAs prevail over CPLDs. They are differentiated further depending on the granularity of the configurable logic.

---

[3] While it is correct to think of alternating cells and wiring channels from a conceptual point of view, you will hardly be able to discern them under a microscope as they are superimposed for the sake of layout density.

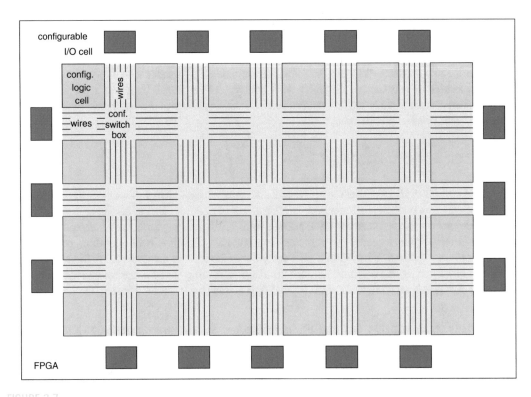

FIGURE 2.7

General architecture of FPGAs.

Fine-grained FPGAs. One speaks of a fine-grained architecture when the configurable cells are so simple that they are capable of implementing no more than a few logic gates and/or one bistable. In the example of fig.2.8a, for instance, each logic cell can be configured into a latch, or a flip-flop, or into almost any 3-input gate.

Coarse-grained FPGAs. Here cells are designed to implement combinational functions of four or more variables and are capable of storing two or more bits at a time. The logic cell of fig.2.8b has 16 inputs and 11 outputs, includes two programmable 4-input lookup tables (LUT), two generic bistables that can be configured either into a latch or a flip-flop, a bunch of configurable multiplexers, a fast carry chain, plus other gates. Of course, the superior functional capabilities offered by a coarse-grained cell are accompanied by a larger area occupation.

The gate-level netlists produced by general synthesis tools map more naturally onto fine-grained architectures because fine-grained FPGAs and semi-custom ICs provide similar primitives. This makes it possible to move back and forth between field- and mask-programmed circuits with little overhead and to defer final commitment until fairly late in the design cycle. Conversely, fine-grained FPGAs tend to be more wasteful in terms of configuration bits, routing resources, and propagation delays.

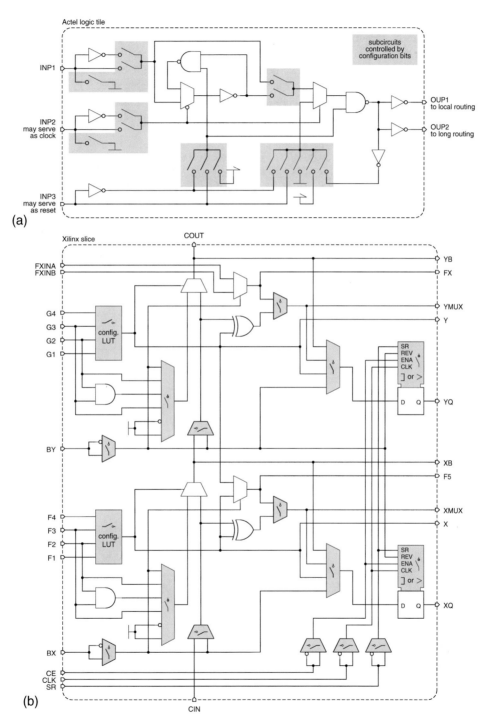

FIGURE 2.8

Fine-grained vs. coarse-grained FPGAs. A small (Actel ProASIC) (a) and a large logic cell (Xilinx Virtex-4, simplified) (b).

Vendors of coarse-grained FPGAs have done a fair bit to overcome the drawbacks of their idiosyncratic FPGA architectures by providing their customers with proprietary software tools that help them make good usage of the available hardware resources.[4] Another reason that contributed to the popularity of coarse-grained FPGAs is that on-chip RAMs come at little extra cost when combined with configuration from static memory. In fact, a reprogrammable LUT is nothing else than a tiny storage array. It is thus possible to string together multiple logic cells such as to make them act collectively like a larger RAM. In the occurrence of fig.2.8b, each of the two larger LUTs in each logic tile contributes another 16 bit of storage capacity.

Fig.2.9 indicates that the optimum trade-off for LUTs has shifted from 4 to 6 inputs over the last couple of generations. A comparison of figs.2.8b (2004, 90 nm) and 2.10 (2009, 40 nm) indeed confirms the trend towards coarser granularities. An excerpt from the datasheet of a competing product (2010, 20 nm) is shown in fig.2.11.[5]

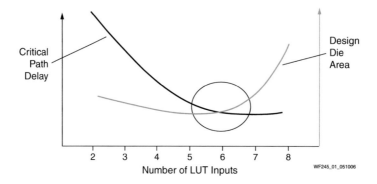

FIGURE 2.9

LUT granularity trade-offs at the 65 nm technology node (source: Xilinx, reprinted with permission).

---

[4] Such as Synplify Pro (Synopsys), Quartus (Altera), Vivado (Xilinx).
[5] FPL vendors refer to configurable logic cells by proprietary names. "VersaTile" is Actel's term for their fine-grained cells while Altera uses the name "adaptive logic module" (ALM) for their coarse-grained building blocks. Xilinx refers to their counterparts as "configurable logic blocks" (CLB). Depending on the product family, one a CLB may be composed of two "slices" each of which includes several LUTs and bistables. Cypress speaks of "universal digital blocks" (UDB).

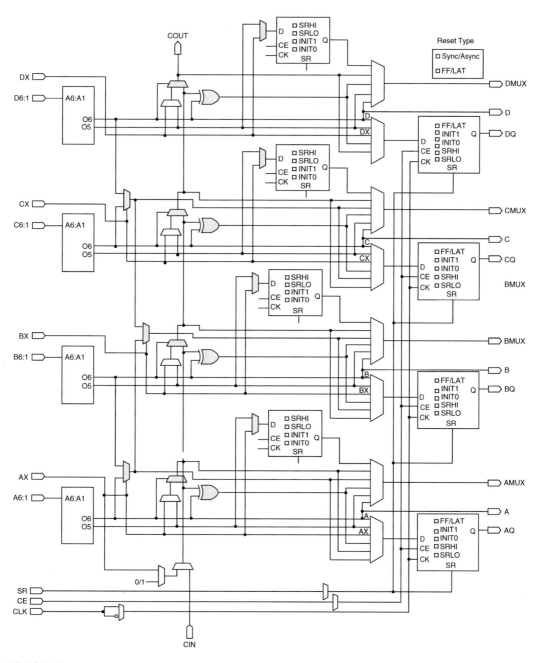

Logic slice from Xilinx Virtex-6, SLICEL with 4 6-input LUTs and 8 bistables (source: Xilinx, reprinted with permission).

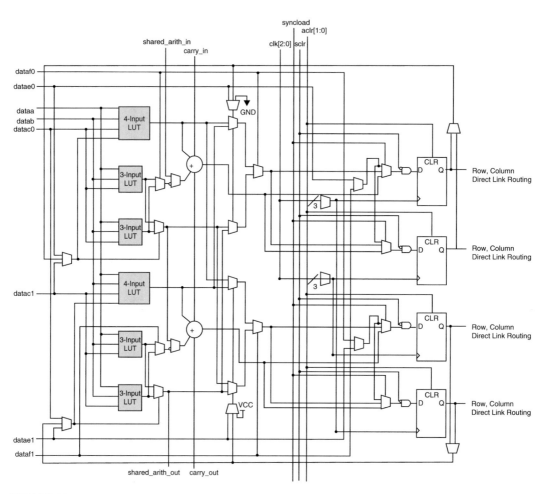

FIGURE 2.11

Adaptive logic module from Altera Stratix V (source Altera, reprinted with permission).

2.4 COMMERCIAL ASPECTS

2.4.1 AN OVERVIEW ON FPL DEVICE FAMILIES

Table 2.2 classifies major commercial CPLD and FPGA device families along the two dimensions configuration technology and hardware organization.

Table 2.2   Field-programmable logic product families.

Configuration technology	Overall organization of hardware resources		
	CPLD	FPGA	
		coarse grained	fine grained
Static memory (SRAM)		Xilinx Virtex, Kintex, Artix, Spartan. Lattice SC, EC, ECP. Altera Stratix, Arria, Cyclone. eASIC Nextreme SL[a]. Achronix Speedster[b].	Atmel AT6000, AT40K.
Flash memory	Xilinx XC9500, CoolRunner-II. Altera MAX. Lattice MACH 1,...,5. Cypress Delta39K, Ultra37000, PSoC 1,...,5LP[f].	Lattice XP[c], MACH XO.	Actel[d] ProASIC3, ProASIC3 nano, Igloo, Fusion[e].
Antifuse (PROM)		QuickLogic Eclipse II, PolarPro.	Actel MX, Axcelerator AX.

[a] Combines RAM-configurable LUTs with e-beam single via-layer customization for interconnect.
[b] Combines synchronous I/O with self-timed clocking inside.
[c] Combines on-chip flash memory with an SRAM-type configuration memory.
[d] Actel has been acquired by Microsemi in late 2010.
[e] Mixed-signal FPGAs.
[f] Mixed-signal CPLDs.

2.4.2 THE PRICE AND THE BENEFITS OF ELECTRICAL CONFIGURABILITY

For obvious reasons, the ability to (re)define a part's functionality long after it has left the fabrication line is extremely valuable in the marketplace. What's more, many applications that mandated a custom ASIC just a few years ago fit into a single FPL device today. Fueled by technological progress and continued price reductions, this trend is bound to carry on.

Yet, FPL is unlikely to rival hardwired logic on the grounds of integration density, unit costs, and energy efficiency unless there is an unforeseen technological breakthrough. This is because FPL must accommodate extra transistors, programmable links, interconnect lines, vias, lithographic masks, and wafer processing steps to provide for configurability. Also, the required and the prefabricated hardware resources never quite match, leaving part of the manufactured gates unused.

In fact, comparisons of FPGAs against cell-based ASICs manufactured with a similar technology have exposed important overheads in terms of area, propagation delays, and power dissipation.

Overhead factors for				
area	timing	power	compared to	source
35	3.4...4.6	14	SRAM-based FPGAs	[9] (2007, 90 nm CMOS)
27	5.1	n.a.	idem	[10] (2013, 130 nm CMOS)

Opting for a reconfigurable FPGA, rather than for a mask-programmed ASIC, is thus likely to inflate the AT-product by more than two orders of magnitude. While antifuse technology, hardwired multipliers, etc. improve the situation, a significant penalty remains. The huge area overhead further explains why large FPGAs continue to be rather expensive, even when bought in substantial quantities. FPL vendors attempt to compensate for this by using the most advanced semiconductor fabrication processes available and have indeed gotten ahead of the average ASIC technology in recent years [11]. Table 2.3 shows what is possible today (2014Q1).

Fig.2.12 puts FPL devices in perspective with semi- and full-custom ICs. We would like to emphasize that this is a simplified overview and that numbers are quoted for illustrative purpose only. It should nevertheless become clear that each technique has its particular niche where it is the best compromise.

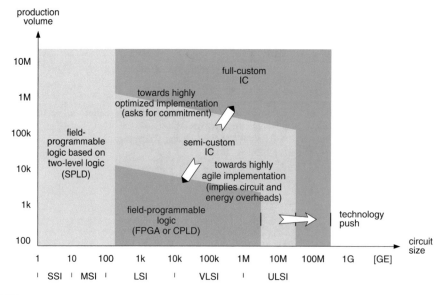

FIGURE 2.12

Application scopes of various implementation techniques as a function of circuit complexity and volume (simplified).

2.5 EXTENSIONS OF THE BASIC IDEA

From the above, it should be clear that configurable logic is best confined to those circuit portions that are subject to frequent changes when unit costs, energy efficiency, or operating speed are critical. Vendors have thus extended their FPL families beyond table 2.2 by combining them with less malleable but more cost-effective and more efficient hardware resources. The idea is to provide just as much configurability as needed to better compete with business rivals such as mask-programmed ICs, signal processors, and with competing FPL products.

Datapath units. Configurable logic cells are designed to implement small look-up tables and random logic functions. When used to implement multiplications on wide data words, for instance, the extensive configurability tends to become overly burdensome. Some FPGA families have, therefore, been equipped with configurable datapath units optimized for multiply-accumulate (MAC) and related arithmetic-logic operations, see fig.2.13. Compared to configurable logic cells, those specialized units come with fairly generous word widths and support digital signal processing applications much more efficiently in terms of hardware usage, throughput and energy provided the synthesis software consistently maps inner products and other suitable operations onto those coarse-grain computing resources.

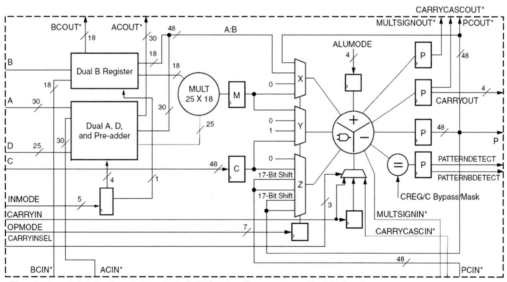

*These signals are dedicated routing paths internal to the DSP48E1 column. They are not accessible via fabric routing resources.

FIGURE 2.13

DSP48E1 slice from Xilinx Virtex-6 (source: Xilinx, reprinted with permission).

Hardwired building blocks. Almost all FPL devices feature hardwired subcircuits on the same die. This is because it makes no sense to tie up precious configurable resources for implementing fixed functions. Typical fixed-function blocks include

- SRAMs, FIFOs, clock recovery circuits, SerDes.
- Industry-standard functions and interfaces
  (such as PCI, USB, FireWire, Ethernet, WLAN, JTAG, LVDS, etc.).
- Analog-to-digital and digital-to-analog converters.
- Entire microprocessor and DSP cores (e.g. PowerPC, ARM).
- Weakly configurable analog subfunctions such as filters or phase locked loops (PLL).

Table 2.3   Maximum resources in two of the most advanced SRAM-based FPGA families. Competing products offer comparable features.

Vendor	Xilinx	
Product	Virtex-7	Virtex Ultrascale
Year introduced	2013	2014
Technology	20 nm CMOS planar	16 nm CMOS finFET
Configurable logic cells[a] [k]	1995	4407
Block RAM [Mbit]	68	115
DSP48 slices	3600	2880
I/O pins	1200	1456
Serial transceivers	96	104
PCI Express blocks	4	6
100G Ethernet blocks	0	7

[a]One Xilinx logic cell roughly corresponds to a 4-input LUT plus a flip-flop.

To give real-world numbers, table 2.3 reproduces selected data of high-end FPGAs from [12]. An extension that goes into a somewhat different direction are the

Field-programmable analog arrays (FPAA). Electrically configurable analog circuits built from OpAmps, capacitors, resistors, and switchcap elements, have begun to appear on the market in the late 1990s. The next logical step was the extension to mixed-signal applications. Advanced products that combine configurable analog building blocks with a micro- or digital signal processor and with analog-to-digital and digital-to-analog converters come quite close to the vision of field-programmable systems on a chip. Vendors of field-programmable analog and mixed-signal arrays include Anadigm, Actel, Cypress, Lattice, and Silego.

Advanced configurable devices that include the right mix of hardwired blocks improve overall energy efficiency, help customers reduce time to market even further than pure FPL parts, and cut their product development and unit costs. Fig.2.14 shows an example for mixed-signal applications. The trend towards (re)configuring larger, more powerful entities (ALUs, datapath units, memories, etc. rather than gates and LUTs) and towards combining (re)configurable logic with processor cores and fixed function blocks is expected to continue. This will naturally lead to platform ICs, a concept that carries electrical configurability further and that is to be discussed in section 3.2.9.

For all enthusiasm about those phantastic capabilities and prospects, note that evaluating FPL devices may be frustrating. As opposed to full-custom ICs, manufactured gates, usable gates, and actual gates are not the same. **Manufactured gates** indicate the total number of GEs that are physically present on a silicon die. A substantial fraction thereof is not usable in practice because the combinational functions in a given design do not fit into the available lookup tables exactly, because an FPL device only rarely includes combinational and storage resources with the desired proportions, and because of limited interconnect resources. The percentage of **usable gates** thus depends on the application. The **actual gate** count, finally, tells how many GEs are indeed put to service by a given design. The three figures frequently get muddled up, all too often in a deliberate attempt to make one product look better than its competitors in advertisements, product charts, and datasheets.

Example

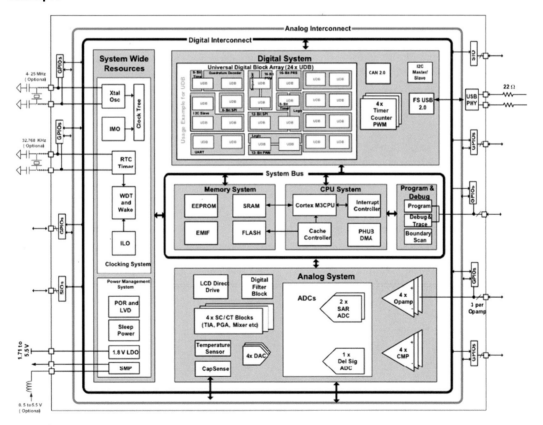

FIGURE 2.14

Block diagram of Cypress mixed-signal PSoC 5LP device (source Cypress, reprinted with permission).
□

That the available resources often get specified using proprietary units other than gate equivalents adds to the confusion. And as exposed in the quote below (after Kevin Morris), this is just the tip of the iceberg:

> Certainly one of the problems with FPGA technology is that you're constantly comparing different things. Apples and oranges, [Xilinx] configurable logic cells and [Altera] adaptive logic modules, field-programmable elements and [largely] hardwired [datapath] units, total negative slack and fastest clock, dynamic power at 20 °C and quiescent power at 85 °C, prices today for quantity 1000 and prices for 9 months from now at quantity 250 000. The list is almost endless, and useful comparison data is virtually impossible to gather.

Hint: It often pays to conduct benchmarks with a few representative designs before undertaking serious cost calculations and making a misguided choice. This also helps to obtain realistic timing figures that take into account interconnect delays.

2.6 THE FPL DESIGN FLOW

Front-end design

The front-end flow is essentially the same as for ASICs. However, depending on the target product, there are a couple of particularities that affect architectural choices. In the occurrence of the popular coarse-grained FPGAs, these include:

± Look-up tables are cheap and typically come in chunks of 64 entries each (6 inputs)

− Routing dominates over gate delay (due to configuration switches and larger dies).
  ↝ Large, deep combinational networks put at a disadvantage.

− Routing resources are limited. ↝ The ideal architecture consists of many loosely connected circuit blocks, each of which fits into one logic cell, logic slice, adaptive logic module, or whatever is the name of the configurable function block.

+ Flip-flops come in generous numbers. ↝ Pipelining is essentially free.

~ Data and/or state coding schemes other than minimum bit encoding sometimes yield better solutions. ↝ Check one-hot, Gray, and Johnson coding, for instance.

+ Parts come with sophisticated on-chip clock preparation circuits (nets, drivers, PLLs), but their number is limited. ↝ A few large clock domains work best.

− Asynchronous reset networks compete for global interconnect resources with clocks in some products. ↝ Synchronous or partial resets tend to facilitate routing.[6]

+ Parts come with sophisticated input/output circuits (adjustable, LVDS, synchronization).

+ Parts include on-chip block RAMs (depending on product).

+ Many parts include weakly configurable datapath units (discussed in section 2.5).

± Hardwired function blocks (such as datapaths, multipliers, adders, and memories) come with fixed word widths. ↝ While it may be difficult to make good use of them, they usually outperform LUT-based alternatives.

+ Some parts are available with one or more on-chip microcontrollers.

− Devices come in thousands of variations. ↝ May be confusing.

− Parts come in fixed sizes. ↝ Circuit complexity matters mostly when up- or downgrading from one size to the next.

− Tools may make suboptimal decisions without designers becoming aware of.

Observation 2.1. *As all resources come with coarser granularities, the cost matrix of FPGAs is not the same as with ASICs. Generally speaking, it pays to be aware of the realities of the target platform before writing RTL synthesis code.*

---

[6] (Re-)configuring an FPL device can be understood as the strongest possible way to re-initialize the circuit's state, making it possible to dispense with a separate reset mechanism in certain applications.

Back-end design

Back-end design for field-programmable logic (FPL) differs considerably from the one depicted in fig.1.13. As FPL parts come with everything prefabricated, there is no need for actually placing cells or for routing interconnect lines. Instead, the gate-level netlist obtained from HDL synthesis is mapped onto the existing configurable cells in the target device and gets reoptimized to make the best possible usage of the logic resources available. EDA software further decides how to configure the switches and line drivers such as to obtain the wiring specified in the netlist. The combined result is then converted into a **configuration bit stream** for download into the FPL device. As FPGA and CPLD products come with many diverse architectures, FPL vendors make available proprietary tools for this procedure.

Apart from short turnaround times, several more factors make the design process simpler and more efficient, and so contribute to the success of FPL.

- Built-in processor cores, interfaces, and other standard functions greatly help to accelerate the FPL design cycle when compared to a custom design where the same functionality must be obtained from macrocells and virtual components.

- Many issues that must be addressed in extenso when designing a custom IC are implicitly solved. There is no need to agonize over subordinate details such as I/O subcircuits, clock and power distribution, embedded memories, testability, and more, see fig.2.15.

- Design tools are more affordable and up-front costs considerably lower than with any other hardware alternative.

- To aid with debugging, vendors provide so-called "signal taps" (Altera) or "chip scopes" (Xilinx) that help monitoring the waveforms on user-selected circuit nodes. These can be thought of as Virtual Components (VCs) that get temporarily inserted into the payload circuitry to sample signals and to store the values in on-chip memories for later inspection.

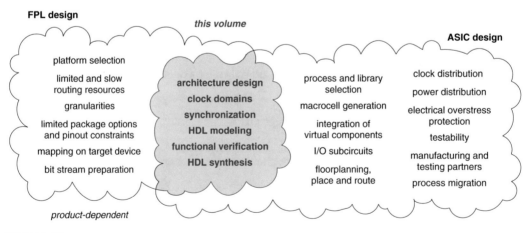

FIGURE 2.15

Primary concerns of FPL customers and full-custom ASIC designers.

Whoever has acquired the skills for designing full-custom ICs is in an excellent position for developing semi-custom ICs and for working with FPL, but not necessarily the other way round. The present volume has been written with all three approaches in mind and begins with those topics that matter for all audiences. If you ultimately plan to limit yourself to configuring FPL devices, you may be satisfied with skipping the sections from 7.2.4 onwards through 7.3. Conversely, you should then pay special attention to section 3.2.9.

Technical details on commercial FPL devices are distributed over thousands of datasheets and white papers prepared by Altera, Xilinx, and other vendors named in table 2.2. Open websites such as [13] [14] help to keep track of products and manufacturers. More condensed background information is available from references such as [15] [16] [17] [18]. [19] specifically discusses FPGAs in image and video processing applications.

2.7 CONCLUSIONS

Field programmable logic is ideal for

- ○ Prototyping and other
- ○ Situations where agility is crucial because specifications are subject to change at any time,
- ○ Products that sell in modest quantities or where time to market is paramount, and for
- ○ Products that need to be reconfigured from remote.

FROM ALGORITHMS TO ARCHITECTURES

3.1 THE GOALS OF ARCHITECTURE DESIGN

VLSI architecture design is concerned with deciding on the necessary hardware resources for carrying out computations from data and/or signal processing and with organizing their interplay such as to meet target specifications defined by marketing.

The foremost concern is to get the desired **functionality** right. The second priority is to meet some given performance target often expressed in terms of data **throughput** or operation rate. A third objective, of economic nature this time, is to minimize **production costs**. Assuming a given fabrication process, this implies minimizing **circuit size** and maximizing fabrication yield such as to obtain as many functioning parts per processed wafer as possible.

Another general concern in VLSI design is **energy efficiency**. Battery-operated equipment, such as hand-held phones, tablet and laptop computers, digital hearing aids, etc. obviously impose stringent limits on the acceptable power consumption. It is perhaps less evident that energy efficiency is also of interest when power gets supplied from the mains. One reason for this is the cost of removing the heat generated by high-performance high-density ICs.

The capability to change from one mode of operation to another in very little time, the flexibility to accommodate evolving needs and/or to upgrade to future standards are other highly desirable qualities and subsumed here under the term **agility**. Last but not least, two distinct architectures are likely to differ in terms of the overall **engineering effort** required to work them out in full detail and, hence also, in their respective time to market.

3.1.1 AGENDA

Driven by dissimilar applications and priorities, hardware engineers have, over the years, devised a multitude of diverse architectural concepts that we will put into perspective in this chapter. Section 3.2, opposes program-controlled and hardwired hardware concepts before showing how their respective strengths can be combined. After the necessary groundwork for architectural analysis has been laid in section 3.3, we then discuss how to select, arrange, and improve the necessary hardware resources in an efficient way with a focus on dedicated architectures. Section 3.4 is concerned with organizing computations of combinational nature, section 3.6 extends our analysis to non-recursive sequential computations before timewise recursive computations are addressed in section 3.7. Finally, section 3.8 generalizes our findings to other than word-level computations on real numbers. Inserted in between is section 3.5 that discusses the options available for temporarily storing data and their implications on architectural decisions. How to make sure functionality is implemented correctly will be the subject of chapter 5.

3.2 THE ARCHITECTURAL SOLUTION SPACE

3.2.1 THE ANTIPODES

Basically, any given computation algorithm can be implemented either as a software program that gets executed an instruction-set computer — such as a microprocessor or a digital signal processor (DSP) — or, alternatively, as a hardwired electronic circuit that carries out the necessary computation steps (Figure 3.1 and Table 3.1). The systems engineer, therefore, has to decide between two fundamentally different options early on in each project.

a) Select a processor-type general-purpose architecture and write program code for it, or
b) Tailor a dedicated hardware architecture for the specific computational needs.

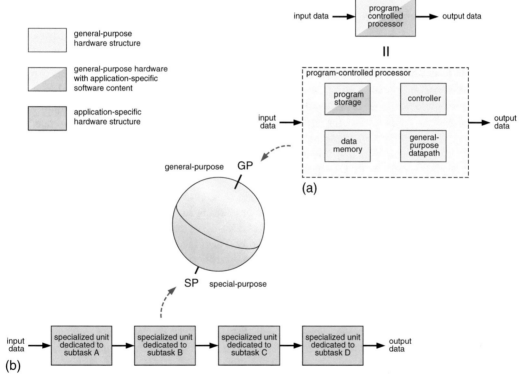

FIGURE 3.1

Program-controlled general-purpose processor (a) and dedicated (special-purpose) hardware structure (b) as architectural antipodes. Table 3.1 gives further details.

Table 3.1   The architectural antipodes compared.

	Hardware architecture	
	General purpose	Special purpose
Algorithm	any, not known a priori	fixed, must be known
Architecture	instruction set processor, von Neumann or Harvard style	dedicated design, no single established pattern
Execution model	fetch-load-execute-store cycle "instruction-oriented"	process data item and pass on "dataflow-oriented"
Datapath	universal operations, ALU(s) plus memory	specific operations only, customized design
Controller	with program microcode	typically hardwired
Performance indicator	instructions per second, run time of various benchmark programs	data throughput, can be anticipated analytically
Paradigm from manufacturing	craftsman in his machine shop working according to different plans every day	division of labor in a factory set up for smooth production of a few closely related goods
Possible hardware implementations	standard µC|DSP components or ASIC with on-chip µC|DSP	ASIC of dedicated architecture or FPL (FPGA|CPLD)
Engineering effort	mostly software design	mostly hardware design
Strengths	highly flexible, immediately available, routine design flow, low up-front costs	room for max. performance, highly energy-efficient, lean circuitry

This is a major decision that has to be made before embarking in the design of a complex circuit. A great advantage of commercial microprocessors is that developers can focus on higher-level issues such as functionality and system-level architecture right away. There is no need for them to address all those exacting chores that burden semi- and — even more so — full-custom design.[1] In addition, there is no need for custom fabrication masks.

Observation 3.1. *Opting for commercial instruction-set processors and/or FPL sidesteps many technical issues that absorb much attention when a custom IC is to be designed instead. Conversely, it is precisely the focus on the payload computations, the absence of programming and configuration overhead together with the full control over every aspect of architecture, circuit, and layout design that make it possible to optimize performance and energy efficiency.*

Circuit examples where dedicated architectures outperform instruction set computers follow.

---

[1] Such as power distribution, clock preparation and distribution, input/output design, physical design and verification, signal integrity, electrical overstress protection, wafer testing, and package selection, all to be discussed in forthcoming chapters. Setting up a working CAE/CAD design flow typically also is a major stumbling block, to say nothing of estimating sales volume, hitting a narrow window of opportunity, finding the right partners, and providing the necessary resources, in-house expertise, and investments. Also note that field-programmable logic (FPL) dispenses developers from dealing with many of these issues too.

Example

Table 3.2   Comparison of architectural alternatives for a Viterbi decoder (code rate $\frac{1}{2}$, constraint length 7, soft decision decoding, Euclidean distance metric). DSPs are at their best for sustained multiply-accumulate operations and offer word widths of 32 bit or so. However, as the Viterbi algorithm can be arranged to make no use of multiplication and to do with word widths of 6 bit or less, DSPs cannot take advantage of these resources. A pipeline of tailored-made stages optimized for branch metric computation, path metric update, and survivor path traceback operations, in contrast, makes it possible to exploit the parallelism inherent in the Viterbi algorithm. Diverse throughput requirements can be accommodated by trading the number of computational units in each stage for throughput. Sophisticated DSPs, such as the C6455, include an extra coprocessor to accelerate path metric update and survivor traceback.

Architecture	General purpose		Special purpose	
Key component	DSP		ASIC	
	TI TMS320C6455		sem03w6	sem05w1
	without	with	ETH	ETH
	Viterbi coprocessor VCP2			
Number of chips	1	1	1	1
CMOS process	90 nm	90 nm	250 nm	250 nm
Program code	187 Kibyte	242 Kibyte	none	none
Circuit size	n.a.	n.a.	73 kGE	46 kGE
Max. throughput	45 kbit/s	9 Mbit/s	310 Mbit/s	54 Mbit/s
@ clock	1 GHz	1 GHz	310 MHz	54 MHz
Power dissipation	2.1 W	2.1 W	1.9 W	50 mW
Year	2005	2005	2004	2006

☐

Example

Table 3.3   Comparison of architectural alternatives for a secret-key block encryption/decryption algorithm (IDEA cipher as shown in fig.3.20, block size 64 bit, key length 128 bit). The clear edge of the VINCI ASIC is due to a high degree of parallelism in its datapath and, more particularly, to the presence of four pipelined computational units for multiplication modulo $(2^{16} + 1)$ designed in full-custom layout that operate concurrently and continuously. The more recent IDEA kernel combines a deep submicron fabrication process with four highly optimized arithmetic units. Full-custom layout was no longer needed to achieve superior performance.

Architecture	General purpose		Special purpose	
Key component	DSP	RISC Workst.	ASSP	ASSP
	Motorola 56001	Sun Ultra 10	VINCI [20]	IDEA Kernel
Number of chips	1 + memory	motherboard	1	1
CMOS process	n.a.	n.a.	1.2 μm	250 nm
Max. throughput	1.25 Mbit/s	13.1 Mbit/s	177 Mbit/s	700 Mbit/s
@ clock	40 MHz	333 MHz	25 MHz	100 MHz
Year	1995	1998	1992	1998

☐

Example

Table 3.4   Comparison of architectural alternatives for lossless data compression with the Lempel-Ziv-77 algorithm that heavily relies on string matching operations [21]. The dedicated hardware architecture is implemented on a reconfigurable coprocessor board built around four field-programmable gate-array components. 512 special-purpose processing elements are made to carry out string comparison subfunctions in parallel. The content-addressed symbol memory is essentially organized as a shift register thereby giving simultaneous access to all entries. Of course, the two software implementations obtained from compiling C source code cannot nearly provide a similar degree of concurrency.

Architecture	General purpose		Special purpose
Key component	RISC Workst. Sun Ultra II	CISC Workst. Intel Xeon	FPGA Xilinx XC4036XLA
Number of chips	motherboard	motherboard	4 + config.
CMOS process	n.a.	n.a.	n.a.
Max. throughput	3.8 Mbit/s	5.2 Mbit/s	128 Mbit/s
@ clock	300 MHz	450 MHz	16 MHz
Year	1997	1999	1999

□

Example

Table 3.5   Comparison of architectural alternatives for a secret-key block encryption/decryption algorithm (AES cipher, block size 128 bit, key length 128 bit). The Rijndael algorithm makes extensive use of a so-called S-Box function and its inverse; the three hardware implementations include multiple lookup tables (LUT) for implementing that function. Also, (de)ciphering and sub-key preparation are carried out concurrently by separate hardware units. On that background, the throughput of the assembly language program running on a Pentium III is indeed impressive. This largely is because the Rijndael algorithm has been designed with the Pentium architecture in mind (MMX instructions, LUTs that fit into cache memory, etc.). Power dissipation remains daunting, though.

Architecture	General purpose		Special purpose		
Key component	RISC Proc. Embedded Sparc	CISC Proc. Pentium III	FPGA Virtex-II Amphion	ASIC CryptoFun ETH	ASIC core only UCLA [22]
Number of chips	motherbo.	motherbo.	1 + config.	1	1
Programming	C	Assembler	none	none	none
Circuit size	n.a.	n.a.	n.a.	76 kGE	173 kGE
CMOS process	n.a.	n.a.	150 nm	180 nm	180 nm
Max. throughp.	133 kbit/s	648 Mbit/s	1.32 Gbit/s	2.00 Gbit/s	1.6 Gbit/s
@ clock	120 MHz	1.13 GHz	n.a.	172 MHz	125 MHz
Power dissipat.	120 mW	41.4 W	490 mW	n.a.	56 mW[a]
@ supply			1.5 V	1.8 V	1.8 V
Year	n.a.	2000	≈ 2002	2007	2002

[a] Most likely specified for core logic alone, that is without I/O circuitry.

□

Upon closer inspection, one finds that dedicated architectures fare much better in terms of performance and/or dissipated energy than even the best commercially available general-purpose processors in some situations, whereas they prove a dreadful waste of both hardware and engineering resources in others.

Algorithms that are very irregular, highly data-dependent, and memory-hungry are unsuitable for dedicated architectures. Situations of this kind are found in **electronic data processing** such as databank applications, accounting, and in **reactive systems**[2] like industrial control,[3] user interfaces, and others. In search of optimal architectures for such applications, one will invariably arrive at hardware structures patterned after instruction set processors. Writing code for a standard microcomputer — either bought as a physical part or incorporated into an ASIC as a megacell or as a virtual component — is more efficient and more economic in this case.

Situations where data streams are to be processed in fairly regular ways offer far more room for coming up with dedicated architectures. Impressive gains in performance and energy efficiency over solutions based on general-purpose parts can so be obtained, see tables 3.2, 3.3, 3.4 and 3.5 among other examples.

Generally speaking, situations that favor dedicated architectures are often found in real-time applications from **digital signal processing** and **telecommunications** such as

- Source coding (i.e. data, audio and video (de)compression),
- (De)ciphering (primarily for secret key ciphers),
- Channel coding (i.e. error correction),
- Digital (de)modulation (for modems, wireless communication, and disk drives),
- Adaptive channel equalization (after transmission over copper lines and optical fibers),
- Filtering (for noise cancellation, preprocessing, spectral shaping, etc.),
- Multipath combiners in broadband wireless access networks (RAKE, MIMO),
- Digital beamforming with phased-array antennas (Radar),
- Computer graphics and video rendering,
- Multimedia (e.g. MPEG, HDTV),
- Packet switching (e.g. ATM, IP),
- Transcoding (e.g. between various multimedia formats),
- Medical signal processing,
- Pattern recognition, and more.

---

[2] A system is said to be **reactive** if it interacts continuously with an environment, at a speed imposed by that environment. The system deals with <u>events</u> and the mathematical formalisms for describing them aim at capturing the complex ordering and causality relations between events that may occur at the inputs and the corresponding reactions — events themselves — at the outputs. Examples: elevators, protocol handlers, antilock brakes, process controllers, graphical user interfaces, operating systems.

As opposed to this, a **transformatorial** system accepts new input values — often at regular intervals —, uses them to compute output values, and then rests until the subsequent data items arrive. The system is essentially concerned with arithmetic/logic processing of <u>data values</u>. Formalisms for describing transformatorial systems capture the numerical dependencies between the various data items involved. Examples: filtering, data compression, ciphering, pattern recognition and other applications colloquially referred to as number crunching but also compilers and payroll programs.

[3] Control in the sense of the German "programmierte Steuerungen" not "Regelungstechnik".

Observation 3.2. *Processing algorithms and hardware architectures are intimately related. While dedicated architectures outperform program-controlled processors by orders of magnitude in many applications of predominantly transformatorial nature, they cannot rival the agility and economy of processor-type designs in others of more reactive nature.*

More precise criteria for finding out whether a dedicated architecture can be an option or not from a purely technical point of view follow in section 3.2.2 while fig.3.2 puts various applications from signal and data processing into perspective.

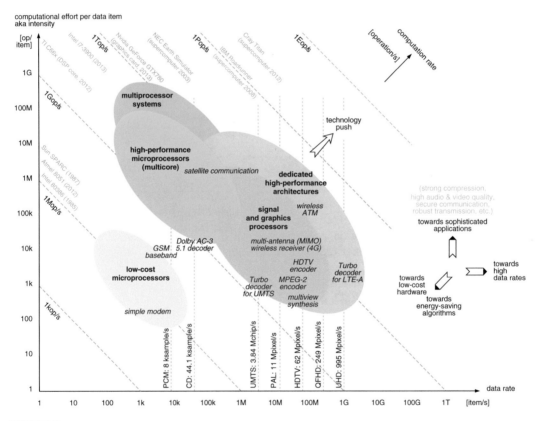

FIGURE 3.2

Computational needs and capabilities (approximate, exact meaning of operation and data item left unspecified, 16 bit-by-16 bit multiply-accumulate (MAC) operations on 16 bit samples are often considered typical in a context of digital signal processing).

3.2.2 WHAT MAKES AN ALGORITHM SUITABLE FOR A DEDICATED VLSI ARCHITECTURE?

Costs are not the same in hardware as in software. As an example, permutations of bits within a data word are time-consuming operations in software as they must be carried out sequentially. In hardware, they reduce to simple wires that cross while running from one subcircuit to the next. Lookup tables (LUT) of almost arbitrary size, on the other hand, have become an abundant and cheap resource in any microcomputer while large on-chip RAMs and ROMs tend to eat substantial proportions of the timing and area budgets of ASIC designs.

In an attempt to provide some guidance, we have collected ten criteria that an information processing algorithm should <u>ideally</u> meet in order to justify the design of a special-purpose VLSI architecture and to take full advantage of the technology. Of course, very few real-world algorithms satisfy all of the requirements listed. It is nevertheless safe to say that designing a dedicated architecture capable of outperforming a general-purpose processor on the grounds of performance and costs will prove difficult when too many of these criteria are violated. The list below begins with the most desirable characteristics and then follows their relative significance.

1. **Loose coupling between major processing tasks.** The overall data processing lends itself to being decomposed into tasks that interact in a simple and unmutable way. Whether those tasks are to be carried out consecutively or concurrently is of secondary importance at this point, what counts is to come up with a well-defined functional specification for each task and with manageable interactions between them. Architecture design, functional verification, optimization, and reuse otherwise become real nightmares.

2. **Simple control flow.** The computation's control flow is simple. This key property can be tracked down to two more basic considerations:

   a) The course of operation does not depend too much on the data being processed; for each loop the number of iterations is a priori known and constant.[4]

   b) The application does not ask for computations to be carried out with overly many varieties, modes of operations, data formats, distinct parameter settings, etc.

   The benefit of a simple control flow is twofold. For one thing, it is possible to anticipate the datapath resources required to meet a given performance goal and to design the chip's architecture accordingly. There is no need for statistical methods in estimating the computational burden nor in sizing data memories and the like. For another thing, datapath control can be handled by counters and by simple finite state machines (FSM) that are small, fast, energy-efficient and — most important — easy to verify.

   An overly complicated course of operations, on the other hand, that involves much data-dependent branching, multitasking, and the like, favors a processor-type architecture that operates under control of stored microcode. Most control operations will then translate into a sequence of machine instructions that take several clock cycles to execute.

---

[4] Put in different terms, the target algorithm is virtually free of branchings and loops such as `if...then[...else]`, `while...do`, and `repeat...until` that include data items in their condition clauses.

3. Regular data flow. The flow of data is regular and their processing is based on a recurrence of a fairly small number of identical operations, there are no computationally expensive operations that are called only occasionally. Regularity opens a door for sharing hardware resources in an efficient way by applying techniques such as iterative decomposition and time sharing, see subsections 3.4.2 and 3.4.5 respectively. Conversely, multiple data streams that are to be processed in a uniform way lend themselves for concurrent processing by parallel functional units. A regular data flow further helps to reduce communications overhead both in terms of area and interconnect delay as the various functional units can be made to exchange data over fixed local links. Last but not least, regularity facilitates reuse and reduces design and verification effort.

As opposed to this, operations that are used infrequently will either have to be decomposed into a series of substeps to be executed one after the other on a general-purpose datapath, which is slow, or will necessitate dedicated functional units bound to sit idle for most of the time, which inflates chip size. Irregular data flow asks for long and flexible communication busses which are at the expense of layout density, operating speed and energy efficiency.

4. Reasonable storage requirements. Overall storage requirements are modest and have a fixed upper bound which precludes the use of dynamic data structures. Memories that occupy an inordinate amount of chip area cannot be incorporated into ASICs in an economic way and must, therefore, be implemented off-chip from standard parts, see subsection 3.5. Massive storage requirements in conjunction with moderate computational burdens tend to place dedicated architectures at a disadvantage.

5. Compatible with finite precision arithmetics. The algorithm is insensitive to effects from finite precision arithmetics. That is, there is no need for floating point arithmetics; fairly small word widths of, say, 16 bit or less, suffice for the individual computation steps. Standard microprocessors and DSPs come with datapaths of fixed and often generous width (24, 32, 64 bit, or even floating point) at a given price. No extra costs arise unless the programmer has to resort to multiple precision arithmetics.

As opposed to this, ASICs and FPL offer an opportunity to tune the word widths of datapaths and on-chip memories to the local needs of computation. This is important because circuit size, logic delay, interconnect length, parasitic capacitances, and energy dissipation of addition, multiplication, and other operations all tend to grow with word width, combining into a burden that multiplies at an overproportional rate.[5]

---

[5] Processor datapaths tend to be fast and area efficient because they are typically hand-optimized at the transistor level (e.g. dynamic logic) and implemented in tiled layout rather than built from standard cells. These are only rarely options for ASIC designers.

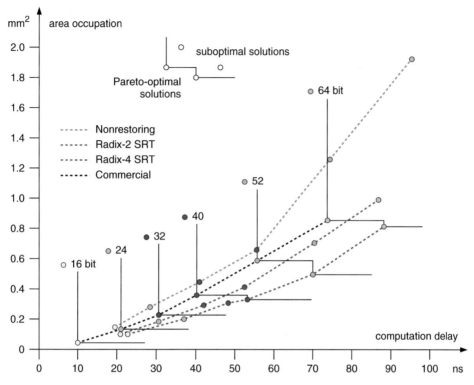

FIGURE 3.3

Comparison of hardware divider architectures in terms of area and delay estimated from pre-layout netlists for a 180 nm CMOS process under worst-case PTV conditions [23]. Note the impact of quotient width on both circuit size and performance for all four architectures. The concept implemented in the commercial component is not disclosed by the vendor. SRT stands for Sweeney, Robertson and Tocher.

6. Non-recursive linear time-invariant computation. The processing algorithm describes a non-recursive linear time-invariant system over some algebraic field.[6] Each of these properties opens a door to reorganizing the data processing in one way or another, see sections 3.4 through 3.9 for details and table 3.11 for an overview. High throughputs, in particular, are much easier to obtain from non-recursive computations as will become clear in section 3.7.

7. No transcendental functions. The algorithm does not make use of roots, logarithmic, exponential, or trigonometric functions, arbitrary coordinate conversions, translations between incompatible number systems, and other transcendental functions as these must either be stored in large lookup

---

[6] Recursiveness is to be defined in section 3.7. **Linear** is meant to imply the principle of superposition: $f(x(t) + y(t)) \equiv f(x(t)) + f(y(t))$ and $f(cx(t)) \equiv cf(x(t))$. **Time-invariant** means that the sole effect of shifting the input in time is a corresponding time shift of the output: if $z(t) = f(x(t))$ is the response to $x(t)$ then $z(t - T)$ is the response to $x(t - T)$. Fields and other algebraic structures are compared in section 3.11.

tables (LUT) or get calculated on-line in lengthy and often irregular computation sequences. Such functions can be implemented more economically provided modest accuracy requirements allow for approximation by way of lookups from tables of reasonable size possibly followed by interpolation steps.[7]

8. **Extensive usage of data operations unavailable from standard instruction sets.** Of course, there exist many processing algorithms that cannot do without costly arithmetic/logic operations. It is often possible to outperform traditional program-controlled processors in cases where such operations need to be assembled from multiple instructions. Dedicated hardware can then be designed to do the same computation in a more efficient way. Examples include finite field arithmetics, add-compare-select operations, many ciphering operations, and, again, CORDIC.[7] It also helps when part of the arguments are constants because this makes it possible to apply some form of preprocessing. Multiplication by a variable is more onerous than by a constant, for instance.[8]

9. **No divisions and multiplications on very wide data words, no matrix inversions.** The algorithm does not make extensive use of multiplications and even less so of divisions as their VLSI implementation is much more expensive than that of addition/subtraction when the data words involved grow wide. Also, depending on the arguments, multiplication and division may vastly expand the numerical range of the results. This gives rise to scaling issues, particularly in conjunction with a fixed-point number system. Matrix inversion is a particularly nasty case in point as it involves division operations and often brings about numerical instability.

10. **Throughput rather than latency is what matters.** Tight latency requirements rule out pipelining, one of the most effective ways to boost the throughput of a datapath. Stringent latency requirements are not in favor of microprocessors either, however, as program-controlled operation cannot guarantee brief and fixed response times, even less so when a complex operating system is involved.

3.2.3 THERE IS PLENTY OF LAND BETWEEN THE ANTIPODES

Most markets ask for performance, agility, low power, and a modest design effort at a time. This requires that the throughput and the energy efficiency of a dedicated VLSI architecture for demanding but highly repetitive computations be combined with the convenience and flexibility of an instruction set processor for more control-oriented tasks. The question is

"How to blend the best of both worlds into a suitable architecture design?"

---

[7] The most popular such technique is known as CORDIC and used to calculate coordinate rotations, trigonometric and hyperbolic functions, see section 3.3.1. Advice on how to approximate other transcendental functions can be found in [24] (sine functions), [25] [26] (logarithms), and table 3.8 (magnitude function) among many others.

[8] Dropping unit factors and/or zero sum terms (both at word and bit levels), substituting integer powers of 2 as arguments in multiplications and divisions, omitting insignificant contributions, special number representation schemes, taking advantage of symmetries, precomputed lookup tables, and distributed arithmetic, see subsection 3.8.3, are just a few popular measures that may help to lower the computational burden in situations where part of the arguments are known ahead of time.

Six approaches for doing so are going to be presented in sections 3.2.4 through 3.2.9 with diagrammatic illustrations in figs.3.4 to 3.9.

3.2.4 ASSEMBLIES OF GENERAL-PURPOSE AND DEDICATED PROCESSING UNITS

The observation below forms the starting point for the conceptually simplest approach.

Observation 3.3. *It is often possible to segregate the needs for computational efficiency from those for flexibility.*

This is because those parts of a system that ask for maximum computation rate are not normally those that are subject to change very often, and vice versa. Examples abound, see table 3.6. The finding immediately suggests a setup where a software-controlled microcomputer cooperates with one or more dedicated hardware units. Separating the quest for computational efficiency from that for agility makes it possible to fully dedicate the various functional units to their respective tasks and to optimize them accordingly. Numerous configurations are possible and the role of the instruction set microcomputer varies accordingly.

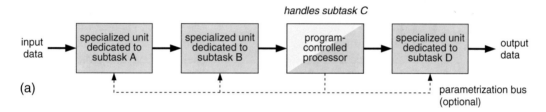

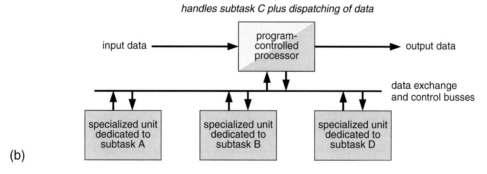

FIGURE 3.4

General-purpose processor and dedicated satellite units working in a chain (a), a host computer with specialized fixed-function blocks or coprocessors (b).

In fig.3.4a, three dedicated and one program-controlled processing units are arranged in a chain. Each unit does its data processing job and passes the result to the downstream unit. While offering ample room for optimizing performance, this structure cannot accommodate much variation if everything is hardwired and tailor-made. Making the specialized hardware units support a limited degree of parametrization (e.g. wrt data word width, filter order, code rate, data exchange protocol, and the like) renders the overall architecture more versatile while, at the same time, keeping the overhead in terms of circuit complexity and energy dissipation fairly low. The term **weakly programmable satellites** has been coined to reflect the idea. An optional parametrization bus suggests this extension of the original concept in fig.3.4a.

Example

Table 3.6   Some digital systems and the computing requirements of major subfunctions thereof.

Application	Subfunctions primarily characterized by	
	irregular control flow and/or need for flexibility	repetitive control flow and need for computing efficiency
Blu-ray player	user interface, track seeking, tray and spindle control, processing of non-video data (directory, title, author, subtitles, region codes)	16-to-8 bit demodulation, error correction, MPEG-2 decompression (DCT), deciphering (AACS AES-128), video signal processing
Smartphone	user interface, SMS/MMS, directory management, battery monitoring, communication protocol, channel allocation, roaming, accounting	intermediate frequency filtering, (de)modul., channel (de)coding, error correction (de)coding, (de)ciphering, speech and video (de)compression, display graphics
Pattern recognition (e.g. as part of a defensive missile)	pattern classification, objects tracking, target acquisition, triggering of actions	image stabilization, redundancy reduction, image segmentation, feature extraction

□

3.2.5 HOST COMPUTER WITH HELPER ENGINES

Fig.3.4b is based on segregation too but differs in how the various components interact. All satellites now operate under command of a software-programmable **host**. A bidirectional bus gives the necessary liberty for transferring data and control words back and forth. Each unit is optimized for a few specific subtasks such as filtering, video and/or audio (de)coding, (de)ciphering, modem operations, display graphics, and the like. These helper engines, as they are fittingly called, may either be hardwired fixed-function blocks or be themselves instruction-set programmable, in which case they are to be viewed as true **coprocessors**.

A helper engine sits idle until it receives a set of input data along with a starting command. As an alternative, the data may be kept in the host's own memory all the time but get accessed by the coprocessor via direct memory access (DMA). Once local computation has come to an end, the helper sets a status flag and/or sends an interrupt signal to the host computer. The host then accepts the processed data and takes care of further action.

As evident from fig.3.5, industrial examples are more complex but the concept of co-operating instruction-set processors and optimized fixed-function blocks remains.

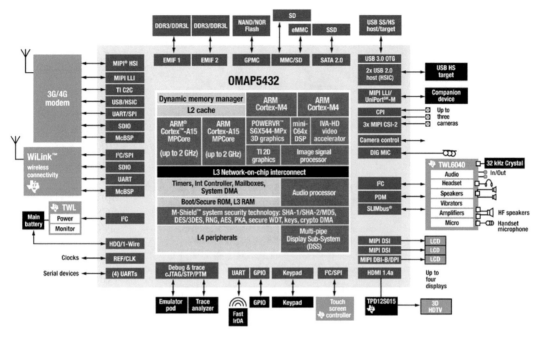

FIGURE 3.5

A system on a chip for smartphones (source Texas Instruments, reprinted with permission).

3.2.6 APPLICATION-SPECIFIC INSTRUCTION SET PROCESSORS

Patterning the overall architecture after a program-controlled processor affords much more flexibility. Application-specific features are largely confined to the data processing circuitry itself. That is, one or more datapaths are designed and hardwired such as to support specific data manipulations while operating under control of a common microprogram. The number of ALUs, their instruction sets, the data formats supported, the capacity of local storage, etc. are tailored to the computational problems to be solved. What's more, the various datapaths can be made to operate simultaneously on different pieces of data thereby providing a limited degree of concurrency. The resulting architecture, sketched in fig.3.6, is that of an application-specific instruction set processor (ASIP).

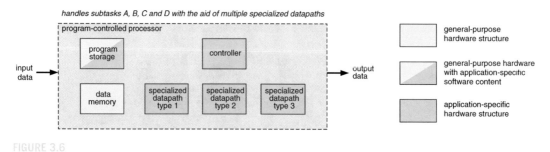

FIGURE 3.6

Application-specific instruction set processor (ASIP).

Example

Table 3.7   An ASIP implementation of the AES (Rijndael) algorithm, compare with table. 3.5

Architecture	ASIP
Key component	Cryptoprocessor core UCLA [27]
Number of chips	1
Programming	Assembler
Circuit size	73.2 kGE
CMOS process	180 nm 4Al2Cu
Throughput	3.43 Gbit/s
@ clock	295 MHz
Power dissipation	86 mW[a]
@ supply	1.8 V
Year	2004

[a]Estimate for core logic alone, that is without I/O circuitry, not a measurement.

□

The hardware organization of an ASIP bears much resemblance to architectural concepts from general-purpose computing. As more and more concurrent datapath units are being added, what results essentially is a **very-long instruction word** (VLIW) architecture. An open choice is that between a **multiple-instruction multiple-data** (MIMD) machine, where an individual field in the overall instruction word is set apart for each datapath unit, and a **single-instruction multiple-data** (SIMD) model, where a bunch of identical datapaths work under control of a single instruction word. Several data items can so be made to undergo the same operation at a time.[9]

---

[9] In an effort to better serve high-throughput video and graphics applications, many vendors have enhanced their microprocessor families in the late 1990's by adding special instructions that provide some degree of concurrency. During each such instruction, the processor's datapath gets split up into several smaller subunits. A datapath of 64 bit can so be made to process four 16 bit data words at a time, for instance, provided the operation is the same for all of them. The technique is best described as **sub-word parallelism**, but is better known under various trademarks such as multimedia extensions (MMX), streaming SIMD extensions (SSE) (Pentium family), Velocity Engine, AltiVec, and VMX (PowerPC family).

Be cautioned that defining a proprietary instruction set makes it impossible to take advantage of existing compilers, debugging aids, assembly language libraries, experienced programmers, and other resources that are routinely available for commercial processors. Industry provides us with such a vast selection of micro- and signal processors that only very particular requirements justify the design of a proprietary CPU. Rather than anywhere else, opportunities can be found in mobile signal processing devices that ask for a combination of specific processing needs, agility, performance, and energy efficiency.

[28] describes an interesting framework for ASIP development whereby assembler, linker, simulator, and RTL synthesis code are generated automatically by system-level software tools. Using LISA 2.0 (Language for Instruction Set Architectures), hardware designers essentially begin by defining the most adequate instruction set for a target application. Transiting through a cycle-accurate model, they can cast their ideas into an RTL-type model which means they remain in full control over the processor's architecture if they wish to do so. With some restrictions, that model then gets translated into VHDL synthesis code. All this happens within the same EDA environment and using the same language, so that the process of evaluating competing architectural options and of working out all the details is greatly expedited. A number of predefined processor templates further helps with the initial modeling steps.[10]

Example
While generally acknowledged to produce more realistic renderings of 3D scenes than industry-standard raster graphics processors, ray tracing algorithms have long been out of reach for real-time applications due to the myriad floating point computations and the immense memory bandwidth they require. Hardwired custom architectures do not qualify either as they cannot be programmed and as ray tracing necessitates many data-dependent recursions and decisions.

Ray tracing may finally find more general adoption in multi-ASIP architectures that combine multiple ray processing units (RPU) into one powerful rendering engine. Working under control of its own program thread, each RPU operates as a SIMD processor that follows a subset of all rays in a scene. The independence of light rays allows for a welcome degree of scalability where frame rate can be traded against circuit complexity. The authors of [29] have further paid attention to define an instruction set for their RPUs that is largely compatible with pre-existing industrial graphics processors.
□

3.2.7 RECONFIGURABLE COMPUTING

Another crossbreed between dedicated and general-purpose architectures has not become viable until the late 1990's but is now being promoted by FPL manufacturers and researchers [30] [31]. The IEEE 1532 standard has also been created in this context. The idea is to reuse the same hardware for implementing subfunctions that are mutually exclusive in time by reconfiguring FPL devices on the fly.

---

[10] A commercial product that has emanated from this research is "Processor Designer" by Synopsys.

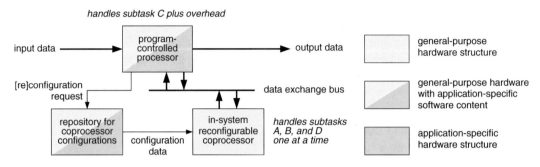

FIGURE 3.7

General-purpose processor with juxtaposed reconfigurable coprocessor.

As shown in fig.3.7, the general hardware arrangement bears some resemblance to the coprocessor approach of fig.3.4b, yet **in-system configurable** (ISC) devices are being used instead of hardwired logic. As a consequence, the course of operations is more sophisticated and requires special action from the hardware architects. For each major subtask, the architects must ask themselves whether the computations involved

- Qualify for being delegated to in-system configurable logic,
- Never occur at the same time — or can wait until the FPL device becomes free —, and
- Whether the time for having the FPL reconfigured in between is acceptable or not.

Typically this would be the case for repetitive computations that make use of sustained, highly parallel, and deeply pipelined bit-level operations. When designers have identified some suitable subfunction, they devise a hardware architecture that solves the particular computational problem with the resources available in the target FPGA or CPLD, prepare a configuration file, and have that stored in a configuration memory. In some sense, they create a large **hardware procedure** instead of programming a software routine in the customary way.

Whenever the host computer encounters a call to such a hardware procedure, it configures the FPL accordingly by downloading the pertaining configuration file. From now on, all the host has to do is to feed the "new" coprocessor with input data and to wait until the computation completes. The host then fetches the results before proceeding with the next subtask.[11]

It so becomes possible to support an assortment of data processing algorithms each with its optimum architecture — or almost so — from a single hardware platform. What often penalizes this approach in practice are the dead times incurred whenever a new configuration is being loaded. Another price to pay is the extra memory capacity for storing the configuration bits for all operation modes. Probably

---

[11] As an extension to the general procedure described here, an extra optimization step can get inserted before the coprocessor is being configured [32]. During this stage, the host would adapt a predefined generic configuration to take advantage of particular conditions of the specific situation at hand. Consider pattern recognition, for instance, where the template remains unchanged for a prolonged lapse of time, or secret-key (de)ciphering where the same holds true for the key. As stated in subsection 3.2.2 item 3.2.2, it is often possible to simplify arithmetic and logic hardware a lot provided that part of the operands do have a fixed value.

the most valuable benefit, however, is the possibility to upgrade information processing hardware to new standards and/or modes of operation even after the system has been fielded.

Examples

Transcoding video streams in real time is a good candidate for reconfigurable computing because of the many formats in existence such as DV, AVI, MPEG-2, DivX and H.264. For each conversion scheme, a configuration file is prepared and stored in local memory from where it is being transferred into the reconfigurable coprocessor on demand. And should a video format or variation emerge that was unknown or unpopular at the time when the system was being developed, extra configuration files can be made available in a remote repository from where they can be fetched much like software plug-ins get downloaded via the Internet.

The results from a comparison between Lempel-Ziv data compression with a reconfigurable coprocessor and with software execution on a processor [21] have been summarized in table 3.4. A related application was to circumvent the comparatively slow PCI bus in a PC [33].
□

3.2.8 EXTENDABLE INSTRUCTION SET PROCESSORS

This offbeat approach pioneered by Stretch Inc. combines ideas from ASIP design and reconfigurable computing. As indicated in fig.3.8, the general architecture includes both a program-controlled processor and electrically reconfigurable logic.

handles subtasks A, B, C, and D with a combination of fixed and configurable datapaths

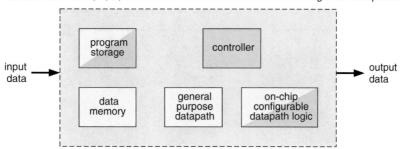

FIGURE 3.8

Combining performance and agility with an Extendable instruction set processor (simplified).

The key innovation is a suite of proprietary EDA tools that allows system developers to focus on writing their application program in C or C++ as if for a regular general-purpose processor. Those tools begin by profiling the software code in order to identify sequences of instructions that are executed many times over. For each such sequence, reconfigurable logic is then synthesized into a dedicated and massively parallel computation network that completes within one clock cycle — ideally at least. Finally, each occurrence of the original instruction sequence in the machine code gets by replaced by a simple function call that activates the custom-made datapath logic.

In essence, the base processor gets unburdened from lengthy code sequences by augmenting its instruction set with a few essential additions that fit the application and that get tailor-made almost on the fly. Yet, the existence of reconfigurable logic and the business of coming up with a suitable hardware architecture are hidden from the system developer. The fact that overall programm execution remains strictly sequential should further simplify the design process. A related concept is discussed in [34].

3.2.9 PLATFORM ICS (DSPP)

The exploding costs of engineering and mask sets are pushing the minimum sales quantity required to economically justify custom silicon to ever higher values, creating a need for malleable platforms that can cover a range of applications with a single design. This calls for a new concept that borrows from all other approaches discussed so far in an attempt to reconcile performance, energy efficiency, agility, and fast turnaround. As illustrated in fig.3.9, a typical platform IC[12] has a heterogeneous architecture that generously marries hardware resources from

- Instruction-set-programmable processors (CPUs, GPUs, ASIPs),
- Hardwired circuits (fixed-function blocks useful for one domain of application),
- Electrically configurable fabrics (FPL), and
- Various types of on-chip memories (SRAM, DRAM, flash).

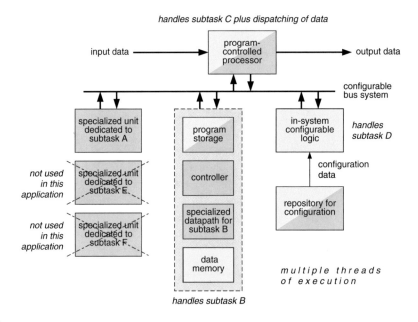

FIGURE 3.9

Domain-specific programmable platform (simplified).

---

[12] There is no established name for the concept yet. While **Domain-Specific Programmable Platform** (DSPP), Domain-Specific Heterogeneous Standard Platform (DSHSP), and Targeted Design Platform capture the spirit, the term Application-Specific Programmable Platform (ASPP) is slightly off the point. We usually prefer the name platform IC for conciseness.

Most of the programming occurs in a domain-specific high-level language.[13] For each subtask, developer tools determine the most adequate execution unit (or units) such as to meet the performance targets with the minimum dissipated energy. The configurable logic is used to extend the datapaths and/or the instruction sets where beneficial in terms of throughput, energy efficiency, updates, etc., but time-consuming on-the-fly reconfigurations are minimized. The degree of concurrency is limited by the circuit resources available. To reduce static power, all subcircuits are turned off when inactive.

While the platform circuitry is enormously complex and includes subcircuits that may never be used in a given application or product, the on-going technological progress tends to make such concerns less and less relevant. What's more, the concept benefits from numerous trends to be explained in forthcoming chapters. Fig.3.10 shows an industrial example that can be viewed as a forerunner. Paraphrasing EE Times, the author believes

CPU and GPU cores are the new gates, and platform ICs are the new gate arrays.

3.2.10 DIGEST

Program execution on a general-purpose processor and hardwired circuitry optimized for one specific flow of computation are two architectural antipodes. Luckily, many useful compromises exist in between as reflected in figs.3.11 and 3.13. A general advice is this:

Observation 3.4. *Rely on dedicated hardware only for those subfunctions that are called many times and are unlikely to change, keep the rest programmable either via software, or via reconfiguration, or both.*

---

[13] Such languages are currently being developed by the supercomputing community to ensure code portability and optimum performance in spite of unlike and more and more heterogeneous computers without having to rewrite the code. Example: Liszt for mesh-based solvers of partial differential equations.

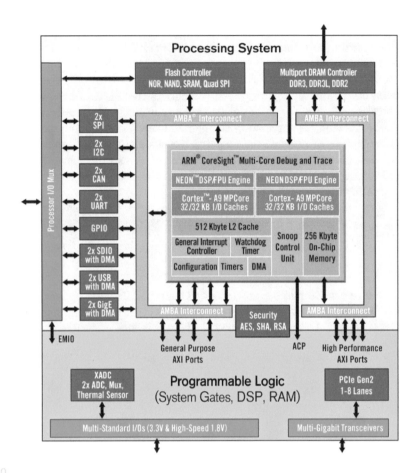

FIGURE 3.10

Zynq All Programmable SoC Architecture (source Xilinx, reprinted with permission).

Fig.3.13 gives rise to an interesting observation. While there are many ways to trade agility for computational efficiency and vice versa, the two seem to be mutually exclusive as we know of no architecture that would meet both goals at a time.

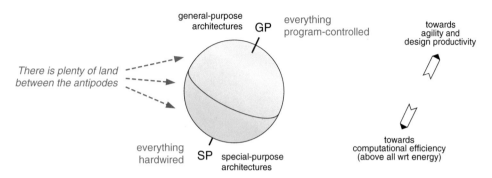

FIGURE 3.11

The architectural solution space viewed as a globe.

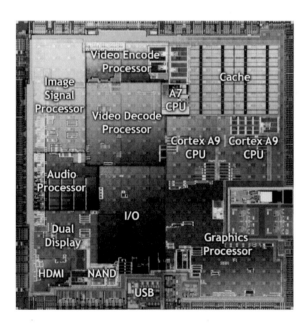

FIGURE 3.12

Coexistence of general-purpose CPUs, ASIPs, and hardwired helper engines in the Tegra 2 SoC (source Nvidia, reprinted with permission).

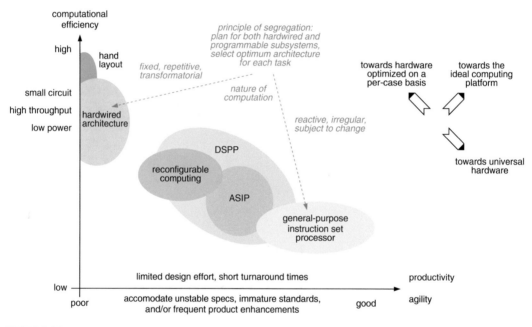

FIGURE 3.13

The basic options of architecture design (simplified).

3.3 DEDICATED VLSI ARCHITECTURES AND HOW TO DESIGN THEM

Let us now turn our attention to the main topic of this chapter:

"How to decide on the necessary hardware resources for solving a given computational problem and how to best organize them?"

Their conceptual differences notwithstanding, many techniques for obtaining high performance at low cost are the same for general- and special-purpose architectures. As a consequence, much of the material presented in this chapter applies to both of them. Yet, the emphasis is on dedicated architectures as the a priori knowledge of a computational problem offers room for a number of ideas that do not apply to instruction set processors.[14]

Observation 3.5. *Most data and signal processing algorithms would lead to grossly inefficient or even infeasible solutions if they were implemented in hardware as they are. Adapting processing algorithms to the technical and economical conditions of large scale integration is one of the intellectual challenges in VLSI design.*

Basically, there is room for remodeling in two distinct domains, namely in the algorithmic domain and in the architectural domain.

3.3.1 THERE IS ROOM FOR REMODELING IN THE ALGORITHMIC DOMAIN ...

In the algorithmic domain, the focus is on minimizing the number of computational operations weighted by the estimated costs of such operations. A given processing algorithm thus gets replaced by a different one better suited to hardware realization in VLSI. Data structures and number representation schemes are also subject to optimizations such as subsampling and/or changing from floating point to fixed point arithmetics. All this implies that alternative solutions are likely to slightly differ in their functionality as expressed by their input-to-output relations.

Six examples

When designing a digital filter, one is often prepared to tolerate a somewhat lower stopband suppression or a larger passband ripple in exchange for a reduced computational burden obtained, for instance, from substituting a lower order filter and/or from filling in zeros for the smaller coefficients. Conversely, a filter structure that necessitates a higher number of computations may sometimes prove acceptable in exchange for less stringent precision requirements imposed on the individual arithmetic operations and, hence, for narrower data words.

In a decoder for digital error-correction, one may be willing to sacrifice 0.1 dB or so of coding gain for the benefit of doing computations in a more economic way. Typical simplifications to the ideal

---

[14] There exists a comprehensive literature on general-purpose architectures including [35] [36]. The historical evolution of the microprocessor is summarized in [37] [38] along with economic facts and trends. [39] [40] [41] emphasize the impact of deep-submicron technology on high-performance microprocessor architectures.

Viterbi algorithm include using an approximation formula for branch metric computation, truncating the dynamic range of path metrics, rescaling them when necessary, and restricting traceback operations to some finite depth.

The autocorrelation function (ACF) has many applications in signal processing, yet it is not always needed in the form mathematically defined.

$$ACF_{xx}(k) = r_{xx}(k) = \sum_{n=-\infty}^{\infty} x(n) \cdot x(n+k) \tag{3.1}$$

Many applications offer an opportunity to relax the effort for multiplications because one is interested in just a small fragment of the entire ACF, because one can take advantage of symmetry, or because modest precision requirements allow for a rather coarse quantization of data values. It is sometimes even possible to substitute the average magnitude difference function (AMDF) that does away with costly multiplication altogether.

$$AMDF_{xx}(k) = r'_{xx}(k) = \sum_{n=0}^{N-1} |x(n) - x(n+k)| \tag{3.2}$$

Code-excited linear predictive (CELP) coding is a powerful technique for compressing speech signals, yet it has long been left aside in favor of regular pulse excitation because of its prohibitive computational burden. CELP requires that hundreds of candidate excitation sequences be passed through a cascade of two or three filters and be evaluated in order to pick the one that fits best. In addition, the process must be repeated every few milliseconds. Yet, experiments have revealed that the usage of sparse (up to 95% of samples replaced with zeros), of ternary $(+1,0,-1)$, or of overlapping excitation sequences has little negative impact on auditory perception while greatly simplifying computations and reducing memory requirements [42].

When having to compute trigonometric functions, lookup tables (LUT) are likely to prove impractical because of size overruns. Executing a lengthy algorithm, on the other hand, may be just too slow, so a tradeoff between circuit size, speed, and precision must be found. The CORDIC (coordinate rotation digital computer) family of algorithms is one such compromise that has been put to service in scientific pocket calculators in the 1960s and that continues to find applications in DSP [43] [44] [45]. Today, CORDIC units are available as virtual components. Note that CORDIC can be made to compute hyperbolic and other transcendental functions too.

Computing the magnitude function $m = \sqrt{a^2 + b^2}$ is a rather costly proposition in terms of circuit hardware. Luckily, there exist at least two fairly precise approximations based on add, shift, and compare operations exclusively, see table 3.8 and problem 1. Better still, the performance of many optimization algorithms used in the context of demodulation, error correction, and related applications does not suffer much when the computationally expensive ℓ^2-norm gets replaced by the much simpler ℓ^1- or ℓ^∞-norm. See [46] for an example.
□

The common theme is that the most obvious formulation of a processing algorithm is not normally the best starting point for VLSI design. Departures from some mathematically ideal algorithm are almost always necessary to arrive at a solution that offers the throughput and energy efficiency requested at economically feasible costs. Most algorithmic modifications alter the input-to-output mapping and so imply an **implementation loss**, that is a minor cut-back in signal-to-noise ratio, coding gain, bit-error-rate, mean time between errors, stopband suppression, passband ripple, phase response, false-positive and false-negative rates, data compression factor, fidelity of reproduction, total harmonic distortion, image and color definition, intelligibility of speech, or whatever figures of merit are most important for the application.

Table 3.8   Approximations for computing magnitudes.

Name	aka	Formula
lesser	$\ell^{-\infty}$-norm	$l = \min(|a|, |b|)$
sum	ℓ^1-norm	$s = |a| + |b|$
magnitude (reference)	ℓ^2-norm	$m = \sqrt{a^2 + b^2}$
greater	ℓ^∞-norm	$g = \max(|a|, |b|)$
Approximation 1		$m \approx m_1 = \frac{3}{8}s + \frac{5}{8}g$
Approximation 2 [47]		$m \approx m_2 = \max(g, \frac{7}{8}g + \frac{1}{2}l)$

Experience tells us that enormous improvements in terms of throughput, energy efficiency, circuit size, design effort, and agility can be obtained by adapting an algorithm to the peculiarities and cost factors of hardware. Optimizations in the algorithmic domain are thus concerned with

"How to tailor an algorithm such as to cut the computational burden, to trim down memory requirements, and/or to speed up calculations without incurring unacceptable implementation losses?"

What the trade-offs are and to what extent departures from the initial functionality are acceptable depends very much on the application. It is, therefore, crucial to have a good command of the theory and practice of the computational problems to be solved.

Observation 3.6. *Digital signal processing programs often come with floating point arithmetics. Reimplementing them in fixed point arithmetics, with limited computing resources, and with minimum memory results in an implementation loss. The effort for finding a good compromise between numerical accuracy and hardware efficiency is often underestimated.*

The necessity to validate trimmed-down implementations for all numerical conditions that may occur further adds to the effort. It is not uncommon to spend as much time on issues of numerical precision as on all subsequent VLSI design phases together.

3.3.2 ... AND THERE IS ROOM IN THE ARCHITECTURAL DOMAIN

In the architectural domain, the focus is on meeting given performance targets for a specific data processing algorithm with a minimum of hardware resources. The key concern is

"How to organize datapaths, memories, controllers, and other hardware resources for implementing some given computation flow such as to optimize throughput, energy efficiency, circuit size, design effort, agility, overall costs, and similar figures of merit while leaving the original input-to-output relationship unchanged except, possibly, for latency?"

As computations are just <u>reorganized</u>, not altered, there is no implementation loss at this point.

Given some data or signal processing algorithm, there exists a profusion of alternative architectures although the number of fundamental options available for reformulating it is rather limited. This is because each such option can be applied at various levels of detail and can be combined with others in many different ways. Our approach is based on reformulating algorithms with the aid of **equivalence transforms**. The remainder of this chapter gives a systematic view on all such transforms and shows how they can be applied to optimize VLSI architectures for distinct size, throughput, and energy targets.

3.3.3 SYSTEMS ENGINEERS AND VLSI DESIGNERS MUST COLLABORATE

Systems theorists tend to think in mathematical categories, so an algorithm from data or signal processing is not much more than a sequence of equations to them. To meet pressing deadlines — or simply for reasons of convenience — they tend to model such algorithms in floating point arithmetics, even when a fairly limited numeric range would amply suffice for the application. This is typically unacceptable in VLSI architecture design, and establishing a lean **bit-true software model** is a first step towards a cost-effective circuit.

Generally speaking, it is always necessary to balance many contradicting requirements to arrive at a working and marketable embodiment of the mathematical or otherwise abstracted initial model of a system. A compromise will have to be found between the theoretically desirable and the economically feasible. So, there is more to VLSI design than accepting a given algorithm and turning that into gates with the aid of some HDL synthesis tool.

Algorithm design is typically carried out by systems engineers whereas VLSI architecture is more the domain of hardware designers. The strong mutual interaction between algorithms and architectures mandates a close and early collaboration between the two groups, see fig.3.14.

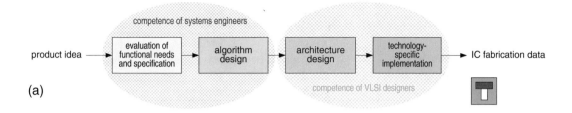

(a)

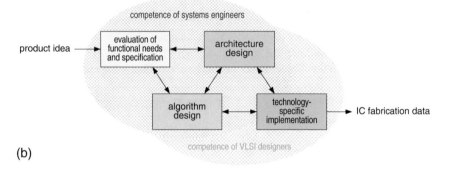

(b)

FIGURE 3.14

Models of collaboration between systems engineers and hardware designers. Sequential thinking doomed to failure (a) versus a networked team more likely to come up with satisfactory results (b).

Observation 3.7. *Finding a good trade-off between the key characteristics of the final circuit and implementation losses asks for an on-going collaboration between systems engineers and VLSI experts during the phases of specification, algorithm development, and architecture design.*

The fact that algorithm design is not covered in this text does not imply that it is of less importance to VLSI than architecture design. As illustrated by the example below, the opposite is true. A comprehensive textbook that covers the joint development of algorithms and architectures is [48], anecdotal observations can be found in [49].

Example

Sequence estimation (SE) is an important subfunction in every EDGE[15] receiver and actually part of the channel estimator - channel equalizer function. Design targets include a soft (i.e. multi valued) output, a processing time of less than $577\,\mu s$ per burst, a small circuit, low power dissipation, and a minimal block error rate at any given signal-to-noise ratio. The key properties of three algorithmic alternatives are summarized below and in fig.3.15.

---

[15] Enhanced Data Rates for GSM Evolution, a standard from wireless telecommunications.

Algorithm	Delayed Decision Feedback	Max-log-MAP	Soft Output Viterbi Equalizer
Soft output	no	yes	yes
Forward recursion	yes	yes	yes
Backward recursion	no	yes	no
Backtracking step	yes	no	no
Memory requirements	1x	50x	0.13x

The DDFSE algorithm does not qualify because of its inferior estimation performance that is due to the hard (i.e. binary) output. While the max-log-MAP and SOVE algorithms perform almost equally well, the backward iteration step of the former requires so many branch metrics to be kept in memory that no architecture-level optimization can compensate the handicap. This explains why the SOVE algorithm was chosen as a starting point in the project from which the above data were taken [50].

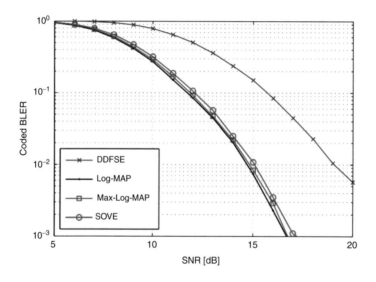

FIGURE 3.15

Block error rate vs. SNR for various sequence estimation algorithms.

□

3.3.4 A GRAPH-BASED FORMALISM FOR DESCRIBING PROCESSING ALGORITHMS

We will often find it useful to capture a data processing algorithm in a **data dependency graph** (DDG) as this visual formalism is suggestive of possible hardware structures. A DDG is a directed graph where vertices and edges have non-negative weights, see fig.3.16. A vertex stands for a memoryless operation and its weight indicates the amount of time necessary to carry out that operation. The precedence of one operation over another is represented as a directed edge. The weight of an edge indicates by how many computation cycles or sampling periods execution of the first operation must precede that of the second.[16] Edge weight zero implies the two operations are scheduled to happen within the same computation or sampling period — one after the other, though —. An edge may also be viewed as expressing the transport of data from one operation to another and its weight as indicating the number of registers included in that transport path.

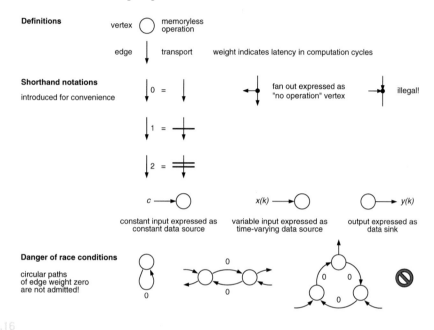

Data dependency graph (DDG) notation.

To warrant consistent outcomes from computation, circular paths of total edge weight zero are disallowed in DDGs.[17] Put differently, any feedback loop shall include one or more latency registers.

---

[16] The term "computation cycle" is to be explained shortly in section 3.3.7.

[17] A **circular path** is a closed walk in which no vertex, except the initial and final one, appears more than once and that respects the orientation of all edges traversed. As the more customary terms "circuit" and "cycle" have other meanings in the context of hardware design, we do prefer "circular path" in spite of its clumsiness. For the same reason, let us use "vertex" when referring to graphs and "node" when referring to electrical networks. A zero-weight circular path in a DDG implies immediate feedback and expresses a self-referencing combinational function. Such zero-latency feedback loops are known to expose the pertaining electronic circuits to unpredictable behavior and are, therefore, highly undesirable, see section 6.4.3 for details.

Example

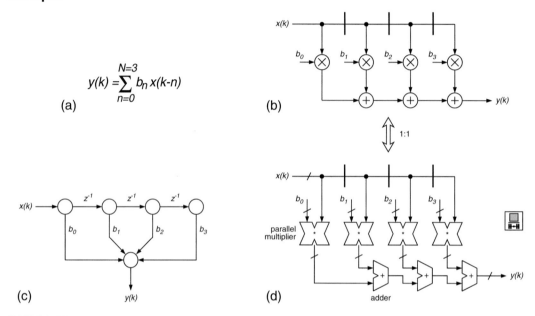

(a)
$$y(k) = \sum_{n=0}^{N=3} b_n\, x(k-n)$$

(b)

(c)

(d)

FIGURE 3.17

Third order ($N = 3$) transversal filter expressed as a mathematical function (a), drawn as data dependency graph (DDG) (b), and implemented with the isomorphic hardware architecture (d). Signal flow graph shown for comparison (c).

□

No matter how one has arrived at some initial proposal, it always makes sense to search for a better hardware arrangement. Inspired VLSI architects let guide themselves by intuition and experience to come up with a few tentative designs before looking for beneficial reorganizations. Yet, for the subsequent discussion and evaluation of the various equivalence transforms available, we need something to compare with. A natural candidate is the isomorphic architecture, see fig.3.17d for an example, where

- Each combinational operation in the DDG is carried out by a hardware unit of its own,
- Each hardware register stands for a latency of one in the DDG,
- There is no need for control because DDG and block diagram are isomorphic,[18] and
- Clock rate and data input/output rate are the same.

---

[18] Two directed graphs are said to be **isomorphic** if there exists a one-to-one correspondence between their vertices and between their edges such that all incidence relations and all edge orientations are preserved. More informally, two isomorphic graphs become indistinguishable when the labels and weights are removed from their vertices and edges. Remember that how a graph is drawn is of no importance for the theory of graphs.

An architecture design as naive as this obviously cannot be expected to utilize hardware efficiently, but it will serve as a reference for discussing both the welcome and the unfavorable effects of various architectural reorganizations. You may also think of the isomorphic architecture as a hypothetical starting point from which any more sophisticated architecture can be obtained by applying a sequence of equivalence transforms.[19]

3.3.6 RELATIVE MERITS OF ARCHITECTURAL ALTERNATIVES

Throughout our analysis, we will focus on the subsequent figures of merit.

Cycles per data item Γ denotes the number of computation cycles that separates the releasing of two consecutive data items, or — which is normally the same — the number of computation cycles between accepting two subsequent data items.

Longest path delay t_{lp} indicates the lapse of time required for data to propagate along the longest combinational path through a given digital network. Path lengths are typically indicated in ns. What makes the maximum path length so important is that it limits the operating speed of a given architecture. For a circuit to function correctly, it must always be allowed to settle to a — typically new — steady state within a single computation period T_{cp}.[20] We thus obtain the requirement $t_{lp} \leq T_{cp}$ where the exact meaning of computation period is to be defined shortly in section 3.3.7.

Time per data item T indicates the lapse of time between releasing two subsequent data items. Depending on the application, T might be stated in μs/sample, ms/frame, s/computation, or time per unit of what you want to have done in the end. $T = \Gamma \cdot T_{cp} \geq \Gamma \cdot t_{lp}$ holds with equality if the circuit gets clocked at the fastest possible rate.

Data throughput $\Theta = \frac{1}{T}$ is the most meaningful measure of computational performance. Throughput gets expressed in terms of data items or operations processed per time unit e.g. in pixel/s, sample/s, frame/s, data record/s, FFT/s, matrix inversion/s, and the like. It is given by

$$\Theta = \frac{f_{cp}}{\Gamma} = \frac{1}{\Gamma \cdot T_{cp}} \leq \frac{1}{\Gamma \cdot t_{lp}} \tag{3.3}$$

for a circuit operated at computation rate f_{cp} or, which is the same, with a computation period T_{cp}.[21] Again, we are most interested in the maximum throughput where $T_{cp} = t_{lp}$.

Latency L indicates the number of computation cycles from a data item being entered into a circuit until the pertaining result becomes available at the output. Latency is zero when the result appears within the very clock cycle during which the input datum was fed in.

---

[19] See problem 6 for a more thorough exposure. Also observe that our transform approach to architecture design bears some resemblance to the theory of evolution.

[20] We neither consider multi-cycle paths, nor wave-pipelined operation, nor asynchronous circuits at this point.

[21] It is sometimes more adequate to express data throughput in terms of bits per time unit. (3.3) must then be restated as $\Theta = w \frac{f_{cp}}{\Gamma}$ where w indicates how many bits make up one data item.

Circuit size A. Depending on how actual hardware costs are best expressed, the designer is free to interpret size as area occupation (in mm^2 or lithographic squares F^2 for ASICs) or as circuit complexity (in terms of GE for ASICs and FPL).

Size-time product AT combines circuit size and computation time to indicate the hardware resources spent to obtain a given throughput. This is simply because $AT = \frac{A}{\Theta}$. The lower the AT-product, the more hardware-efficient a circuit.

Energy per data item E is meant to quantify the amount of energy dissipated in carrying out some given computation on a data item. As examples, consider indications in pJ/MAC, nJ/sample, μJ/datablock or mWs/videoframe.

This quantity can also be understood as power-per-throughput ratio $E = \frac{P}{\Theta}$ measured in units like $\frac{\text{mW}}{\text{Mbit/s}}$ or $\frac{\text{W}}{\text{Gop/s}}$ because $\frac{\text{energy}}{\text{data item}} = \frac{\text{energy per second}}{\text{data item per second}} = \frac{\text{power}}{\text{throughput}}$. When it comes to evaluating microprocessors, the inverse figure of merit expressed in terms of $\frac{\text{Mop/s}}{\text{mW}}$ or $\frac{\text{Gop/s}}{\text{W}}$ is more popular, however.

Energy per data item is closely related to the **power-delay product** (PDP) $pdp = P \cdot t_{lp}$, a quantity often used for comparing standard cells and other transistor-level circuits. The difference is that our definition accounts for slack and for multi-cycle operations because $E = PT = P \cdot \Gamma \cdot T_{cp} \geq P \cdot \Gamma \cdot t_{lp} = \Gamma \cdot pdp$.

Energy-time product ET indicates how much energy gets spent for achieving a given throughput since $ET = \frac{E}{\Theta} = \frac{P}{\Theta^2}$ which also explains the synonym "energy-per-throughput ratio". ET may be expressed in $\frac{\mu J}{\text{datablock/s}}$ or $\frac{\text{mWs}}{\text{videoframe/s}}$, for instance, and is a useful yardstick whenever a better performance must be bought at the expense of energy efficiency. The energy-delay product (EDP) is to ET what the PDP is to E.

Example
In the occurrence of the architecture shown in fig.3.17d, one easily finds the quantities below

$$A = 3A_{reg} + 4A_* + 3A_+ \tag{3.4}$$

$$\Gamma = 1 \tag{3.5}$$

$$t_{lp} = t_{reg} + t_* + 3t_+ \tag{3.6}$$

$$AT = (3A_{reg} + 4A_* + 3A_+)(t_{reg} + t_* + 3t_+) \tag{3.7}$$

$$L = 0 \tag{3.8}$$

$$E = 3E_{reg} + 4E_* + 3E_+ \tag{3.9}$$

where indices $*$, $+$ and reg refer to a multiplier, an adder, and a data register respectively.

$\square$

A word of caution is due here. Our goal in using formulae to approximate architectural figures of merit is not so much to obtain numerical values for them but to explain how the various equivalence transforms available to VLSI architects tend to affect them.[22] For illustration purposes, we will repeatedly use a graphical representation that suggests hardware organization, circuit size, longest path length, data throughput, and latency in a symbolic way, see figs 3.18b and c for a first example.

3.3.7 COMPUTATION CYCLE VERSUS CLOCK PERIOD

So far, we have been using the term computation period without defining it. In synchronous digital circuits, a calculation is broken down into a series of shorter computation cycles the rhythm of which gets imposed by a periodic **clock** signal. During each computation cycle, fresh data emanate from a register, propagate through combinational circuitry where they undergo various arithmetic, logic, and/or routing operations before the result gets stored in the next analogous register (same clock, same active edge).

Definition 1. A computation period T_{cp} is the time span that separates two consecutive computation cycles.

For the moment being, it is safe to assume that computation cycle, computation period, clock cycle, and clock period are all the same, $T_{cp} = T_{clk}$, which is indeed the case for all those circuits that adhere to single-edge-triggered one-phase clocking.[23] The inverse, that is the number of computation cycles per second, is referred to as **computation rate** $f_{cp} = \frac{1}{T_{cp}}$.

---

[22] As an example, calculation of the long path delay t_{lp} is grossly simplified in (3.6). For one thing, interconnect delays are neglected which is an overly optimistic assumption. For another thing, the propagation delays of the arithmetic operations are simply summed up which sometimes is a pessimistic assumption, particularly in cascades of multiple ripple-carry adders (RCA) where all operands arrive simultaneously. Synthesis followed by place and route often is the only way to determine overall path delays with sufficient accuracy.

[23] As an exception, consider dual-edge-triggering where each clock period comprises two consecutive computation periods so that $T_{cp} = \frac{1}{2}T_{clk}$. Details are to follow in section 7.2.3.

3.4 EQUIVALENCE TRANSFORMS FOR COMBINATIONAL COMPUTATIONS

A computation that depends on the present arguments exclusively is termed combinational. A sufficient condition for combinational behavior is a DDG which is free of circular paths and where all edge weights equal zero.

Consider some fixed but otherwise arbitrary combinational function $y(k) = f(x(k))$. As suggested by the dashed edges in fig.3.18a, both input $x(k)$ and output $y(k)$ can include several subvectors. No assumptions are made about the complexity of f which, in principle, can range from a two-bit addition, over an algebraic division, to the Fast Fourier Transform (FFT) operation on a data block, see fig.3.27, and beyond. In practice, architecture designers primarily focus on those operations that heavily impact chip size, performance, power dissipation, etc.

The isomorphic architecture simply amounts to a hardware unit that does nothing but evaluate function f, a rather expensive proposition if f is complex such as in the FFT example. Three options for reorganizing and improving this unsophisticated arrangement exist.[24]

1. **Decomposing** function f into a sequence of subfunctions that get executed one after the other in order to reuse the same hardware as much as possible.
2. **Pipelining** of the functional unit for f to improve computation rate by cutting down combinational depth and by working on multiple consecutive data items simultaneously.
3. **Replicating** the functional unit for f and having all units work concurrently.

It is intuitively clear that replication and pipelining both trade circuit size for performance while iterative decomposition does the opposite. This gives rise to questions such as

"Does it make sense to combine pipelining with iterative decomposition in spite of their contrarian effects?"
"How do replication and pipelining compare?"
"Are there situations where one should be preferred over the other?"

which we will try to answer in the following subsections.

---

[24] Of course, many circuit alternatives for implementing a given arithmetic or logic function also exist at the gate level. However, within the general context of architecture design, we do not address the problem of developing and evaluating such options as this involves lower-level considerations that strongly depend on the specific operations and on the target library. The reader is referred to the specialized literature on computer arithmetics and on logic design.

DDG for some combinational function f (a). A symbolic representation of the reference hardware configuration (b) with its key characteristics highlighted (c).

The architectural arrangement that will serve as a reference for comparing various alternative designs is essentially identical to the isomorphic configuration of fig.3.18a with a register added at the output to allow for the cascading of architectural chunks without their longest path delays piling up. The characteristics of the reference architecture then are

$$A(0) = A_f + A_{reg} \tag{3.10}$$

$$\Gamma(0) = 1 \tag{3.11}$$

$$t_{lp}(0) = t_f + t_{reg} \tag{3.12}$$

$$AT(0) = (A_f + A_{reg})(t_f + t_{reg}) \tag{3.13}$$

$$L(0) = 1 \tag{3.14}$$

$$E(0) = E_f + E_{reg} \tag{3.15}$$

where subscript $_f$ stands for the datapath hardware that computes some given combinational function f and where subscript $_{reg}$ denotes a data register. For the sake of simplicity, the word width w in the datapath is assumed to be constant throughout. The quotient A_f/A_{reg} relates the size of the datapath hardware to that of a register, and t_f/t_{reg} does the same for their respective time requirements.[25] Their product $\frac{A_f}{A_{reg}} \frac{t_f}{t_{reg}}$ thus reflects the computational complexity of function f in some sense. (3.16) holds whenever logic function f is a fairly substantial computational operation. We will consider this the typical, although not the only possible case.

$$A_{reg}t_{reg} \ll A_f t_f \tag{3.16}$$

Many of the architectural configurations to be discussed require extra circuitry for controlling datapath operation and for routing data items. Two additive terms A_{ctl} and E_{ctl} are introduced to account for this where necessary. As it is very difficult to estimate the extra hardware without detailed knowledge of the specific situation at hand, the only thing that can be said for sure is that A_{ctl} is on the order of

---

[25] Typical size A and delay figures t for a number of logic and arithmetic operations are given as illustrative material in appendix 3.12.

A_{reg} or larger for most architectural transforms. Control overhead may in fact become significant or even dominant when complex control schemes are brought to bear as a result from combining multiple transforms.

As for energy, we will focus on the dynamic contribution that gets dissipated in charging and discharging electrical circuit nodes as a consequence of fresh data propagating through gate-level networks. Any dissipation due to static currents or due to idle switching is ignored.

Throughout our architectural comparisons, we further assume all electrical and technological conditions to remain the same.[26] A comparison of architectural alternatives on equal grounds is otherwise not possible as a shorter path delay or a lower energy figure would not necessarily point to a more efficient design alternative.

3.4.2 ITERATIVE DECOMPOSITION

The idea behind iterative decomposition — or decomposition, for short — is nothing else than **resource sharing** through *step-by-step execution*. The computation of function f is broken up into a sequence of d subtasks which are carried out one after the other. From a dataflow point of view, intermediate results are recycled until the final result becomes available at the output d computation cycles later, thereby making it possible to reuse a single hardware unit several times over. A configuration that reuses a multifunctional datapath in a time-multiplex fashion to carry out f in $d = 3$ subsequent steps is symbolically shown in fig.3.19. Note the addition of a control section that pilots the datapath on a per-cycle basis over a number of control lines.

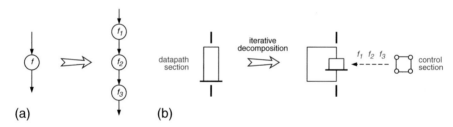

(a)   (b)

FIGURE 3.19

Iterative decomposition. DDG (a) and hardware configuration for $d = 3$ (b).

Performance and cost analysis

Assumptions:

1. The total size requirement for implementing the various subfunctions into which f is decomposed ranges between $\frac{A_f}{d}$ and A_f.
2. The decomposition is lossless and balanced, i.e. it is always possible to break up f into d subfunctions the computations of which require a uniform amount of time $\frac{t_f}{d}$.

---

[26] This includes supply voltage, cell library, transistor sizes, threshold voltages, fabrication process, and the gate-level structure of arithmetic units.

As a first-order approximation, iterative decomposition leads to the following figures of merit:

$$\frac{A_f}{d} + A_{reg} + A_{ctl} \leq A(d) \leq A_f + A_{reg} + A_{ctl} \tag{3.17}$$

$$\Gamma(d) = d \tag{3.18}$$

$$t_{lp}(d) \approx \frac{t_f}{d} + t_{reg} \tag{3.19}$$

$$d(A_{reg} + A_{ctl})t_{reg} + (A_{reg} + A_{ctl})t_f + A_f t_{reg} + \frac{1}{d}A_f t_f \leq AT(d) \leq$$

$$d(A_f + A_{reg} + A_{ctl})t_{reg} + (A_f + A_{reg} + A_{ctl})t_f \tag{3.20}$$

$$L(d) = d \tag{3.21}$$

$$E(d) \gtrsim E_f + E_{reg} \tag{3.22}$$

Let us confine our analysis to situations where the control overhead can be kept small so that $A_{reg} \approx A_{ctl} \ll A_f$ and $E_{reg} \approx E_{ctl} \ll E_f$. A key issue in interpreting the above results is whether size $A(d)$ tends more towards its lower or more towards its upper bound in (3.17). While iterative decomposition can, due to (3.16), significantly lower the AT-product in the former case, it does not help in the latter.

The lower bounds hold in (3.17) and (3.20) when the chunk's function f makes repetitive use of a single subfunction because the necessary datapath is then essentially obtained from cutting the one that computes f into d identical pieces only one of which is implemented in hardware. A monofunctional processing unit suffices in this case.

At the opposite end are situations where computing f asks for very disparate subfunctions that cannot be made to share much hardware resources in an efficient way. Iterative decomposition is not an attractive option in this case, especially if register delay, control overhead, and the difficulty of meeting assumption 2 are taken into consideration.

Multiplication, for instance, can be broken into repeated shift & add operations. Chains of additions and subtractions in either fixed point or floating point arithmetics also lend themselves well for being combined into a common computational unit. Similarly, angle rotations, trigonometric, and hyperbolic functions can all be computed with essentially the same CORDIC datapath.

As for energy efficiency, there are two mechanisms that counteract each other. On the one hand, iterative decomposition entails register activity not present in the original circuit. The extra control and data recycling logic necessary to implement step-by-step execution further inflate dissipation.

On the other hand, we will later find that long register-to-register signal propagation paths tend to foster transient node activities, aka glitches. Cutting such propagation paths often helps to mitigate glitching activities and the associated energy losses. Such second order effects are not accounted for in the simplistic unit-wise additive model introduced in (3.15), however, making it difficult to apprehend the impact of iterative decomposition on energy before specific circuit details become available.

Example

A secret-key block cipher operated in electronic code book (ECB) mode is a highly expensive combinational function. ECB implies a memoryless mapping $y(k) = c(x(k), u(k))$ where $x(k)$ denotes the plaintext, $y(k)$ the ciphertext, $u(k)$ the key, and k the block number or time index. What most block ciphers, such as the Data Encryption Standard (DES), the International Data Encryption Algorithm (IDEA), and the Advanced Encryption Standard (AES) Rijndael have in common is a cascade of several rounds, see fig.3.20 for the IDEA algorithm [51]. The only difference between the otherwise identical rounds is in the values of the subkeys used that get derived from $u(k)$. What is referred to as output transform is nothing else than a subfunction of the previous rounds.

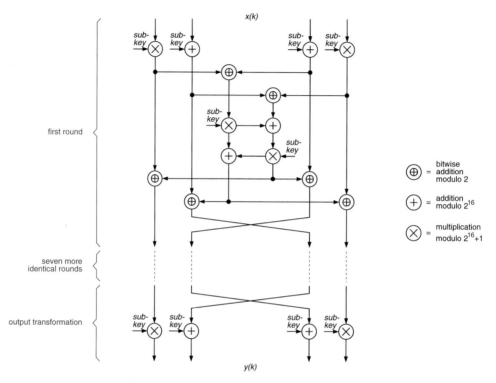

FIGURE 3.20

DDG of the block cipher IDEA.

If we opt for iterative decomposition, a natural choice consists in designing a datapath for one round and in recycling the data with changing subkeys until all rounds have been processed. As control is very simple, the circuit's overall size is likely to stay close to the lower bound in (3.17) after this first step of decomposition. When continuing in the same direction, however, benefits will diminish because the operations involved (bitwise addition modulo 2, addition modulo 2^{16} and multiplication modulo $(2^{16} + 1)$) are very disparate. In addition, the impact of control on the overall circuit size would be felt. □

A more radical approach is to decompose arbitrary functions into sequences of arithmetic and/or logic operations from a small but extremely versatile set and to provide a single ALU instead. The datapath of any **microprocessor** is just a piece of universal hardware that arose from the general idea of step-by-step computation, and the **reduced instruction set computer** (RISC) can be viewed as yet another step in the same direction. While iterative decomposition together with programmability and time sharing, see 3.4.5, explains the outstanding Flexibility and hardware economy of this paradigm, it also accounts for its modest performance and poor energy efficiency when compared to more focussed architecture designs.

Examples

Examples of ASICs the throughputs of which exceeded that of contemporary high-end general-purpose processors by orders of magnitude are given in sections 3.2 and 3.7.3.
□

3.4.3 PIPELINING

Pipelining aims at increasing throughput by cutting combinational depth into several separate stages of approximately uniform computational delays by inserting registers in between.[27] The combinational logic between two subsequent **pipeline registers** is designed and optimized to compute one specific subfunction. As an ensemble, the various stages cooperate like specialist workers on an *assembly line*. Fig.3.21 sketches a functional unit for f subdivided into $p = 3$ pipeline stages by $p - 1$ extra registers. Note the absence of any control hardware.

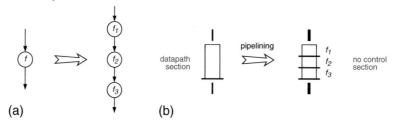

(a)   (b)

FIGURE 3.21

Pipelining. DDG (a) and hardware configuration for $p = 3$ (b).

Performance and cost analysis

Assumptions:

1. The combinational logic for implementing function f is not affected by the number of pipeline stages introduced. Its overall size A_f, therefore, remains constant.
2. The pipeline is lossless and balanced, i.e. similarly to decomposition it is always possible to partition the logic into p stages such that all have identical delays $\frac{t_f}{p}$.
3. The size penalty of pipelining can be expressed by an additive term A_{reg} for each register accounting for the silicon area occupied by storage elements.
4. At each pipeline stage a performance penalty results from introducing a register delay t_{reg} which includes the delay caused by the storage element.

Pipelining changes performance and cost figures as follows:

$$A(p) = A_f + pA_{reg} \tag{3.23}$$

$$\Gamma(p) = 1 \tag{3.24}$$

$$t_{lp}(p) \approx \frac{t_f}{p} + t_{reg} \tag{3.25}$$

---

[27] For a more formal discussion see subsection 3.6.1.

$$AT(p) \approx pA_{reg}t_{reg} + (A_{reg}t_f + A_f t_{reg}) + \frac{1}{p}A_f t_f \qquad (3.26)$$

$$L(p) = p \qquad (3.27)$$

$$E(p) \underset{\text{coarse grain}}{\overset{\text{fine grain}}{\gtrless}} E_f + E_{reg} \qquad (3.28)$$

Both performance and size grow monotonically with pipeline depth. The same holds true for latency. What is more interesting, a modest number of pipeline stages each of which has a substantial depth dramatically lowers the AT-product due to (3.16). This regime is referred to as **coarse grain pipelining**.

Example

Equation (3.25) relates combinational delay to register delay. Another popular way to quantify the degree of pipelining is to express the delay on the longest path as a multiple of fanout-of-4 (FO4) inverter delays.[28]

CPU	year	clock freq. [MHz]	FO4 inverter delays per pipeline stage
Intel 80386	1989	33	≈ 80
Intel Pentium 4	2003	3200	12...16
Core 2 Duo	2007	2167	≈ 40
Core i7 980X	2011	3333...3600	42...46
IBM POWER5	2004	1650...1900	22
IBM POWER6	2007	3500...5000	13
IBM Cell Processor	2006	3200	11

□

Continuing along this line, one may want to insert more and more pipeline registers. However, (3.25) reveals that the benefit fades when the combinational delay per stage $\frac{t_f}{p}$ approaches the register delay t_{reg}. For large values of p the area-delay product gets dominated by the register delay rather than by the payload function. A natural question for this type of deep or **fine grain pipelining** is to ask

"What is the maximum computation rate for which a pipeline can be built?"

The fastest logic gates that can do any data processing are 2-input NAND and NOR gates.[29] Even if we are prepared to profoundly redesign a pipeline's logic in an attempt to minimize the longest path t_{lp}, we must leave room for at least one such gate between two consecutive registers.

$$T_{cp} \geq \min(t_{lp}) = \min(t_{gate}) + t_{reg} = \min(t_{nand}, t_{nor}) + t_{suff} + t_{pdff} \qquad (3.29)$$

thus represents a lower bound for (3.25). Practical applications that come close to this theoretical minimum are limited to tiny subcircuits, however, mainly because of the disproportionate number of registers required, but also because meeting assumptions 1 and 2 is difficult with fine grained pipelines. Even in

---

[28] Comparing circuit alternatives in terms of FO4 inverters makes sense because fanout-of-4 inverters commonplace in buffer trees driving large loads and because the delays of other static CMOS gates have been found to track well with those of FO4 inverters.

[29] This is because binary NAND and NOR operations (a) form a complete gate set each and (b) are efficiently implemented from MOSFETs, see section A.2.10.

high-performance datapaths, energy efficiency and economic reasons preclude pipelining below 12 or so FO4 inverter delays per stage. As the above numbers illustrate, Intel has in fact reversed the historical trend towards ever deeper pipelines when transiting from the Pentium 4 to the Core architecture to reclaim energy efficiency [39]. Fig.3.22 confirms that this is indeed a general industry trend.

Equation (3.29) further indicates that register delay is critical in high speed design. In fact, a typical relation is $t_{reg} \approx 3...5 \min(t_{gate})$. As a consequence, it takes twenty or so levels of logic between subsequent registers before flip-flop delays are relegated to insignificant proportions. A high-speed cell library must, therefore, not only include fast combinational functions but also provide bistables with minimum insertion delays.[30]

Example
Plugging into (3.29) typical numbers for a 2-input NOR gate and a D-type flip-flop with no reset from a 130 nm CMOS standard cell library, one obtains $T_{cp} \geq t_{NOR2D1} + t_{DFFPB1} = 18\,\text{ps} + 249\,\text{ps} \approx 267\,\text{ps}$ which corresponds to a maximum computation rate of about 3.7 GHz.
□

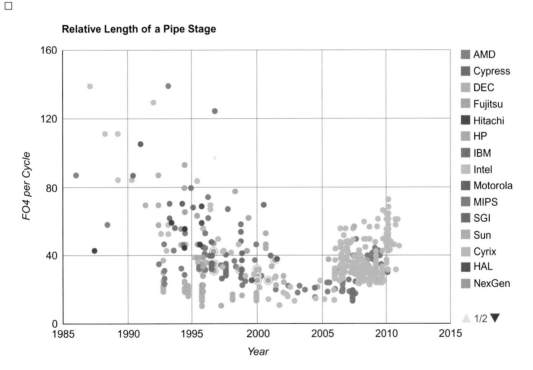

FIGURE 3.22

Evolution of pipeline depth over the years (source Stanford CPU database).

---

[30] Function latches where bistables and combinational logic get merged into a single library cell in search of better performance are to be discussed in section 7.2.6.

"How many stages yield optimum pipeline efficiency?"

Optimum hardware efficiency means minimum size-time product

$$AT(p) = \min \tag{3.30}$$

which is obtained for

$$p_0 = \sqrt{\frac{A_f t_f}{A_{reg} t_{reg}}} \tag{3.31}$$

Beyond this point, adding more pipeline registers causes the size-time product to deteriorate even though performance is still pushed further. It also becomes evident from (3.31) that, in search of an economic solution, the more complex a function, the more pipelining it supports. In practice, efficiency is likely to degrade before p_0 is reached because our initial assumptions 1 and 2 cannot be entirely satisfied. Also, shallow logic just a few gates deep is more exposed to on-chip variations (OCV).

"How does pipelining affect energy efficiency?"

The additional registers suggest that any pipelined datapath dissipates more energy than the reference architecture does. This is certainly true for fine grain pipelines where the energy wasted by the switching of all those extra subcircuits becomes the dominant contribution.

For coarse grain designs, the situation is more fortunate. Experience shows that pipeline registers tend to reduce the unproductive switching activity associated with glitching in deep combinational networks, a beneficial side effect neglected in a simple additive model.

Interestingly, our finding that throughput is greatly increased makes it possible to take advantage of coarse grain pipelining for improving energy efficiency, albeit indirectly. Recall that the improved throughput is a result from cutting the longest path while preserving a processing rate of one data item per computation cycle. The throughput of the isomorphic architecture is thus readily matched by a pipelined datapath implemented in a slower yet more energy-efficient technology, e.g. by operating CMOS logic from a lower supply voltage or by using mostly minimum size transistors. Our model cannot reflect this opportunity because we have decided to establish energy figures under the assumption of identical operating conditions and cell libraries. Another highly welcome property of pipelining is the absence of energy-dissipating control logic.

Pipelining in the presence of multiple feedforward paths

Although pipelining can be applied to arbitrary feedforward computations, there is a reservation of economic nature when a DDG includes many parallel paths. In order to preserve overall functionality, any latency introduced into one of the signal propagation paths must be balanced by inserting an extra register into each of its parallel paths. Unless those **shimming registers** help cut combinational depth there, they bring about substantial size and energy penalties, especially for deep pipelines where p is large.

Example

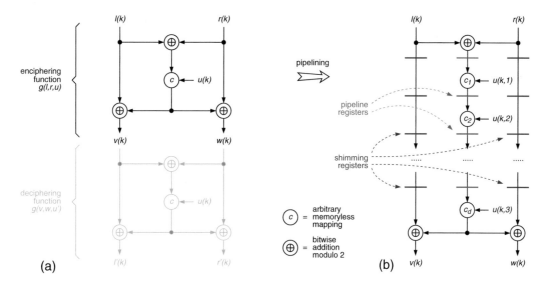

FIGURE 3.23

Involutory cipher algorithm. DDG before (a) and after pipelining (b).

With simplifications, fig.3.23a reproduces the block cipher IDEA. Variable k stands for the block index, $l(k)$ and $r(k)$ each denote half of a 64 bit plaintext block while $v(k)$ and $w(k)$ do the same for a 64 bit ciphertext block. $u(k)$ and $u'(k)$ stand for the keys used during enciphering and deciphering operations respectively. Provided the two keys are the same, i.e. $u'(k) = u(k)$, the net result is $l'(k) = l(k)$ and $r'(k) = r(k)$, which implies that the plaintext is recovered after calling g twice. Note that this involution property[31] is totally independent of function c which, therefore, can be designed such as to maximize cryptographic security.

Extensive pipelining seems a natural way to reconcile the computational complexity of c with ambitious performance goals. Yet, as a consequence of the two paths bypassing c, every pipeline register entails two shimming registers, effectively tripling the costs of pipelining, see fig.3.23b. This is the reason why pipeline depth had to be limited to eight stages per round in a VLSI implementation of the IDEA cipher in spite of stringent throughput requirements [52].
□

---

[31] A function g is said to be **involutory** iff $g(g(x)) \equiv x$, $\forall x$. As trivial examples, consider multiplication by -1 in classic algebra where we have $-(-x) \equiv x$, the complement function in Boolean algebra where $\bar{\bar{x}} \equiv x$, or a mirroring operation from geometry. Involution is a welcome property in cryptography since it makes it possible to use exactly the same equipment for both enciphering and deciphering.

3.4.4 REPLICATION

Replication is a brute force approach to performance: If one functional unit does not suffice, allow for several of them. Concurrency is obtained from providing q instances of identical functional units for f and from having each of them process one out of q data items in a cyclic manner. To that end, two synchronous q-way switches distribute and recollect data at the chunk's input and output respectively. An arrangement where $q = 3$ is shown in fig.3.24.[32] Overall organization and operation is reminiscent of a *multi-piston pump*.

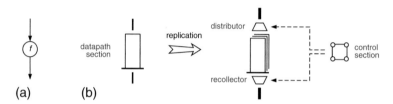

(a)   (b)

FIGURE 3.24

Replication. DDG (a) and hardware configuration for $q = 3$ (b).

Performance and cost analysis

Assumptions:

1. Any size penalties associated with distributing data to replicated functional units and with recollecting them are neglected.
2. Any energy dissipated in data distribution and recollection is ignored.

The above assumption hold fairly well provided the circuitry for computing f is much larger than the one for data distribution and recollection. The key characteristics of replication then become:

$$A(q) = q(A_f + A_{reg}) + A_{ctl} \tag{3.32}$$

$$\Gamma(q) = \frac{1}{q} \tag{3.33}$$

$$t_{lp}(q) \approx t_f + t_{reg} \tag{3.34}$$

$$AT(q) \approx (A_f + A_{reg} + \frac{1}{q} A_{ctl})(t_f + t_{reg}) \approx (A_f + A_{reg})(t_f + t_{reg}) \tag{3.35}$$

$$L(q) = 1 \tag{3.36}$$

$$E(q) \approx E_f + E_{reg} + E_{ctl} \tag{3.37}$$

---

[32] Multiple processing units that work in parallel are also found in situations where the application naturally provides data in parallel streams, each of which is to undergo essentially the same processing. In spite of the apparent similarity, this must not be considered as the result of replication, however, because DDG and architecture are isomorphic. This gets reflected by the fact that no data distribution and recollection mechanism is required in this case. Please refer to section 3.4.5 for the processing of multiple data streams.

As everyone would expect, replication essentially trades area for speed. Except for the control overhead, the AT-product remains the same. Pipelining, therefore, is clearly more attractive than replication as long as circuit size and performance do not get dominated by the pipeline registers, see fig.3.25 for a comparison.

Energywise, replication is indifferent except for the contributions for datapath control and data distribution/recollection. Also note, by the way, that replication does not shorten the computation period which contrasts with iterative decomposition and pipelining.[33]

A more accurate evaluation of replication versus pipelining would certainly require revision of some of the assumptions made here and does depend to a large extent on the actual DDG and on implementation details. Nevertheless, it is safe to conclude that neither fine grain pipelining nor replication are as cost-effective as coarse grain pipelining.

Its penalizing impact on circuit size confines replication to rather exceptional situations in ASIC design. A megacell available in layout form exclusively represents such a need because adding pipeline registers to a finished layout would ask for a disproportionate effort. Replication is limited to high performance circuits and always combined with generous pipelining.

Superscalar and multicore microprocessors are two related ideas from computer architecture.[34] Several factors have pushed the computer industry towards replication: CMOS technology offered more room for increasing circuit complexity than for pushing clock frequencies higher. The faster the clock, the smaller the region on a semiconductor die that can be reached within a single clock period.[35] Fine grain pipelines dissipate a lot of energy for relatively little computation. Reusing a well-tried subsystem benefits design productivity and lowers risks. A multicore processor can still be of commercial value even if one of its CPUs is found to be defective.

Example

Consider a simple network processor that handles a stream of incoming data packets and does some address calculations before releasing the packets with a modified header. Let that processor be characterized by the subsequent cost figures: $A_f = 60w$ GE, $A_{reg} = 6w$ GE, where w is an integer relating to datapath width, $t_f = 12$ ns, $t_{reg} = 1.2$ ns and $\min(t_{gate}) = 0.3$ ns.

---

[33] Observe that the entirety of functional units must be fed with q data items per computation cycle and that processed data items emanate at the same rate. Only the data distribution and recollection subcircuits must be made to operate at a rate q times higher than the computational instances themselves. High data rates are obtained from configuring data distribution/recollection networks as heavily pipelined binary trees. Maximum speed is, again, determined by (3.29). Yet, circumstances permitting, it may possible to implement data distribution and recollection using a faster technology than the one being used in the body of the processing chunk ("superfast distribution and recollection"). Also see [53] for further information on fast data distribution/recollection circuitry.

[34] A **superscalar** processor combines multiple execution units, such as integer ALUs, FPUs, load/store units, and the like, into one CPU. The architecture enables the processor to concurrently fetch and process multiple instructions from a thread of execution. **Multicore** architectures go one step further in that they replicate entire CPUs on a single chip and so enable a processor to work on two or more threads of execution at a time.

[35] For a rationale, refer to section 7.3 that discusses delay in interconnect lines without and with repeaters.

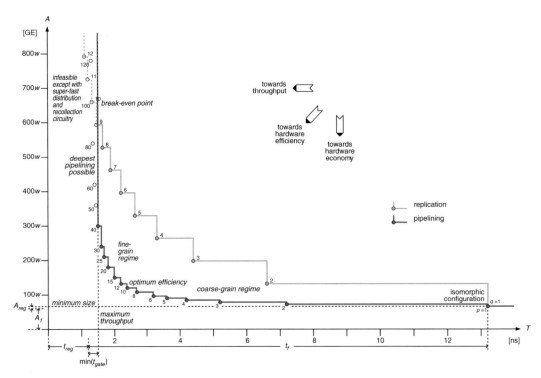

FIGURE 3.25

AT-characteristics of pipelining and replication compared. Simplified by assuming perfect balancing in the occurrence of pipelining and by abstracting from control, data distribution and recollection associated with replication.

□

3.4.5 TIME SHARING

So far we have been concerned with the processing of a single data stream as depicted in fig.3.18. Now consider a situation where a number of parallel data streams undergo processing as illustrated in fig.3.26, for instance. Note that the processing functions f, g and h may or may not be the same. The isomorphic architecture calls for a separate functional unit for each of the three operations in this case. This may be an option in applications such as image processing where a great number of dedicated but comparatively simple processing units are repeated along one or two dimensions, where data exchange is mainly local, and where performance requirements are very high.

More often, however, the costs of fully parallel processing are unaffordable and one seeks to cut overall circuit size. A natural idea is to pool hardware by having a single functional unit process the parallel data streams one after the other in a cyclic manner. Analogously to replication, a synchronous *s*-way switch at the input of that unit collects the data streams while a second one redistributes the processed data at the output. While the approach is known as time sharing in computing, it is more often referred to as **multiplexing** or as **resource sharing** in the context of circuit design.[36] What it requires is that the circuitries for computing the various functions involved all be combined into a single datapath of possibly multifunctional nature. A *student sharing his time between various subjects* might serve as an analogy from every-day live.

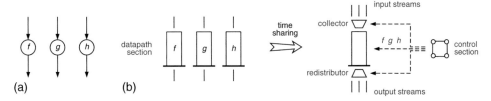

FIGURE 3.26

Time sharing. DDG with parallel data streams (a) and hardware configuration for $s = 3$ (b).

Performance and cost analysis

Assumptions:

1. The size of a circuit capable of implementing functions f, g and h with a single computational unit ranges between $\max\limits_{f,g,h}(A) = \max(A_f, A_g, A_h)$ and $\sum_{f,g,h} A = (A_f + A_g + A_h)$.

2. The time for the combined computational unit to evaluate any of the functions f, g and h has a fixed value $\max\limits_{f,g,h}(t) = \max(t_f, t_g, t_h)$.

3. As for replication, any size and energy penalties associated with collecting and redistributing data are neglected.

4. The energy spent for carrying out functions f, g and h (all together) with one shared unit is closer to $s\max\limits_{f,g,h}(E) = s\max(E_f, E_g, E_h)$ than to $\sum_{f,g,h} E = E_f + E_g + E_h$.

Time sharing yields the following circuit characteristics:

$$\max\limits_{f,g,h}(A) + A_{reg} + A_{ctl} \leq A(s) \leq \sum_{f,g,h} A + A_{reg} + A_{ctl} \tag{3.38}$$

$$\Gamma(s) = s \tag{3.39}$$

$$t_{lp}(s) \approx \max\limits_{f,g,h}(t) + t_{reg} \tag{3.40}$$

$$s(\max\limits_{f,g,h}(A) + A_{reg} + A_{ctl})(\max\limits_{f,g,h}(t) + t_{reg}) \leq AT(s) \leq$$

$$s(\sum_{f,g,h} A + A_{reg} + A_{ctl})(\max\limits_{f,g,h}(t) + t_{reg}) \tag{3.41}$$

$$L(s) = s \tag{3.42}$$

$$E(s) \approx s\max\limits_{f,g,h}(E) + E_{reg} + E_{ctl} \tag{3.43}$$

---

[36] This is our second resource sharing technique after iterative decomposition introduced in section 3.4.2.

Similarly to what was found for iterative decomposition, the question whether size $A(s)$ tends more towards its lower or more towards its upper bound in (3.38) greatly depends on how similar or dissimilar the individual processing tasks are.

The most favorable situation occurs when one monofunctional datapath proves sufficient because all streams are to be processed in exactly the same way. In our example we then have $f \equiv g \equiv h$ from which $\max_{f,g,h}(A) = A_f = A_g = A_h$ and $\max_{f,g,h}(t) = t_f = t_g = t_h$ follow immediately. Apart from the overhead for control and data routing, $AT(s)$ equals the lower bound $s(A_f + A_{reg})(t_f + t_{reg})$ which is identical to the isomorphic architecture with its s separate computational units. It is in this best case exclusively that time sharing leaves the size-time product unchanged and may, therefore, be viewed as antithetic to replication.

The contrary condition occurs when f, g and h are very dissimilar so that little or no savings can be obtained from concentrating their processing into one multifunctional datapath. Time sharing will then just lower throughput by a factor of s thereby rendering it an unattractive option. Provided speed requirements are sufficiently low, a radical solution is to combine time sharing with iterative decomposition and to adopt a processor-style architecture as already mentioned in subsection 3.4.2.

The energy situation is very similar. If the processing functions are all alike and if the computation rate is kept the same, then the energy spent for processing actual data also remains much the same.[37] Extra energy is then spent only for controlling the datapath and for collecting and redistributing data items. More energy is going to get dissipated in a combined datapath when the functions markedly differ from each other. As time sharing has no beneficial impact on glitching activity either, we conclude that such an architecture necessarily dissipates more energy than a comparable non-time-shared one.

By processing s data streams with a single computational unit, time sharing deliberately refrains from taking advantage of the parallelism inherent in the original problem. This is of little importance as long as performance goals are met with a given technology. When in search of more performance, however, a time-shared datapath will have to run at a much higher speed to rival the s concurrent units of the isomorphic architecture which implies that data propagation along the longest path must be substantially accelerated. Most measures suitable to do so, such as higher supply voltage, generous transistor sizing, usage of high-speed cells and devices, adoption of faster but also more complex adder and multiplier schemes, etc., tend to augment the amount of energy spent for the processing of one data item even further.

Example

The Fast Fourier Transform (FFT) is a rather expensive combinational function, see fig.3.27. Luckily, due to its regularity, the FFT lends itself extremely well to various reorganizations that help reduce hardware size. A first iterative decomposition step cuts the FFT into $\log_2(n)$ rounds where n stands for the number of points. When an in-place computation scheme is adopted, those rounds become identical

---

[37] Consider (3.43) and note that the equation simplifies to $sE_f + E_{reg} + E_{ctl}$ when f, g and h are the same. The partial sum $sE_f + E_{reg}$ then becomes almost identical to $s(E_f + E_{reg})$ of the reference architecture. The apparent saving of $(s-1)E_{reg}$ obtained from doing with a single register does not materialize in practice because of the necessity to store data items from all streams.

except for their coefficient values.[38] Each round so obtained consists of $\frac{n}{2}$ parallel computations referred to as **butterflies** because of their structure, see fig.3.27b. The idea of sharing one or two computational units between the butterfly operations of the same round is very obvious at this point.[39]

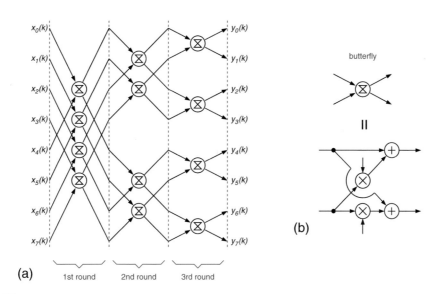

FIGURE 3.27

DDG of 8-point FFT (a) and of butterfly operator (b) (reduced for simplicity).

DDGs as regular as this offer ample room for devising a range of architectures that represent diverse compromises between a single-ALU microprocessor and a hardwired data pipeline maximized for throughput. Providing a limited degree of scalability to accommodate FFTs of various sizes, is not overly difficult either. Favorable conditions similar to these are found in many more applications including, among others, transversal filters (repetitive multiply-accumulate operations), correlators (idem), lattice filters (identical stages), and block ciphers (cascaded rounds).

□

---

[38] For a number of computational problems, it is a logical choice to have two memories that work in a **ping-pong** fashion. At any moment of time, one memory provides the datapath with input data while the other accommodates the partial results presently being computed. After the evaluation of one round is completed, their roles are swapped. As simple as it is, this approach unfortunately requires twice as much memory than needed to store one set of intermediate data. A more efficient technique is **in-place computation** where part of the input data are immediately overwritten by the freshly computed values. In-place computation may cause data items to get scrambled in memory, though, which necessitates corrective action. Problems amenable to in-place computation combined with memory unscrambling include the FFT and the Viterbi algorithm.

[39] Alternative butterfly circuits and architectures have been evaluated in [54] with emphasis on energy efficiency. As opposed to the above, [55] discusses systolic radix-4 high-performance FFT architectures. Also, in fig.3.27, we have assumed input samples to be available as eight concurrent data streams. FFT processors often have to interface to one continuous word-serial stream, however. Architectures that optimize hardware utilization for this situation are being discussed in [56].

So far, we have come up with four equivalence transforms, namely

- Iterative decomposition,
- Pipelining,
- Replication, and
- Time sharing.

Fig.3.28 puts them into perspective in a grossly simplified but also very telling way. More comments will follow in section 3.4.8.

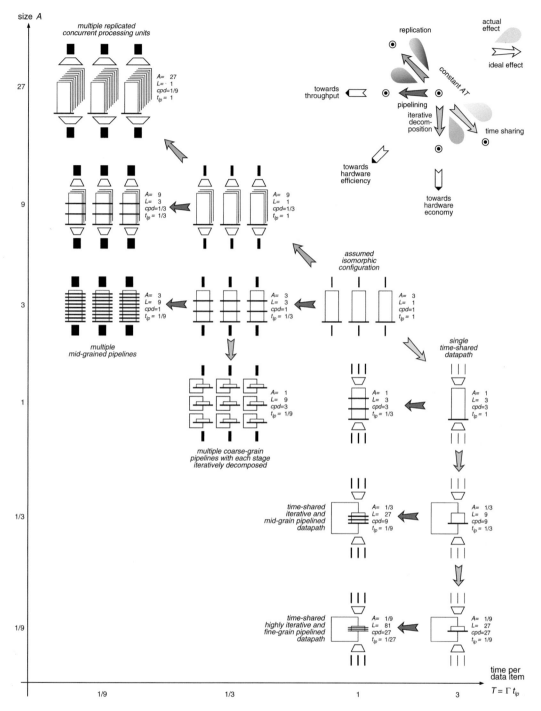

FIGURE 3.28

A roadmap illustrating the four universal transforms for tailoring combinational hardware. Only a subset of all possible architectural configurations is shown, see problem 6. Greatly simplified by

- abstracting from register overhead ($A_{reg} = 0$, $t_{reg} = 0$), which also implies
- not making any difference between RAMs and flip-flops ($A_{RAM} = A_{ff} \cdot \#_{bits}$, $t_{RAM} = t_{ff}$),
- assuming ideal iterative decomposition and ideal time sharing (lower bounds in (3.17) and (3.38)), and by
- ignoring any overhead associated with control and/or data distribution and collection.

3.4.6 ASSOCIATIVITY TRANSFORM

All four architectural transforms discussed so far have one thing in common. Whether and how to apply them for maximum benefit can be decided from a DDG's connectivity and weights alone, no matter what operations the vertices stand for. In the sequel, we will call any architectural reorganization that exhibits this property a **universal transform**.

The practical consequence is that any computational flow qualifies for reorganization by way of universal transforms. This also implies that any two computations the DDGs of which are isomorphic can be solved by the same architecture just with the vertices interpreted differently.

   More on the negative side, universal transforms have a limited impact on the flow of computation because the number and precedence of operations are left unchanged.[40] As many computational problems ask for more specific and more profound forms of reorganization in order to take full advantage of the situation at hand, one cannot expect to get optimum results from universal transforms alone. Rather, one needs to bring in knowledge on the particular functions involved and on their algebraic properties. Architectural reorganizations that do so are referred to as operation-specific or **algebraic transforms**.

Probably the most valuable algebraic property from an architectural point of view is the associative law. Associativity can be capitalized on to

   o Convert a chain structure into a tree or vice versa, see example below,
   o Reorder operations such as to accommodate input data that arrive later than others do,
   o Reverse the order of execution in a chain as demonstrated in section 3.6, or to
   o Relocate operations from within a recursive loop to outside the loop, see section 3.7.

This explains why the associativity transform is also known as **operator reordering** and as **chain/tree conversion**.

Example
Consider the problem of finding the minimum among I input values

$$y(k) = \min(x_i(k)) \quad \text{where} \ \ i = 0, 1, ..., (I - 1) \tag{3.44}$$

Assuming the availability of 2-way minimum operators, this immediately suggests a chain structure such as the one depicted in fig.3.29a for $I = 8$. The delay along the longest path is $(I - 1)t_{min}$ and increases linearly with the number of terms. As the 2-way minimum function is associative, the DDG lends itself to being rearranged into a balanced tree as shown in fig.3.29b. The longest path is thereby shortened from $I - 1$ to $\lceil \log_2 I \rceil$ operations which makes the tree a much better choice, especially for large values of I. The number of operations and the circuit's size remain the same.
□

---

[40] While it is true that the number of DDG vertices may change, this is merely a consequence of viewing the original operations at a different level of detail.

The conversion of a chain of operations into a tree, as in the above example, is specifically referred to as **tree-height minimization**. As a side effect, this architectural transform often has a welcome impact on energy efficiency. This is because glitches die out more rapidly and are more likely to neutralize when all data propagation paths are of similar lengths. In addition, we observe the same indirect benefit as with pipelining, in that a shortened longest path makes it possible to use a slower yet more energy-efficient circuit style or a reduced supply voltage if circumstances permit.

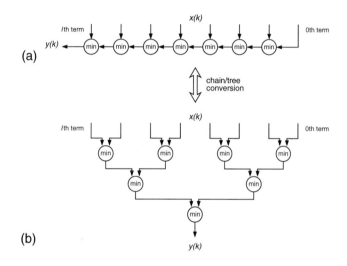

FIGURE 3.29

8-way minimum function. Chain-type DDG (a), tree-type DDG (b).

3.4.7 OTHER ALGEBRAIC TRANSFORMS

It goes without saying that many more algebraic laws can be put to use for improving dedicated architectures. **Distributivity** helps to replace the computation of $(a^2 - 2ax + x^2)$ by the more economic form of $(a - x)^2$, for instance, and is instrumental in exploiting known **symmetries** in coefficient sets of correlators, (matched) filters, and the like. Together with **commutativity**, distributivity is also at the heart of distributed arithmetic to be introduced in subsection 3.8.3. **Horner's scheme** serves to evaluate polynomials with a minimum number of multiplications while the **method of finite differences** can calculate any number of equidistant values with no multiplication at all. **Karatsuba multiplication** reduces the effort for very wide data words, the principle of **superposition** holds in linear systems, **De Morgan's theorem** helps in optimizing Boolean networks, and so on. See problem 5 for yet another operation-specific alteration. As a rule, always ask yourself what situation-specific properties might be capitalized on. The transforms discussed in this text just represent the more common ones and are by no means exhaustive.

3.4.8 DIGEST

- Iterative decomposition, pipelining, replication, and time sharing are based on the DDG as a graph and make no assumptions on the nature of computations carried out in its vertices, which is why they are qualified as <u>universal</u>. Associativity transform, in contrast, is said to be an <u>algebraic</u> transform because it depends on the operations involved being identical and associative.

- Iterative decomposition, pipelining, replication, and a variety of algebraic transforms make it possible to tailor combinational computations on a single data stream to given size and performance constraints without affecting functionality. Time sharing is another option in the presence of inherently parallel data streams and operations.

- For iterative decomposition to be effective, complex functions must be amenable to being broken into similar subfunctions such as to make it possible to reuse the same circuitry. Much the same reasoning also holds for time sharing in that parallel functions must not be too diverse to share a single functional unit in a fairly efficient way.

- Pipelining is generally more efficient than replication, see fig.3.25. While coarse grain pipelining improves throughput dramatically, benefits decline as more and more stages are included. When in search of utmost performance, begin by designing a pipeline the depth of which yields close-to-optimum efficiency. Only then — if throughput is still insufficient — consider replicating the pipelined functional unit a few times, see problem 2 for an example. This approach also makes sense in view of design productivity because duplicating a moderately pipelined unit may be easier and quicker than pushing pipelining to the extreme.

- A theoretical upper bound on throughput, expressed as data items per time unit, that holds for any circuit technology and architecture is $\Theta \leq \frac{1}{\min(t_{gate}) + t_{reg}}$.[41]

- Pipelining and iterative decomposition are complementary in that they both can contribute to lowering the size-time product AT. While the former acts to improve performance, the latter cuts circuit size by sharing resources. Combining them indeed makes sense, within certain bounds, in order to obtain a high throughput from a small circuit.

- Starting from the isomorphic configuration, a great variety of architectures is obtained from applying equivalence transforms in different order. Combining several of them is typical for VLSI architecture design. Figure 3.28 gives an idealized overview of the design space spanned up by the four universal transforms.[42] Which configuration is best in practice cannot be decided without fairly detailed knowledge of the application at hand, of the performance requirements, and of the target cell library and technology.

---

[41] Further improvements are possible only by processing larger data chunks at a time i.e. by packing more bits, pixels, samples, characters, or whatever into one data item. Note this is tantamount to opting for a larger w in the sense of footnote 21.

[42] As a more philosophical remark, observe from fig.3.28 that there exists no single transform that leads towards optimum hardware efficiency. To move in that direction, designers always have to combine two or more transforms much as a yachtsman must tack back and forth to progress windward with his sailboat.

- Program-controlled microprocessors follow an architectural concept that pushes iterative decomposition and time sharing to the extreme and that combines them with pipelining, and often with replication too. Developers of general-purpose hardware cannot take advantage of algebraic transforms as their application requires detailed knowledge about the data processing algorithm.

- It can be observed from fig.3.28 that lowering the size-time product AT always implies cutting down the longest path t_{lp} in the circuit. This comes with no surprise as better hardware efficiency can only be obtained from keeping most hardware busy for most of the time by means of a higher computation rate f_{cp}.

- Important power savings are obtained from operating CMOS logic with a supply voltage below its nominal value. Clever architecture design must compensate for the loss of speed that is due to the extended gate delays. Suggestions are not only given throughout this chapter, but also in the forthcoming material on energy efficiency.

120   CHAPTER 3 FROM ALGORITHMS TO ARCHITECTURES

3.5 OPTIONS FOR TEMPORARY STORAGE OF DATA

Except for trivial SSI/MSI circuits, any IC includes some form of memory. If the original data processing algorithm is of sequential nature and, therefore, mandates the temporary storage of data, we speak of <u>functional</u> memory. If storage gets introduced into a circuit as a consequence from architectural transformations, the memory is sometimes said to be of <u>nonfunctional</u> nature.

The major options for temporary storage of data are as follows:

- On-chip registers built from individual bistables (flip-flops or latches),
- On-chip memory (SRAM macrocell or — possibly — embedded DRAM), or
- Off-chip memory (SRAM or DRAM catalog part).[43]

There are important differences from an implementation point of view that matter from an architectural perspective and that impact high-level design decisions.

3.5.1 DATA ACCESS PATTERNS

Standard single-port RAMs provide access to data words one after the other.[44] This is fine in sequential architectures as obtained from iterative decomposition and time sharing that process data in a step-by-step fashion. Program-controlled microprocessors with their "fetch, load, execute, store" processing paradigm are a perfect match for RAMs.

High-throughput architectures obtained from pipelining, retiming and loop unfolding,[45] in contrast, keep data moving in every computation cycle which mandates the usage of registers as only those allow for simultaneous access to all of the data words stored.

Incidentally, also keep in mind that the contents of DRAMs needs to be periodically refreshed which dissipates electrical power even when no data accesses take place.

3.5.2 AVAILABLE MEMORY CONFIGURATIONS AND AREA OCCUPATION

Next compare how much die area gets occupied by registers and by on-chip RAMs. While register files allow for any conceivable memory configuration in increments of one data word of depth and one data bit of width, their area efficiency is rather poor. In the occurrence of fig.3.30, registers occupy an order of magnitude more area than a single-port SRAM for capacities in excess of 5000 bits. This is because registers get assembled from individual flip-flops or latches with no sharing of hardware resources.

Due to their simple and extremely compact bit cells, RAMs make much better use of area in spite of the shared auxiliary subcircuits they must accommodate. In a typical commodity DRAM, for instance, roughly 60% of the die is occupied by storage cells, the rest by address decoders, switches, precharge

---

[43] Please refer to section A.4 if totally unfamiliar with semiconductor memories.
[44] Dual-port RAMs can access two data words at a time, multi-port RAMs are rather uncommon.
[45] Retiming and loop unfolding are to be explained in sections 3.6 and 3.7 respectively.

circuitry, sense amplifiers, internal sequencers, I/O buffers, and padframe. Yet, such circuit overheads tend to make RAMs less attractive for storing smaller data quantities which is also evident from fig.3.30. A more serious limitation is that macrocells are available in a very limited number of configurations only. Adding or dropping an address bit alters memory capacity by a factor of two, and fractional cores are not always supported.

Always keep in mind that such effects have been ignored in the cost and performance analyses carried out in sections 3.4.2 through 3.4.5 where A_{reg} had been assumed to be fixed with no distinction between registers and RAMs. More specifically, this also applies to fig.3.28.

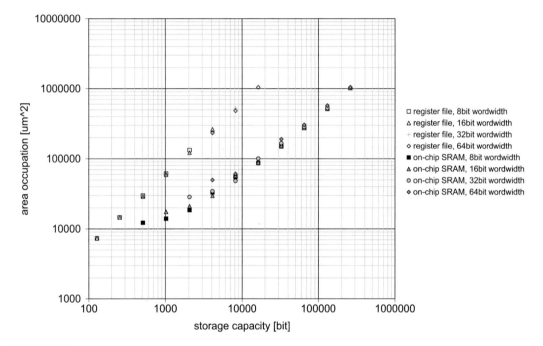

FIGURE 3.30

Area occupation of registers and on-chip RAMs for a 130 nm CMOS technology.

3.5.3 STORAGE CAPACITIES

Embedded memories cannot rival the copious storage capacities offered by commodity RAMs. The densest memory chips available are DRAMs built from one-transistor cells whereas the macrocells intended for being embedded within ASICs typically get assembled from six-transistor SRAM storage cells, see table 3.9.[46]

---

[46] DRAMs further take advantage of three-dimensional trench capacitors and other area-saving structures made possible by dedicated fabrication steps and process options unavailable to macrocells that are to be compatible with a baseline CMOS process. Also, processes for commodity memories are often ahead of ASIC processes in terms of feature size. Finally, the layout of memory chips is highly optimized by hand.

The economic disparity between on-chip memories and cheap commodity DRAMs, that are cost-optimized, fabricated in large quantities, and — all too often — subject to ruinous competition has long discouraged the on-chip storage of very large amounts of data within ASICs. The design of true "systems on a chip" requires this gap being closed. In fact, some integrated VLSI manufacturers are capable of embedding high-density DRAMs within their designs, but the approach is slow to catch on with the ASIC community and, more particularly, with the providers of ASIC design kits.

Examples

Embedded DRAMs occupy a large part of the market for 3D graphics accelerator chips for laptops because higher performance and lower power dissipation are key value propositions [57]. The so-called "Emotion Engine" superscalar multimedia processor chip designed and produced by Toshiba for Sony's PlayStation 2 is another popular example.

□

3.5.4 WIRING AND THE COSTS OF GOING OFF-CHIP

On the negative side, off-chip memories add to pin count, package count, and board space. Communicating with them involves a profusion of parasitic capacitances and delays that cause major bottlenecks in view of operating speed, performance, and energy efficiency.

In addition, most commodity RAMs feature bidirectional data pins in an effort to keep costs and pin counts as low as possible. They so impose the adoption of a bidirectional bus on any IC that is going to exchange data with them. Yet, note that bidirectional on-chip busses and even more so bidirectional pads require special attention during circuit design and test.

- Not only stationary but even transient drive conflicts must be avoided because of the strong drivers and important currents involved.
- Automated test equipment (ATE) must be made to alternate between read and write modes with no physical access to any control signal within the chip.
- Test patterns must be prepared for verifying bidirectional operation and high-impedance states during circuit simulation and testing.
- Electrical and timing measurements become more complicated.

3.5.5 LATENCY AND TIMING

RAM-type memories further differ from registers in terms of latency, paging, and timing. Firstly, some RAMs have latency while others do not. In a read operation, we speak of latency zero if the content of a memory location becomes available at the RAM's data output in the very clock cycle during which its address has been applied to the RAM's address port. This is also the behavior of a register bank.

As opposed to this, we have a latency of one if the data word does not appear before an active clock edge has been applied. Latency is even longer for memories that operate in a pipelined manner internally. Latency may have a serious impact on architecture design and certainly affects HDL coding.[47]

---

[47] A trick that may help to conceal latency is to operate the RAM's memory address register on the opposite clock edge than the rest of the circuit. Note that this introduces extra timing conditions with respect to the supposedly inactive clock edge that do not exist in a single-edge triggered circuit, however.

Secondly, commodity DRAMs have their row and column addresses multiplexed over the same pins to cut down package size and board-level wiring. Latency then depends on whether a memory location shares the row address with the one accessed before, in which case the two are said to sit on the same page, or not. Paged memories obviously affect architecture design.

Thirdly, address decoding, precharging, the driving of long word and bit lines, and other internal suboperations inflate the access times of both SRAMs and DRAMs. RAMs thus impose a comparatively slow clock that encompasses many gate delays per computation period whereas registers are compatible with much higher clock frequencies.[48]

3.5.6 DIGEST

Most on-chip RAMs available for ASIC design

- $\pm$ are of static nature (SRAMs),
- $+$ have their views obtained automatically using a macrocell generator,
- $-$ offer a limited choice of configurations (in terms of $\#_{words}$ and w_{data}),
- $+$ occupy much less area than flip-flops,
- $-$ but do so only above some minimum storage capacity,
- $+$ greatly improve speed and energy efficiency over off-chip RAMs,
- $-$ but cannot compete in terms of capacity,
- $-$ restrict data access to one read or write per cycle,
- $-$ impose rigid constraints on timing, minimum clock period, and latency.

---

[48] It is sometimes possible to multiply maximum RAM access rate by recurring to **memory interleaving**, a technique whereby a small number of memories operate in a circular fashion such as to emulate a single storage array of shorter cycle time.

Table 3.9   Key characteristics of register-based and RAM-based data storage compared.

architectural option	on-chip				off-chip
	bistables		embedded		commodity
	flip-flop	latch	SRAM	DRAM	DRAM
fabrication process	compatible with logic				optimized
devices in each cell	20...30T	12...16T	6T	1T1C	1T1C
cell area per bit [F^2]^a	1700...2800	1100...1800	135...170	18...30^b	6...8
extra circuit overhead	none		$1.3 \leq$ factor ≤ 2		off-chip
memory refresh cycles	n o n e				y e s
extra package pins	none		none		addr. & data bus
nature of wiring	multitude of local lines		on-chip busses		package & board
bidirectional busses	none		optional		mandatory
access to data words	all at a time		one at a time		
available configurations	any		restricted		
energy efficiency	good^c		fair	poor	very poor
latency and paging	none		no fixed rules		yes
impact on clock period	minor		substantial		severe

^{a}Area of one bit cell in multiples of F^2 where F^2 stands for the area of one lithographic square.
^{b}As low as 6...8 for processes that accomodate 3D capacitors at the price of 4 to 6 extra masks [58].
^{c}Depending on the register access scheme, conditional clocking may or may not be an option.

Examples

Cu-11 is the name of an ASIC technology by IBM that has a drawn gate length — and hence also a minimum feature size — of 110 nm and that operates at 1.2 V. The process combines copper interconnect with low-k interlevel dielectric materials. As part of the Cu-11 design library, IBM offers an SRAM macrocell generator for memories ranging from 128 bit to 1 Mibit as well as embedded DRAM megacells of trench capacitor type with up to 16 Mibit. A 1 Mibit eDRAM has a cycle time of 15 ns which is equivalent to 555 times the nominal delay of a 2-input NAND gate. eDRAM bit cell area is 0.31 μm^2 which corresponds to 25.6F^2. A 1 Mibit eDRAM occupies an area of 2.09 mm^2 (with an overhead factor 1.84) and its 16 Mibit counterpart 14.1 mm^2 (overhead factor 1.63).

Actel's ProASIC$\underline{^{PLUS}}$ flash-based FPGA family makes embedded SRAMs available in chunks of 256x9 bit. The APA1000 part includes 88 such blocks which corresponds to 198 Kibit of embedded RAM if fully used.
□

Flash memories have not been addressed here as they do not qualify for short-time random-access storage. This is primarily because data must be erased in larger chunks before it becomes possible to rewrite individual words. The comparatively low speed and limited endurance are other limitations that make flash more suitable for longer-term storage applications such as retaining FPL configurations as explained in section 2.2.

Just for comparison, the physical bit cell area of flash technology is a mere 4 to $12F^2$ and, hence, comparable to DRAM rather than SRAM. What's more, by using four voltage levels instead of two, two bits can be stored per flash cell bringing down the minimum area to just $2F^2$ per bit. Endurance is on the order of 100 000 write&erase cycles for flash cells that hold one bit (two states) and of 10 000 cycles for those that hold two bits (four states). Still higher numbers are made possible by wear-leveling schemes implemented in special memory controllers.

Observation 3.8. *Only registers allow for simultaneous access to all data but occupy a lot of die area per bit. SRAMs serve to temporarily hold more significant quantities of data with access times that are slower than registers but faster than DRAMs. DRAM and Flash memories are cost-efficient for large data quantities. Flash is used for permanent storage, speed is much lower than with RAMs, though. Off-chip commodity memories offer virtually unlimited capacities at low costs, but are is associated with speed, energy and other penalties.*

As further details of the various memory technologies are of little importance here, the reader is referred to the literature [59] [60] [61] [62]. An excellent introduction to flash memory technology is given in [63], [64] elaborates on improvements towards high-density storage and high-speed programming, while [65] and [66] focus on the role of NAND flash for mass storage and on the system and packaging aspects of removable media respectively.

3.6 EQUIVALENCE TRANSFORMS FOR NON-RECURSIVE COMPUTATIONS

Unlike combinational computations, the outcome of sequential computations depends not only on present but also on past values of its arguments. Architectures for sequential computations must therefore include memory. In the DDG this gets reflected by the presence of edges with weights greater than zero. However, as non-recursiveness implies the absence of feedback, the DDG remains free of circular paths. The storage capacity required by the isomorphic architecture is referred to as **memory bound** because no other configuration exists that could do with less.[49] Table 3.9 allows for approximate translation from memory bits to chip area.

3.6.1 RETIMING

The presence of registers in a circuit suggests a new type of reorganization known as retiming or as register balancing, whereby registers get relocated such as to allow for a higher computation rate without affecting functionality [67] [68]. The goal is to equalize computational delays between any two registers, thereby shortening the longest path that bounds the computation period from below. Referring to a DDG one must therefore know

"In what way is it possible to modify edge weights without altering the original functionality?"

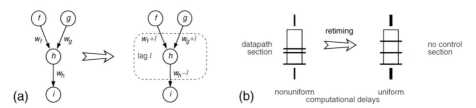

Retiming. DDG (a) and a hardware configuration for $l = 1$ (b).

Let us follow an intuitive approach to find an answer.[50] Consider a DDG and pick a vertex, say h in fig.3.31, for instance. Now suppose the operation of vertex h is made to lag behind those of all others by adding latency to every edge pointing towards that vertex, and by removing the same amount of latency from every edge pointing away from that vertex. Conversely, any vertex could be made to lead the others by transferring latency from its incoming to its outgoing edges. Since any modifications made to the incoming edges are compensated for at the outgoing edges, nothing changes when the DDG is viewed from outside. The retimed DDG is, therefore, functionally equivalent to the initial one. As it is always possible (a) to think of an entire subgraph as one supervertex, and (b) to repeat the operation with changing vertices and supervertices, the idea paves the way for significant hardware reorganizations.

---

[49] For combinational computations, the memory bound is obviously zero.
[50] A more formal treatise is given in [69].

Not all DDGs obtained in this way are legal, though, because the general requirements for DDGs stated in subsection 3.3.4 impose a number of restrictions on edge weights or — which is the same — on latency registers. Any legal retiming must observe the rules below.

1. Neither data sinks (i.e. outputs) nor sources of time-varying data (i.e. variable inputs) may be part of a supervertex that is to be retimed. Sources of constant data (i.e. fixed inputs) do not change in any way when subject to retiming.
2. When a vertex or a supervertex is assigned a lag (lead) by l computation cycles, the weights of all its incoming edges are incremented (decremented) by l and the weights of all its outgoing edges are decremented (incremented) by l.
3. No edge weight may be changed to assume a negative value.
4. Any circular path must always include at least one edge of strictly positive weight.[51]

Interestingly, rule 4 does not need to be taken into consideration explicitly — provided it was satisfied by the initial DDG — because the weight along a circular path will never change when rule 2 is observed. The proof is trivial.

As a direct consequence of rule 1, retiming does not affect latency. Retiming necessitates neither control logic nor extra data storage facilities but may alter the total number of registers in a circuit depending on the structure of the DDG being subject to the transformation.[52]

Energywise, it is difficult to anticipate the overall impact of retiming as switching activities, fanouts, and node capacitances all get altered. Yet, much as for pipelining, the reduced long path delay either allows for compensating a lower supply voltage or can be invested in doing with a slower but more energy-efficient circuit style or technology. The fact that retiming does not normally burden a circuit with much overhead renders it particularly attractive.

3.6.2 PIPELINING REVISITED

Recall from section 3.4.3 that pipelining introduces extra registers into a circuit and necessarily increases its latency, which contrasts with what we have found for retiming. Pipelining can in fact be interpreted as a transformation of edge-weights that differs from retiming in that rule 1 is turned into its opposite, namely

1. Any supervertex to be assigned a lag (lead) must include all data sinks (all time-varying data sources).

What retiming and pipelining have in common is that they allow one to operate a circuit at a higher computation rate. Most high-speed architectures therefore combine the two.

---

[51] Although irrelevant here due to the absence of circular paths, this stipulation does apply in the context of recursive computations.

[52] For many applications it is important that a sequential circuit assumes its operation from a well-defined start state. As a rule, initial state does matter for controllers but is often irrelevant for datapath circuitry. If so, a mechanism must be provided for forcing the circuit into that state. Finding the start state for a retimed circuit is not always obvious. The problem of systematically computing the start state for retimed circuits is addressed in [70].

Example

Consider the problem of computing

$$y(k) = \sum_{n=0}^{N} h(c_n, x(k-n)) \tag{3.45}$$

i.e. a time-invariant N-th order correlation where $h(c,x)$ stands for some unspecified — possibly nonlinear — function. Think of $h(c,x)$ as some distance metric between two DNA fragments c and x, for instance, in which case $y(k)$ stands for the overall dissimilarity between the DNA strings $(c_0, c_1, ..., c_N)$ and $(x(k), x(k-1), ..., x(k-N))$ of length N.

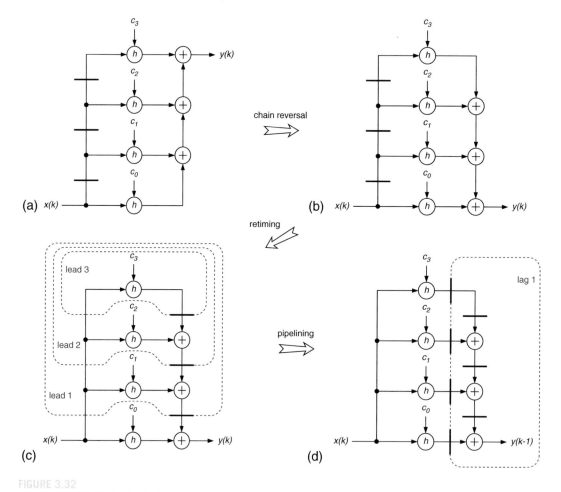

Nonlinear time-invariant third order correlator. Original DDG (a), with adder chain reversed by associativity transform (b), after retiming (c), with pipelining added on top such as to obtain a systolic architecture (d).

The DDG of a third order correlator is shown in fig.3.32 together with its stepwise reorganization. For the sake of concreteness, let us assume that a register delay is $t_{reg} = 0.5$ns, that computing one distance metric takes $t_h = 3$ns, and that adding up two items costs $t_+ = 2$ns.

(a) Original DDG as obtained from straight interpretation of (3.45). The delay along the longest path is stated in the table below, note that is grows with correlation order N. There is no retiming that would relocate the existing registers in a useful way. Although the configuration is amenable to pipelining, reformulating it first will eventually pay off.

(b) Same as (a) with the adder chain reversed by an associativity transform. Maximum delay and register count remain unchanged, but the computation has now become suitable for retiming. Also refer to problem 3.

(c) Functional registers transferred into the adder chain by retiming of (b). The three vertices and supervertices enclosed by dashed lines have been assigned leads of 1, 2, and 3 computation cycles respectively. Long path delay is substantially reduced with no registers added. Also notice that the maximum operating speed is no longer a function of correlation order N.

(d) Retiming complemented by pipelining. The supervertex enclosed by a dashed line, which includes the data sink, has been assigned a lag of 1. The longest path is further shortened at the price of extra registers and of an increased latency. Observe that it is not possible to improve performance any further unless one is willing to intervene into the adders on a lower level of circuit detail, also see section 3.8.1.

Key characteristics	Architectural variant			
	(a)	(b)	(c)	(d)
arithmetic units	$(N+1)A_h + NA_+$	idem	idem	idem
functional registers	NA_{reg}	idem	idem	idem
nonfunctional registers	0	idem	idem	$(N+1)A_{reg}$
cycles per data item Γ	1	idem	idem	idem
longest path delay t_{lp}	$t_{reg} + t_h + N t_+$	idem	$t_{reg} + t_h + t_+$	$t_{reg} + \max(t_h, t_+)$
for $N = 3$ [ns]	9.5	idem	5.5	3.5
for $N = 30$ [ns]	63.5	idem	5.5	3.5
latency L	0	idem	idem	1

☐

Make sure you understand there is a fundamental difference between the architectural transforms used in the above example. While retiming and pipelining are universal transforms that do not depend on the operations involved, changing the order of execution in the algebraic transform that leads from (a) to (b) insists in the operations concerned being identical and associative. The practical significance is that the sequence of reorganizations that has served to optimize the nonlinear correlator example also applies to transversal filters which are of linear nature, for instance, but not to DDGs of similar structure where addition is replaced by some non-associative operation.

3.6.3 SYSTOLIC CONVERSION

Both pipelining and retiming aim at increasing computation rate by reducing and by equalizing register-to-register delays. For a given granularity, maximum speed is obtained when there is no more than one combinational operation between any two registers. This is the basic idea behind systolic computation.

A DDG is termed **systolic** if the edge weight between any two vertices is one or more. The architecture depicted in fig.3.32d is in fact systolic, and the ultimate pipeline in the sense of (3.29) is now also recognized as a circuit that is systolic at the gate level. Please refer to [71] for a more comprehensive discussion of systolic computation and to [72] for details on systolic conversion, that is on how to turn an arbitrary non-recursive computation into a systolic one.

3.6.4 ITERATIVE DECOMPOSITION AND TIME SHARING REVISITED

Applying the ideas of decomposition and time sharing to sequential computations is straightforward. Clearly, only combinational circuitry can be multiplexed whereas functional memory requirements remain the same as in the isomorphic architecture.

Example

In the isomorphic architecture of a transversal filter, see fig.3.17d, each filter tap is being processed by its own multiplier. All calculations associated with one sample are thus carried out in parallel and completed within a single computation cycle. Nevertheless, the architecture is slow because the longest path traverses the entire adder chain thereby mandating a long computation period. Also, hardware costs are immoderate, at least for higher filter orders N.

A more economic alternative that handles one filter tap after the other follows naturally, see fig.3.33. This architecture manages with a single multiplier that gets time-shared between taps. A single adder iteratively sums up the partial products until all taps that belong to one sample have been processed after $N + 1$ computation cycles. An accumulator stores the intermediate sums. Coefficients may either be kept in a hardwired lookup table (LUT), in a ROM, or in some sort of writable memory. Datapath control is fairly simple. An index register that counts $n = N, N - 1, ..., 0$ addresses one coefficient at any time. The very same register also selects the samples from the input shift register, either by way of a multiplexer, of a three-state bus, or by arranging the shift register in circular fashion and by having data there make a full cycle between any two subsequent samples. An output register maintains the end result while computation proceeds with the next sample.

For filters of higher order, one would want to substitute a RAM for the input shift register. While this requires some extra circuitry for addressing, it does not fundamentally change the overall architecture.

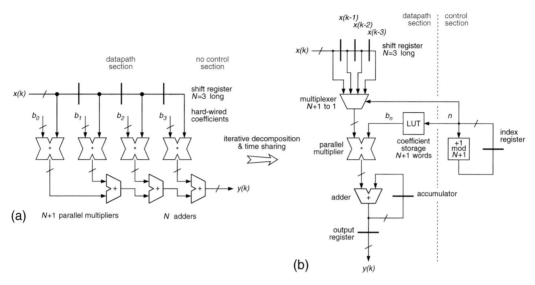

Third order transversal filter. Isomorphic architecture (a) and a more economic alternative obtained by combining time sharing with iterative decomposition (b) (simplified).

□

3.6.5 REPLICATION REVISITED

The concept of replication introduced in section 3.4.4 cannot normally be applied to sequential computations as the processing of one data item is dependent on previous data items. A notable exception exists in the form of so-called **multipath filters** (aka N-path filters) designed to implement sequential computations of linear time-invariant nature. With $H_1(z)$ denoting the transfer function of a single path, all of which are identical, and with q as replication factor, a composite transfer function

$$H(z) = H_1(z^q) \tag{3.46}$$

is obtained [73], which implies that the elemental transfer function $H_1(z)$ is compressed and repeated along the frequency axis by a scaling factor of q. Due to the resulting extra passbands the usefulness of this approach is very limited. An extended structure capable of implementing general FIR and IIR[53] transfer functions is proposed in [73] under the name of delayed multipath structures. Regrettably, the number of necessary operations is found to grow with q^2.

---

[53] FIR stands for finite impulse response, IIR for infinite impulse response.

3.6.6 DIGEST

- The throughput of arbitrary feedforward computations can be improved by way of retiming, by way of pipelining, and by combining the two. Replication is, in general, not a viable alternative.

- The associative law is often helpful in rearranging a DDG prior to iterative decomposition, pipelining, and especially retiming in order to take full advantage of these transforms.

- Much as for combinational computations, iterative decomposition and time sharing are the two options available for reducing circuit size for feedforward computations.

- Highly time-multiplexed architectures are found to dissipate energy on a multitude of ancillary activities that do not directly contribute to data computation, however.

3.7 EQUIVALENCE TRANSFORMS FOR RECURSIVE COMPUTATIONS

A computation is said to be timewise recursive — or recursive for short — if it somehow depends on an earlier outcome from that very computation itself, a circumstance that gets reflected in the DDG by the presence of a feedback loop. Yet, recall that circular paths of weight zero have been disallowed to exclude the risk of race conditions. Put differently, circular paths do exist but each of them includes at least one latency register.

This section examines equivalence transforms that apply specifically to recursive computations. We will find that such transforms are not universal. This is why we address linear time-invariant, linear time-variant, and nonlinear computations separately.

3.7.1 THE FEEDBACK BOTTLENECK

Consider a linear time-invariant first-order recursive function

$$y(k) = ay(k-1) + x(k) \qquad (3.47)$$

which, in the z domain, corresponds to transfer function

$$H(z) = \frac{Y(z)}{X(z)} = \frac{1}{1 - az^{-1}} \qquad (3.48)$$

The corresponding DDG is shown in fig.3.34a. Many examples for this and similar types of computations are found in IIR filters, DPCM[54] data encoders, servo loops, etc.

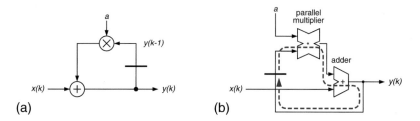

(a)   **(b)**

FIGURE 3.34
Linear time-invariant first-order feedback loop. DDG (a) and isomorphic architecture with longest path highlighted (b).

Recursion demands that the result from the actual computation cycle be available no later than the next input sample. The longest path defined by all computations that are part of the loop must, therefore, not exceed one sampling interval. In the occurrence

$$\sum_{loop} t = t_{reg} + t_* + t_+ = t_{lp} \le T_{cp} \qquad (3.49)$$

---

[54] DPCM is an acronym for differential pulse code modulation.

As long as this **iteration bound** is satisfied by the isomorphic architecture of fig.3.34b implemented using some available and affordable technology, there is no out-of-the-ordinary design problem. As an example, we could easily provide a sustained computation rate of 100 MHz if the three delay figures for the actual word width were of 0.5 ns, 5 ns, and 2 ns respectively.

When in search of some higher throughput, say 200 MHz, recursiveness becomes a real bottleneck since there is no obvious way to make use of replication or to insert pipeline registers without altering the overall transfer function and input-to-output mapping. Retiming does not help either as the weight along a circular path is always preserved. So the problem is

"How to allow more time for those computations that are part of the recursion loop?"

3.7.2 UNFOLDING OF FIRST-ORDER LOOPS

The key idea is to relax the timing constraint by inserting additional latency registers into the feedback loop while preserving the original transfer function. In other words, a tentative solution for a first-order loop must look like

$$H(z) = \frac{Y(z)}{X(z)} = \frac{N(z)}{1 - a^p z^{-p}}$$
(3.50)

where an unknown expression $N(z)$ is here to compensate for the changes that are due to the new denominator $1 - a^p z^{-p}$. Recalling the sum of geometric series we easily establish $N(z)$ as

$$N(z) = \frac{1 - a^p z^{-p}}{1 - az^{-1}} = \sum_{n=0}^{p-1} a^n z^{-n}$$
(3.51)

The new transfer function can then be completed to become

$$H(z) = \frac{\sum_{n=0}^{p-1} a^n z^{-n}}{1 - a^p z^{-p}}$$
(3.52)

and the new recursion in the time domain follows as

$$y(k) = a^p y(k - p) + \sum_{n=0}^{p-1} a^n x(k - n)$$
(3.53)

The modified equations correspond to a cascade of two sections. A first section, represented by the numerator of (3.52), is a DDG that includes feedforward branches only. This section is amenable to pipelining as discussed in section 3.4.3. The denominator stands for the second section, a simple feedback loop which has been widened to include p unit delays rather than one as in (3.47).

Using retiming, the corresponding latency registers can be redistributed into the combinational logic for computing the loop operations such as to serve as pipeline registers there. Neglecting, for a moment, the limitations to pipelining found in section 3.4.3, throughput can in fact be multiplied by an arbitrary positive integer p through this unfolding technique, several variations of which are discussed in [74].

Unless p is prime, it is further possible to simplify the DDG — and hence the implementing circuit — by factoring the numerator into a product of simpler terms. Particularly elegant and efficient solutions exist when p is an integer power of 2 because of the lemma

$$\sum_{n=0}^{p-1} a^n z^{-n} = \prod_{m=0}^{\log_2 p - 1} (a^{2^m} z^{-2^m} + 1) \qquad p = 2, 4, 8, 16, \ldots \tag{3.54}$$

The feedforward section can then be realized by cascading $\log_2 p$ subsections each of which consists of just one multiplication and one addition.

Example

The linear time-invariant first-order recursive function (3.47) takes on the following form after unfolding by a factor of $p = 4$

$$y(k) = a^4 y(k-4) + a^3 x(k-3) + a^2 x(k-2) + ax(k-1) + x(k) \tag{3.55}$$

which corresponds to transfer function

$$H(z) = \frac{1 + az^{-1} + a^2 z^{-2} + a^3 z^{-3}}{1 - a^4 z^{-4}} \tag{3.56}$$

Making use of (3.54) the numerator can be factorized into such as to obtain

$$H(z) = \frac{(1 + az^{-1})(1 + a^2 z^{-2})}{1 - a^4 z^{-4}} \tag{3.57}$$

The DDG corresponding to this equation is shown in figure 3.35a. Note that the configuration is not only simple but also highly regular. Further improvements are obtained from pipelining in conjunction with retiming and shimming where necessary. The final architecture, shown in fig.3.35b, is equivalent except for latency. Incidentally, also note that threefold instantiation of one pipelined multiply-add building block favors further optimizations at lower levels of detail.

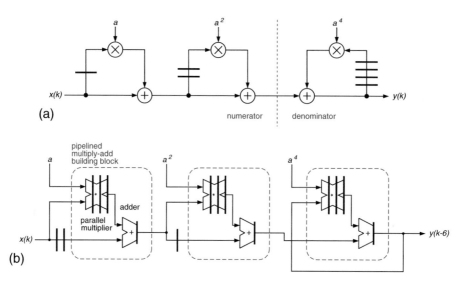

(a)

numerator   denominator

(b)

FIGURE 3.35
Linear time-invariant first-order feedback loop. DDG after unfolding by a factor of $p = 4$ (a) and high-performance architecture with pipelining and retiming on top (b).
◻

Performance and cost analysis

In the occurrence of optimally efficient configurations, where p is an integer power of 2, a lower bound for total circuit size can be given as follows

$$A(p) \geq (\log_2 p + 1)A_f + p(\log_2 p + 1)A_{reg} \tag{3.58}$$

In the above example, we count three times the original arithmetic logic plus 14 extra (nonfunctional) registers. In return for an almost fourfold throughput, this is finally not too bad.

Analogously to what was found for pipelining in section 3.4.3, the speedup of loop unfolding tends to diminish while the difficulties of balancing delays within the combinational logic tend to grow when unfolding is pushed too far, $p \gg 1$.

A hidden cost factor associated with loop unfolding is due to finite precision arithmetics. For the sake of economy, datapaths are designed to do with minimum acceptable word widths, which implies that output and intermediate results get rounded or truncated somewhere in the process. In the above example, for instance, addition would typically handle only part of the bits that emanate from multiplication. Now, the larger number of roundoff operations that participate in the unfolded loop with respect to the initial configuration leads to more quantization errors, a handicap which must be offset by using somewhat wider data words [75].

Loop unfolding greatly inflates the amount of energy dissipated in the processing of one data item because of the extra feedforward computations and the many latency registers added to the unfolded circuitry. More on the positive side, the shortened longest path may bring many recursive computations into the reach of a relatively slow but power-efficient technology or may allow for a lower supply voltage.

The idea of loop unfolding demonstrated on a linear time-invariant first-order recursion can be extended into various directions, and this is the subject of the forthcoming three subsections.

3.7.3 HIGHER-ORDER LOOPS

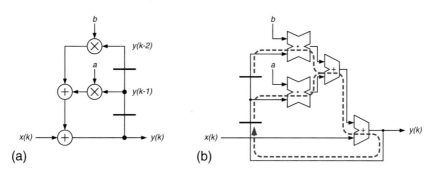

FIGURE 3.36

Linear time-invariant second-order feedback loop. DDG (a) and isomorphic architecture with longest paths highlighted (b).

Instead of unfolding loops of arbitrary order directly, we make use of a common technique from digital filter design that consists in factoring a higher-order transfer function into a product of second- and first-order terms. The resulting DDG takes the form of cascaded second- and first-order sections. High-speed IIR filters of arbitrary order can be constructed by pipelining the cascade so obtained. As an added benefit, cascade structures are known to be less sensitive to quantization of coefficients and signals than direct forms. We will, therefore, limit the discussion to second-order recursive functions here

$$y(k) = ay(k-1) + by(k-2) + x(k) \tag{3.59}$$

which correspond to the DDG depicted in fig.3.36. The equivalent in the z domain is

$$H(z) = \frac{Y(z)}{X(z)} = \frac{1}{1 - az^{-1} - bz^{-2}} \tag{3.60}$$

After multiplying numerator and denominator by a factor of $(1 + az^{-1} - bz^{-2})$ the transfer function becomes

$$H(z) = \frac{1 + az^{-1} - bz^{-2}}{1 - (a^2 + 2b)z^{-2} + b^2 z^{-4}} \tag{3.61}$$

which matches the requirements for loop unfolding by a factor of $p = 2$.
Analogously we obtain for $p = 4$

$$H(z) = \frac{(1 + az^{-1} - bz^{-2})(1 + (a^2 + 2b)z^{-2} + b^2 z^{-4})}{1 - ((a^2 + 2b)^2 - 2b^2)z^{-4} + b^4 z^{-8}} \tag{3.62}$$

Example

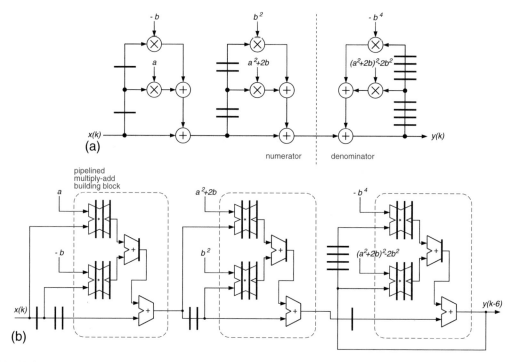

(a)

numerator ┆ denominator

(b)

FIGURE 3.37

Linear time-invariant second-order feedback loop. DDG after unfolding by a factor of $p = 4$ (a) and high-performance architecture with pipelining and retiming on top (b).

Figure 3.37 shows a DDG and a block diagram that implement the second order recursion (3.62). Except for finite precision effects, the transfer function remains exactly the same as in (3.59), but the arithmetic operations inside the loop can now be carried out in four rather than one computation period. The same pipelined hardware block is instantiated three times.

A high-speed fourth-order ARMA[55] filter chip that includes two sections similar to fig.3.37b has been reported back in 1992 [76]. Pipelined multiply-add units have been designed as combinations of consecutive carry-save and carry-ripple adders. The circuit, fabricated in a 0.9 μm CMOS technology, has been measured to run at a clock frequency of 85 MHz and spits out one sample per clock cycle, so $\Gamma = 1$. Overall computation rate roughly is 1.5 Gop/s.[56] The authors write that one to two extra data bits had to be added in the unfolded datapath to maintain similar roundoff and quantization characteristics as in the initial configuration. Circuit size is approximately 20 kGE. At full speed the chip dissipates

---

[55] ARMA is an acronym for "auto recursive moving average" used to characterize IIR filters that comprise both recursive (AR) and non-recursive computations (MA).
[56] Multiply-add operations, in the occurrence, taking into account all of the filter's AR and MA sections.

2.2 W from a 5 V supply. While performance and power data are obsolete today, loop unfolding allows to push out the need for fast but costly fabrication technologies such as GaAs, then and now.

☐

Performance and cost analysis

When compared to the first-order case, the number of pipeline registers per subsection is doubled while the other figures remain unchanged. Hence, size estimation yields

$$A(p) \geq (\log_2 p + 1)A_f + (2p(\log_2 p + 1))A_{reg} \tag{3.63}$$

3.7.4 TIME-VARIANT LOOPS

Here, the feedback coefficient a is no longer constant but varies as a function of time $a(k)$

$$y(k) = a(k)y(k-1) + x(k) \tag{3.64}$$

The unfolded recursions are derived in the time domain. Substituting $y(k-1)$ in (3.64) yields

$$y(k) = a(k)a(k-1)y(k-2) + a(k)x(k-1) + x(k) \tag{3.65}$$

which computes $y(k)$ from $y(k-2)$ directly, so the unfolding factor is $p = 2$. Repeating this operation leads to a configuration with $p = 3$ where

$$y(k) = a(k)a(k-1)a(k-2)y(k-3) + a(k)a(k-1)x(k-2) + a(k)x(k-1) + x(k) \tag{3.66}$$

and once more to $p = 4$

$$y(k) = a(k)a(k-1)a(k-2)a(k-3)y(k-4)+$$

$$+ a(k)a(k-1)a(k-2)x(k-3) + a(k)a(k-1)x(k-2) + a(k)x(k-1) + x(k) \tag{3.67}$$

As for the time-invariant case, the process of unfolding can be continued to widen the recursive loop by an arbitrary positive integer p as expressed by

$$y(k) = (\prod_{n=0}^{p-1} a(k-n)) \cdot y(k-p) + \left(\sum_{n=1}^{p-1} (\prod_{m=0}^{n-1} a(k-m)) \cdot x(k-n) \right) + x(k) \tag{3.68}$$

However, because precomputation is not applicable here, all necessary coefficient terms must be calculated on-line, which requires extra hardware. Depending on how the terms of (3.68) are combined, various DDGs can be obtained. One of them derived from (3.67) is depicted in fig.3.38.

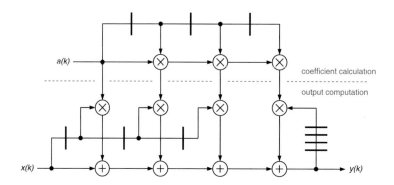

FIGURE 3.38

Linear time-variant first-order feedback loop. DDG after unfolding by a factor of $p = 4$.

Performance and cost analysis

The count of adders and multipliers is proportional to the number of subsections p. Each subsection requires approximately $2p$ pipeline registers as both multipliers must be pipelined. Together with shimming registers, many of which are needed in this configuration due to the numerous parallel branches, roughly $4p^2$ registers are needed.

3.7.5 NONLINEAR OR GENERAL LOOPS

A nonlinear difference equation implies that the principle of superposition does not hold. The most general case of a first-order recursion is described by

$$y(k) = f(y(k-1), x(k)) \qquad (3.69)$$

and can be unfolded an arbitrary number of times. For simplicity we will limit our discussion to a single unfolding step, i.e. to $p = 2$ where

$$y(k) = f(f(y(k-2), x(k-1)), x(k)) \qquad (3.70)$$

The associated DDG of fig.3.39c shows that loop unfolding per se does not relax the original timing constraint, the only difference is that one can afford two cycles for two operations f instead of one cycle for one operation. As confirmed by fig.3.39d, there is no room for any meaningful retiming in this case.

Yet, the unfolded recursion can serve as a starting point for more useful reorganizations. Assume function f is known to be associative. Following an associativity transform the DDG is redrawn as shown in fig.3.39e. The computation so becomes amenable to pipelining and retiming, see fig.3.39f, which cuts the longest path in half when compared to the original architecture of fig.3.39b. Even more speedup can be obtained from higher unfolding degrees, the price to pay is multiplied circuit size and extra latency, though. In summary, architecture, performance, and cost figures resemble those found for linear computations.

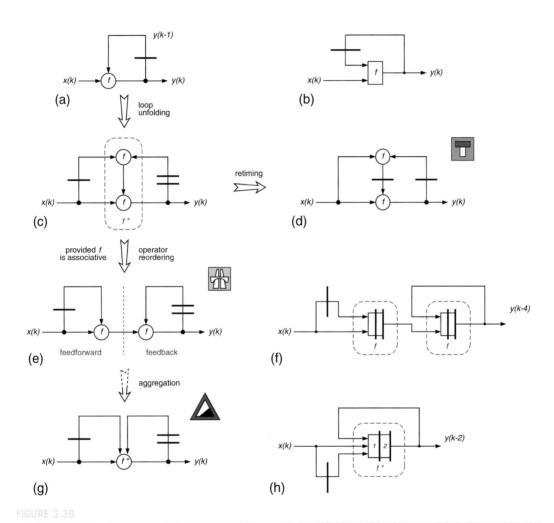

FIGURE 3.39

Architectural alternatives for nonlinear time-variant first-order feedback loops. Original DDG (a) and isomorphic architecture (b), DDG after unfolding by a factor of $p = 2$ (c), same DDG with retiming added on top (d). DDG reorganized for an associative function f (e), pertaining architecture after pipelining and retiming (f), DDG with the two functional blocks for f combined into f'' (g), pertaining architecture after pipelining and retiming (h).

The situation is definitely more difficult when f is not associative. Still, it is occasionally possible to relax the loop constraint to some extent by playing a trick. Reconsider fig.3.39c and think of the two occurrences of f being combined into an **aggregate computation**

$$y(k) = f''(y(k-2), x(k-1), x(k)) \tag{3.71}$$

as sketched in fig.3.39g. If that aggregate computation can be made to require less than twice as much time as the original computation, then the bottleneck gets somewhat alleviated. This is because it should then be possible to insert a pipeline register into the datapath unit for f'' so that the maximum path length in either of the two stages becomes shorter than the longest delay in a datapath that computes f alone.

$$t_{lpf''} = \max(t_{lpf''_1}, t_{lpf''_2}) < t_{lpf} \tag{3.72}$$

More methods for speeding up general time-variant first-order feedback loops are examined in [77]. One technique, referred to as **expansion** or **look-ahead**, is closely related to aggregate computation. The idea is to process two or more samples in each recursive step so that an integer multiple of the sampling interval becomes available for carrying out the necessary computations. In other terms, the recursive computation is carried out at a lower pace but on wider data words. This approach should be considered when the combinational logic is not amenable to pipelining, for example because it is implemented as table lookup in a ROM. The limiting factor is that the size of the lookup table (LUT) tends to increase dramatically.

Yet another approach, termed **concurrent block technique**, groups the incoming data stream into blocks of several samples and makes the processing of these blocks independent from each other. While data processing within the blocks remains sequential, it so becomes possible to process the different blocks concurrently.

The **unified algebraic transformation** approach promoted in [78] combines both universal and algebraic transforms to make the longest path independent of problem size in computations such as recursive filtering, recursive least squares algorithm, and singular value decomposition.

Any of the various architectural transforms that helps to successfully introduce a higher degree of parallel processing into recursive computations takes advantage of algorithmic properties such as linearity, associativity, fixed coefficients, limited word width, or of a small set of register states. If none of these applies, we can't help but agree with the authors of [77].

Observation 3.9.   *When the state size is large and the recurrence is not a closed-form function of specific classes, our methods for generating a high degree of concurrency cannot be applied.*

Example
Cryptology provides us with a vivid example for the implications of nonlinear nonanalytical feedback loops. Consider a block cipher that works in **electronic code book** (ECB) mode as depicted in fig.3.41a. The algorithm implements a combinational function $y(k) = c(x(k), u(k))$ where $u(k)$ denotes the key and k the block number or time index. However complex function c, there is no fundamental obstacle to pipelining or to replication in the datapath.

Unfortunately, ECB is cryptologically weak as two identical blocks of plaintext result in two identical blocks of ciphertext because $y(k) = y(m)$ if $x(k) = x(m)$ and $u(k) = u(m)$. If a plaintext to be encrypted contains sufficient repetition, the ciphertext necessarily carries and betrays patterns from the original plaintext. Fig.3.40 nicely illustrates this phenomenon.

(a)   (b)   (c)

FIGURE 3.40

A computer graphics image in clear text, ciphered in ECB mode, and ciphered in CBC-1 mode (from left to right, Tux by Larry Ewing).

To prevent this from happening, block ciphers are typically used with feedback. In **cipher block chaining** (CBC) mode, the ciphertext gets added to the plaintext before encryption takes place, see fig.3.41b. The improved cipher algorithm so becomes $y(k) = c(x(k) \oplus y(k-1), u(k))$ and is sometimes referred to as CBC-1 mode because $y(k-1)$ is being used for feedback.

From an architectural point of view, however, this first-order recursion is awkward because it offers little room for reorganizing the computation. This is particularly true in ciphering applications where the nonlinear functions involved are chosen to be complex, labyrinthine, and certainly not analytical. The fact that word width (block size) is on the order of 64 or 128 bit makes everything worse. Inserting pipeline registers into the computational unit for c does not help since this would alter algorithm and ciphertext. Throughput in CBC mode is thus limited to a fraction of what is obtained in ECB mode.[57]

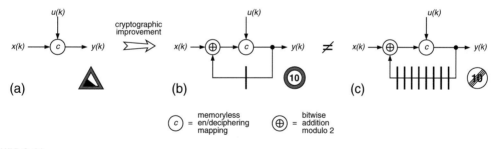

FIGURE 3.41

DDGs for three block ciphering modes. Combinational operation in ECB mode (a) vs. time-variant nonlinear feedback loop in CBC mode (b), and CBC-8 operation b (c).

□

---

[57] Higher data rates must then be bought by measures on lower levels of abstraction, that is by optimizing the circuitry at the arithmetic/logic level, by recurring to transistor-level design in conjunction with full-custom layout, and/or by adopting a faster target technology, all of which measures ask for extra effort and come at extra cost. Only later have cryptologists come up with a better option known as **counter mode** (CTR) that manages without feedback and still avoids the leakage of plaintext into ciphertext that plagues ECB.

It has repeatedly been noted in this section that any attempt to insert an extra register into a feedback loop with the idea of pipelining the datapath destroys the equivalence between original and pipelined computations unless its effect is somehow compensated. After all, circuits c and b of fig.3.41 behave differently. Although this may appear a futile question, let us ask

"What happens if we do just that to a first-order recursion?"

Adding an extra latency register to (3.69) results in the DDG of fig.3.42a and yields

$$y(k) = f(y(k-2), x(k)) \tag{3.73}$$

Observe that all indices are even in this equation. As k increments with time $k = 0, 1, 2, 3, ...$ indices do in fact alternate between even and odd values. It thus becomes possible to restate the ensuing input-to-output mapping as two separate recursions with no interaction between "even" data items $x(k = 0, 2, 4, ..., 2n, ...)$ and "odd" items $x(k = 1, 3, 5, ..., 2n+1, ...)$.

$$y(2n) = f(y(2n-2), x(2n)) \tag{3.74}$$

$$y(2n+1) = f(y(2n-1), x(2n+1)) \tag{3.75}$$

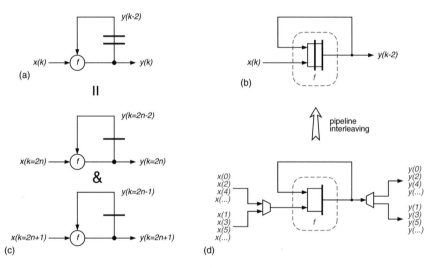

FIGURE 3.42

Pipeline interleaving. DDG of nonlinear time-variant first-order feedback loop with one extra register inserted (a) and isomorphic architecture (b). Interpretation as two interleaved data streams each of which gets processed exactly as specified by the original nonlinear first-order recursion of fig.3.39a (c,d).

This pair of equations says that the original data processing recursion of (3.69) now gets applied to two separate data streams as depicted in fig.3.42c. From a more general perspective, it is indeed possible to cut the combinational delay in any first-order feedback loop down to $\frac{1}{p}$ by inserting $p-1$ pipelining registers, yet the computation then falls apart into the processing of p interleaved but otherwise independent data streams. More often than not this is undesirable. However, practical applications exist where it is possible to take advantage of this effect.

Examples

Cipher block chaining (CBC) implements the recursion $y(k) = c(x(k) \oplus y(k-1), u(k))$. What counts from a cryptographic point of view is that patterns from the plaintext do not show up in the ciphertext. Whether this is obtained from feeding back the immediately preceding block of ciphertext $y(k-1)$ (CBC-1 mode) or some prior block $y(k-p)$ where $2 \leq p \in \mathbb{N}$ (CBC-p mode) is of minor importance. Some cryptochips, therefore, provide a fast but nonstandard CBC-8 mode in addition to the regular CBC-1 mode, see fig.3.41c. In the occurrence of the IDEA chip described in [79], maximum throughout is 176 Mbit/s both in pipelined ECB mode and in pipeline-interleaved CBC-8 mode as compared to just 22 Mbit/s in nonpipelined CBC-1.

Fig.3.43 shows a high-level block diagram of a sphere decoder, a key subfunction in a MIMO OFDM (orthogonal frequency division multiplex) receiver. Sphere decoding is essentially a sophisticated tree-traversal algorithm that achieves close-to-minimum error rate performance at a lower average search complexity than an exhaustive search. Pipelining the computation in search of throughput is not an option because of the (nonlinear) first-order recursion. Instead, the facts that (a) OFDM operates on many subcarriers at a time (typically 48 to 108) and that (b) each such subcarrier poses an independent tree-search problem, make sphere decoding an ideal candidate for pipeline interleaving. This and many other refinements to sphere decoding have been reported in [80] from which fig.3.44 has been taken.

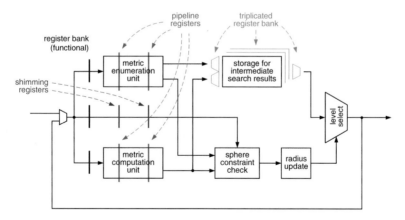

FIGURE 3.43

Sphere decoder. The dashed arrows point to the extra circuitry required to handle three individual subcarriers in an interleaved fashion (simplified).

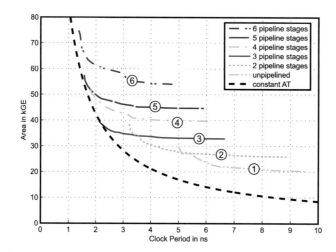

FIGURE 3.44

The beneficial impact of pipeline interleaving on area and throughput of a sphere decoder circuit (diagram courtesy of Dr. Markus Wenk).

For a much simpler example, consider some image processing algorithm where rows of pixels are dealt with independently from each other. Rather than scanning the image row by row, pixels from p successive rows are entered one by one in a cyclic manner before the process is repeated with the next column, and so on. All processing can so be carried out using a single pipelined datapath of p stages [81].
□

Pipeline interleaving does obviously not qualify as an equivalence transform. Still, it yields useful architectures for any recursive computation — including nonlinear ones — provided that data items arrive as separate time-multiplexed streams that are to be processed independently from each other, or can be arranged to do so. From this perspective, pipeline interleaving is easily recognized as a clever and efficient combination of time sharing with pipelining.

3.7.7 DIGEST

- When in search of high performance for recursive computations, reformulating a high-order system as a cascade of smaller-order sections in order to make the system amenable to coarse grain pipelining should be considered first. As a by-product, the reduced orders of the individual recursion loops offer additional speedup potential.

- Throughput of low-order recursive computations can be significantly improved by loop unfolding in combination with fine grain pipelining. This may bring computations into the reach of static CMOS technology that would otherwise ask for faster but also more expensive alternatives such as SiGe, BiCMOS, or GaAs.

- Whether the inflated latency is acceptable or not depends on the application. Also, the rapid growth of overall circuit size tends to limit economically practical unfolding degrees to fairly low values, say $p = 2...8$, especially when the system is a time-variant one.

- The larger number of roundoff operations resulting from loop unfolding must be compensated for by using longer word widths, which increases the cost of loop unfolding beyond the sole proliferation of computational units and intermediate registers.

- Nonlinear feedback loops are, in general, not amenable to throughput multiplication by applying unfolding techniques. A notable exception exists when the loop function is associative.

- Pipeline interleaving is highly efficient for accelerating recursive computations because it does not depend on any specific properties of the operations involved. Yet, as it modifies the input-to-output mapping, it is not an option unless the application admits that multiple data streams undergo the same processing independently from each other.

3.8 GENERALIZATIONS OF THE TRANSFORM APPROACH

3.8.1 GENERALIZATION TO OTHER LEVELS OF DETAIL

As stated in section 3.3.4, DDGs are not concerned with the granularity of operations and data. Recall, for instance, figs.3.20 and 3.41a that describe the same block cipher at different levels of detail. As a consequence, the techniques of iterative decomposition, pipelining, replication, time sharing, algebraic transform, retiming, loop unfolding, and pipeline interleaving are not limited to any specific level of abstraction although most examples so far have dealt with operations and data at the word level, see table 3.10.

Table 3.10   An excerpt from the VLSI abstraction hierarchy.

Level of abstraction	Granularity	Relevant items	
		Operations	Data
Architecture	◯	subtasks, processes	time series, pictures, blocks
Word	○	arithmetic/logic operations	words, samples, pixels
Bit	·	gate-level operations	individual bits

Architecture level

Things are pretty obvious at this higher level where granularity is coarse. As an example, fig.3.45 gives a schematic overview of a visual pattern recognition algorithm. Four subtasks cooperate in a cascade with no feedback, namely preprocessing, image segmentation, feature extraction, and object classification.

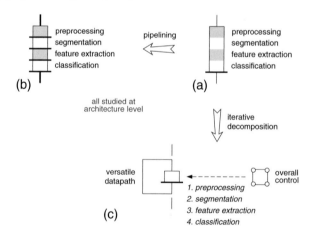

FIGURE 3.45

Overall architectural alternatives for a pattern recognition system. Isomorphic architecture (a), pipelined operation (b), and iteratively decomposed computation flow (c).

In a real-time application, one would definitely begin by introducing pipelining because four processing units are so made to work concurrently at negligible cost. In addition, each unit is so dedicated to a specific subtask and can be optimized accordingly.

The option of replicating the entire datapath would most likely get discarded at this point because it cannot compete with pipelining in terms of hardware efficiency. Replication of selected functional units could become an interesting alternative in later transforms, however, when the actual performance bottlenecks are known.

Iterative decomposition would only be considered if the desired throughput were so modest that it could be obtained from a single processing unit.

Bit level

Equivalence transformations can also be beneficial at low levels of abstraction. Take addition, for instance, which is an atomic operation when considered at the word level, see fig.3.46a. When viewed at the gate level, however, the same function appears as a composite operation that can be implemented from bit-level operations in many ways, the most simple of which is a ripple-carry adder shown in fig.3.46b. This detailed perspective clearly exposes the longest path and opens new opportunities for reorganizing the DDG that remain hidden from a word or higher level of abstraction.

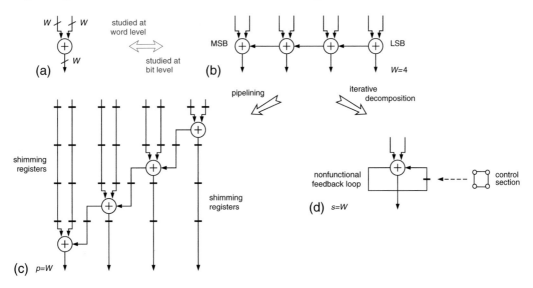

FIGURE 3.46

4-bit addition at the register transfer level (a), broken down into a ripple-carry adder (b) before being pipelined (c) or iteratively decomposed (d).

A gate-level pipelined version makes it possible to operate the circuit at a computation rate many times higher than with word-level pipelining alone, see fig.3.46c. As this amounts to fine-grain pipelining, the price to pay in terms of circuit size is likely to be excessive, however. In the above example, better solutions are obtained from more sophisticated arithmetic schemes such as carry-save, carry-skip, carry-select, or carry-lookahead adders, possibly in combination with moderate pipelining. Incidentally, note that modern synthesis tools support automatic retiming of gate-level networks.

Conversely, the structure shown in fig.3.46d follows when the W-bit addition is decomposed into one-bit suboperations. Computation starts with the LSB. A flip-flop withholds the carry-out bit for the next computation period where it serves as carry-in to the next more significant bit. Obviously, the flip-flop must be properly initialized in order to process the LSB and any potential carry input in a correct way. Although this entails some extra control overhead, substantial hardware savings may nevertheless result when the words being processed are sufficiently wide.

3.8.2 BIT-SERIAL ARCHITECTURES

The idea underlying fig.3.46d gives rise to an entirely different family of architectures known as bit-serial computation [82]. While most datapaths work in a bit-parallel fashion in that word-level operations are executed one after the other with all bits of a data word being processed simultaneously, organizing computations the other way round is also possible. Here, the overall structure remains isomorphic with the DDG, but the various word-level operations are decomposed into steps that are carried out one bit after the other.

Example

A bit-serial implementation of a third order transversal filter is shown in fig.3.47, also see fig.3.17c for the DDG. The w-bit wide input samples $x(k)$ enter serially with the LSB first whereas coefficients $b_n(k)$ must be applied in a parallel format. The circuit is operated at a computation rate s times higher than the sampling frequency. Evaluation proceeds from the LSB to the MSB with computation periods numbered $w = s - 1, ..., 0$.

The first computation period ingests the LSB of the actual input sample $x(k)$, evaluates the LSBs of all samples $x(k), ..., x(k - 3)$, and sees the LSB of the result $y(k)$ emerge at the output. The second period then handles the next significant bit and so on.[58] Shifts and carry-overs from one bit to the next higher position are obtained from using latency registers. As these registers may contain carries from previous additions after all W bits of the input have been processed, extra computation periods are required to bring out the MSB of $y(k)$, so that $s > W$.

Note that iterative decomposition has led to <u>nonfunctional</u> feedback loops in the architecture of a transversal filter, although the DDG is free of circular paths by definition. As this kind of feedback is confined to within the multiply and add units, the filter as a whole remains amenable to pipelining provided computations inside the loops are not affected.

---

[58] DSP applications frequently deal with input data scaled such that $|x_k| < 1$ and coded with a total of W bits in 2's-complement format, see (3.77). In this particular case, computation periods $w = s - 1, ..., 1$ process the input bits of weight 2^{-w} respectively, while the last computation period with $w = 0$ is in charge of the sign bit.

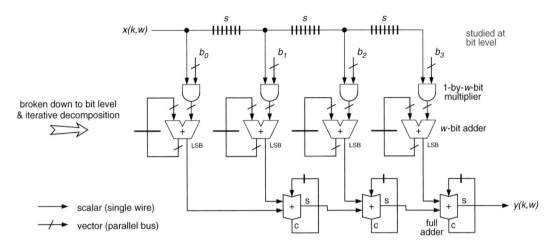

FIGURE 3.47

Bit-serial implementation of a third order transversal filter functionally equivalent to the bit-parallel architecture of fig.3.33 (simplified).

□

Closer examination of bit-serial architectures reveals the following properties:

+ Control overhead is small when compared to bit-parallel alternatives because the isomorphism of DDG and architecture is maintained.

− As the DDG is hardwired into the datapath with no explicit control instance, changes to the processing algorithm, switching between different modes of operation, and exception handling are awkward to deal with.

+ The shallow combinational depth in conjunction with a high computation rates helps to keep all computational units busy for most of the time.

+ All non-local data communication is via serial links that operate close to the maximum rate supported by the target technology which cuts down on-chip wiring requirements.

+ Much of the data circulation is local which contrasts favorably with the data items travelling back and forth between datapath and storage in bit-parallel architectures.

+ As FPL devices provide only limited wiring resources, the two previous assets tend to favor bit-serial architectures when designing for FPGAs or CPLDs.

− The word width must be the same throughout a serial architecture. Parts of the computation that do not make use of the full precision typically cause the hardware to idle for a number of computation periods.

+ Conversely, arbitrary precision can be accommodated with the same hardware when execution time is allowed to grow with word length.

— Bit-serial computation is incompatible with the storing of data in word-oriented RAMs and ROMs. Extra format conversion circuitry is required whenever such memories are preferred for their high storage densities.

— Division, data-dependent decisions, and many other functions are ill-suited to bitwise iterative decomposition and pipelining. While it is possible to incorporate bit-parallel functions, provided their interfaces are serialized and that they exhibit a fixed latency, the resulting hybrid structures often prove unsatisfactory and inefficient.

— Algorithms based on successive approximation naturally operate with the MSB first and the same applies to picking the maximum or minimum value from a set of numbers. It is sometimes possible to reconcile LSB-first and MSB-first operations at the price of recurring to redundant number representation schemes.

In summary, bit-serial architectures are at their best for unvaried real-time computations that involve fixed and elementary operations such as addition and multiplication by a constant. The reader is referred to the specialized literature [82] [83] for case studies and for further information on bit-serial design techniques.

3.8.3 DISTRIBUTED ARITHMETIC

Bit-serial architectures have been obtained from breaking down costly word-level operations into bit-level manipulations followed by universal transforms such as iterative decomposition and pipelining. Another family of serial architectures results from making use of algebraic transforms at the bit level too. Consider the calculation of the following inner product

$$y = \sum_{k=0}^{K-1} c_k x_k \tag{3.76}$$

where each c_k is a fixed coefficient and where each x_k stands for an input variable. Fig.3.48a shows the architecture that follows from routinely applying decomposition at the word level. Computation works by way of repeated multiply-accumulate operations, takes K computation periods per inner product, and essentially requires a hardware multiplier plus a look up-table for the coefficients.

Now assume that the inputs are scaled such that $|x_k| < 1$ and coded with a total of W bits in 2's-complement format.[59] We then have

$$x_k = -x_{k,0} + \sum_{w=1}^{W-1} x_{k,w}\, 2^{-w} \tag{3.77}$$

---

[59] This is by no means a necessity. We simply assume $|x_k| < 1$ for sake of convenience and 2's-complement format because it is the most common representation scheme for signed numbers in digital signal processing.

with $x_{k,0}$ denoting the sign bit and with $x_{k,w}$ standing for the bit of weight 2^{-w} in the input word x_k. By combining (3.76) and (3.77) the desired output y can be expressed as

$$y = \sum_{k=0}^{K-1} c_k (-x_{k,0} + \sum_{w=1}^{W-1} x_{k,w} 2^{-w}) \tag{3.78}$$

With the aid of the distributive law and the commutative law of addition, the computation now gets reorganized into the equivalent form below where the order of summation is reversed.

$$y = \sum_{k=0}^{K-1} c_k (-x_{k,0}) + \sum_{w=1}^{W-1} (\sum_{k=0}^{K-1} c_k x_{k,w}) 2^{-w} \tag{3.79}$$

The pivotal observation refers to the term in parentheses

$$\sum_{k=0}^{K-1} c_k x_{k,w} = p(w) \tag{3.80}$$

For any given bit position w, calculating the sum of products takes one bit from each of the K data words x_k which implies that the result $p(w)$ can take on no more than 2^K distinct values. Now, as coefficients c_k have been assumed to be constants, all those values can be precomputed and kept in a lookup table (LUT) instead of being calculated from scratch whenever a new data set x_k arrives at the input. A ROM is typically used to store the table. It must be programmed such as to return the partial product $p(w)$ when presented with the address w, i.e. with a vector obtained from concatenating all bits of weight 2^{-w} across all input variables x_k. Playing this trick, and noting that $\sum_{k=0}^{K-1} c_k (-x_{k,0})$ is nothing else than $p(0)$ with the sign reversed, (3.79) takes on the utterly simple form

$$y = -p(0) + \sum_{w=1}^{W-1} p(w) 2^{-w} \tag{3.81}$$

While the isomorphic architecture calls for W LUTs with identical contents, a much smaller architecture can be obtained from decomposing the evaluation of (3.81) into a series of W consecutive steps. The new architecture manages with a single lookup table but requires a nonfunctional register for accumulating the partial products, see fig.3.48b. Calculation proceeds one bit position at a time, starting with the LSB in computation period $w = W - 1$ and processing the sign bit in the final cycle where $w = 0$.

A minor complication comes from the fact that the term $-p(0)$ has a sign opposite to all other contributions to y. A simple solution consists of using an adder-subtractor working under control of a "sign-bit cycle" signal from a modulo W counter that acts as a controller. The same counter is also in charge of releasing the fully completed result and of clearing the accumulator at the end of the last computation period (two details not shown in fig.3.48b). In addition, it guides the selection of the appropriate input bits unless the x_ks can be made to arrive in a bit-serial format LSB first.

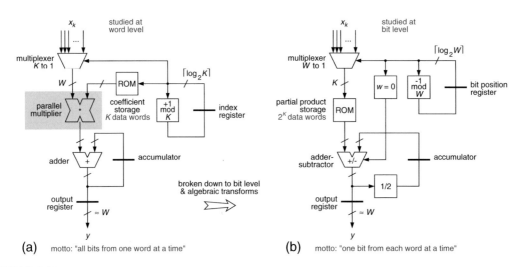

FIGURE 3.48

Architectures for computing a sum of products by way of repeated multiply-accumulate operations (a) and with distributed arithmetic (b) (simplified).

The most striking difference between the two architectural alternatives of fig.3.48 is the absence of any multiplier in the second design. Rather than being concentrated in a single hardware unit, multiplication is spread over the circuit which is why such architectures were given the name **distributed arithmetic**.

A limitation of distributed arithmetic is that memory size is proportional to 2^K where K is the order of the inner product to be computed. Although a more sophisticated coding scheme makes it possible to introduce a symmetry into the lookup table which can then be exploited to halve its size [84], the required storage capacity continues to grow exponentially with K. More impressive memory savings are obtained from reformulating (3.80) in the following way

$$\sum_{k=0}^{K-1} c_k x_{k,w} = \sum_{k=0}^{H-1} c_k x_{k,w} + \sum_{k=H}^{K-1} c_k x_{k,w} \tag{3.82}$$

where $0 < H < K$. Instead of having all K bits address a single ROM, they are split into two subsets of H and $K - H$ bits respectively, each of which drives its own LUT. The total storage requirement is so reduced from 2^K data words to $2^H + 2^{K-H}$ which amounts to $2^{\frac{K}{2}+1}$ when input bits are split such that $H = \frac{1}{2}K$. The price to pay is an extra adder for combining the outputs from the two tables. Clearly, the idea can be extended to more than two tables.

Memory requirements may sometimes be slashed further by taking advantage of properties of the coefficient values at hand such as symmetries, repetitions, and relations between their binary codes, also see problem 8. Memory size and circuit structure thereby become highly dependent on the particular coefficient values, though, which makes it difficult to accommodate modifications.

In conclusion, distributed arithmetic should be considered when coefficients are fixed, when the number of distinct coefficient values is fairly small, and when lookup tables (LUT) are available at little cost compared to bit-parallel multipliers. This explains why this approach has recently regained popularity in the context of DSP applications with LUT-based FPGAs [85] [86]. Please refer to the literature for tutorials [84] and further VLSI circuit examples [87] if distributed arithmetic appears to be an option for your filtering problem.

3.8.4 GENERALIZATION TO OTHER ALGEBRAIC STRUCTURES

So far we have mostly been dealing with the infinite field $(\mathbb{R}, +, \cdot)$ formed by the set of all real numbers together with addition and multiplication.[60] Accordingly, most examples have been taken from digital signal processing where this type of computation is commonplace. Now, as all algebraic fields share a common set of axioms, any algebraic transform that is valid for some computation in $(\mathbb{R}, +, \cdot)$ must necessarily hold for any other field.[61]

Finite fields

Galois fields such as GF(2), GF(p), and GF(p^n) have numerous applications in data compression (source coding), error correction (channel coding), and information security (ciphering). Thus, when designing high-speed telecommunications or computer equipment, it sometimes proves useful to know that the loop unfolding techniques discussed for linear systems in $(\mathbb{R}, +, \cdot)$ directly apply to any linear computation in any Galois or other finite field too.

Semirings

The analysis of recursive computations in section 3.7 has revealed that almost all successful and efficient loop unfolding techniques are tied to linear systems over a field. That computation be performed in a field is a sufficient but not a necessary condition, however, as will become clear shortly. Recall how loop unfolding was derived for the time-variant linear case in (3.64) through (3.68). Substituting the generic operator symbols $\boxplus$ for $+$ and $\boxdot$ for $\cdot$ we can write

$$y(k) = a(k) \boxdot y(k-1) \boxplus x(k) \tag{3.83}$$

After the first unfolding step, i.e. for $p = 2$, one has

$$y(k) = a(k) \boxdot a(k-1) \boxdot y(k-2) \boxplus a(k) \boxdot x(k-1) \boxplus x(k) \tag{3.84}$$

and for arbitrary integer values of $p \geq 2$

$$y(k) = \left(\prod_{n=0}^{p-1} a(k-n)\right) \boxdot y(k-p) \boxplus \sum_{n=1}^{p-1}\left(\prod_{m=0}^{n-1} a(k-m)\right) \boxdot x(k-n) \boxplus x(k) \tag{3.85}$$

where $\sum$ and $\prod$ refer to operators $\boxplus$ and $\boxdot$ respectively. The algebraic axioms necessary for that derivation were closure under both operators, associativity of both operators, and the distributive law of $\boxdot$ over $\boxplus$. The existence of identity or inverse elements is not required. Also, we have never made use of commutativity of operator $\boxdot$ which implies (a) that the result also holds for other than commutative

---

[60] See appendix 3.11 for a summary on algebraic structures.
[61] Universal transforms remain valid anyway as they do not depend on algebraic properties.

operators $\square$, in which case (b) the above order of "multiplication" is indeed mandatory. The algebraic structure defined by these axioms is referred to as a semiring.

The practical benefit is that recursive computations of seemingly nonlinear nature when formulated in the field $(\mathbb{R}, +, \cdot)$ — or in some other field — become amenable to loop unfolding, provided it is possible to restate them as linear computations in a ring or semiring [88]. A number of problems related to finding specific paths through graphs are amenable to reformulation in this way. Suitable algebraic systems that satisfy the axioms of a semiring are listed in [89] under the name of path algebras and in appendix 3.11.

Example
Convolutional codes find applications in telecommunication systems for error recovery when data get transmitted over noisy channels. While a convolutional (en)coder is simple, the computational effort for decoding at the receiver end is more substantial. The most popular decoding method is the **Viterbi algorithm** [90] [91], a particular case of dynamic programming[62] for finding the shortest path through a trellis graph. The subsequent discussion is about the central step of fig.3.49.

Figure 3.50 shows a couple of architectural options for an 8-state convolutional code. The trellis graph of fig.3.50a can be understood as an abstracted DDG where each node stands for an **add-compare-select** (ACS) operation. ACS operations occur in pairs of two and each pair is said to form a butterfly.[63] Figure 3.50d specifically suggests a datapath that includes the necessary hardware to compute one set of path metrics from the previous set in a single clock cycle and to store the result in a bank of registers. This obvious solution represents a dramatic improvement over the highly inefficient and, hence, impractical isomorphic architecture of fig.3.50a. Yet, there is room for more beneficial reorganizations depending on application requirements.

---

[62] **Dynamic programming** is an odd name — chosen by Richard Bellman for promotional reasons — for a broad class of optimization algorithms that decompose the search for a global optimum into a sequence of simpler decision problems at a local level. All decisions must obey Bellman's principle of optimality which states that the globally optimum solution includes no suboptimal local decision. This is a very welcome property because it allows to prune inferior candidates early during the search process. Dynamic programming finds applications in fields as diverse as telecommunications, speech recognition, video coding, watermark detection, flight trajectory planning, and genome sequencing.

   For an anecdotal introduction, think of the following situation. During the darkness of night, a group of four has to cross a fragile suspension bridge that can carry no more than two persons at a time. The four persons take 5, 10, 20 and 25 minutes respectively for traversing the bridge. A torch must be carried while on the bridge, the torch available will last for exactly one hour. The problem is how to organize the operation. Draw a graph where each state of affair gets represented by a vertex and each traversal of the bridge by an edge. By solving this quiz in a systematic way, you are bound to discover the ideas behind dynamic programming yourself. The solution is available on the book's companion website.

[63] This is in fact not the only commonality with the FFT. Much as the computations there, the path metric update step in the Viterbi algorithm is amenable to in-place computation [92].

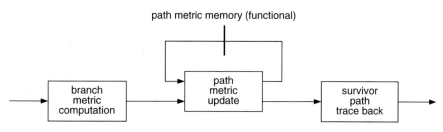

path metric memory (functional)

FIGURE 3.49

The three major steps of the Viterbi algorithm.

Smaller circuit. Apply time sharing and then again iterative decomposition as suggested earlier in the context of the FFT (in section 3.4.5). This route essentially trades throughput for economy and ultimately leads to a processor-type datapath built around an ALU as portrayed in fig.3.50e.

Reduced clock. The longest path in fig.3.50d includes no more than two adders, a multiplexer, a register, and the pertaining wiring. As the remainder of the circuit is likely to impose a much longer clock period, there will be a mismatch in many situations. Besides, the maximum clock rate may be undesirably high. The architecture suggested by fig.3.50c may then prove more adequate as it yields roughly the same throughput with half the clock. The more complex datapath — combinational logic gets approximately doubled — together with the inflated AT-product are the downsides of this modification.

Still higher throughput. As a consequence from iterative decomposition, each butterfly participates in a recursive computation illustrated in fig.3.51a that goes

$$y_1(k) = \min(a_{11}(k) + y_1(k-1), a_{12}(k) + y_2(k-1)) \tag{3.86}$$

$$y_2(k) = \min(a_{21}(k) + y_1(k-1), a_{22}(k) + y_2(k-1)) \tag{3.87}$$

As all other computations in the Viterbi algorithm are of feedforward nature, the maximum achievable throughput of the decoding process gets indeed limited by this nonlinear recursion.

Now consider a semiring where

- Set of elements: $S = \mathbb{R} \cup \{\infty\}$,
- Algebraic addition: $\boxplus = \min$, and
- Algebraic multiplication: $\boxdot = +$.

The new and linear — in the semiring — formulation of the ACS operation goes

$$y_1(k) = a_{11}(k) \boxdot y_1(k-1) \boxplus a_{12}(k) \boxdot y_2(k-1) \tag{3.88}$$

$$y_2(k) = a_{21}(k) \boxdot y_1(k-1) \boxplus a_{22}(k) \boxdot y_2(k-1) \tag{3.89}$$

which, making use of vector and matrix notation, can be rewritten as

$$\vec{y}(k) = A(k) \boxdot \vec{y}(k-1) \tag{3.90}$$

By replacing $\vec{y}(k-1)$ in (3.90) one gets the unfolded recursion for $p = 2$

$$\vec{y}(k) = A(k) \boxdot A(k-1) \boxdot \vec{y}(k-2) \tag{3.91}$$

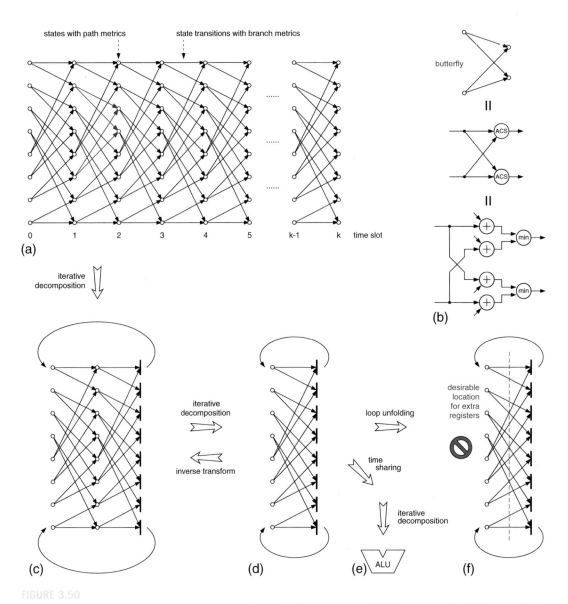

FIGURE 3.50

Architectural options for the Viterbi algorithm. Abstracted trellis-type DDG for path metric computation (a) with details for one butterfly (b). Datapath architectures obtained from different degrees of iterative decomposition (c,d,e). Doomed attempt to boost throughput by inserting extra latency registers into the nonlinear first-order feedback loop (f) (simplified).

To take advantage of this unfolded form, the product $B(k) = A(k) \boxdot A(k-1)$ must be computed outside the loop. Resubstituting the original operators and scalar variables we finally obtain the recursion

$$y_1(k) = \min(b_{11}(k) + y_1(k-2), b_{12}(k) + y_2(k-2)) \tag{3.92}$$

$$y_2(k) = \min(b_{21}(k) + y_1(k-2), b_{22}(k) + y_2(k-2)) \tag{3.93}$$

which includes the same number and types of operations as the original formulation but allows for twice as much time. The price to pay is the extra hardware required to perform the non-recursive computations below in a heavily pipelined way.

$$b_{11}(k) = \min(a_{11}(k) + a_{11}(k-1), a_{12}(k) + a_{21}(k-1)) \tag{3.94}$$

$$b_{12}(k) = \min(a_{11}(k) + a_{12}(k-1), a_{12}(k) + a_{22}(k-1)) \tag{3.95}$$

$$b_{21}(k) = \min(a_{21}(k) + a_{11}(k-1), a_{22}(k) + a_{21}(k-1)) \tag{3.96}$$

$$b_{22}(k) = \min(a_{21}(k) + a_{12}(k-1), a_{22}(k) + a_{22}(k-1)) \tag{3.97}$$

□

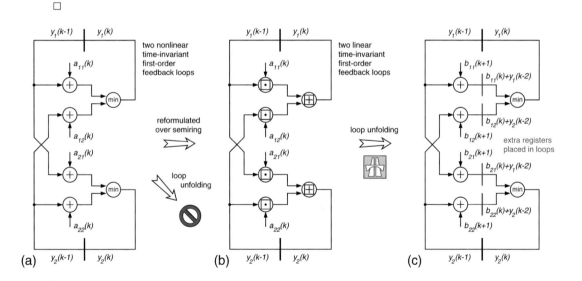

The first-order recursion of the Viterbi algorithm before (a) and after being reformulated over a semiring (b), and with loop unfolding added on top (c) (simplified, only one butterfly shown).

The remarkable hardware structure so obtained demonstrates that taking advantage of specific properties of an algorithm and of algebraic transforms has more potential to offer than universal transforms alone. Some computations can be accelerated by creating concurrencies that did not exist in the original formulation which opens a door to solutions that would otherwise have remained off-limits.

3.8.5 DIGEST

- The transform approach to architecture design promoted in this text has been found to yield useful solutions at any level of granularity. Some of the resulting architectures are truly surprising.

- Both bit-serial architectures and distributed arithmetic follow quite naturally when arithmetic/logic operations are dissected into bit-level manipulations before the various equivalence transforms are being applied. They are worthwhile to be considered when fixed and data-independent computations are to be carried out with limited hardware resources and moderate performance. After having sunk into oblivion for many years, the two techniques have had a comeback for filtering and other DSP applications with LUT-based FPGAs.

- All universal <u>and</u> algebraic transforms that apply to computations on the field of reals also apply to computations on Galois fields, of course.

- While loop unfolding is applicable to any linear computation in the field of reals, this is not a necessary condition. In case a recursion forms a bottleneck when in pursuit of higher performance, check whether it is possible to restate or modify the computations within the feedback loop such as to make them linear over a semiring.

3.9 CONCLUSIONS

3.9.1 SUMMARY

We have begun this chapter by comparing instruction set processors with dedicated architectures. It was found that general-purpose computing asks for a high degree of flexibility that only program-controlled processors can provide. However, the ability to execute an arbitrary sequence of instructions on an unknown range of data types brings about numerous inefficiencies and largely precludes architectural optimizations. For well-defined computational tasks, much better performance and energy efficiency can be obtained from hardwired architectures with resources tailored to the specific computational needs of the target application. Segregation, weakly-programmable satellites, ASIPs, and configurable computing have been found to form useful compromises.

Next, we have investigated a number of options for organizing datapath hardware. Our approach was based on reformulating a given data processing algorithm such as to preserve its input-to-output relationship except, possibly, for latency, while improving on performance, circuit size, energy efficiency, and the like. Findings on how to best rearrange combinational, non-recursive, and recursive computations were given in sections 3.4.8, 3.6.6, and 3.7.7 respectively. The approach was then generalized in terms of granularity and algebraic structure with the results summarized in section 3.8.5. The essences of these insights is collected in tables 3.12 and 3.11.

Table 3.11 **Options available for reorganizing datapath architectures. Upper-case letters denote transforms generally available whereas lower-case letters indicate some preconditions must be satisfied by the application and/or type of computation to make this a viable option.**

		Type of computation		
		combinational (aka memoryless)	sequential (aka memorizing)	
			non-recursive	recursive
	Data flow	feedforward	feedforward	feedback
	Memory	no	yes	yes
	Data dependency graph	DAG^a with all edge weights zero	DAG with some or all edge weights non-zero	Directed cyclic graph with no circular path of weight zero
	Response length	$M = 1$	$1 < M < \infty$	$M = \infty$
Nature of system	linear time-invariant	D,P,Q,S,a	D,P,q,S,a,R	D,S,a,R,i,U
	linear time-variant	D,P,Q,S,a	D,P,S,a,R	D,S,a,R,i,U
	nonlinear	D,P,Q,S,a	D,P,S,a,R	D,S,a,R,i,u
	Discussed in section	3.4	3.6	3.7

D : Iterative decomposition
P : Pipelining
Q : Replication
q : Multipath filtering as special case of replication
     provided the resulting repetitive transfer function is acceptable
S : Time sharing
a : Associativity transform provided operations are identical and associative
R : Retiming
i : Pipeline interleaving, i.e. pipelining in conjunction with time sharing,
     provided a number of data streams can be processed separately from each other
U : Loop unfolding
u : Loop unfolding provided computation is linear over a semiring

---

^{a}DAG is an acronym for directed acyclic graph, i.e. for a directed graph with no circular path.

Table 3.12   Summary of the most important architectural transforms and their characteristics.

Architectural transform	Decom-position	Pipe-lining	Repli-cation	Time sharing	Associa-tivity	Retiming	Loop unfolding
Kind	universal	universal	universal	universal	algebraic	universal	algebraic
Applicable to	combinational computations					sequential computations	
Impact on						nonrecurs.	recursive
A	$-... =$	$= ...+$	$+$	$-... =$	$=$	$=$	$+$
Γ	$+$	$=$	$-$	$+$	$=$	$=$	$=$
t_{lp}	$-$	$-$	$=$, mux $-$	$=$	$-...+$	$-$	$-$
$T = \Gamma \cdot t_{lp}$	$=$	$-$	$-$	$+$	$-...+$	$-$	$-$
AT	$-... =$	$-... =$	$=$	$= ...+$	$-...+$	$-$	$+$
L	$+$	$+$	$=$, mux $+$	$+$	$=$	$=$	$+$
E	$-...+$	$-...+$	$=$	$= ...+$	$-...+$	$=$	$+$
Extra hardware overhead	recy. and cntl.	none	distrib., recoll., and cntl.	collect., redist., and cntl.	none	none	extra word width
Helpful for indirect energy saving	no	coarse grain yes	possibly yes	no	yes	yes	possibly yes
Compatible storage type	any	register	register	any	any	register	register
Discussed in subsection(s)	3.4.2 3.6.4	3.4.3 3.6.2	3.4.4	3.4.5 3.6.4	3.4.6	3.6.1 3.7.2	3.7.2

A :	circuit size	$=$:	approximately constant
Γ :	cycles per data item	$+$:	tends to increase
t_{lp} :	longest path length	$-$:	tends to decrease
T :	time per data item	$...$:	in less favorable situations
L :	latency in computation periods		
E :	energy per data item		

auxiliary circuitry for
recy.: data recycling
cntl.: datapath control
dist.: data distribution
coll.: data collection

As energy efficiency depends on so many parameters, the pertaining entries of table 3.12 deserve further clarification. Assuming fixed energy costs per operation and ignoring any static currents, most architectural transforms discussed inflate the energy dissipated on a given calculation as conveyed by table entries E and "Extra hardware overhead". Put in other words, cutting circuit size and boosting throughput typically are at the expense of energy efficiency.

The picture is more favorable when there is room for cutting the energy spent per computational operation by playing with voltages, transistor sizes, circuit style, fabrication process, and the like. The most effective way to do so in CMOS is to lower the supply voltage since the energy dissipated per operation augments quadratically with voltage whereas a circuit's operating speed does not. The long paths through a circuit are likely to become unacceptably slow, though. A suitable architecture transform may then serve to trim these paths such as to compensate for the loss of speed. This is precisely what the attribute "Helpful for indirect energy saving" in table 3.12 refers to. Retiming, chain/tree conversion, and coarse grain pipelining are the most promising candidates as they entail no or very little overhead. Note that the net gain of any such technique must be examined in detail on a per case basis, however.

3.9.2 THE GRAND ARCHITECTURAL ALTERNATIVES FROM AN ENERGY POINT OF VIEW

Over the first decade of the 21th century, power efficiency has become as important as die size, so let us re-examine the fundamental architectural alternatives from an energy point of view. Program-controlled processors heavily rely on subcircuits and activities such as

- General-purpose multi-operation ALUs,
- Generic register files of generous capacity,
- Multi-driver busses, bus switches, multiplexers, and the like,
- Program and data memories along with address generation,
- Controllers or program sequencers,
- Instruction fetching and decoding,
- Stack operations and interrupt handling,
- Dynamic reordering of operations,
- Branch prediction and speculative execution,
- Data shuffling between main memory and multiple levels of cache, and
- Various mechanisms for maintaining cache and data consistency.

From a purely functional point of view, all of this is a tremendous waste of energy because none of it contributes to payload data processing. The welcome agility of instruction set processors is thus paid for with an overhead in terms of control operations and a formidable inflation of switching activity.

The isomorphic architecture, in contrast, does not carry out any computations or data transfers unless mandated by the original data processing algorithm itself. There are no instruction to fetch and decode. There is no addressing and no accessing of memories either as all data are kept in registers. Data transfers are local, there is no data shuffling between registers, cache, and main memory. There are no busses with their important load capacitances to drive.

"Does this imply the isomorphic architecture is the most energy-efficient option then?"

Somewhat surprisingly, this is not so because of glitching and leakage. Glitching is a phenomenon observed in digital circuits that causes redundant signal transients to occur on top of those stipulated by the computation per se. Glitch-induced switching is particularly intense when data recombine in combinational logic after having travelled along propagation paths of largely disparate lengths because node voltages are then likely to rock up and down several times before settling.[64] By cutting overlong propagation paths, moderate pipelining and iterative decomposition tend to abate glitching and so help to improve overall energy efficiency.

Leakage refers to static transistor currents that flow irrespective of whether they sit idle or are busy carrying out computations. Everything else being equal, a smaller circuit, as obtained from iterative decomposition and time sharing, tends to have fewer leakage paths.

Back to the main point, it is not unusual to find that a program-controlled processor dissipates two or three orders of magnitude as much energy as an application-specific architecture does for the same computation.[65] A quote from Steve Jobs (2010) underlines the point.

> "To achieve long battery life when playing video, mobile devices must decode the video in hardware (on the GPU); decoding it in software (on the CPU) uses too much power. ... The difference is striking: on an iPhone 4, for example, H.264 videos play for up to 10 h, while videos decoded in software play for less than 5 h before the battery is fully drained."

General-purpose processors operate with data words of uniform and often oversized width throughout an entire algorithm.[66] As opposed to this, dedicated architectures make it possible to fine-tune the number of bits in every register and logic block to individual requirements as there is no compelling need to combine subfunctions with largely different precision requirements into a single datapath sized for worst-case requirements. The overriding concern is to avoid switching activities that are not relevant to the final result.

Last but not least, the impressive throughputs of general-purpose processors have been bought at the price of operating CMOS circuits under conditions that are far from optimal from an energy efficiency perspective (fine-grain pipelining, extremely fast clock, large overdrive factors i.e. comparatively high supply voltage U_{dd} combined with low MOSFET threshold voltages U_{th} and, hence, significant leakage).

Observation 3.10. *Increasing performance in power-constrained applications (almost all today), requires that the amount of energy spent per payload operation be lowered because power dissipation is nothing else than the product of operations per second and energy per operation.*

A key challenge of architecture design is to

- minimize redundant switching activities,
- provide as just as much flexibility as required,
- keep the effort for design and verification within reasonable bounds,

---

[64] You may want to refer to appendix A.5 to learn more about the causes of glitching.
[65] [93] estimates the gap to be up to four orders of magnitude over direct-mapped architectures and growing.
[66] The so-called multimedia instructions can provide programmers with an opportunity to process fewer bits per data item. Yet, not all instruction sets include them and not all algorithms lend themselves to taking advantage of sub-word parallelism.

and all this at a time. In most cases, this will require a clever combination of hardwired units with program-controlled processors. Energy considerations thus tend to give architecture design new momentum over the unimaginative usage of general-purpose microcomputers.

3.9.3 A GUIDE TO EVALUATING ARCHITECTURAL ALTERNATIVES

In spite of our efforts to present a systematic overview on dedicated datapath architectures, we must admit that architecture design is more art than science. Many practical constraints and technical idiosyncrasies make it impossible to obtain a close-to-optimum solution by analytical means alone. The common procedure, therefore, is to come up with a variety of alternative ideas, to devise the corresponding architectures to a reasonable level of detail, and to evaluate their respective merits and drawbacks before decisions are taken. This approach — which is typical for many engineering activities — asks for creativity, methodology, and endeavor. It is nevertheless hoped that the material in this chapter gives some insight into the options available and some directions on when to prefer what option for tailoring VLSI architectures to specific technical requirements. What follows are some practical guidelines.

1. Begin by analyzing the algorithm. Section by section, identify the data flow and the nature of the essential computations. Estimate the necessary datapath resources by giving quantitative indications for
   • the word widths (check [94] for references),
   • the data rates between all major building blocks,
   • the memory bounds, access rates and access patterns in each building block, and
   • the computation rates for all major arithmetic operations.[67]

2. Make sure the potential for simplifications and optimizations in the algorithmic domain has been fully tapped, see section 3.3.1 for suggestions.

3. Identify the controllers that are required to govern the flow of computation along with its interplay with the external world. Examine the control flow in terms of data dependencies, overall complexity, and flexibility requirements. Find out where to go for a hard-wired dedicated architecture, where for a program-controlled processor, and where to look for a compromise.

4. Rather than starting from a hypothetical isomorphic architecture, let your intuition come up with a number of preliminary architectural concepts. Establish a rough **block diagram** for each of them. Make the boundaries between major subfunctions coincide with registers as you would otherwise have to trace path delays across circuit blocks during timing verification and optimization.

---

[67] Watch out if you are given source code from some prior implementation, such as C code for a 32 bit DSP, for instance. You are likely to find items solely mandated by the resources available there or by software engineering considerations. Typical examples include operations related to (un)packing and (re)scaling, usage of computationally expensive data types, arithmetic operations substituted for bit-level manipulations, multitudes of nicely named variables that unnecessarily occupy distinct memory locations, and more.

5. It is always a good idea to prepare a comprehensive and, hence, fairly large table that opposes the different architectures under serious consideration.

The rows serve to describe the hardware resources. Each major subfunction occupies a row of its own. Each such subfunction is then hierarchically decomposed into ever smaller subcircuits on a number of subsequent rows until it becomes possible to give numerical estimates for A, t_{lp}, and, possibly, for E as well. Once a subcircuit has been broken down to the RTL level, one can take advantage of HDL synthesis to obtain those figures with a good degree of precision. Finite state machines, in particular, are difficult to estimate otherwise.

For each architectural variant, a few adjacent columns are reserved to capture A, t_{lp}, E and Γ. An extra column is set aside for n, a natural number that indicates how many times the hardware resource is meant to be instantiated for the architecture being considered. Γ stands for the number of cycles required to obtain one processed data item with the hardware resources available. Of course, this quantity tends to diminish with n but it is not possible to state the exact dependency in general terms.

6. Estimating the overall circuit size, cycles per data item, latency, and dissipated energy for each architecture now essentially becomes a matter of bookkeeping that can be carried out with the aid of spreadsheet software. Path delays are more tricky to deal with as logic and interconnect delays are subject to vary significantly as a function of lower-level details.[68] It is thus quite common to code, synthesize, place, and route the most time-critical portions of a few competing architectures merely for the purpose of evaluating $\max(t_{lp})$ and of extrapolating clock frequency, overall computation rate, and overall throughput.

7. Analysis of the figures so obtained will identify performance bottlenecks and inacceptably burdensome subfunctions in need of more efficient implementations. This is the point where the equivalence transforms discussed in this chapter come into play.

8. Compare the competing architectural concepts against the requirements. Narrow down your choice before proceeding to more detailed analyses and implementations.

Example

The table below shows results from exploring the design space for AES encryption with a key length of 128 bit [95]. The available options for trading datapath resources for computation time are evident. The narrower datapaths require extra circuitry for storing and routing intermediate results which inflates complexity and adds to path delays. What all variants have in common is that the ten cipher rounds are carried out by a single datapath as a result from iterative decomposition of the AES algorithm. Also, none of the architectures makes use of pipelining which results in latency and cycles per data item being the same. SubBytes refers to the cipher's most costly operation from a hardware point of view. While the figures include control logic and have been obtained from actual synthesis, simplifications have been made to obtain reasonably accurate estimates for the key figures of merit without having to establish the HDL code for each architectural alternative in full detail.

---

[68] Please refer to footnote 22 for a comment on the limitations of anticipating path delays.

Datapath width [bit]	8	16	32	64	128
Parallel SubBytes units	1	2	4	8	16
Circuit complexity [GE]	5052	6281	7155	11 628	20 410
Area A (normalized)	1	1.27	1.47	2.43	4.27
Cycles per data item Γ	160	80	40	20	10
Longest path delay t_{lp} (normalized)	1.35	1.34	1.21	1.13	1
Time per data item T (normalized)	21.6	10.7	4.83	2.23	1
Size-time product AT	21.6	13.6	7.10	5.42	4.27

☐

Designers of large VLSI chips running at elevated clock rates, such as high-performance uniprocessors, inevitably run into interconnect delay as another limiting factor. This is because it is no longer possible to transmit data from one corner of a chip to the opposite one within a single clock cycle. While this important aspect has been left aside in this chapter, more will be said in fig.7.18.

As a concluding remark, we would like to recall once more that good solutions call for analysis and reorganization of data processing algorithms at all levels of details, including architecture (process/block), register transfer (arithmetic/word), and logic (gate/bit) levels. It has been shown on numerous occasions that viewing a problem from a totally different angle can pave the way to unexpected architectural solutions that feature uncommon characteristics. Also, the possibilities for replacing a given algorithm by a truly different suite of computations that is equivalent for any practical purpose of the application at hand, but better suited to VLSI, should always be investigated first.

1. ∗∗∗ Computationally efficient approximations for the magnitude function $\sqrt{a^2 + b^2}$ have been presented in table 3.8. (a) Show that approximation 2 remains within ±3% of the correct result for any values of a and b. (b) Give three alternative architectures that implement the algorithm and compare them in terms of datapath resources, cycles per data item, longest path, and control overhead. Assume input data remain valid as long as you need them, but plan for a registered output. Begin by drawing the DDG.

2. ∗∗ Discuss the idea of combining replication with pipelining. Using fig.3.25 and the numbers that come along with it as a reference, take a pipelined datapath before duplicating it. Sketch the result in the AT-plane for various pipeline depths, e.g. for $p = 2, 3, 4, 5, 6, 8, 10$. Compare the results with those of competing architectures that achieve similar performance figures (a) by replicating the isomorphic configuration and (b) by extending the pipeline approach beyond the most efficient depth. How realistic are the various throughput figures when data distribution/recollection is to be implemented using the same technology and cell library as the datapath?

3. ∗∗ Reconsider the third order correlator of fig.3.32a. (a) To boost performance, try to retime and pipeline the isomorphic architecture without prior reversal of the adder chain. How does the circuit so obtained compare with fig.3.32d. Give estimates for datapath resources, cycles per data item, longest path, latency, and control overhead. (b) Next assume your prime concern is area occupation. What architectures qualify?

4. ∗∗ Fig.3.33 shows a viable architecture for a transversal filter. Before this architecture can be coded using an HDL, one must work out the missing details about clocking, register clear, register enable, and multiplexed control signals. Establish a schedule that lists clock cycle by clock cycle what data items the various computational units are supposed to work on, what data items or states that the various registers are supposed to hold, and what logic values the various control signals must assume to marshal the interplay of all those hardware items. Samples are to be processed as specified by fig.3.17a.

5. ∗∗ Arithmetic mean $\bar{x}$ and standard deviation σ are defined as

$$\bar{x} = \frac{1}{N} \sum_{n=1}^{N} x_n \qquad\qquad \sigma^2 = \frac{1}{N-1} \sum_{n=1}^{N} (x_n - \bar{x})^2 \qquad\qquad (3.98)$$

Assume samples x_n arrive sequentially one at a time. More specifically, each clock cycle sees a new w-bit data item appear. Find a dedicated architecture that computes $\bar{x}$ and σ^2 after N clock cycles and where N is some integer power of two, say 32. Definitions in (3.98) suggest one needs to store up to $N - 1$ past values of x. Can you do with less? What mathematical properties do you call on? What is the impact on datapath word width? This is actually an old problem the solution of which has been made popular by early scientific pocket calculators such as the HP-45, for instance. Yet, it nicely shows the difference between a crude and a more elaborate way of organizing a computation.

6. ∗ Most locations in the map of fig.3.28 can be reached from the isomorphic configuration on more than one route. Consider the location where $A = 1/3$ and $T = 1$, for instance. Possible routes include
o (time share → decompose → pipeline) as shown on the map,
o (time share → pipeline → decompose),
o (pipeline → decompose → time share),
o (pipeline → time share → decompose), and
o (decompose → decompose).
Architectures obtained when following distinct routes typically differ. Fig.3.28 indicates only one possible outcome per location and is, therefore, incomplete. Adding the missing routes and datapath configurations is left as a pastime to the reader. Purely out of academic interest, you may want to find out which transforms form commutative pairs.

7. ∗ Fig.3.28 shows a kind of compass that expresses the respective impact of iterative decomposition, pipelining, replication, and time sharing. Include the impact of the associativity transform in a similar way.

8. ∗∗∗ Calculating the convolution of a two-dimensional array with a fixed two-dimensional operator is a frequent problem from image processing. The operator $c_{x,y}$ is moved over the entire original image $p(x, y)$ and centered over one pixel after the other. For each position X, Y the pertaining pixel of the convoluted image $q(x, y)$ is obtained from evaluating the inner product

$$q(X, Y) = \sum_{y=-w}^{+w} \sum_{x=-w}^{+w} c_{x,y}\, p(X + x, Y + y) \tag{3.99}$$

Consider an application where $w = 2$. All pixels that contribute to (3.99) are then confined to a 5-by-5 square with the current location in its center. An uninspired implementation with distributed arithmetics would thus call for a lookup table with 2^{25} entries which is exorbitant. A case study by an FPGA manufacturer explains how it is possible to cut this requirement down to one lookup table of a mere 16 words. Clearly, this remarkable achievement requires a couple of extra adders and flip-flops and is dependent on the particular set of coefficients given below. It combines putting together multiple occurrences of identical weights, splitting of the lookup table, taking advantage of nonoverlapping 1s across two coefficients, and clever usage of the carry input for the handling unit weights. Try to reconstruct the architecture. How close can you come to the manufacturer's result published in [85]?

$c_{x,y}$		x				
		-2	-1	0	1	2
	2	-16	-7	-13	-7	-16
	1	-7	-1	12	-1	-7
y	0	-13	12	160	12	-13
	-1	-7	-1	12	-1	-7
	-2	-16	-7	-13	-7	-16

9. ∗ It has been claimed in section 3.9.2 that dedicated architectures can carry out computations with a small fraction of the energy that a general purpose microprocessor would dissipate. However, in the iPhone example given there, Steve Jobs found the impact on battery run time to be just a factor of 2. Are these two statements in contradiction?

3.11 APPENDIX I: A BRIEF GLOSSARY OF ALGEBRAIC STRUCTURES

Any algebraic structure is defined by a set of elements S and by one or more operations. The nature of the operations involved determines which of the axioms below are satisfied.

Consider a first binary operation $\boxplus$

1. Closure wrt $\boxplus$: if a and b are in S then $a \boxplus b$ is also in S.
2. Associative law wrt $\boxplus$: $(a \boxplus b) \boxplus c = a \boxplus (b \boxplus c)$.
3. Identity element wrt $\boxplus$: There is a unique element e such that $a \boxplus e = e \boxplus a = a$ for any a, (e is often referred to as "zero" element).
4. Inverse element wrt $\boxplus$: For every a in S there is an inverse $-a$ such that $a \boxplus -a = -a \boxplus a = e$.
5. Commutative law wrt $\boxplus$: $a \boxplus b = b \boxplus a$.

Consider a second binary operation $\boxdot$ that always takes precedence over operation $\boxplus$

6. Closure wrt $\boxdot$: if a and b are in S then $a \boxdot b$ is also in S.
7. Associative law wrt $\boxdot$: $(a \boxdot b) \boxdot c = a \boxdot (b \boxdot c)$.
8. Identity element wrt $\boxdot$: There is a unique element i such that $a \boxdot i = i \boxdot a = a$ for any a, (i is often referred to as "one" or "unity" element).
9. Inverse element wrt $\boxdot$: For every a in S there is an inverse a^{-1} such that $a \boxdot a^{-1} = a^{-1} \boxdot a = i$, the only exception is e for which no inverse exists.
10. Commutative law wrt $\boxdot$: $a \boxdot b = b \boxdot a$.
11. Distributive law of $\boxdot$ over $\boxplus$: $a \boxdot (b \boxplus c) = a \boxdot b \boxplus a \boxdot c$ and $(a \boxplus b) \boxdot c = a \boxdot c \boxplus b \boxdot c$.
12. Distributive law of $\boxplus$ over $\boxdot$: $a \boxplus b \boxdot c = (a \boxplus b) \boxdot (a \boxplus c)$ and $a \boxdot b \boxplus c = (a \boxplus c) \boxdot (b \boxplus c)$.
13. Complement: For every a in S there is a complement $\bar{a}$ such that $a \boxplus \bar{a} = i$ and $a \boxdot \bar{a} = e$.

Name of algebraic structure	Opera-tions	1	2	3	4	5	6	7	8	9	10	11	12	13
Set														
Semigroup	$\boxplus$	1	2											
Monoid	$\boxplus$	1	2	3										
Group	$\boxplus$	1	2	3	4									
Abelian or commutative group	$\boxplus$	1	2	3	4	5								
Abelian semigroup	$\boxplus$	1	2			5								
Abelian monoid	$\boxplus$	1	2	3		5								
Ring	$\boxplus\boxdot$	1	2	3	4	5	6	7				11		
Ring with unity	$\boxplus\boxdot$	1	2	3	4	5	6	7	8			11		
Division ring aka skew field	$\boxplus\boxdot$	1	2	3	4	5	6	7	8	9		11		
Field	$\boxplus\boxdot$	1	2	3	4	5	6	7	8	9	10	11		
Commutative ring	$\boxplus\boxdot$	1	2	3	4	5	6	7			10	11		
Commutative ring with unity	$\boxplus\boxdot$	1	2	3	4	5	6	7	8		10	11		
Semiring	$\boxplus\boxdot$	1	2			5	6	7				11		
Commutative semiring	$\boxplus\boxdot$	1	2			5	6	7			10	11		
Boolean algebra	$\boxplus\boxdot$	1	2	3		5	6	7	8		10	11	12	13

Examples with one operation

Let S_{DNA} stand for the set of all possible DNA sequences of non-zero length with characters taken from the alphabet {A,T,C,G}. This set together with the binary operation of string concatenation denoted as $\smile$[69] forms an infinite **semigroup** $(S_{DNA}, \smile)$.

A **monoid** $(S_{DNA} \cup \{\epsilon\}, \smile)$ is obtained iff the empty sequence ϵ is also admitted.

All possible permutations of a given number of elements make up a **group** when combined with binary composition of functions[70] as sole operation. For a practical example, consider all six distinct rearrangements of three elements, shown below, and let us refer to them as set S_3.

1.	2.	3.		1.	2.	3.		1.	2.	3.		1.	2.	3.		1.	2.	3.		1.	2.	3.
↓	↓	↓		↓	↓	↓		↓	↓	↓		↓	↓	↓		↓	↓	↓		↓	↓	↓
1.	2.	3.		2.	3.	1.		3.	1.	2.		3.	2.	1.		1.	3.	2.		2.	1.	3.

This set of permutations along with binary composition $\circ$ forms a finite group $(S_3, \circ)$.

The set of all positive integers $\mathbb{N}^+ = \{1, 2, 3, ...\}$ together with addition as sole operation constitutes an infinite **Abelian semigroup** $(\mathbb{N}^+, +)$.

When supplemented with 0, the above structure turns into an **Abelian monoid** $(\mathbb{N}, +)$.

The set of all integers $\mathbb{Z}$ together with addition forms an infinite **Abelian group** $(\mathbb{Z}, +)$.

Examples with two operations

A **commutative ring with unity** $(\mathbb{Z}, +, \cdot)$ results when multiplication is added as a second operation to the aforementioned Abelian group. The number of elements is obviously infinite.

$(\mathbb{N}, +, \cdot)$, in contrast, is merely a **commutative semiring** because natural numbers have no additive inverses. Yet, in addition to the necessary requirements for a semiring, identity elements with respect to both operations also happen to exist in this example.

The set of all rational numbers $\mathbb{Q}$ together with addition as a first and multiplication as a second operation forms the **field**[71] $(\mathbb{Q}, +, \cdot)$. Other popular fields with infinitely many elements are $(\mathbb{R}, +, \cdot)$ over the real numbers, and $(\mathbb{C}, +, \cdot)$ over the complex numbers.

The set of all quotients $\frac{P(x)}{Q(x)}$ of two polynomials $P(x)$ and $Q(x)$ with real-valued coefficients together with addition and multiplication makes up for yet another infinite field.

For any square matrix M_{nxn} with coefficients taken from a field, there exist identity elements with respect to both addition and multiplication. Any such matrix has an additive inverse, but a multiplicative only if the determinant $||M_{nxn}|| \neq 0$. Also, matrix multiplication is not commutative. The algebraic structure

---

[69] Alternative symbols for the concatenation operator include . (mathematics) and & (computer science).

[70] Binary composition of functions means that two functions are invoked one after the other $(...) \circ f \circ g \equiv g(f(...))$. Also keep in mind that a permutation is just a particular kind of function.

[71] The German term for a field is "(Zahlen)körper", the French "corps", and the Italian "campo".

formed by the set of all square matrices together with matrix addition and matrix multiplication thus is an infinite **ring with unity**.[72]

Any subset of integers $S = \{0, 1, ..., p - 1\}$ forms a field together with addition modulo p and multiplication modulo p iff p is a prime number. Any such **finite field** is called a **Galois field** GF(p). For $p = 5$, for instance, we obtain the GF(5) ($\{0,1,2,3,4\}, + \bmod 5, \cdot \bmod 5$) with the addition and multiplication tables given below.

$\boxplus = + \bmod 5$	0	1	2	3	4
0	0	1	2	3	4
1	1	2	3	4	0
2	2	3	4	0	1
3	3	4	0	1	2
4	4	0	1	2	3

$\boxdot = \cdot \bmod 5$	0	1	2	3	4
0	0	0	0	0	0
1	0	1	2	3	4
2	0	2	4	1	3
3	0	3	1	4	2
4	0	4	3	2	1

The best-known finite field is the GF(2) ($\{0,1\}, \oplus, \wedge$). Observe that the additive inverse in GF(2) of a $-a$ is a itself and that the multiplicative inverse of 1 1^{-1} is 1 while 0 has no multiplicative inverse.

As opposed to this, algebraic structures of type ($\{0,1,...,m - 1\}, + \bmod m, \cdot \bmod m$) where m is not prime merely form finite commutative rings with unity as 0 is not the only element that lacks a multiplicative inverse. In the occurrence of ($\{0,1,2,3,4,5,6,7,8\}, + \bmod 9, \cdot \bmod 9$), this also applies to elements 3 and 6.

Cardinalities of finite fields are not confined to prime numbers. A so-called **extension field** GF(p^n) can be defined for any power p^n provided $2 \leq n \in \mathbb{N}^+$. All polynomials $P(x)$ of degree 0, 1, ..., $n - 1$ with coefficients from GF(p) constitute the set of elements. The first operation is addition modulo $M(x)$ and the second one multiplication modulo $M(x)$ where $M(x)$ an irreducible polynomial[73] of degree n with coefficients from GF(p). GF(3^2), for instance, consists of the nine elements $\{0, 1, 2, x, x + 1, x + 2, 2x, 2x + 1, 2x + 2\}$, the operations are $+ \bmod (x^2 + 1)$ and $\cdot \bmod (x^2 + 1)$ with $M(x) = x^2 + 1$ being an irreducible polynomial.

The factors of 30 together with operations least common multiple (lcm) and greatest common divisor (gcd) constitute a **Boolean algebra** of eight elements ($\{1,2,3,5,6,10,15,30\}$, lcm, gcd). It necessarily follows that taking the complement $\bar{a}$ is tantamount to computing $\frac{30}{a}$ for any a.

(S, $\cup$, $\cap$) is a Boolean algebra with union and intersection as binary operations iff S is a power set $\mathfrak{P}$. Consider a set of three elements $\Omega = \{a,b,c\}$, for instance. The set of all sets that can be composed from these three elements, that is $\{ \emptyset, \{a\}, \{b\}, \{c\}, \{a,b\}, \{a,c\}, \{b,c\}, \{a,b,c\} \}$, is called the power set of Ω and denoted as $\mathfrak{P}(\Omega)$. ($\mathfrak{P}(\Omega)$, $\cup$, $\cap$) then forms a Boolean algebra. Two particular elements of $\mathfrak{P}(\Omega)$, namely the empty set $\emptyset$ and the universal set Ω, act as identity elements e and i for the first and the second operation respectively. Each structure element $x \in \mathfrak{P}(\Omega)$ has a complement $\bar{x} = \Omega - x$.

---

[72] Incidentally note that all concepts of linear algebra (matrices, inverses, determinants, etc.) apply to matrices with coefficients from any field.

[73] An irreducible polynomial cannot be expressed as a product of non-trivial polynomials of lower degree.

The above structure is readily extended to an infinite Boolean algebra when elements and sets are chosen such that $|\mathfrak{P}(\Omega)| = \infty$ which, in turn, is obtained from making $|\Omega| = \infty$. As an example, assume Ω is made up of all DNA sequences of arbitrary length.

The well-known **switching algebra** $(\{0,1\}, \vee, \wedge)$ is a Boolean algebra with just two elements. The complement of an element is its logic inverse and denoted with the negation operator $\neg$.

With no more than six axioms, the class of **semirings** is very broad. It encompasses but is not limited to the embodiments tabulated below:

Constituent	S	$\boxplus$	$\boxdot$
the commutative semiring of natural numbers	$\mathbb{N}$	$+$	$\cdot$
the commutative ring with unity of integers	$\mathbb{Z}$	$+$	$\cdot$
the "ordinary" fields	$\mathbb{Q}$	$+$	$\cdot$
	$\mathbb{R}$	$+$	$\cdot$
	$\mathbb{C}$	$+$	$\cdot$
all Galois fields, e.g.	$\{0,1\}$	$\oplus$	$\wedge$
all other fields, e.g.	$\frac{P(x)}{Q(x)}$	$+$	$\cdot$
the switching algebra	$\{0,1\}$	$\vee$	$\wedge$
other finite Boolean algebras, e.g.	$\{0,1\}$	$\wedge$	$\vee$
or	$\{1,2,3,4,6,12\}$	lcm	gcd
all other Boolean algebras, e.g.	$\mathfrak{P}(\Omega)$	$\cup$	$\cap$
the path algebras	$\{0,1\}$	max	min
	$\mathbb{R} \cup \{\infty\}$	min	$+$
	$\mathbb{R} \cup \{-\infty\}$	max	$+$
	$\{x \in \mathbb{R} \mid 0 \leq x \leq 1\}$	max	$\cdot$
	$\{x \in \mathbb{R} \mid x \geq 0\} \cup \{\infty\}$	max	min
the matrix algebras for every $n \in \mathbb{N}^+$	$\mathsf{M}_{n \times n}$	$+$	$\cdot$

3.12 APPENDIX II: AREA AND DELAY FIGURES OF VLSI SUBFUNCTIONS

with contributions by Beat Muheim

This appendix gives real-world numbers for common subfunctions such as logic gates, bistables, adders, and multipliers. All data refer to commercial high-performance cell libraries in static CMOS technology under typical operating conditions.[74] They extend over multiple process generations and nicely document the benefits of geometric downscaling and other improvements.[75]

Process generations and figures of merit compared

Process generation	M1 min. half-pitch F [nm]	Lithographic square F^2 [nm^2]	Number of metals	Supply [V]
250 nm	320	102 400	5	2.5
180 nm	240	57 600	6	1.8
130 nm	160	25 600	up to 8	1.2
90 nm	120	14 400	up to 9	1.2
65 nm	90	8100	up to 12	1.2
45 nm	70	4900	up to 11	1.1
28 nm	50	2500	up to 11	1.0

Quantity A states the area occupied by the circuitry required to implement the target functionality. Except for individual standard cells, the intercell wiring has been completed and the associated overhead included. Parameter t_{id} denotes the insertion delay.[76]

---

[74] Numbers for individual library cells relate to a variant with simple output strength (1x drive) loaded with four standard inverters (FO4). In the occurrence of bistables, delay figures refer to the non-inverting output.

[75] Do not jump to conclusions from the numbers alone, discontinuities exist. Firstly, changes in the industry have made it impossible to compile the tables from data sets of a single source. A horizontal line indicates where the foundry or the cell library vendor alters. Secondly, the lack of a universal standard for library characterization is a major source of inconsistencies. Thirdly, transistor-level circuits, transistor geometries, and threshold voltages for any given function may shift from one library generation to the next.

[76] Insertion delay reflects the lapse of time that a subcircuit takes to pass on a data item from its input to the output and is defined in section A.6. As a reminder, $t_c = t_{id\,c} = \max(t_{pd\,c})$ for combinational functions, $t_{ff} = t_{id\,ff} = t_{su\,ff} + t_{pd\,ff}$ for flip-flops, and $t_{lc} = t_{id\,lc} = t_{su\,lc} + t_{pd\,lc}$ for latches.

Bistable storage functions

Table 3.13   Selected flip-flops and latches (D flip-flops with no reset are found in pipelines).

D flip-flop with no reset	A [μm^2]	[F^2]	t_{id} [ps]	D scan flip-flop with async. reset	A [μm^2]	[F^2]	t_{id} [ps]
250 nm	97.9	956	n.a.	250 nm	121.0	1181	n.a.
180 nm	50.0	868	321	180 nm	65.6	1139	648
130 nm	25.6	1000	209	130 nm	35.8	1400	435
90 nm	13.3	924	154	90 nm	18.8	1306	286
65 nm	6.8	840	129	65 nm	10.1	1247	246
45 nm	4.1	837	180	45 nm	5.7	1163	273
28 nm[a]	3.8	1520	118	28 nm	4.4	1760	174

[transparent] Latch with async. reset	A [μm^2]	[F^2]	t_{id} [ps]
250 nm	63.4	619	n.a.
180 nm	43.7	759	343
130 nm	23.0	898	200
90 nm	11.8	819	138
65 nm	5.8	716	195
45 nm	2.9	592	194
28 nm		n.a.	

[a] Figures refer to a scanable cell here as the 28 nm library includes no non-scan flip-flops.

Elementary logic functions

Table 3.14   Selected logic gates.

Inverter	A		t_{id}	C_{inp}
	$[\mu m^2]$	$[F^2]$	[ps]	[fF]
250 nm	11.5	113	n.a.	5.6
180 nm	6.2	108	71	3.8
130 nm	3.8	148	55	2.0
90 nm	2.4	167	29	2.0
65 nm	1.1	136	26	1.0
45 nm	0.5	102	29	0.6
28 nm	0.3	120	11	0.6

Full adder	A		t_{id}
	$[\mu m^2]$	$[F^2]$	[ps]
250 nm	144.0	1410	n.a.
180 nm	78.1	1356	341
130 nm	41.0	1600	173
90 nm	25.9	1800	133
65 nm	7.6	938	128
45 nm	4.3	878	137
28 nm	2.6	1040	46

2-input NAND	A		t_{id}
	$[\mu m^2]$	$[F^2]$	[ps]
250 nm	17.3	169	n.a.
180 nm	9.4	162	75
130 nm	5.1	200	65
90 nm	3.1	215	31
65 nm	1.4	173	39
45 nm	0.7	143	36
28 nm	0.5	200	14

2-input NOR	A		t_{id}
	$[\mu m^2]$	$[F^2]$	[ps]
250 nm	17.3	169	n.a.
180 nm	9.4	162	129
130 nm	5.1	200	92
90 nm	3.1	215	52
65 nm	1.4	173	61
45 nm	0.7	143	51
28 nm	0.5	200	21

2-input XOR	A		t_{id}
	$[\mu m^2]$	$[F^2]$	[ps]
250 nm	51.8	506	n.a.
180 nm	28.1	488	179
130 nm	14.1	550	146
90 nm	7.8	542	68
65 nm	3.6	444	93
45 nm	1.8	367	70
28 nm	1.1	440	21

2-to-1 MUX	A		t_{id}
	$[\mu m^2]$	$[F^2]$	[ps]
250 nm	46.1	450	n.a.
180 nm	21.9	379	164
130 nm	11.5	450	121
90 nm	7.1	493	79
65 nm	3.6	444	76
45 nm	2.0	408	71
28 nm	1.5	600	25

Arithmetic functions Tables 3.15 and 3.16 refer to unpipelined adders and multipliers respectively. They include the approximate area for intercell wiring as estimated by Synopsys DesignCompiler. Synthesis results have been obtained by instantiating the appropriate DesignWare component followed by optimization with no timing constraint.

Table 3.15 **2's complement adders with carry-in and carry-out. Note that delay data do not grow smoothly throughout but exhibit steps.**

ripple-carry adder DW01_add		A		t_{id}	carry-lookahead DW01_addsub		A		t_{id}
		$[\mu m^2]$	$[F^2]$	$[ps]$			$[\mu m^2]$	$[F^2]$	$[ps]$
90 nm	8 bit	207	14 400	529	90 nm	8 bit	267	18 600	578
90 nm	16 bit	414	28 800	1016	90 nm	16 bit	627	43 600	629
90 nm	24 bit	621	43 100	1512	90 nm	24 bit	880	61 100	933
90 nm	32 bit	828	57 500	2008	90 nm	32 bit	1231	85 500	933

ripple-carry adder DW01_add		A		t_{id}	carry-lookahead DW01_addsub		A		t_{id}
		$[\mu m^2]$	$[F^2]$	$[ps]$			$[\mu m^2]$	$[F^2]$	$[ps]$
28 nm	8 bit	20.9	8 360	296	28 nm	8 bit	40.1	16 100	226
28 nm	16 bit	41.8	16 700	580	28 nm	16 bit	93.2	37 300	270
28 nm	24 bit	62.7	25 100	865	28 nm	24 bit	131.2	52 500	388
28 nm	32 bit	83.6	33 400	1150	28 nm	32 bit	185.2	74 100	377

Table 3.16 **2's complement multipliers.**

carry-save multiplier DW02_mult		A		t_{id}
		$[\mu m^2]$	$[F^2]$	$[ps]$
90 nm	8 bit x 8 bit	1 780	123 400	1315
90 nm	16 bit x 16 bit	7 180	498 300	2480
90 nm	24 bit x 24 bit	16 290	1 131 100	3750
90 nm	32 bit x 32 bit	29 230	2 029 600	4680

carry-save multiplier DW02_mult		A		t_{id}
		$[\mu m^2]$	$[F^2]$	$[ps]$
28 nm	8 bit x 8 bit	214	85 800	665
28 nm	16 bit x 16 bit	817	326 600	1200
28 nm	24 bit x 24 bit	1834	733 600	1740
28 nm	32 bit x 32 bit	3252	1 301 000	2080

CIRCUIT MODELING WITH HARDWARE DESCRIPTION LANGUAGES

4

4.1 MOTIVATION AND BACKGROUND

4.1.1 WHY HARDWARE SYNTHESIS?

VLSI designers find themselves in a difficult situation. On the one hand, buyers ask for microelectronic products that integrate more and more functions on a single chip. Following Moore's law, fabrication technology has always supported this aspiration by quadrupling the achievable circuit complexity every three years or so. Market pressure, on the other hand, vetoes a proportional dilation of product development times. Worse than this, time to market is even expected to shrink. As a consequence, design productivity must constantly improve. Hardware description languages (HDL) and design automation come to the rescue in four ways:

- Exonerate designers from low-level details by moving design entry to more abstract levels.
- Allow designers to focus on functionality as automatic synthesis tools generate circuits.
- Facilitate design reuse by capturing a circuit description in a parametrized technology- and platform-independent form (as opposed to schematic diagrams, for instance).
- Making functional verification more efficient by supporting stimuli generation, automatic response checking, assertion-based verification, and related techniques.

The transition from structural to physical is largely automated in digital VLSI design today. The transition from behavioral to structural has not yet reached the same maturity, but HDL synthesis is routinely used for turning register transfer level (RTL) descriptions into gate-level networks that are then processed further with the aid of cell-based design automation software.

4.1.2 AGENDA

After an overview on the HDLs currently available, section 4.1.6 introduces the key concepts that set HDLs apart from software languages. The text then bifurcates. Section 4.2 discusses VHDL language constructs and explains how they relate to the underlying concepts while section 4.3 does the same for SystemVerilog. Readers are free to chose one or the other depending on their needs. Both threads reconverge in section 4.4 that focusses on circuit synthesis. How to organize and code simulation testbenches will have to wait until section 5.5. Appendix 4.7 provides side-by-side comparisons of the two HDLs and lists textbooks and syntax references. Appendix 4.8 is about various VHDL language extensions. Last but not least, the book's companion website includes more substantial code examples than those printed in the text.

4.1.3 ALTERNATIVES FOR MODELING DIGITAL HARDWARE

SystemVerilog shares most key concepts with **VHDL**. To a somewhat lesser extent this also applies to **Verilog** the precursor of SystemVerilog, making the differences between RTL synthesis models captured using those three languages largely a matter of syntax and coding style. Beyond that, SystemVerilog offers a better support for functional verification as it has inherited various mechanisms from specialized languages. It is, therefore, sometimes dubbed the first hardware description and verification language (HDVL). That said, VHDL and SystemVerilog have their pros and cons, see below and have a look at table 4.12 for technical details.

Criterion	VHDL	Verilog	SystemVerilog
Synthesis support	yes	yes	growing
Parametrization & abstract modeling	good	poor	good
Type checking & scoping rules	strong	none	loose
Deterministic event queue mechanism	yes	not really	not really
Modeling of electric phenomena	9-valued	4-valued	4-valued
High-level verification support	limited	poor	excellent

Many companies currently use VHDL for synthesis and SystemVerilog for system-level verification. The future will tell whether SystemVerilog will one day supersede Verilog <u>and</u> VHDL, reconcile their user communities, and so bring an unfortunate schism to an end.

SystemC is not really a language but a C++ class library that includes the necessary extensions for hardware modeling plus a simulation kernel. This means that a model written in SystemC can be executed with only a standard GNU compiler. SystemC is intended for use in software/hardware co-design and co-simulation, but does not qualify for gate-level simulation and timing verification because of its inaptitude for modeling detailed timing behavior.[1]

4.1.4 THE GENESIS OF VHDL AND SYSTEMVERILOG

Providing spare parts over many years for industrial products that include ASICs and other custom-designed state-of-the-art electronic components proves painful as technology evolves and as companies restructure. In search of a standard format for documenting digital ICs and for exchanging design data other than layout polygons, the US Department of Defense (DoD) in 1983 commissioned IBM, Intermetrics and Texas Instruments to define an HDL. Ada was taken as a starting point. As the project

---

[1] While SystemC adds clocking information to C++ functions, it does not support any timing finer than one clock cycle. Synthesis path is via translation to RTL VHDL or [System]Verilog by way of automatic allocation, scheduling and binding, which makes RTL expertise unavoidable for VLSI designers.

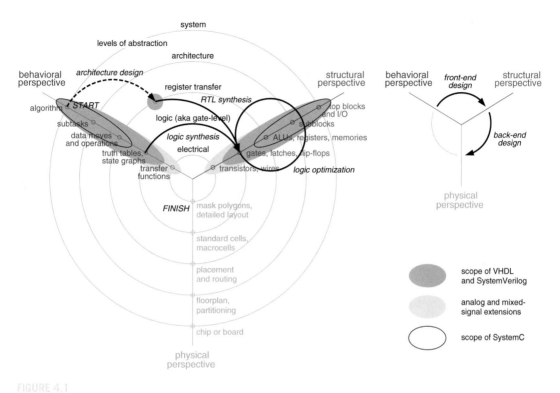

FIGURE 4.1

HDLs shown in the Y-chart.

had originated from DoD's Very High Speed Integrated Circuits (VHSIC) program, the new language was given the acronym VHDL.

After a number of revisions and after military restrictions had been lifted, the proposal was accepted as **IEEE 1076**[-87] standard in 1987. IEEE bylaws require any standard to be revised on a periodical basis and a first re-examination led to the IEEE 1076-93 revision which found wide adoption. Approximately at the same time IEEE also passed the IEEE 1164 standard, a nine-valued logic system used in conjunction with VHDL. Following a series of relatively minor improvements and clarifications over the subsequent years, the latest release IEEE 1076-2008 has brought about a wider set of modifications as documented in [96].

The first software tools built around HDLs were compilers and simulators. Around 1984, Gateway Design Automation had in fact developed a logic simulator along with a proprietary language for capturing the circuits to be simulated. That language was Verilog, and its syntax was very much inspired by the C programming language. It became more prevalent in the early 1990s when Cadence acquired Gateway, made the language an open standard, and ultimately pushed it through the IEEE standardization process in 1995.

Only later did people want to come up with automatic synthesis tools that would accept behavioral specifications stated using an HDL and churn out gate-level netlists implementing them. Synopsys pioneered synthesis technology, supporting both VHDL and Verilog for design entry. As will become evident later in this chapter, current language versions continue to suffer from the fact that synthesis issues were ignored when the languages were originally defined.

Verilog had not much to offer in terms of abstract data types, model parametrization, and functional verification. As these limitations became obvious, the Accellera industry consortium set out to create a successor, incorporating useful constructs from VHDL, powerful assertions for verification purposes, plus other features that have proven highly effective for writing advanced testbenches, notably from languages such as OpenVera, PSL, and Superlog.[2] The result essentially collects three languages under the SystemVerilog brand and was accepted as **IEEE 1800**[-05] standard. On the other hand, the consortium chose not to clean up the messier parts of the Verilog language for reasons of backward code compatibility.[3] SystemVerilog and Verilog officially co-existed until 2009 when Verilog got absorbed into the IEEE 1800-09 standard that brought further enhancements, mostly for verification. [98] is the 1315-page language reference manual for the current IEEE 1800-12 SystemVerilog release.

4.1.5 WHY BOTHER LEARNING HARDWARE DESCRIPTION LANGUAGES?

As shown in fig.4.1, VHDL and SystemVerilog encompass behavioral and structural views but not physical ones. The levels of abstraction covered range from purely algorithmic descriptions down to logic design.[4] That said, it seems a tempting idea to view HDLs as nothing more than intermediate formats for exchanging data between electronic system-level (ESL) tools and VLSI CAE/CAD suites.

"Why not skip learning hardware description languages and have electronic system-level tools generate HDL code from specifications automatically?"

There indeed exists a variety of software packages that deal with systems at levels of abstraction well above those addressed during actual VLSI design, see fig.1.12. Many of them deal with transformatorial systems as found in signal processing and telecommunications that typically get pieced together from high-level function blocks such as oscillators, modulators, filters, phase shifters, delay lines, frequency dividers, phase locked loops (PLL), synchronizers, etc.

Other high-level tools model system behavior on the basis of condition/action pairs captured as state graphs, state charts, Petri nets, and the like. The graphical design capture and animation facilities provided are intuitive and very helpful for defining, checking, debugging, and improving the functionality of reactive systems.

---

[2] Dynamic data structures, object-oriented programming (classes), constrained random stimuli generation, and functional coverage metering, plus a new programming language interface (DPI).

[3] Such as the byzantine syntax, the overly indulgent typing scheme, and the perplexing scheduler. You may want to see [97] for a collection of language misfeatures that tend to breed bugs.

[4] The inability to model time-continuous phenomena and in terms of electrical quantities confines both languages to digital circuits. Extensions for analog and mixed-signal circuits have been standardized as VHDL-AMS and Verilog-AMS respectively. The basic ideas will be briefly reviewed in section 4.8.2.

More EDA packages are geared towards some field of application such as the analysis of communication channels, source and channel coders, data transfer networks, image and video processing, pattern recognition, optimization of instruction set computers, and so forth.

The common theme is that EDA tools working above the HDL level tend to focus on fairly specific problem classes. Though probably unavoidable, this fragmentation is unfortunate in the context of VLSI design where building blocks of transformatorial and reactive nature coexist on the same chip.

ESL tools typically include code generators that produce software code for popular microcomputers or DSPs. Many of them are also capable of producing HDL code. All too often, this code is or was nothing else than a translation of the processor code, however. While this may have allowed for simulation under certain conditions, it is clearly not acceptable for synthesis, and the results so obtained remained unsatisfactory.

Genuine high-level synthesis tools, on the other hand, typically work on the basis of resource allocation, scheduling, and binding.[5] Finding a good overall solution implies exploring a vast solution space that involves both algorithmic and architectural issues. Yet, today's tool suites are often restricted to a few predefined hardware patterns with limited optimization capabilities.

As a result, it will take a couple of years before true system-level synthesis pervades the industrial production environment [99]. Developing a circuit model actually goes more like this:

A behavioral model established early on serves to validate the intended functionality. Software languages, MATLAB, and other general purpose tools qualify as little attention needs to be paid to architectural issues and to limitations of synthesis technology at this point.

Later in the design cycle, a more detailed model is written — mostly at the RTL level — as a starting point for synthesis. Simulation here serves to make sure this complex piece of HDL code is correct before it gets synthesized into a gate-level netlist.

A third simulation round is then carried out on the netlist so obtained, and a fourth round with the same netlist following place and route (P&R), layout extraction, and back-annotation. In the end, four circuit models have been used to capture the identical functionality at distinct levels of detail as summarized in table 4.1.

---

[5] The tools essentially accept an algorithmic description in C and a series of pragmas or other human input that outlines the hardware resources to be made available. The output is RTL code that describes a VLIW ASIP operating under control of either a stored microprogram or a hardwired finite state machine. Final implementation is with FPGAs or as a cell-based ASIC. Catapult C by Mentor Graphics is a commercial example.

Table 4.1   **Circuit models and HDLs encountered during a typical VLSI design flow.**

Design stage & model	Main purpose	Level of abstraction	Timing	Predominant languages
1. Algorithmic model	system-level simulation	behavioral	none	C, Matlab
			tentative	VHDL, SysVer
2. RTL model	simulation	register transfer	optional fake delays	VHDL, SysVer
	synthesis		constraints in Tcl	
3. Post-synthesis netlist	simulation, timing analysis place & route	gate level	estimated with wire load models	Verilog, (VHDL&VITAL)
4. Post-layout netlist	simulation, timing analysis, sign-off	gate level	extracted from layout and back-annotated	Verilog, (VHDL&VITAL)

Even where ESL tools do generate a useful HDL (sub)model, manual interventions in the source code may be required to parametrize, adapt, interface, or optimize the circuit or the model. Finally, HDLs are indispensable for modeling library cells and virtual components.

Observation 4.1. *For the foreseeable future, hardware description languages such as VHDL and SystemVerilog are bound to remain prominent hubs for all VLSI design activities.*

4.1.6 A FIRST LOOK AT VHDL AND SYSTEMVERILOG

In sections 4.2 and 4.3, we shall study the VHDL and SystemVerilog languages respectively by asking ourselves

"What features are required to model digital electronic circuits for simulation and synthesis?"

In anticipation of our findings, we will identify a multitude of needs that can be collected into the six broad categories listed below. The language concepts addressing those needs will be introduced accordingly. The discussion of basic concepts that HDLs share with modern programming languages has been postponed to the end of our presentations based on the assumption that readers have had some exposure to software engineering.

Observation 4.2. *In a nutshell, HDLs can be characterized as follows:*

			VHDL	SystemVerilog
HDL	=	*structured programming language*	4.2.6	4.3.6
	+	*circuit hierarchy and connectivity*	4.2.1	4.3.1
	+	*interacting concurrent processes*	4.2.2	4.3.2
	+	*a discrete replacement for electrical signals*	4.2.3	4.3.3
	+	*an event-driven scheme of execution*	4.2.4	4.3.4
	+	*model parametrization facilities*	4.2.5	4.3.5
	+	*verification aids*		chapter 5

A few more remarks are due before we start with our discussions.

- The entire chapter puts emphasis on the concepts behind HDLs and on applying them to hardware modeling. There will be no comprehensive exposure to syntax or grammar. To become proficient in writing circuit models of your own, you will need a more detailed documentation on your preferred HDL. See table 4.13 for an annotated bibliography.

- A look at table 4.10 will facilitate the understanding of the upcoming code examples.

- More substantial VHDL and SystemVerilog code examples have been collected on the book's companion website to show the usage of important language constructs in context. Newcomers are strongly encouraged to go through the listings there as well to develop an understanding of HDL coding styles and options.

- This text discusses two HDLs as defined by various international standards. Commercial EDA tools occasionally deviate in terminology and implementation. Some recent additions may not be supported by all tools.

- Our discussion of HDL concepts is accompanied by a series of illustrations that include figs.4.2, 4.3, 4.6, 4.8, and that culminates with the full picture in fig.4.12. Refer back to these synoptical drawings when in danger of getting lost in minor details.

- Throughout this text, we will use "process" as a broad and generic term for any kind of concurrent activity, i.e. as a shorthand for "concurrent process", "parallel process", or "thread of execution".[6]

- Please observe a linguistic ambiguity in the context of hardware modeling:

Meaning of "sequential" with reference to	Synonym	Antonyms
○ program execution during simulation	step-by-step	concurrent, parallel
○ nature of circuit being modeled	memorizing	combinational, memoryless

---

[6] And as a translation of the German "nebenläufiger Prozess".

4.2 KEY CONCEPTS AND CONSTRUCTS OF VHDL

4.2.1 CIRCUIT HIERARCHY AND CONNECTIVITY

The need for supporting modularity and hierarchical composition

Consider a motherboard from a personal computer, for instance. At the top level, you will discern a CPU, a graphics processor, a ROM or two, and several memory modules composed of multiple RAM chips. Then there are all sorts of peripheral circuits plus a variety of passive components. If you could look into the ICs, you would find datapaths, controllers, storage arrays, and the like. And each such block consists of thousands of logic gates and bistables, that are in turn assembled from the most elementary devices, from transistors.

Electronic systems are organized into multiple layers of hierarchy because it is entirely impractical to specify, understand, design, fabricate, test, and document circuits with millions of gates as flat collections of transistors. Instead, larger entities are hierarchically composed from subordinate entities that are interconnected with the aid of busses and individual wires. Abstraction, modularity, and repetition further help to arrive at manageable circuit descriptions. Any HDL that wants itself useful must support these techniques, see fig.4.2 for a first impression.

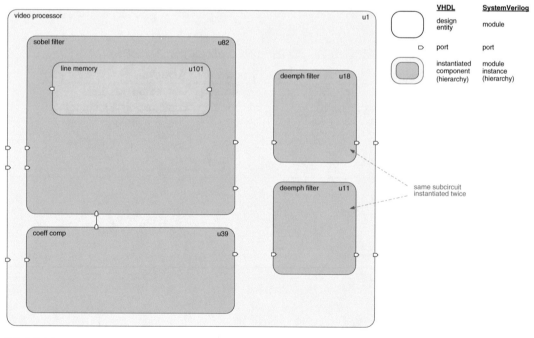

FIGURE 4.2

HDL capabilities (I): Hierarchical composition ...

Design entity

The `entity` is the basic building block of a VHDL design. Its primary purpose is to encapsulate the data, functionality and timing of a circuit or subcircuit. VHDL makes a strict distinction between a (sub)circuit's external and internal views. The interface is specified in the entity declaration whereas the details of its implementation are captured in a language element called architecture body, or architecture for short.

Entity declaration and ports

The entity declaration specifies the external interface of an `entity`. Its most important part is the **port clause** that enumerates those circuit nodes that are visible from outside. Put differently, each `signal` that is part of the `port` clause corresponds to a connector on the icon of that subcircuit as illustrated in fig.4.13. An example of an entity declaration is given below.

```
-- entity declaration
entity lfsr4 is
   port (
      Clk_CI :  in std_logic;
      Rst_RBI : in std_logic; -- reset is active low
      Ena_SI :  in std_logic;
      Oup_DO :  out std_logic );
end lfsr4;
```

Hint: VHDL is case-insensitive, e.g. `clk_ci` = `CLK_CI`. *An exception are the so-called "extended identifiers" written between two backslashes, e.g.* `\clk_ci\` ≠ `\CLK_CI\`.

Hint: Naming a `signal` *or a* `port` `In` *or* `Out` *is all too tempting, yet these are reserved words in VHDL. We recommend the use of* `Inp` *and* `Oup` *instead.*

Architecture body, structural view

An architecture body is the place where the internal technicalities of a design entity are being described. Any `entity` is permitted to contain **instances** of other design entities, thereby creating hierarchy. A code example follows in listing 4.1. Although you are probably not yet in a position to understand everything, it should become clear that the circuit is composed of five logic gates and four flip-flops. As an exercise, draw a schematic diagram for the circuit.

LISTING 4.1

Structural view of a linear feedback shift register (LFSR) of length 4.

```
-- architecture body
architecture structural of lfsr4 is

   -- declare the components to be used
   component GTECH_FD2     -- D-type flip-flop with reset
      port (
```

```
        D, CP, CD : in  std_logic;
        Q : out std_logic );
   end component;
   component GTECH_FD4     -- D-type flip-flop with set
      port (
        D, CP, SD : in  std_logic;
        Q : out std_logic );
   end component;
   component GTECH_MUX2    -- 2-input multiplexer
      port (
        A, B, S : in std_logic;
        Z : out std_logic );
   end component;
   component GTECH_XOR2    -- 2-input XOR gate
      port (
        A, B: in std_logic;
        Z : out std_logic );
   end component;

   -- declare a signal for each inner node
   signal State_DP : std_logic_vector(1 to 4);
   signal n11, n21, n31, n41, n42 : std_logic;

begin

   -- instantiate components and connect them by listing port maps
   u10 : GTECH_FD2
      port map( D => n11, CP => Clk_CI, CD => Rst_RBI, Q => State_DP(1) );
   u20 : GTECH_FD2
      port map( D => n21, CP => Clk_CI, CD => Rst_RBI, Q => State_DP(2) );
   u30 : GTECH_FD2
      port map( D => n31, CP => Clk_CI, CD => Rst_RBI, Q => State_DP(3) );
   u40 : GTECH_FD4
      port map( D => n41, CP => Clk_CI, SD => Rst_RBI, Q => State_DP(4) );
   u11 : GTECH_MUX2
      port map( A => State_DP(1), B => n42, S => Ena_SI, Z => n11 );
   u21 : GTECH_MUX2
      port map( A => State_DP(2), B => State_DP(1), S => Ena_SI, Z => n21 );
   u31 : GTECH_MUX2
      port map( A => State_DP(3), B => State_DP(2), S => Ena_SI, Z => n31 );
   u41 : GTECH_MUX2
      port map( A => State_DP(4), B => State_DP(3), S => Ena_SI, Z => n41 );
```

```
      u42 : GTECH_XOR2
         port map( A => State_DP(3), B => State_DP(4), Z => n42 );

      -- connect state bit of rightmost flip-flop to output port
      Oup_DO <= State_DP(4);

end structural;
```

How to compose a circuit from components

How do you proceed when asked to fit a circuit board with components? You think of the exact name of a part required, go and fetch a copy of it, and solder the terminals of that one copy in a well-defined manner to metal pads interconnected by narrow lines on the board. The **component instantiation statement** of VHDL does exactly this, albeit with virtual `components` and `signals` instead of physical parts and wires.

In the above code example, nine components get instantiated following the keyword `begin`. As multiple copies of the same component must be told apart, each instance is assigned a unique identifier; `u10`, `u20`, ... , `u42`, in the occurrence. Further observe that the operator `=>` in the **port map** clauses does not indicate any assignment. Rather, it is an association between two `signals` that stands for an electrical connection made between the instance terminal to its left and a node in the superordinate circuit the name of which is indicated to the right.

The first four statements in the `lfsr4 architecture` body are **component declarations**. VHDL requires that the names and ports of all `component` models be known prior to instantiation.[7] Signals running back and forth between instances must be declared as well. Those connecting to the outside world are automatically known from the `port` clause and need not be declared a second time. Inner nodes, in contrast, must be defined in a series of **signal declarations** just before the keyword `begin`. More on `signals` is to follow shortly.

---

[7] There are essentially two ways for declaring a subcircuit model, yet the difference is a subtlety that can be skipped for a first reading. Assume you are describing a circuit by way of hierarchical composition in a **top-down** fashion, that is, beginning with the top-level design entity. In doing so, you must anticipate what subcircuits you will need. All that is really required for the moment are the complete port lists of those subcircuits-to-be that you are going to instantiate. Their implementations can wait until work proceeds to the next lower level of hierarchy. Declaring the external interfaces of such prospective subcircuits locally, that is within the current `architecture` body, is exactly what the `component` declaration statement is intended for.

Now consider the opposite **bottom-up** approach. You begin with the lowest-level subcircuits by capturing the interface in an `entity` declaration and the implementation in an `architecture` body for each subcircuit. These models are then instantiated at the next-higher level of the design hierarchy, and so on. Instantiation always refers to an existing design entity which explains why this type of instantiation is said to be direct. No `component` declarations are required in this case, yet the `component` instantiation statement are complemented with the extra keyword `entity` and with an optional `architecture` identifier as follows.

```
u6756 : entity lfsr4 (behavioral)
   port map( Clk_CI => n18, Rst_RBI => n8, Ena_SI => n199, Oup_DO => n4 );
```

For **direct instantiation** to work, design entities must be made visible with a `use work.all` clause. `use` clauses and `configuration` specification statements are to be introduced in sections 4.2.6 and 4.2.5 respectively.

Observation 4.3. *VHDL can describe the hierarchical composition of a digital circuit by instantiating* `components` *or* `entities` *and by interconnecting them with the aid of* `signals`.

A model that describes a circuit as a bunch of interconnected components is qualified as **structural**. It essentially holds the same information as the circuit netlist does. Manually coding `architecture` bodies in this way is not particularly attractive. Indeed, most structural models are obtained from register-transfer-level (RTL) models by automatic synthesis.

4.2.2 INTERACTING CONCURRENT PROCESSES

The need for modeling concurrent activities

While we have learned how to capture a circuit's hierarchical composition, our model remains devoid of life up to this point as we have no means for expressing circuit behavior, precluding both simulation and synthesis. So there must be more to VHDL.

The most salient feature of any electronic system is the concurrent operation of its subcircuits, just think of all those ICs on a typical circuit board, or of the many thousands of logic gates and storage cells within each such chip. This inherent parallelism contrasts sharply with the line-by-line execution of program code on a computer. Another innate trait is the extensive communication that permanently takes place between subcircuits and that is physically manifest in the multitude of wires that run across

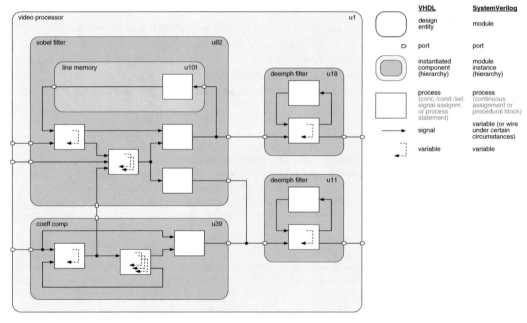

FIGURE 4.3

HDL capabilities (II): ... plus behavior emulated by concurrent processes ...

chips and boards. This is simply because there can be no cooperation between subcircuits without on-going exchange of data.

Now assume you wanted to write a software model that imitates the behavior of a substantial circuit using some traditional programming language such as Pascal or C. You would soon get frustrated because of the absence of constructs and mechanisms to handle simultaneous operation and interaction. Hardware description languages extend the expressive power of software languages by supporting concurrent processes and means for exchanging information between them, see fig.4.3 for a first idea. This bears some resemblance to real-time languages and, in fact, Ada has been taken as a starting point for defining VHDL.

Signal

The vehicle for exchanging dynamic information between concurrent processes is the `signal`, there exists no other mechanism in VHDL for this purpose.[8] Any `signal` declared within an `architecture` body is strictly confined to that body and remains inaccessible from outside. A `signal` declaration must specify the name and the data type of the `signal`; an optional argument can be used to assign an initial value.[9]

Example of a `signal` declaration	`signal ThisMonth_D : month;`
Second example	`signal Error_D, Actual_D, Wanted_D : integer := 0;`

> *Hint: HDL code is easier to read when* `signals` *can be told from* `variables` *by their visual appearance. As part of an elaborate naming convention to be presented in section 6.7, we make it a habit to append an underscore followed by an informative suffix of a few upper-case letters to* `signals` *(but not to* `variables`*).*[10]

How to describe combinational logic behaviorally

Many arithmetic and logic computations are straightforward and can be expressed in one instruction with no need for branching. These are best captured in a **concurrent signal assignment**, the most simple VHDL construct for defining the value of a `signal`.

Example with no operation	`ThisMonth_D <= AUGUST;`
Example with logic operations	`Oup_D <= Aa_D xor (Bb_D and not Cc_D);`
Example with an arithmetic operation	`Error_D <= Actual_D - Wanted_D;`

A concurrent signal assignment is nothing else than a process in a single statement. Whenever an operand on the right-hand side changes value, the expression is (re-)evaluated and the `signal` on the left-hand side gets updated accordingly. The assignment operator for `signals` is `<=` which choice is somewhat unfortunate because the same symbol also serves as relational operator (in lieu of $\leq$).

---

[8] "Dynamic" means that the data are free to evolve over time, i.e. to change value as simulation progresses. The purpose of protected shared variables is a totally different one, see section 4.8.1.

[9] Why this is inadequate for modeling a hardware reset will be explained in observation 4.17.

[10] Attentive readers may notice that certain listings and schematic diagrams in this text feature identifiers that do not adhere to this scheme, e.g. CP instead of Cp_CI. This applies to standard cell connectors where the naming gets decided by the library vendor rather than by VLSI designers and HDL code writers.

Situations abound where concurrent signal assignments do not suffice to express the desired functionality, e.g. because conditional execution is required. This is where two more elaborate constructs enter the picture. The **selected signal assignment** is reminiscent of a multiplexer (MUX) or data switch: One out of multiple possible values gets assigned to a signal under control of a selecting expression.

```
with ThisMonth_D select
    ThisQuarter_D <= Q1ST when JANUARY | FEBRUARY | MARCH,
                     Q2ND when APRIL | MAY | JUNE,
                     Q3RD when JULY | AUGUST | SEPTEMBER,
                     Q4TH when others;
```

The **conditional signal assignment** is very similar to the selected signal assignment but a bit more liberal in formulating the branching condition. An example follows.

```
Spring_D <= true when (ThisMonth_D=MARCH and ThisDay_D>=21) or
                      ThisMonth_D=APRIL or ThisMonth_D=MAY or
                      (ThisMonth_D=JUNE and ThisDay_D<=20)
                      else false;
```

Where this is still insufficient to capture a desired input-to-output mapping in an adequate way, designers can recur to the **process statement**, a more powerful — but also rather tricky — construct for expressing a concurrent process. What sets it apart from the three signal assignment statements discussed above are essentially

- Its capability to update two or more `signals` at a time,
- The fact that the instructions for doing so are captured in a sequence of statements that are going to be executed one after the other,
- The liberty to make use of `variables` for temporary storage, plus
- A more detailed control over the conditions for activating the process.[11]

The `process` statement is best summed up as being concurrent outside and sequential inside.[12] `Process` statements cannot be nested but may call subprograms. The example given next is semantically and functionally identical to the conditional signal assignment given before.[13]

---

[11] Several of these items will be clarified in section 4.2.4.

[12] "Sequential" here refers to code execution during simulation, not to the nature of the circuit being modeled.

[13] The (`all`) term is an enhancement introduced with the VHDL-2008 standard revision. It saves programmers from having to enumerate all signals that appear on the right-hand side of assignments in the body of the process statement. For those who must work with earlier language revisions, the VHDL mode of the Emacs editor offers an "auto-update sensitivity list" function to help with this.

LISTING 4.2

Code example of a combinational operation captured in a process statement.

```
memless1: process (all)
begin
   Spring_D <= false;    -- execution begins here
   if ThisMonth_D=MARCH and ThisDay_D>=21 then Spring_D <= true; end if;
   if ThisMonth_D=APRIL                   then Spring_D <= true; end if;
   if ThisMonth_D=MAY                     then Spring_D <= true; end if;
   if ThisMonth_D=JUNE  and ThisDay_D<=20 then Spring_D <= true; end if;
end process memless1;    -- process suspends here
```

You can think of `process` statement used to model a combinational (sub)circuit as a container for a small program that generates the truth table. Interestingly, a `process` statement may as well model sequential circuit behavior. This depends on how the code is organized and the criteria will be detailed later, in observation 4.15. In the above code example, the identifier `memless1` serves to express the designers's intention. Note, however, this is just an optional free-choice label that has no impact on simulation or synthesis whatsoever.

How to describe a register behaviorally

As there exists no particular VHDL language element for modeling storage functions such as flip-flops, latches, registers, and the like, this must be achieved by organizing the `process` statement in a different way, see listing 4.3. Note that the present state of the process is kept from one activation to the next in a `signal`-type vector named `State_DP`.

LISTING 4.3

Code example for an edge-triggered register that features an asynchronous reset, a synchronous load, and an enable. Actual designs are unlikely to combine all three mechanisms in a single register, so a subset of the clauses shown will most often do.

```
p_memzing : process (Clk_C,Rst_RB)
begin
   -- activities triggered by asynchronous reset
   if Rst_RB='0' then
      State_DP <= (others => '0');    -- shorthand for all bits zero
   -- activities triggered by rising edge of clock
   elsif Clk_C'event and Clk_C='1' then
      -- when synchronous load is asserted
      if Lod_S='1' then
         State_DP <= (others => '1');    -- shorthand for all bits one
      -- else assume new value iff enable is asserted
      elsif Ena_S='1' then
         State_DP <= State_DN;    -- admit next state into state register
      end if;
   end if;
end process p_memzing;
```

Architecture body, behavioral view

Most `architecture` bodies include a collection of concurrent processes that together make up the `entity`'s overall functionality. Such models are called **behavioral** because they specify how a design `entity` is to react in response to changing inputs. Potential reactions include the updating of outputs, the updating of the `entity`'s current state, the checking of compliance with some predefined timing conditions, or simply ignoring the new input.

Listing 4.4 shows a behavioral `architecture` body for the LFSR circuit of listing 4.1. Taking fig.4.3 as a pattern, make a small drawing that illustrates the processes and the `signals` being exchanged. Find out what hardware item each process stands for and compare the drawing to that established earlier. What liberties do you have in coming up with a schematic diagram?

LISTING 4.4

Behavioral view of a linear feedback shift register (LFSR) of length 4.

```
-- architecture body
architecture behavioral of lfsr4 is
   signal State_DP, State_DN : std_logic_vector(1 to 4);
   -- for present and next state respectively
begin

   -- computation of next state using concatenation of bits
   State_DN <= (State_DP(3) xor State_DP(4)) & State_DP(1 to 3);

   -- updating of state
   process (Clk_CI,Rst_RBI)
   begin
      -- activities triggered by asynchronous reset
      if Rst_RBI='0' then
         State_DP <= "0001";
      -- activities triggered by rising edge of clock
      elsif Clk_CI'event and Clk_CI='1' then
         if Ena_SI='1' then
            State_DP <= State_DN; -- admit next state into state register
         end if;
      end if;
   end process;

   -- updating of output
   Oup_DO <= State_DP(4);

end behavioral;
```

Observation 4.4. *In VHDL, the behavior of a digital circuit typically gets described by a collection of concurrent processes that execute simultaneously, that communicate via* signals, *and where each such process represents some subfunction.*

Hardware modeling styles compared

Except for physical descriptions, VHDL supports various circuit modeling styles.

A **procedural model** essentially describes functionality in a sequence of steps much as a piece of conventional software code. A circuit is captured in one process statement and its behavior gets implemented with the aid of sequential statements there.[14]

A **dataflow model** describes the behavior as a collection of concurrent signal assignments that get executed under the coordination of the signals exchanged.

A **structural model** describes the composition of a circuit by way of component instantiation statements along with the interconnections in between. It is equivalent to a netlist.

Listing 4.5 juxtaposes three architecture bodies, each coded from a different perspective while fig.4.5 illustrates their differences and commonalities.[15] Make sure you understand the profound difference between the procedural and dataflow models in spite of their apparent similarity.

Observation 4.5. *VHDL allows for procedural, dataflow, and structural modeling styles to be freely combined, and circuit models almost always take advantage of this.*

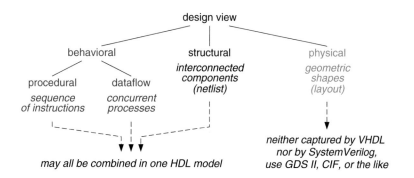

FIGURE 4.4

Modeling styles and their relationships to HDLs and other EDA languages.

---

[14] Procedural, dataflow, and structural are just user-defined terms, not reserved words of the language.
[15] A full-adder has been chosen in this example for its simplicity and obviousness. Adders are normally synthesized from algebraic expressions that include a + operator.

LISTING 4.5

Procedural, dataflow, and structural modeling styles compared.

```
entity fulladd is
   port (
      Aa_DI, Bb_DI, Cc_DI : in std_logic;
      Sum_DO, Carry_DO : out std_logic );
end fulladd;
--------------------------------------------------------------------------
-- compute result in a series of sequential steps
architecture procedural1 of fulladd is
begin
   process (Aa_DI,Bb_DI,Cc_DI)
      variable loc1, loc3, loc4 : std_logic;
   begin
      loc1      := Aa_DI xor  Bb_DI;
      Sum_DO    <= Cc_DI xor  loc1;
      loc3      := Cc_DI nand loc1;
      loc4      := Aa_DI nand Bb_DI;
      Carry_DO <= loc3  nand loc4;
   end process;
end procedural1;
--------------------------------------------------------------------------
-- spawn a concurrent signal assignment for each logic operation
architecture dataflow1 of fulladd is
   signal Loc1_D, Loc3_D, Loc4_D : std_logic;
begin
   Loc1_D    <= Aa_DI  xor  Bb_DI;
   Sum_DO    <= Cc_DI  xor  Loc1_D;
   Loc3_D    <= Cc_DI  nand Loc1_D;
   Loc4_D    <= Aa_DI  nand Bb_DI;
   Carry_DO <= Loc3_D nand Loc4_D;
end dataflow1;
--------------------------------------------------------------------------
-- describe circuit network as a bunch of interconnected std cells
-- note: cells from Synopsys' generic library are used here
architecture structuralgtech of fulladd is

   component GTECH_XOR2
      port ( A, B : in std_logic;
             Z : out std_logic );
   end component;
   component GTECH_NAND2
      port ( A, B : in std_logic;
```

```
                      Z : out std_logic );
        end component;


        signal Loc1_D, Loc3_D, Loc4_D : std_logic;

begin
        u1: GTECH_XOR2
            port map ( A=>Bb_DI,  B=>Aa_DI,  Z=>Loc1_D );
        u2: GTECH_XOR2
            port map ( A=>Cc_DI,  B=>Loc1_D, Z=>Sum_DO );
        u3: GTECH_NAND2
            port map ( A=>Cc_DI,  B=>Loc1_D, Z=>Loc3_D );
        u4: GTECH_NAND2
            port map ( A=>Aa_DI,  B=>Bb_DI,  Z=>Loc4_D );
        u5: GTECH_NAND2
            port map ( A=>Loc3_D, B=>Loc4_D, Z=>Carry_DO );
end structuralgtech;
```

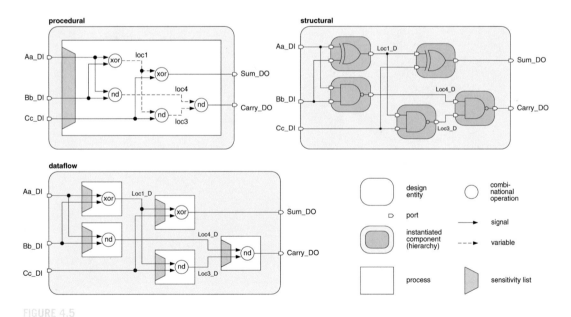

FIGURE 4.5

The modeling styles of listing 4.5 illustrated.

4.2.3 A DISCRETE REPLACEMENT FOR ELECTRICAL SIGNALS

The need for multiple logic values to describe a circuit node

An innocent approach to hardware modeling would be to use one binary digit per circuit node. VHDL actually provides two predefined data `types` for describing 2-valued data:

`bit`	which can take on value 0 or 1.
`boolean`	which can take on value `false` or `true`.

Digital circuits exhibit traits that cannot be captured with a 2-valued abstraction, however. Just think of transients, indeterminate states following power-up, three-state outputs, multiple buffers driving a common node with an inherent potential for conflicts, and the like.

Observation 4.6. *Distinguishing between logic 0 and 1 is inadequate for modeling the binary signals found in digital circuits. A more elaborate multi-valued logic system must be sought that is capable of capturing the effects of both node voltage and source impedance.*

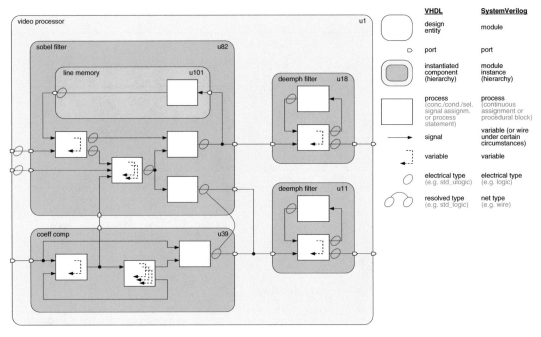

FIGURE 4.6

HDL capabilities (III): ... plus data types for modeling electrical phenomena ...

VHDL offers two more sophisticated and closely related data types named `std_ulogic` and `std_logic` respectively. Using either of these data types, the electrical conditions of a circuit node are condensed into one **logic value** at any time as shown in table 4.2 and fig.4.7a.

A number of clarifications are due at this point.

- Node voltage is quantized into three **logic states**:

  low   logic low, that is below U_l.

  high   logic high, that is above U_h.

  unknown   may be either "low", "high", or anywhere in the forbidden interval in between, e.g. as a result from a short between two conflicting drivers.

Table 4.2   The IEEE 1164 standard MVL-9 (multiple value logic) system.

logic value → ↓	logic state		
	low	unknown	high
uninitialized		U	
strength strong	0	X	1
weak	L	W	H
high-impedance	Z	Z	Z
don't care		–	

No difference is made between a drive conflict, the outcome of which is truly unknown, and a ramping node, the voltage of which is known to assume values between thresholds U_l and U_h for some time. Either condition is modeled as "unknown".

- The amount of current that a subcircuit can sink or source — which is inversely related to the source impedance — gets mapped onto three discrete **drive strengths**:

  strong   the low impedance value commonly exhibited by a driving output.

  high-impedance   the almost infinite impedance exhibited by a disabled three-state output.

  weak   an impedance somewhere between "strong" and "high-impedance", e.g. as exhibited by a passive pull-up/-down resistor or a snapper.

  No difference is made between "charged high" and "charged low". All high-impedance conditions are merged into a single condition of undetermined state or voltage.

- Two extra values have been added, namely:

  uninitialized   has never been assigned any value, e.g. the internal state of a storage element immediately after power-up, distinguished from "unknown" as the latter can arise from causes other than failed initialization (applicable to simulation only).

  don't care   whether the node is "low" or "high" is considered immaterial, used by designers to leave the choice to the logic optimization tool (applicable to synthesis only).

- Data types `std_ulogic` and `std_logic` are not part of the VHDL language IEEE 1076 itself, but have been defined in separate standard IEEE 1164 and made available in a package named `ieee.std_logic_1164`.[16] The difference is to be explained shortly.

---

[16] Incorporating a logic system into VHDL would have biased the language towards a specific circuit technology such as CMOS, TTL, ECL or GaAs, and would preclude its evolution towards unforeseen technologies in the future. Keeping the MVL-9 system apart makes it is possible to replace it by some user-defined alternative at any time should the necessity occur.

Some logic values get collapsed during synthesis

Not all nine options make sense from a synthesis point of view. The semantics of 0 and 1 are obvious. A don't care symbol - on the right hand-side of an assignment implies the outcome is of no importance in which case logic optimization is free to select either a 0 or a 1 such as to minimize gate count, path delay, and/or power. z calls for a driver with built-in three-state capability. Values U, X and W capture situations that may occur during simulation, but have no sensible interpretation for synthesis. L and H are not honored either. Most synthesis tools collapse meaningless (to them) values to more sensible ones, e.g. L to 0, H to 1, and X or W to -.

> *Hint: For the sake of clarity and portability, do not use logic values other than 0, 1, z and - in VHDL source code that is intended for synthesis.*

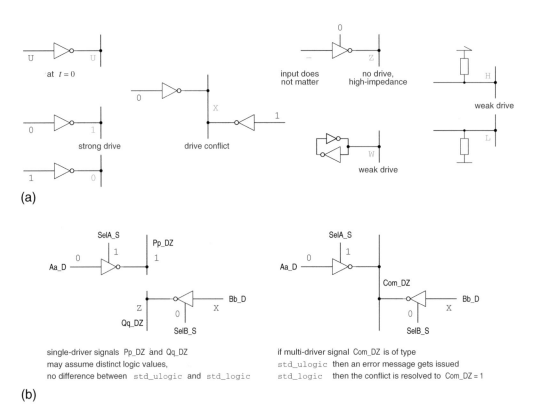

(a)

(b)

The IEEE 1164 MVL-9 illustrated (a), type `std_logic` vs. `std_ulogic` (b).

How to model three-state outputs and busses

How do we tell we want a node to be released, i.e. reverted to an undriven condition? With the aid of the MVL-9 system, the answer is straightforward: just assign logic value z.

Example   `Oup_DZ <= Inp_D when Ena_S='1' else 'Z';`

As illustrated in fig.4.6, many digital circuits include **multi-driver nodes** that operate under control of multiple processes. How to model them is now obvious. In the code fragment below, the common node Com_DZ is left floating, that is in a high-impedance condition, when neither of the two drivers is enabled. In the occurrence of a parallel **bus**, assigning the shorthand form `{(others = > 'Z')}` will do so for all bits, no matter how wide the bus is.

```
signal Com_DZ, Aa_D, Bb_D, SelA_S, SelB_S : std_logic;
.....
Com_DZ <= not Aa_D when SelA_S='1' else 'Z';
.....
Com_DZ <= not Bb_D when SelB_S='1' else 'Z';
.....
```

A multi-driver node may give rise to drive conflicts when something goes wrong, however.[17] It is, therefore, important to know how such situations are dealt with during simulation. Built on top of std_ulogic sits a subtype called std_logic that shares the same set of nine values. As shown in fig.4.7b, the difference matters in the presence of multiple drivers.

If the signal is of type std_ulogic, some form of error message will be generated during compilation or simulation.[18] On a std_logic-type signal, in contrast, any conflict between diverging values is tacitly solved at simulation time by calling a **resolution function** that determines the most plausible outcome. One such function is part of the IEEE 1164 standard, its name is resolved, see listing 4.6. As an example, this function specifies that simulation is to continue with a 1 should an attempt be made to drive a node to z, w and 1 at the same time.

Observation 4.7. Signals *of type* std_logic *can accommodate multiple drivers whereas those of type* std_ulogic *cannot. In more general terms, a* signal *is allowed to be driven from multiple processes iff a resolution function is defined that determines the outcome.*

The IEEE 1076 standard insists that a resolution function be available for any signal that is being driven from multiple sources but places the details under the designer's control. Function resolved is fine for fully complementary static CMOS logic.[19] Make sure to understand that there can be no such thing as a resolution function for variables. The same applies to bit, bit_vector, integer, real, and similar data types.

---

[17] A **drive conflict** implies that two or more processes attempt to drive a signal to incompatible logic values.

[18] The details depend on the product being used. ModelSim categorically rejects any unresolved signals that are driven from multiple processes whereas Synopsys DesignCompiler accepts them with warnings issued at elaboration time.

[19] By programming his own resolution functions, the designer can indicate how to solve driver conflicts in other situations, e.g. for open-collector and open-drain outputs with their wired-AND operations.

LISTING 4.6

IEEE 1164 standard resolution function.

```
---------------------------------------    -------------------------
-- resolution function "resolved"
---------------------------------------------------------------------
constant resolution_table : stdlogic_table := (
--      ----------------------------------------------------------
--      | U    X    0    1    Z    W    L    H    -      |  |
--      ----------------------------------------------------------
        ( 'U', 'U', 'U', 'U', 'U', 'U', 'U', 'U', 'U' ), -- | U |
        ( 'U', 'X', 'X', 'X', 'X', 'X', 'X', 'X', 'X' ), -- | X |
        ( 'U', 'X', '0', 'X', '0', '0', '0', '0', 'X' ), -- | 0 |
        ( 'U', 'X', 'X', '1', '1', '1', '1', '1', 'X' ), -- | 1 |
        ( 'U', 'X', '0', '1', 'Z', 'W', 'L', 'H', 'X' ), -- | Z |
        ( 'U', 'X', '0', '1', 'W', 'W', 'W', 'W', 'X' ), -- | W |
        ( 'U', 'X', '0', '1', 'L', 'W', 'L', 'W', 'X' ), -- | L |
        ( 'U', 'X', '0', '1', 'H', 'W', 'W', 'H', 'X' ), -- | H |
        ( 'U', 'X', 'X', 'X', 'X', 'X', 'X', 'X', 'X' )  -- | - |
    );
```

Selecting adequate data types

Data types `std_ulogic` and `std_logic` emulate the electrical behavior of digital circuits in a much more realistic way than type `bit` does, and should be used whenever one wants to model an actual circuit node (as opposed to an auxiliary variable in a process statement, for instance).[20] Usually `std_logic` prevails.[21]

---

[20] In theory, any multi-valued data type occupies more memory space than its two-valued equivalent does, demands a higher computational effort and, hence, slows down simulation. In practice, everything depends on how the simulator software actually stores the bit, bit_vector, std_logic, and std_logic_vector types. It is, therefore, difficult to give general and qualified advice on whether it pays to begin with architecture-level simulations on the basis of bit types before upgrading to std_logic types during RTL design. What is for sure, however, is that you will not want to use MVL-9 data types for items other than actual circuit nodes.

[21] Simulating with unresolved std_ulogic and std_ulogic_vector types is more conservative than simulating with their resolved counterparts because an error message will tell should any of those accidentally get involved in a drive or naming conflict. Yet, the IEEE 1164 standard recommends: "For scalar ports and signals, the developer may use either std_ulogic or std_logic type. For vector ports and signals, the developer should use std_logic_vector type." Two reasons are given for this surprising advice: Concerns expressed by EDA vendors that they might not be able to optimize simulator performance for both data types, and interoperability of circuit and testbench models from different sources.

data type	bit	std_ulogic	std_logic
defined in	VHDL	ieee.std_logic_1164	
value set per binary digit	2	9	
for simulation purposes			
modeling of power-up phase	no	yes	yes
modeling of weakly driven nodes	no	yes	yes
modeling of multi-driver nodes	no	yes	yes
handling of drive conflicts	n.a.	reported	resolved
for synthesis purposes			
three-state drivers	no	yes	yes
don't care conditions	no	yes	yes

Data types for modeling multi-bit signals

It goes by itself that scalars of the same type can be collected into a vector, known as bit_vector, std_ulogic_vector, and std_logic_vector respectively. Unfortunately, the original standards did not support arithmetic operations on them. To fill the gap, two packages have been accepted as IEEE standard 1076.3. Both define two extra data types called unsigned and signed that are overloaded for standard VHDL arithmetic operators as much as possible. Objects of type unsigned are interpreted as unsigned integer numbers, and objects of type signed as signed integer numbers coded in 2's complement (2'C) format. The table below compares the various types available that comprise multiple data bits.

data type(s)	integer, natural, positive	bit_vector	std_logic _vector	signed, unsigned	signed, unsigned
defined in	VHDL	VHDL	ieee.std_logic_1164	ieee.numeric_bit	ieee.numeric_std
value set per binary digit	2	2	9	2	9
word width	32 bit	at the programmer's discretion			
arithmetic operations	yes	no	no	yes	yes
logic operations	no	yes	yes	yes	yes
access to subwords or bits	no	yes	yes	yes	yes
modeling of electrical effects	no	no	yes	no	yes

The difference between the two packages is that ieee.numeric_bit is composed of bit-type elements, whereas ieee.numeric_std operates on std_logic elements. As they otherwise define identical data types and functions, only one of the two packages can be used at a time.

LISTING 4.7

Examples of VHDL multi-bit data types.

```
integer                    32 bit, signed, 2-valued
natural                 -- 32 bit, range 0...(2^31)-1, 2-valued
positive                -- 32 bit, range 1...(2^31)-1, 2-valued
bit_vector(11 downto 0)  -- 12 bit, no numerical interpretation, 2-valued
std_logic_vector(5 downto 0) --  6 bit, no numerical interpretation, 9-valued
unsigned(11 downto 0)    -- 12 bit, unsigned, from ieee.numeric_bit -> 2-valued
signed(5 downto 0)       --  6 bit, signed,   from ieee.numeric_bit -> 2-valued
unsigned(11 downto 0)    -- 12 bit, unsigned, from ieee.numeric_std -> 9-valued
signed(5 downto 0)       --  6 bit, signed,   from ieee.numeric_std -> 9-valued
```

VHDL is a strongly typed language, which implies that every object has a type and that extensive type checking is performed. Types must be converted before an assignment or a comparison becomes possible across distinct types, see section 4.8.6 for conversion functions.

Orientation of binary vectors

When encoding a number using a positional number system, there is a choice between spelling the data word with the MSB or the LSB first. In VHDL, this primarily applies to data types unsigned and signed.[22] Any misinterpretation will cause severe malfunctions.

Hint: Any vector that contains a data item coded in some positional number system should consistently be declared as (i_{MSB} downto i_{LSB}) where 2^i is the weight of the binary digit with index i. The MSB will so have the highest index assigned to it and will appear in the customary leftmost position because $i_{MSB} \geq i_{LSB}$.

Example   signal Hour_D : unsigned(4 downto 0) := "10111";

Data types unsigned and signed are intended to represent integer numbers with the radix point immediately following the LSB. We can thus illustrate their formats as $\boxed{\texttt{ii...i.}}$ and $\boxed{\texttt{si...i.}}$ respectively, where s stands for the sign bit and each i for one binary digit.

Data types for modeling fractional and floating point numbers

The IEEE 1076-2008 revision saw the arrival of data types for representing fixed and floating point numbers. Note that each of those types comes in a resolved and an unresolved variety.

---

[22] bit_vector, std_logic_vector and std_ulogic_vector are concerned as well if used to encode numbers.

type prefix	unresolved_			unresolved
data type	ufixed	sfixed	float	resolved
defined in	fixed_ generic_ pkg		float_ generic_ pkg	
arithmetics	fixed point unsigned | signed		floating point	
word width	at the programmer's discretion			
arithmetic operations	yes			
logic operations	yes			
access to subwords or bits	yes			
modeling of electrical effects	yes (resolved)			

Example of a fractional number `signal HourWithQuarter_D : ufixed(4 downto -2) := "1011111";`
($\boxed{\text{iiiii.ff}}$ $i_{MSB}=4$, $i_{LSB}=-2$, $w=i_{MSB}-i_{LSB}+1=7$ with a range from 0 to $11111.11_2 = 31.75_{10}$ in increments of $\frac{1}{4}$ and an initial value of $10111.11_2 = 23.75_{10}$)

Second example `variable JustIntegers : ufixed(7 downto 2) := "010110";`
($\boxed{\text{iiiiii}}$ 00. $i_{MSB}=7$, $i_{LSB}=2$, $w=6$ assumes integer values from 4 to 1020 in steps of 4)

Third example `variable JustFractions : ufixed(-4 downto -7) := "1000";`
(.000 $\boxed{\text{ffff}}$ $i_{MSB}=-4$, $i_{LSB}=-7$, $w=4$ with a range from $\frac{1}{128}$ to $\frac{15}{128}$ and initialized to $\frac{1}{16}$)

It goes by itself that the 2'C format is applied to represented fractional numbers with a sign.

Example `signal HourWithQuarter_D : sfixed(4 downto -2) := "1011111";`
($\boxed{\text{siiii.ff}}$ $i_{MSB}=4$, $i_{LSB}=-2$, $w=7$ with a range from $10000.00_2 = -16.00_{10}$ to $01111.11_2 = 15.75_{10}$ in increments of $\frac{1}{4}$ and with an initial value of $10111.11_2 = -9.75_{10}$)

Floating point numbers include a sign bit and an exponent by definition. Their formats adhere to the principles of the IEEE 754 standard $\boxed{\text{see...e.ff...f}}$ except that the number of bits reserved to accommodate exponent and mantissa, $\#e$ and $\#f$ respectively, are not fixed but can be defined by the user in a type declaration.[23] The mantissa is a fractional number normalized to the interval [1...2) and coded in 2'C format.[24] The exponent is coded in offset-binary (O-B) format with an offset of $2^{\#e-1}-1$.

Example `signal ToyFloat_D : float(5 downto -8);`
($\boxed{\text{seeeee.ffffffff}}$ $\#e = 5$ and $\#f = 8$ with an offset of $+15$ which implies that the true unbiased exponent is calculated by subtracting 15 from $\boxed{\text{eeeee}}$)

---

[23] The default widths are $\#e=8$ and $\#f=23$ respectively for a total word width of $w=32$ bit.
[24] As a consequence of the normalization, the binary digit of the mantissa immediately preceding the radix point is always 1 and, hence, devoid of information. The mantissa field is thus limited to the fractional bits to the right of the radix point with an (invisible) 1 implied to its left. See section A.1.2 for further details.

To render the fixed and floating point packages as versatile as possible, several aspects relating to arithmetic behavior are kept user-adjustable with the aid of `generics`.

- Rounding behavior (round $\approx$ vs. truncate $\downarrow$)
- Overflow behavior (saturate _/‾ vs. wrap around /\/\/)
- Number of guard bits for division operation[25]

4.2.4 AN EVENT-DRIVEN SCHEME OF EXECUTION

The need for a mechanism that schedules process execution

Recall the concurrent VHDL constructs introduced in section 4.2.2.

- o Concurrent signal assignment (simplest).
- o Selected signal assignment.
- o Conditional signal assignment.
- o The `process` statement (most powerful).

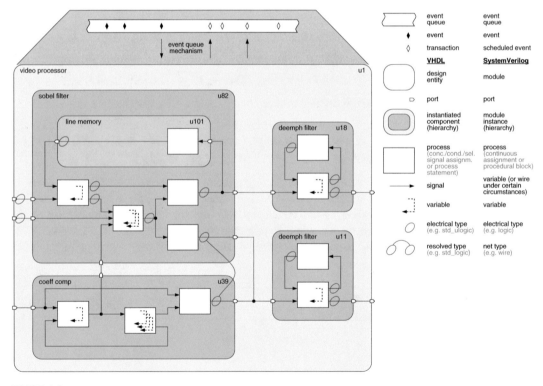

FIGURE 4.8

HDL capabilities (IV): ... plus an event queue mechanism that governs process activation ...

---

[25] Calculating a result to a certain precision w_r may require carrying out intermediate steps at some higher precision w_i as roundoff errors can accumulate. Guard bits is the name for the extra $w_r - w_i$ binary digits.

Simulation must, of course, yield the same result as if the many processes present in a circuit model were operating simultaneously, although no more than a few processor cores are normally available for running the simulation code. What is obviously required then is a mechanism that schedules processes for sequential execution and that combines their effects such as to perfectly mimic concurrency. This mechanism that always sits in the background of HDL models is the central theme of this section. A good comprehension of how a model's concurrent processes are being scheduled during simulation is essential for writing synthesis code.

Simulation time versus execution time

Simulation time is to an HDL model what physical time is to the hardware described by that model. The simulator software maintains a counter that is set to zero when a new simulation run begins and that registers the progress of simulation time from then on. This counter can be likened to a stopwatch, and any event that occurs during simulation can be thought of being stamped with the time currently displayed by that clock.

Execution time, aka wall clock, refers to the time a computer takes to execute statements from the HDL code during simulation. It is of little interest to circuit designers as long as their simulation runs complete within an acceptable lapse of time.

The benefits of a discretized model of time

Assume you wanted to model a digital circuit using some software language. Capturing the functionality of gates and registers poses no major problem, but how about taking into account their respective propagation delays? How would you organize a simulation run? You would find that no computations are required unless a circuit node switches. For the sake of efficiency, you would consider time as being discrete and devise some data structure that activates the relevant circuit models when they have to (re-)evaluate their inputs. These are precisely the ideas underlying event-driven simulation.

Observation 4.8. *In VHDL simulation, the continuum of time gets subdivided by events each of which occurs at a precise moment of simulation time. An* **event** *is said to happen whenever the value of a* `signal` *changes.*

Event-driven simulation

The key element that handles events and that invokes processes is called **event queue** and can be thought of as a list where entries are arranged according to their time of occurrence, see fig.4.9. An entry is referred to as a **transaction**.

Event-driven simulation works in cycles where three stages alternate:

1. Advance simulation time to the next transaction in the event queue thereby making it the current one.[26]
2. Set all `signals` that are to be updated at the present moment of time to the target value associated with the current transaction.

---

[26] Multiple entries may be present for the same moment of time, but the general procedure remains the same.

3. Invoke all processes that need to respond to the new situation and have them (re-)evaluate their inputs. Every signal assignment supposed to modify a `signal`'s value causes a transaction to be entered into the event queue at that point in the future when that `signal` is anticipated to take on its new value. This stage comes to an end when all processes invoked suspend after having finished to schedule `signal` updates in response to their current input changes.

After completing the third stage, a new simulation cycle is started. Simulation stops when the event queue becomes empty or when simulation time reaches some predefined final value.

As nothing happens between transactions, an event-driven simulator essentially skips from one transaction to the next. No computational resources are wasted while models sit idle. Parallel processes and event-queue together form a powerful mechanism for modeling the behavior of discrete-time systems.[27] Refer to fig.4.14 for a wider perspective on the simulation cycle.

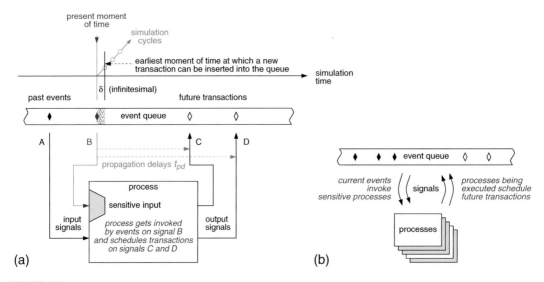

(a) (b)

FIGURE 4.9

Event-driven simulation in VHDL. Interactions between the event queue and a process (a), actions that repeat during every simulation cycle (b).

Transaction versus event

It is important to note that any `signal` update actually occurs in two steps. Execution of the assignment statement causes a transaction to be entered into the event queue but has no immediate impact. The update is to become effective only later when simulation time has reached the scheduled time for that transaction.

---

[27] Incidentally, note that the event queue mechanism is by no means confined to electronic hardware but is also being used for simulating land, air and data traffic, for evaluating communication protocols, for planning fabrication and logistic processes, and in many other discrete-time applications.

By the same token, not every signal assignment that is being carried out necessarily causes a `signal` to toggle. All too often, a process gets evaluated in response to some event on one of the wake-up `signals` just to find out that the result is the same as before. Consider a four-input NAND gate or an E-type flip-flop, for instance. Also, the effect of a first transaction may get nullified by a second transaction inserted into the event queue afterwards. This is why transactions and events are not the same. A transaction that does not alter the value of a `signal` is still a transaction but it does not give rise to an event.[28]

Observation 4.9. *Events are observable from the past evolution of a* `signal`*'s value up to the present moment of simulation time whereas transactions merely reflect future plans that may or may not materialize.*

Delay modeling

In the context of simulation, the lapse of time between an event at the input of a process and the ensuing transaction at the output reflects the delay of the piece of hardware being modeled. Circuit delays are typically expressed as part of the signal assignment statement with an `after` clause.[29] The first statement below models the propagation delay of an adder, by scheduling a transaction on its output t_{pd} after an event at either input. The second statement also accounts for contamination delay t_{cd}.

Example
Second example

```
                                    Oup_D <= InpA_D + InpB_D after TPD;
              Oup_D <= 'X' after TCD, InpA_D + InpB_D after TPD;
```

The δ delay

For obvious reasons, a process cannot be allowed to schedule `signal` updates for past or present moments of time. It is, therefore, natural to ask

"What is the earliest point in time at which a new transaction can be entered into the queue?"

In the occurrence of VHDL, the answer is δ time later where δ does not advance simulation time but requires going through another simulation cycle. Put differently, δ can be thought of as an infinitesimally small lapse of time greater than zero. This refinement to the basic event queue mechanism serves to maintain a consistent order of transactions when the simulation involves models that are supposed to respond with delay zero. Without the δ time step, there would be no way to order zero-delay transactions and simulation could, therefore, not be guaranteed to yield meaningful and reproducible results. Although simulation time does not progress in regular intervals, δ may, in some sense, be interpreted as the timewise resolution of the simulator.

---

[28] An event queue resembles very much an agenda in everyday life. Transactions are analogous to entries there. `Signals` reflect the evolution of the state of our affairs such as current location and occupation, health condition, social relations, material possessions, and much more. An entry in the agenda stands for some specific intention as anticipated today. At any time, an event, such as a phone call, may force us to alter our plans, i.e. to add, cancel or modify intended activities to adapt to a new situation. Some of our activities remain in vain and do not advance the state of affairs, very much as part of the transactions do not turn into events. Finally, in retrospect, an agenda also serves as a record of past events and bygone states.

[29] Related language constructs that also express time intervals are `wait for` and `reject`. A `wait for` statement causes a process to suspend for the time indicated before being reactivated. The `reject` clause helps to describe rejection phenomena on narrow pulses in a more concise way.

"How does a simulator handle signal assignments with no `after` clause?"

The answer is that delay is assumed to be zero exactly as if the code would read ... `after 0 ns` .... The transaction is then scheduled for the next simulation cycle or, which is the same, one δ delay later. Omitting the `after` clauses is typical in RTL synthesis models because physically meaningful delay data are unavailable at the time when such models are being established. Much the same applies to behavioral models at the algorithmic level.

> *Hint: When simulating models with no delays other than the infinitesimally small δs, it becomes difficult to tell apart cause and effect in the output waveforms as the respective events appear to coincide. A trick is to artificially postpone transactions by a tiny amount of time in otherwise delayless signal assignments. To allow for quick adjustments, a constant of type time is best declared in a package and referenced throughout a model hierarchy. Note that the largest sum of fake delays must not exceed one clock period, though.*

> *Example*   `Oup_D <= InpA_D + InpB_D after FAKEDELAY;`

Signal versus variable

We now are in a good position to understand what separates `signals` from `variables`. While the difference in terms of scope has already been illustrated in fig.4.3, fig.4.10 exposes those particularities that relate to time. A `variable` has no time dimension attached which is to say that it merely holds a present value. Neither transactions nor events are involved. The effect of a variable assignment is thus felt immediately, in the next statement, exactly as with any software programming language.

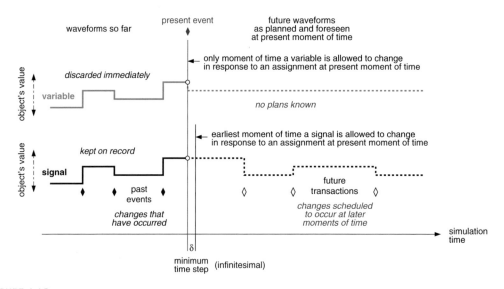

FIGURE 4.10

The past, present and future of VHDL `variables` and `signals`.

A `signal`, in contrast, is a typical element of VHDL. It is defined over time, which implies that a `signal` not only holds a present value, but also past values, plus those values that are anticipated to become manifest in the future. The effect of a signal assignment is not felt before the delay specified in the `after` clause has expired. Minimum delay, and default value in the absence of an `after` clause, is δ. As another particularity, it is possible to schedule multiple transactions in a single signal assignment statement.

Example   `ThisMonth_D <= AUGUST, SEPTEMBER after 744 hr, OCTOBER after 1464 hr;`

Observation 4.10. *VHDL* `signals` *convey time-varying information between processes via the event queue. They are instrumental in process invocation which is directed by the same mechanism.* `Variables`, *in contrast, are confined to within a* `process` *statement or a subprogram and do not interact with the event queue in any way.*

Ignoring the observation below is the cause of frequent misconceptions.

Observation 4.11. *A signal assignment does not become effective before the delay specified in the* `after` *clause has expired. In the absence of an explicit indication, there is a delay of one simulation cycle, so the effect can never be felt in the next statement. This sharply contrasts with a variable assignment the effect of which is felt immediately.*

To constantly remind programmers of this vital difference, two distinct symbols have been set aside in the VHDL syntax for the variable assignment `:=` and for the signal assignment `<=` .

Event-driven simulation revisited

A process is either active or suspended at any time. Simulation time is stopped while the code of the processes presently active is being carried out which implies that
(a) all active processes are executed concurrently with respect to simulation time, and
(b) all sequential statements inside a `process` statement are executed in zero simulation time.

The order of process invocation with respect to execution time is undetermined.

Observation 4.12. *As opposed to software languages where execution strictly follows the order of statements in the source code, there is no fixed ordering for carrying out processes (including concurrent signal assignments and assertion statements) in VHDL. When to invoke a process gets determined solely by events on the* `signals` *that run back and forth between processes.*

Sensitivity list

Each process has its own set of `signals` that cause it to get (re-)activated whenever an event occurs on one or more of them. The entirety of such `signals` are aptly qualified as its wake-up or trigger signals although this is not official VHDL terminology. What are the wake-up signals of a given process? The answer depends on the type of process.

In concurrent, selected, and conditional signal assignments, specifying wake-up signals is neither necessary nor legal as these are sensitive to any `signal` that appears on the right-hand side of the assignment operator `<=` by definition. In the example below, `ThisMonth_D` and `ThisDay_D` act as wake-up signals.

```
Spring_D <= true when  (ThisMonth_D=MARCH and ThisDay_D>=21) or
                        ThisMonth_D=APRIL or ThisMonth_D=may or
                        (ThisMonth_D=JUNE and ThisDay_D<=20)
                        else false;
```

The `process` statement is more liberal in that it accepts a special clause that collects the wake-up signals. This sensitivity list is included within parentheses to the right of the keyword `process`. Upon activation by any of the signals there, instructions get executed one after the other until the `end process` statement is reached. The `process` then reverts to its suspended state. The example below is semantically identical to the conditional signal assignment above.

```
memless2: process (ThisMonth_D, ThisDay_D)
-- an event on any signal listed activates the process
begin
    Spring_D <= false;     -- execution begins here
    if ThisMonth_D=MARCH and ThisDay_D>=21 then Spring_D <= true; end if;
    if ThisMonth_D=APRIL                   then Spring_D <= true; end if;
    if ThisMonth_D=may                     then Spring_D <= true; end if;
    if ThisMonth_D=JUNE  and ThisDay_D<=20 then Spring_D <= true; end if;
end process memless2;    -- process suspends here
```

This feature gives the engineer more freedom but also more responsibility as including or omitting a `signal` from a sensitivity list profoundly modifies `process` behavior. Imagine `ThisDay_D` is dropped. What does that change? Events on `ThisDay_D` are unable to activate the `process` and, hence, no longer update `signal Spring_D`. Only an event on `signal ThisMonth_D` causes the logic value of `ThisDay_D` to get (re-)evaluated. The state of `Spring_D` thus depends on past values of `ThisDay_D` — rather than just on the present one — which implies memory!

Observation 4.13. *Depending on its sensitivity list, the same process statement may or may not model sequential circuit behavior.*

Wait statement [30]

An alternative syntax for indicating where execution of a `process` statement is to suspend and when it is to resume, is to include a `wait` statement. Note that the two forms are mutually exclusive. That is, no `process` statement is allowed to include both a sensitivity list and `wait`s.

As the name suggests, process execution suspends when a `wait` statement is reached. It resumes with the subsequent instruction as soon as a condition specified is met and continues until a next wait is

---

[30] Note to the reader: For a first reading, you may want to skip this paragraph and continue with "How to safely code sequential circuits for synthesis".

encountered, and so on. The `wait` statement comes in four flavors that differ in the nature of the condition for process reactivation:

Statement	Wake-up condition
`wait on ...`	an event (value change) on any of the `signals` listed here
`wait until ...`	*idem* plus the logic conditions specified here
`wait for ...`	a predetermined lapse of time as specified here
`wait`	none, sleep forever as no wake-up condition is given

The code below is functionally interchangeable with process `memless2` shown above but uses a `wait on` statement instead of a sensitivity list.

```
memless3: process    -- no sensitivity list because a wait statement is used
begin
   Spring_D <= false;    -- execution begins here
   if ThisMonth_D=MARCH and ThisDay_D>=21 then Spring_D <= true end if;
   if ThisMonth_D=APRIL                   then Spring_D <= true end if;
   if ThisMonth_D=MAY                     then Spring_D <= true end if;
   if ThisMonth_D=JUNE  and ThisDay_D<=20 then Spring_D <= true end if;
   wait on ThisMonth_D, ThisDay_D;    -- process suspends here until reactivated
                                      -- by an event on any of these signals
end process memless3;    -- execution continues with first statement
```

Execution does not terminate with the `end process` statement but resumes at the top of the `process` body. In a `process` statement with a single `wait`, execution thus necessarily makes a full turn through the code each time the process gets (re-)activated.[31] As opposed to this, only a fragment of the code gets executed in a process with multiple `wait`s, which also implies that there can be no equivalent `process` with a sensitivity list in this case.

Last but not least, take note of the limitation below. As they are of general nature, the reasons behind are going to be explained in section 4.4.3 jointly for VHDL and SystemVerilog.

Observation 4.14. *Process statements that include multiple* `wait`s *are not amenable to synthesis. Only* `process` *statements with a single* `wait` *or with a sensitivity list are.*

What exactly is it that makes a process statement exhibit sequential behavior?
The criteria relate to the organization of the source code and are as follows:

Observation 4.15. *A* `process` *statement implies memory whenever one or more of the conditions below apply. Conversely, memoryless behavior is being modeled iff none of them holds.*

o The `process` statement includes multiple `wait on` or `wait until` statements.
o The `process` statement evaluates input `signals` that have no wake-up capability.
o The `process` statement includes `variables` that get assigned no value before being used.
o The `process` statement fails to assign a value to its output `signals` for every possible combination of values of its inputs.

---

[31] The reason why the `wait` is placed at the end — rather than at the beginning — in the `memless3` code is that all processes get activated once until they suspend as part of the initialization phase at simulation time zero.

Observation 4.16. *VHDL knows of no specific language construct that could tell a sequential model from a combinational one. Similarly, there are no reserved words to indicate whether a piece of code is intended to model a synchronous or an asynchronous circuit, or whether a finite state machine is of Mealy, Moore or Medvedev type.*

What makes the difference is the detailed construction of the source code.

> *Hint: If a* process *statement is to model combinational logic, use the* (all) *notation introduced in listing 4.2. Further make sure to assign a value to each output for all possible combinations of input values.*

Failure to assign a don't care to a signal — or to a variable declared within a process statement — in situations where the object's value does not seem to matter is a frequent mistake. After all, not assigning anything means that the object's present value shall be maintained and, as a consequence, implies memory.

> *Hint: As it is not immediately obvious whether a given* process *statement actually models memorizing or memoryless behavior, it is a good idea to make this information explicit in the source code, either by adding a comment or by choosing some meaningful name for the optional process label.*

This habit not only makes code easier to understand, but also helps to check the presence, nature, and number of bistables that are obtained from synthesis against the code writer's intentions. This is important because a minor oversight during VHDL coding may turn a memoryless process statement into a memorizing one, and hence the associated circuit as well.

How to safely code sequential circuits for synthesis

As we have just learned, coding a process statement can be tricky as sensitivity list, wait statement(s), and wake-up condition(s) must be kept consistent. Listing 4.8 shows a proven template for any (sub)circuit with memory that is supposed to exhibit edge-triggered behavior, such as a register or counter, for instance. A clock signal in the sensitivity list is mandatory. One optional signal is accepted for implementing an asynchronous reset function. No other signals are permitted to (re-)activate the process.

LISTING 4.8

Skeleton of a safe and synthesizable process statement with memory.

```
process (Clk_C, Rst_R) <--------- sensitivity list, no more signals accepted!
begin

   <--------- no other statement allowed here!
   -- activities triggered by asynchronous active-high reset
   if Rst_R='1' then
      PresentState_DP <= STARTSTATE;
      .....
```

```
    -- activities triggered by rising edge of clock
  elsif Clk_C'event and Clk_C='1' then <--------- no more term allowed here!
    <--------- extra subconditions, if any, accepted here.
    PresentState_DP <= NextState_DN; -- admit next state into state register
    .....
  <--------- no further elsif or else clause allowed here!
  end if;
  <--------- no statement allowed here!
end process;
```

The reason for marking various items as disallowed is that their presence would render the model's behavior inconsistent with single-edge-triggered clocking. Safe and predictable behavior are not the only merits of this scheme, it also ensures that the code is universally accepted for synthesis. An actual example has been given in listing 4.3.

Depending on how the code is written, a `process` statement can be made to capture almost anything from a humble piece of wire up to an entire image compression circuit, for instance.

Hint: For the sake of modularity, legibility, and smooth synthesis, do not cram too much functionality into a concurrent process. As a rule, prefer concurrent, selected and conditional signal assignments for describing combinational operations while packing any kind of data storage into separate `process` *statements.*

Initial values cannot replace a reset mechanism

As discussed earlier, the VHDL syntax supports assigning an initial value to a `signal` as part of its declaration statement, and the same applies for `variables`.

Observation 4.17. *The initial value given to a* `signal` *or* `variable` *defines the objects's state at $t = 0$, just before the simulator enters the first cycle. A hardware reset, in contrast, remains ready to reconduct the circuit into a predetermined start state at any time $t \geq 0$. This asks for distributing a dedicated signal to the bistables, both in the circuit and in its HDL model.*

Make sure you understand these are two totally different things. An initialized `signal` or `variable` does not model a hardware reset facility and will, therefore, not synthesize into one. This explains the presence of a reset clause in listings 4.8 and 4.3.

Detecting clock edges and other signal events

In listing 4.8, note that a signal attribute termed `'event` ... is used to trigger action on rising clock edges. During automatic circuit synthesis, this will instruct the software to instantiate (edge-triggered) flip-flops when the netlist is being built.

Alternatively, you may prefer to call upon function `rising_edge(...)` defined in the IEEE 1164 standard along with its counterpart `falling_edge(...)`. Both functions work for signals of type `std_logic` and `std_ulogic` and make use on the `'event` attribute internally.

Example of clock edge detection   `if Clk_C'event and Clk_C='1' then ... endif;`
Alternative syntax[32]   `if rising_edge(Clk_C) then ... endif;`

---

[32] Behavior would differ in the occurrence of a Z condition, but clocks typically have no three-state capability.

Signal attributes

Broadly speaking, a signal attribute is a named characteristic of a `signal`. The most important signal attribute is `'event`, a boolean that is true iff an event has occurred during the current simulation cycle. In addition, VHDL knows of `'transaction`, `'driving`, `'last_value`, `'delayed()`, `'stable()`, and five more. Users are free to declare their own signal attributes on top of those predefined in the IEEE 1076 standard. Not all signal attributes are supported by synthesis, though. In fact, `'event` typically is the only one. As we will see shortly, `'stable` is most useful in simulation models.

How to check timing conditions

Latches, flip-flops, RAMs, and all other sequential subcircuits impose specific timing requirements such as setup and hold times on data, and minimum pulse widths on clock inputs. Should any of these timing conditions get violated, their behavior becomes unpredictable. Checking for compliance is thus absolutely essential for meaningful simulations.

Observation 4.18. *The simulation model of a (sub)circuit is in charge of two things:*
1. *Check whether input waveforms indeed conform with its timing requirements (if any).*
2. *Evaluate input data and (re-)compute the equations to update outputs and state (if any).*

Actually, timing checks must be completed before logic evaluation can begin. They are essentially carried out by having the simulator inspect the event queue for events on the relevant signals. A timing condition is considered as respected if no events are found within the time span during which a given signal is required to remain stable. VHDL supports this idea with signal attributes such as `'stable` and with concurrent assertion statements.

Concurrent assertion statements

Much like a concurrent signal assignment, a concurrent assertion statement is a process that gets (re-)activated by an event on any `signal` present in the assert expression. The difference is that an assertion is neither intended to schedule any new transaction nor capable of doing so. As they cannot update any signal values, concurrent assertion statements are qualified as **passive processes**. They are allowed at any level of abstraction in a design hierarchy and typically used to check user-defined properties and for collecting statistical data during simulation runs.

The `assert` keyword is followed by a boolean expression. Upon activation, the process checks the value of that expression and takes no further action if it evaluates to `true`. If not so, the string following the keyword `report` is sent to the output device. The **severity level** — one of `note`, `warning`, `error` or `failure` — is reported to the simulator and defines where to send that message and whether to proceed or to abort the current simulation run. Listing 4.9 shows an example.[33]

---

[33] The assertion statements are located in the `architecture` body here along with other concurrent processes. They might as well be included in the `entity` declaration, however, following the `port` clause and preceded by a `begin` keyword. Further note that the assertion statement also comes in a sequential version for inclusion in `process` statements and subprograms.

LISTING 4.9

Setup and hold time checks in a D-type flip-flop model

```vhdl
-- simulation model of a single-edge-triggered flip-flop with hardcoded timing
entity setff is
   port (
      Clk_CI :  in std_logic;
      Rst_RBI : in std_logic;
      Dd_DI :   in std_logic;
      Qq_DO :   out std_logic );
end setff;

architecture behavioral of setff is
   signal State_DP : std_logic;    -- state signal
begin

   assert (not (Clk_CI'event and Clk_CI='1' and not Dd_DI'stable(1.09 ns)))
      report "setup time violation" severity warning;
   assert (not (Dd_DI'event and Clk_CI='1' and not Clk_CI'stable(0.60 ns)))
      report "hold time violation" severity warning;

   memzing: process (Clk_CI, Rst_RBI)
   begin
      if Rst_RBI='0' then
         State_DP <= '0';
      elsif Clk_CI'event and Clk_CI='1' then
         State_DP <= Dd_DI;
      end if;
   end process memzing;

   Qq_DO <= State_DP after 0.92 ns;

end behavioral;
```

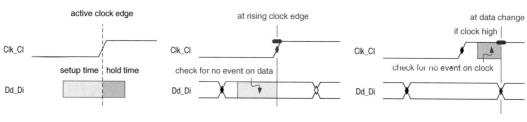

FIGURE 4.11

Checking setup and hold conditions by searching the event queue for past events.[34]

Observation 4.19. *Any inspection of the event queue for compliance with timing requirements must necessarily look backward in time as forthcoming transactions might be added at any time and change the anticipated future evolution of the signal being examined.*

4.2.5 FACILITIES FOR MODEL PARAMETRIZATION

The need for supporting parametrized circuit models

Imagine you have devised a synthesis model for some datapath unit that includes 16 data registers and that is capable of carrying out seventeen distinct arithmetic and logic operations on data words of 32 bit. While continuing on your project, you find out that you need a similar unit for address computations. There are significant differences, however. Addresses are just 24 bit wide, no more than five registers are required, and a subset of eight ALU operations suffices for the purpose. How do you handle such a situation?

It would be fairly easy to derive a separate model by pruning the existing HDL code and by downsizing certain index ranges there. But what if you later needed a third and a fourth model? What if new requirements asked for substantial extensions to the existing model? The problem lies not so much in the initial effort of creating yet another model. Rather, it is the maintenance of a multitude of largely identical source codes that renders this approach so onerous.

A truly reusable model, in contrast, should be written in a parametrized form such as to accommodate distinct choices and parameter settings within a single piece of code. The most salient features of VHDL towards this goal are `generic` parameters, `configurations`, and the `generate` statement that allows for conditional or repeated spawning of processes and that does the same for instantiation. Fig.4.12 shows how these fit into the general picture.

---

[34] Setup and hold times are assumed to be positive in the figure. The modeling of sequential subcircuits that feature a negative timing condition becomes possible by adding fictitious input delays and by adjusting their values such as to make both setup and hold times positive. In order to preserve the original delay figures, all delays added at the input must be compensated for at the output. This is how negative timing condition are to be handled according to the IEEE 1076.4 VITAL standard, at any rate.

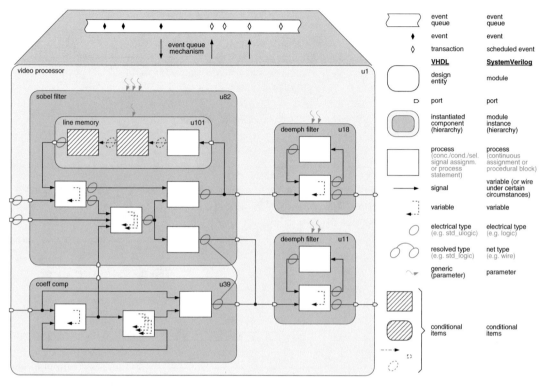

FIGURE 4.12

HDL capabilities (V): ... plus parametrization with adjustable quantities and conditional items.

Generics

As stated earlier, `signals` carry dynamic, i.e. time-varying, information between processes and indirectly also between design entities. `Generics`, in contrast, serve to disseminate static, i.e. time-invariant, details to design entities. Just think of

- Word width,
- Active-low or -high signaling on inputs and outputs,
- Output drive capability,
- Functional options (e.g. details about an instruction set),
- Timing quantities (propagation and contamination delay, setup and hold time), and
- Capacitive load figures.

A `generic`'s name is made part of the `entity` declaration in a so-called `generic` clause in much the same way as signals in a `port` clause. Indicating a default value is optional. As opposed to `ports`, `generics` do not have any direct hardware counterpart. An example follows.

```
component parityoddw    -- w-input odd parity gate
   generic (
      WIDTH : natural range 2 to 32;    -- number of inputs with supported range
      TCD : time := 0 ns,         -- contamination delay with default value
      TPD : time := 1.0 ns );     -- propagation delay with default value
   port (
      Inp_DI : in  std_logic_vector(WIDTH-1 downto 0);
      Oup_DO : out std_logic );
end component;
```

The values of the `generics` get nailed down when the `component` is instantiated, overriding any defaults specified in the `entity` declaration. The syntax of this `generic map` roughly follows that of the `port map`. Note that no semicolon separates the two.

```
constant NUMBITS : natural = 12;
.....
-- component instantiation statement
u173: parityoddw
   generic map ( WIDTH => NUMBITS, TCD => 0.05 ns, TPD => (NUMBITS * 0.1 ns) )
   port map ( Inp_DI => DataVec_D , Oup_DO => Parbit_D );
.....
```

The generate statement used to conditionally spawn concurrent processes

Situations exist where it is impossible to know and freeze the number of processes at the time when the source code is being written. Consider a datapath organized into multiple slices or other array-type circuits where some elemental subfunction is repeated along one or more dimensions. When each such subfunction is captured in a concurrent process of its own, the total number of processes is bound to vary as a function of array size. Now, it is just not possible to write a parametrized behavioral model in such a situation using the constructs of a software programming language that govern execution at run time.

HDLs, therefore, provide the `generate` statement, a mechanism for producing processes under control of `constants` and `generics`. The code fragment below implements the Game of Life.[35]

---

[35] Game of Life by John Horton Conway dates back to 1970. It is not really a game, in fact it is a 2-dimensional cellular automaton where each cell has eight neighbors. Each cell is either dead or alive, and its birth, survival and death depend on how many living neighbors the cell has. A living cell survives if it has two or three live neighbors, and a dead cell is turned into a living one if exactly three of its neighbors are alive (S23/B3).

```
.....
-- spawn a process for each cell in the array
row : for ih in HEIGHT-1 downto 0 generate    -- repetitive generation
   cell : for iw in WIDTH-1 downto 0 generate    -- repetitive generation
      memzing: process(Clk_C)
         subtype live_neighbors_type is integer range 0 to 8;
         variable live_neighbors : live_neighbors_type;
      begin
         if Clk_C'event and Clk_C='1' then
            live_neighbors := live_neighbors_at(ih,iw);
            if State_DP(ih,iw)='0' and live_neighbors=3  then
               State_DP(ih,iw) <= '1';    -- birth
            elsif State_DP(ih,iw)='1' and live_neighbors<=1 then
               State_DP(ih,iw) <= '0';    -- death from isolation
            elsif State_DP(ih,iw)='1' and live_neighbors>=4 then
               State_DP(ih,iw) <= '0';    -- death from overcrowding
            end if;
         end if;
      end process memzing;
   end generate cell;
end generate row;
.....
```

Note that the `generate` statement not only comes in a `for` form but also in an `if` form.

Example   `if i=WIDTH-1 generate ... end generate;`

The generate statement used to conditionally instantiate components

Many circuits are obtained by replicating subcircuits in a more or less regular way. It then makes sense to come up with an algorithm that instantiates and connects components such as to obtain the desired circuit pattern, effectively writing a parametrized **netlist generator**. Straight `component` instantiation cannot cope with such situations; what is needed is a controllable mechanism. The `for ... generate` and `if ... generate` statements indeed fill this role too. The major difference is that instantiation statements rather than concurrent processes are placed between the `generate` and `end generate` keywords this time.

```
-- architecture body
architecture structural of binary2gray is

   -- component declarations
   component xnor2_gate
      port (A1, A2 : in bit;
            ZN : out bit);
   end component;
   component inverter_gate
      port (I : in bit;
            ZN : out bit);
   end component;


   -- signal declarations
   signal Inode_D : bit_vector(WIDTH-1 downto 0);

begin

   -- assemble logic network by instantiating and interconnecting components
   any_slice : for i in 0 to WIDTH-1 generate  -- replicative generate
      -- a row of eqvs except for the MSB which requires an inverter
      less_significant : if i<WIDTH-1 generate  -- conditional generate
         uxn : xnor2_gate  -- component instantiation
            port map (A1=>Inp_D(i), A2=>Inp_D(i+1), ZN=>Inode_D(i));
      end generate;
      most_significant : if i=WIDTH-1 generate  -- conditional generate
         uin : inverter_gate  -- component instantiation
            port map (I=>Inp_D(i), ZN=>Inode_D(i));
      end generate;
      -- a final row of inverters
      uin : inverter_gate  -- component instantiation
         port map (I=>Inode_D(i), ZN=>Oup_D(i));
   end generate;

end structural;
```

Observation 4.20. Generate *statements get interpreted as part of the so-called elaboration phase*[36] *that precedes actual circuit simulation and synthesis.* variables *and* signals *are obviously not acceptable as conditions or as loop boundaries since they are subject to vary at run time, only* constants *and* generics *are.*

---

[36] To be explained shortly, see fig.4.14 for a first idea.

The need to accommodate multiple models for one circuit block

Please recall from table 4.1 that up to four HDL models may occur during a VLSI design cycle to capture a circuit-to-be at distinct levels of detail. Designers also experiment with alternative circuit architectures to compare them in terms of gate count, longest path delay, energy efficiency, and other figures of merit. VHDL accommodates all this need by allowing a design `entity` to have more than one `architecture` body, see fig.4.13.

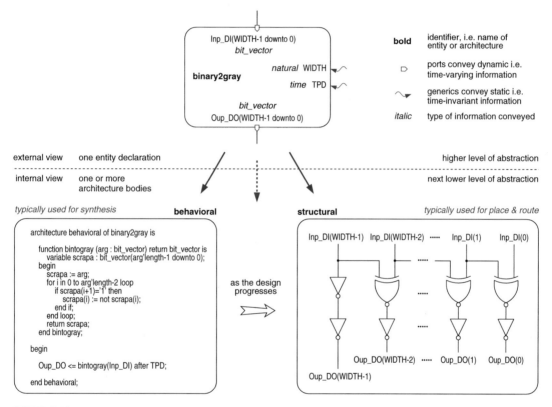

FIGURE 4.13

`Entity` declaration versus `architecture` body and `port` versus `generic`.

> *Warning: Do not pack two functionally distinct behaviors into two architecture bodies that belong to the same entity declaration as this is extremely confusing!*

Configuration specification and binding

In the presence of multiple `architecture` bodies for one `entity` declaration, there must be a way to indicate which `architecture` to consider during simulation and synthesis. This is the configuration specification statement. The mechanism is more general, however, in that it makes it possible to specify a binding between any one instance of a `component` and some `entity-architecture` pair. Put differently, a `component` instantiated under one name can be bound to an `entity` with a different name, and this binding does not need to be the same for all instances drawn from that `component`. Configuration specification statements must appear in the design unit where the `components` concerned are being instantiated.

Prior to simulation and synthesis, the VHDL code must be analyzed for configuration specification statements and the `architecture` bodies indicated must be brought to bear. This preparatory step is also referred to as binding.

The code fragment below specifies that instance `u113` of `entity` or `component` `binary2gray` is to be implemented by the `architecture` body `behavioral` of entity `binary2gray` whereas the body `structural` is to be used in the occurrence of `u188`.

```
.....
for u113: binary2gray use entity binary2gray(behavioral);
for u188: binary2gray use entity binary2gray(structural);
.....
```

The code fragment below specifies that all instances of EQV gates and inverters that have been instantiated in the previous example are to be modeled by the respective `entity-architecture` pairs indicated.

```
.....
for all: xnor2_gate use entity GTECH_XNOR2(behavioral);
for all: inverter_gate use entity GTECH_NOT(behavioral);
.....
```

Hint: Whenever you have multiple architecture bodies for one entity declaration, explicitly indicate which one to use as different EDA tools may otherwise pick different architecture bodies.

Observation 4.21. *VHDL provides a range of complementary constructs that are instrumental in writing parametrized circuit models. More particularly, it is possible to establish a model without committing the code to any specific number of processes and/or instantiated components.*

Elaboration

The fact that the numbers of instances, processes and `signals` in an HDL model are not a priori fixed but obtained under control of the code itself necessitates a preparatory step before simulation or synthesis can begin, see fig.4.14. During that phase, termed elaboration, multiple instantiations get expanded and `generate` statements unrolled so that the final inventory of instances, processes and `signals` can be established. No new instances, processes or `signals` can be created after elaboration is completed.

In preparation to simulating a design, all lowest-level design `entities` (leaf cells) must become available as behavioral models, either as part of the current VHDL design file itself or from a separate components library. Once elaboration is completed, the memory space necessary to hold the entirety of `signals` can be reserved. As processes work concurrently, memory space also needs to be set aside for the `variables` associated with every single process.

In preparation to synthesizing a design, the software has to find out where to resort to actual synthesis and where to simply assemble a netlist from library components. Any elemental `component` instantiated as part of a structural model must be available from the cell library targeted. The numbers of occurrences for all elemental and non-elemental `components` are frozen when elaboration completes.

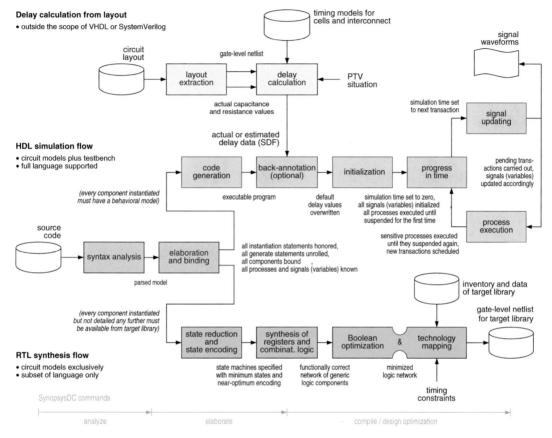

FIGURE 4.14

Major steps in HDL simulation and synthesis (simplified).[37] [38]

Simulation and synthesis essentially follow the same course outlined in fig.4.14 for SystemVerilog and for VHDL, except for the syntax analysis step which is most obviously language-specific. The conceptual commonalities underlying the two HDLs has enabled industry to develop EDA tools that support VHDL-SystemVerilog co-simulation.

---

[37] It may seem counterintuitive that stimuli and expected responses appear nowhere in the simulation part of the drawing. The reason is that the testbench that feeds the event queue with the necessary transactions is itself being described using an HDL and thus enters the processing chain from the left much as the source code of the model under test (MUT).

[38] Organization and vocabulary of commercial products may deviate considerably from our generic figure, take DesignCompiler by Synopsys, for instance. While the "analyze" command does what its name suggests, the subsequent "elaborate" encompasses elaboration, binding, parts of synthesis, plus a preliminary mapping to logic gates from the GTECH generic library. Completing synthesis, timing-driven logic optimization, and technology mapping are then handled by another command named "compile".

4.2.6 CONCEPTS BORROWED FROM PROGRAMMING LANGUAGES

Structured flow control statements

Structured programming is universally accepted by the programming community. The IEEE 1076 standard defines a set of flow control statements that is consistent with this discipline and that supports exception handling in nested loops. Constructs include `if...then...else`, `case`, `loop`, `exit`, `next` the semantics of which are self-explanatory.

Object

A data object, or simply an object, is either a `constant`, a `variable`, a `signal`, or a `file`.

Constant

A constant holds a fixed value that gets assigned where the constant is declared.

Example of a constant declaration   `constant FERMAT_PRIME_4 : integer := 65537;`

Variable

A `variable` holds a changeable value. The scope is limited to the subprogram or `process` where the `variable` is declared, global variables do not exist.[39] As an option, an initial value may be indicated where the variable is declared. `Variables` declared in a subprogram are (re-)initialized whenever the subprogram is called, whereas `variables` declared in a `process` retain their value from one process activation to the next, that is until they are assigned a new value. The assignment operator for `variables` is `:=` whereas `<=` must be used when assigning to a `signal`.[40]

Example of a variable declaration   `variable Brd : real := 2.48678E5;`
Example of a variable assignment   `Brd := Brd + Ddr;`

User-defined data types

In VHDL, `type` is synonym for data type. Basically, a data type defines a set of values and a set of operations that can be performed on them. Types `real`, `integer`, and `time` are part of the IEEE 1076 standard itself, additional types for modeling electrical signals have been introduced in section 4.2.3. Users may declare their own data types or use predefined ones. Enumerated types are supported, and the predefined types `character`, `boolean`, and `bit` are in fact enumerated types. More data types added on top of the VHDL language for the modeling of various electrical phenomena have been discussed in section 4.2.3.

Example of a type declaration   `type audiosample is signed(23 downto 0);`
Enumerated type declaration   `type month is (JANUARY, FEBRUARY, ... , DECEMBER);`

---

[39] So-called protected shared variables have been added later, please refer to section 4.8.1 for their usage.
[40] Further insight on how `variables` relate to `signals` and time can be obtained from section 4.2.4.

Subtypes

A subtype shares the operations with its parent type, but differs in that it takes on a subset of data values only.[41]

Example of a subtype declaration   `subtype day is integer range 1 to 31;`

> *Hint: It is good engineering practice to indicate an upper and a lower bound when using* integers *for the purpose of hardware modeling. A simulation tool can then incorporate on-line checks to ascertain that data values indeed remain within their legal range, while synthesis is put in a position to selectively cut down the number of hardware bits rather than using the default word width of 32 bit.*

Arrays and records

Scalar, i.e. atomic, data types can be lumped together to form composite data types. An array is a collection of elements all of which are of the same type.

Example of an array type declaration   `type byte is bit_vector(7 downto 0);`

As opposed to the homogeneous composition of an array, a record may be assembled from elements of different types which is why it is qualified as heterogeneous. Another difference is that each element is identified by a name unique within the record. An example follows.

```
type date is record
   date_year : integer;
   date_month : month;
   date_day : day;
end record;
```

> *Hint: Records come in handy when a set of* signals *connects to many* entities *or traverses multiple levels of hierarchy. Leaving clock and reset out, use a* record *to collect the entire set into one wholesale quantity that can be referenced as such. Adding or dropping a* signal *or changing the cardinality of a vector then becomes just a matter of modifying one* record *declaration in one* package. *Designers are so dispensed from rewriting the* port *clauses in all* entity *declarations.*

Type attributes and array attributes

A type attribute is a named characteristic of a data type or data value. Type attributes make it possible to recover information such as the range of a type or subtype, the position number, the successor, or the predecessor of a given value in an enumerated type, and the like. An example for the usage of type attributes 'left and 'right is given further down in function nextmonth.

Array attributes operate in a similar way on array types and array objects to obtain their bounds and cardinalities. In the example below, dateandtime'range returns 0 to 5 whereas asking for dateandtime'length yields a value of 6.

```
type sixtupel is array (0 to 5) of integer;   -- type declaration
variable dateandtime : sixtupel;               -- variable declaration
```

---

[41] It is perfectly legal to declare a subtype with an improper subset, i.e. with a data set identical to that of its parent type.

Subprogram, function, and procedure

A subprogram is either a `function` or a `procedure`. A function returns a value and has no side effects, whereas the opposite is true for a procedure. Any subprogram is dynamic, which is to say that it does not persist beyond its current invocation. A first example of a function, named `nextmonth`, is given below, two more follow in listing 4.10.

```
-- package declaration
package calendar is
    type month is (JANUARY, FEBRUARY, MARCH, APRIL, MAY, JUNE, JULY,
                    AUGUST, SEPTEMBER, OCTOBER, NOVEMBER, DECEMBER);
    subtype day is integer range 1 to 31;
    function nextmonth (given_month : month) return month;
    function nextday (given_day : day) return day;
end calendar;

-- package body
package body calendar is

    function nextmonth (given_month : month) return month is
    begin
        if given_month=month'right then return month'left;
        else return month'rightof(given_month);
        end if;
    end nextmonth;

    function nextday (given_day : day) return day is
    .....
    end nextday;


end calendar;
```

Package

A `package` is a named collection of widely used `constants`, `types`, subprograms, and/or `component` declarations. Packages make it possible to sidestep the waste and perils of repeating supposedly identical declarations at multiple places. The contents of a package must be referenced in a `use` clause to make them accessible from some other design unit. This is shown in listings 4.10 and 4.21, the former of which defines two functions that are being called in the latter.

LISTING 4.10

Package holding a pair of Gray code ↔ binary conversion functions.

```
-- Mission: illustrate the use of a package in VHDL.
-- Functionality: Gray code <-> binary code conversion functions.
-- Author: H.Kaeslin.
-------------------------------------------------------------------------


library ieee;
use ieee.std_logic_1164.all;
-------------------------------------------------------------------------


-- package declaration
package grayconv is
   function bintogray (arg : std_logic_vector) return std_logic_vector;
   function graytobin (arg : std_logic_vector) return std_logic_vector;
end grayconv;


-------------------------------------------------------------------------------
--
--                         |     |     |     |     |
-- binary to Gray          |--.  |--.  |--.  |--.  |          # of X-ops on
-- conversion              |  '--X  '--X  '--X  '--X          longest paths
-- for GRAYWIDTH=5         |     |     |     |     |          = 1
--                         v     v     v     v     v
-- bit positions           4     3     2     1     0          X = XOR
--                         |     |     |     |     |
-- Gray to binary          |  ,--X ,--X  ,--X ,--X            # of X-ops on
-- conversion              |--'  |--'  |--'  |--'  |          longest path
-- for GRAYWIDTH=5         |     |     |     |     |          = GRAYWIDTH-1
--                         v     v     v     v     v
--
-------------------------------------------------------------------------------


-- package body
package body grayconv is

   -- purpose: converts binary code into Gray code
   -- by way of in-place computation on a scratchpad variable
   function bintogray (arg : std_logic_vector) return std_logic_vector is
      variable scrapa : std_logic_vector(arg'length-1 downto 0);
   begin
      scrapa := arg;
      for i in 0 to arg'length-2 loop       -- MSB remains unchanged
```

```
            if scrapa(i+1)='1' then
                scrapa(i) := not scrapa(i);
            end if;
        end loop;
        return scrapa;
    end bintogray;


    -- purpose: converts Gray code into binary code
    -- by way of in-place computation on a scratchpad variable
    function graytobin (arg : std_logic_vector) return std_logic_vector is
        variable scrapa : std_logic_vector(arg'length-1 downto 0);
    begin
        scrapa := arg;
        for i in arg'length-2 downto 0 loop    -- MSB remains unchanged
            if scrapa(i+1)='1' then
                scrapa(i) := not scrapa(i);
            end if;
        end loop;
        return scrapa;
    end graytobin;


end grayconv;
```

VHDL applies the principles of **information hiding** to packages by separating package declaration from package body.[42]

All data types and subtypes of VHDL are actually defined in this package along with the pertaining logic and arithmetic operations and a few more features. As package standard comes with the language, always gets precompiled into design library std, and is made available there by default, users do not normally need to care much about it.

There is no point in circuit simulation unless designers can document results and control simulation flow. To facilitate the coding of testbenches and other programs that interface with text files, type declarations and subprograms related to the reading and writing of ASCII files are collected in a special

---

[42] Information hiding is an established principle in software engineering whereby a piece of software is divided into a declaration module or interface that must be made accessible to the caller, and a separate implementation module or body that is deliberately withheld. It is based on the observation that the interactions between software entities remain the same regardless of their implementation details, as long as their interfaces and overall functionalities do not change. Note the similarity with the **black box** concept from electrical engineering.

package named `textio`. Obviously, file I/O code is not for synthesis. Package `textio` routinely gets precompiled into library `std` as well. Yet, to make its definitions immediately available within a design unit, the pertaining source code must include the line `use std.textio.all;`.

Design unit and design file

Incremental compilation is based on the separate processing of individual program modules. The VHDL term for a language construct that is amenable to successful compilation[43] independently from others is design unit. VHDL provides five kinds of design units, namelyn

- `package` declaration,
- `package` body,
- `entity` declaration (see section 4.2.2),
- `architecture` body (see sections 4.2.2 and 4.2.1), and
- `configuration` declaration.

One or more design units are stored in a design file.[44] VHDL compilers, aka VHDL analyzers, accept one design file at a time and store the output in a design library, see fig.4.15.

Observation 4.22. *VHDL supports information hiding and incremental compilation.*

Design library

A design library is a named repository for a collection of design units after compilation on a host computer, which has two major implications. Firstly a design library is not normally portable but specific for a platform. That is, the result depends on the computer for which it has been compiled and on the software tool being used (manufacturer, product, simulator or synthesizer). Secondly, a design library can accommodate many design files and design units. As a consequence, a library is typically referenced under a logical name that differs from the original name(s) of the design file(s) included. Also note that a VHDL design library can hold a program library, a component library, or both.

---

[43] Compilation here loosely refers to the early processing steps that are necessary to simulate and/or synthesize a circuit from VHDL source code, also see fig.4.14. Incidentally, one can distinguish three operating principles in VHDL simulation. A first category translates the original VHDL source into some **pseudo code** which then gets interpreted by a simulation kernel. A second category begins by translating VHDL into C. This C code is then translated into the host's machine code by the local C compiler before being linked with the simulation kernel. Simulation is by executing this machine code. The third category avoids the detour and compiles from VHDL to machine code directly which is why the approach is termed **native compiled code**. As we focus on the language itself, we will not differentiate between such software implementation issues in this text.

[44] Incidentally, note that `design_file` serves as **start symbol** in formal definitions of VHDL.

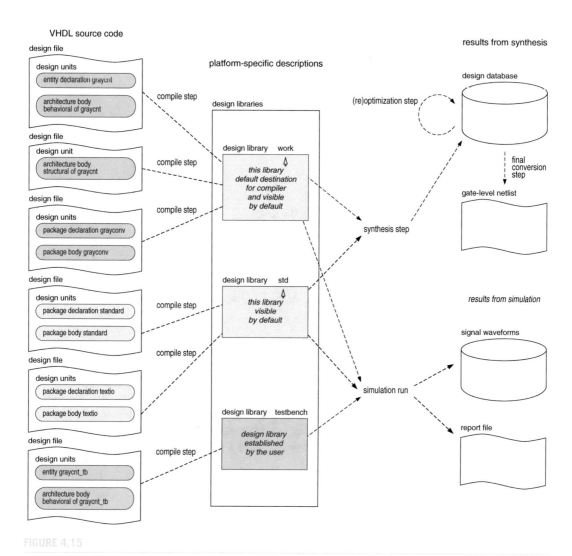

FIGURE 4.15

Organization of VHDL source code and intermediate data (simplified).

Library and use clauses

As most designs make use of VHDL code that is distributed over several design libraries, a mechanism is needed for importing outside subprograms, types, `entities`, and `components`. VHDL provides the `library` statement and the **use clause** for that.

In the code fragment below, the first statement makes the design library `testbench` visible. That is, subprogram `PutSimulationReportLine` from `package graycnt_tb`, for instance, could then be referenced as `testbench.graycnt_tb.PutSimulationReportLine` provided the `package` had been previously compiled into that very design library. Much the same holds for other items declared in the `package`.

The subsequent `use` clause goes one step further in that it dispenses with the need of writing the full library and `package` names each time an item from `package graycnt_tb` is being referenced. In the occurrence, a simple `PutSimulationReportLine` suffices.[45] A second example is shown in listings 4.10 and 4.21 where the code of a Gray counter has been distributed over two design units to illustrate the idea of separate compilation.

```
-- library and use clauses
library testbench;
use testbench.graycnt_tb.all;
.....
```

Special libraries work and std

The design library named `work` differs from all other libraries in that VHDL compilers direct their output to that particular library unless explicitly instructed to do otherwise. Another special library termed `std` accommodates the compilation results from the two packages `standard` and `textio` that come along with VHDL. In addition, libraries `work` and `std` are universally visible because the two statements below are tacitly included in any design unit.

```
library std, work;
use std.standard.all;
```

Note that `standard` is the only package that is immediately available by default, all others ask for an explicit `use` clause. In the example of fig.4.15, the programmer has to include a statement such as `use work.grayconv.all;` in the source code of `graycnt` in order to make the code of package `grayconv` available for reference there.

---

[45] The improved convenience also opens the door for ambiguous references, however, if the same name happens to exist more than once across the various design libraries being made visible in this way.

4.3 KEY CONCEPTS AND CONSTRUCTS OF SYSTEMVERILOG

with contributions by Beat Muheim and Frank Gürkaynak

4.3.1 CIRCUIT HIERARCHY AND CONNECTIVITY

The need for supporting modularity and hierarchical composition

Consider a motherboard from a personal computer, for instance. At the top level, you will discern a CPU, a graphics processor, a ROM or two, and several memory modules composed of multiple RAM chips. Then there are all sorts of peripheral circuits plus a variety of passive components. If you could look into the ICs, you would find datapaths, controllers, storage arrays, and the like. And each such block consists of thousands of logic gates and bistables, that are in turn assembled from the most elementary devices, from transistors.

Electronic systems are organized into multiple layers of hierarchy because it is entirely impractical to specify, understand, design, fabricate, test, and document circuits with millions of gates as flat collections of transistors. Instead, larger entities are hierarchically composed from subordinate entities that are interconnected with the aid of busses and individual wires. Abstraction, modularity, and repetition further help to arrive at manageable circuit descriptions. Any HDL that wants itself useful must support these techniques, see fig.4.2 for a first impression.

Module, structural view

The `module` construct is the basic building block of a SystemVerilog design. Its primary purpose is to encapsulate the data, functionality and timing of a circuit or subcircuit. Each `module` consists of header and body that describe the circuit's external and internal views respectively.

The most important part of the header is the **port list** that enumerates those circuit nodes that are visible from outside between parentheses. Put differently, each signal that is part of the port list corresponds to a connector on the icon of the circuit. Fig.4.13 illustrates this with VHDL code, but you get the idea.

The circuit's internal technicalities are being detailed in the body, syntactically separated from the header by a semicolon. A `module` is permitted to contain **instances** of other `modules`, thereby creating hierarchy. A code example follows in listing 4.11. Although you are probably not yet in a position to understand everything, it should become clear that the circuit is composed of five logic gates and four flip-flops. As an exercise, draw a schematic diagram for the circuit.

LISTING 4.11

Structural view of a linear feedback shift register (LFSR) of length 4.

```
// module header with external interface
module lfsr4
  ( output logic Oup_DO,
    input logic Clk_CI, Rst_RBI, Ena_SI ) ; // reset is active low

// module body with circuit structure
```

```
// declare a variable for each inner node
logic [1:4] State_DP;
logic n11, n21, n31, n41, n42;

// instantiate cells and connect them by listing port maps
// (cells are from Synopsys' generic library in this example)
GTECH_FD2 u10 ( .D(n11), .CP(Clk_CI), .CD(Rst_RBI), .Q(State_DP[1]) );
GTECH_FD2 u20 ( .D(n21), .CP(Clk_CI), .CD(Rst_RBI), .Q(State_DP[2]) );
GTECH_FD2 u30 ( .D(n31), .CP(Clk_CI), .CD(Rst_RBI), .Q(State_DP[3]) );
GTECH_FD4 u40 ( .D(n41), .CP(Clk_CI), .SD(Rst_RBI), .Q(State_DP[4]) );
GTECH_MUX2 u11 ( .A(State_DP[1]), .B(n42),          .S(Ena_SI), .Z(n11) );
GTECH_MUX2 u21 ( .A(State_DP[2]), .B(State_DP[1]),  .S(Ena_SI), .Z(n21) );
GTECH_MUX2 u31 ( .A(State_DP[3]), .B(State_DP[2]),  .S(Ena_SI), .Z(n31) );
GTECH_MUX2 u41 ( .A(State_DP[4]), .B(State_DP[3]),  .S(Ena_SI), .Z(n41) );
GTECH_XOR2 u42 ( .A(State_DP[3]), .B(State_DP[4]),              .Z(n42) );

// connect state bit of rightmost flip-flop to output port
assign Oup_DO = State_DP[4];

endmodule
```

Hint: SystemVerilog is case-sensitive, e.g. `clk_ci` $\neq$ `CLK_CI`.[46]

Hint: Naming a signal `input` *and* `output` *is all too tempting, yet these are reserved words in SystemVerilog, We recommend the use of* `Inp` *and* `Oup` *instead.*

How to compose a circuit from components

How do you proceed when asked to fit a circuit board with components? You think of the exact name of a part required, go and fetch a copy of it, and solder the terminals of that one copy in a well-defined manner to metal pads interconnected by narrow lines on the board. The **module instantiation statement** of SystemVerilog does exactly this, albeit with virtual `modules` and variables[47] instead of physical parts and wires.

In the above code example, nine components get instantiated. As multiple copies of the same component must be told apart, each instance is assigned a unique identifier; `u10, u20, … , u42`, in the occurrence.

---

[46] No rule without exceptions. The letters d, h, o and b that indicate the base in decimal, hexadecimal, octal and binary numbers, the hex digits A through F, and the logic values X and Z are all case-insensitive. As an example, the casing `16'hFE39` for a 16-bit hexadecimal number takes advantage of this for maximum legibility.

[47] Verilog traditionally required that modules be connected by `wires` but this is no longer mandatory. Generally speaking, SystemVerilog variables can assume many roles that necessitated a `wire` in classic Verilog. For simplicity, we prefer to work with variables in this text without emphasizing where `wires` might do the same job. `wires` continue to be compulsory for multi-driver nodes and for bidirectional `inout` ports, though.

Each instantiation statement includes a **port map**, i.e. a series of associations where an electrical connection is indicated between an instance terminal (preceded by a dot .) and a node in the superordinate circuit the name of which is given in parentheses. Named port connections become overly verbose, however, when the instance ports and the connecting circuit nodes carry identical names anyway.

Tedious example of named port connections `.InpA(InpA), .InpB(InpB), .InpC(InpC)`

SystemVerilog comes to rescue with two shorthand notations for port mapping.[48] A first construct does a name-based association for the port(s) specified, while the second does so for all ports of matching names (and leaves the others, if any, for regular named port connections).

Example of dot name port connections `.InpA, .InpB, .InpC`
Example of dot star port connections `.*`

SystemVerilog requires that the signals running back and forth between instances be declared. Those connecting to the outside world are automatically known from the port list and need not be declared a second time. Inner nodes, in contrast, are defined in a series of variable declaration statements that precede the module instantiation statements.

Observation 4.23. *SystemVerilog can describe the hierarchical composition of a digital circuit by instantiating* `modules` *and by interconnecting them with the aid of wires that are normally modeled as variables.*

Special constructs for modeling busses

Larger systems often make use of busses to connect multiple circuits. Using `ports` to describe a bus typically leads to duplication of code, making the code cumbersome to maintain and prone to error. The **interface** construct unique to SystemVerilog collects a set of signals into a single port without indicating a direction like input or output; bidirectional lines are also supported.

In real life, not all signals will necessarily connect to each bus partner, however. Also, a bus master will very likely impose other signal directions than a slave. And a clock signal, if part of the bus, is driven from a single place. The **modport** command is here to capture such individual details, allowing engineers to differentiate between the description of the actual interface and the corresponding module connections. As an `interface` must get instantiated like a `module`, it is best thought of as a cable with multiple wires, and each `modport` as a specific type of connector for that cable. While both constructs are highly useful, giving syntax details and meaningful examples is well beyond this text, please refer to [100] [101] [98].

A model that describes a circuit as a bunch of interconnected components is qualified as **structural**. It essentially holds the same information as the circuit netlist does. Manually coding `module` bodies in this way is not particularly attractive. Indeed, most structural models are obtained from register-transfer-level (RTL) models by automatic synthesis.

---

[48] For reasons of backward code compatibility, SystemVerilog continues to support a fourth scheme whereby ports get mapped according to their positions in the list, but this is not a recommended practice.

4.3.2 INTERACTING CONCURRENT PROCESSES

The need for modeling concurrent activities

While we have learned how to capture a circuit's hierarchical composition, our model remains devoid of life up to this point as we have no means for expressing circuit behavior, precluding both simulation and synthesis. So there must be more to SystemVerilog.

The most salient feature of any electronic system is the concurrent operation of its subcircuits, just think of all those ICs on a typical circuit board, or of the many thousands of logic gates and storage cells within each such chip. This inherent parallelism contrasts sharply with the line-by-line execution of program code on a computer. Another innate trait is the extensive communication that permanently takes place between subcircuits and that is physically manifest in the multitude of wires that run across chips and boards. This is simply because there can be no cooperation between subcircuits without on-going exchange of data.

Now assume you wanted to write a software model that imitates the behavior of a substantial circuit using some traditional programming language such as Pascal or C. You would soon get frustrated because of the absence of constructs and mechanisms to handle simultaneous operation and interaction. Hardware description languages extend the expressive power of software languages by supporting concurrent processes and means for exchanging information between them, see fig.4.3 for a first idea.

How to describe combinational logic behaviorally

Many arithmetic and logic computations are straightforward and can be expressed in one instruction with no need for branching. These are best captured in a **continuous assignment**, the most simple SystemVerilog construct for defining the value of a variable (or of a `wire`).

Example with no operation	`assign ThisMonth_D = AUGUST;`
Example with logic operations	`assign Oup_D = Aa_D^(Bb_D &~Cc_D);`
Example with an arithmetic operation	`assign Error_D = Actual_D - Wanted_D;`

A continuous assignment is nothing else than a process in a single statement. Whenever an operand on the right-hand side changes value, the expression is (re-)evaluated and the variable on the left-hand side gets updated accordingly. Continuous assignments may call functions but no tasks.[49]

Situations abound where continuous assignments do not suffice to express the desired functionality, e.g. because conditional execution is required. A condition operator can then be included. The syntax goes "conditional_expression ? then_expression : else_expression".

Example with a condition   `assign Oup_D = Add_S ? (Aa_D + Bb_D) : (Aa_D - Bb_D);`

Where this is still inadequate for capturing the desired input-to-output mapping, designers can recur to a **procedural block**, a more powerful construct for expressing a concurrent process that comes in several varieties: `always_comb`, `always_ff`, `always_latch`, `always`, `initial`, and `final`. What sets all of them apart from the continuous assignment is this:

---

[49] The difference will be explained in section 4.3.6.

- The capability to update two or more variables at a time,
- The fact that the instructions for doing so are captured in a sequence of statements that are going to be executed one after the other,
- The liberty to make use of [local] variables for temporary storage,
- A more detailed control over the conditions for activating the process.[50]

Procedural blocks are best summed up as being concurrent outside and sequential inside.[51] The SystemVerilog construct for capturing combinational behavior in a sequence of statements is the **always_comb** block. You can think of an `always_comb` block as a container for a small program that generates the truth table for the (sub)circuit being modeled.

```
always_comb
   begin
      Spring_D = 0;    // execution begins here
      if (ThisMonth_D==MARCH & ThisDay>=21) Spring_D = 1;
      if (ThisMonth_D==APRIL)               Spring_D = 1;
      if (ThisMonth_D==MAY)                 Spring_D = 1;
      if (ThisMonth_D==JUNE  & ThisDay<=20) Spring_D = 1;
   end   // process suspends here
```

Note that the keywords `begin` and `end` are mandatory here because the `always_comb` block includes more than one statement; this will be different in the subsequent example.

How to describe a register behaviorally

SystemVerilog provides designers with another procedural block specifically for modeling flip-flops and other edge-triggered circuits. This is the **always_ff** block, see listing 4.12. Note that the present state of the process is kept from one activation to the next in a vector named `State_DP`.

LISTING 4.12

Code example for an edge-triggered register that features an asynchronous reset, a synchronous load, and an enable. Actual designs are unlikely to combine all three mechanisms in a single register, so a subset of the clauses shown will most often do.

```
always_ff @(posedge Clk_C, negedge Rst_RB)   // sensitivity list
   // activities triggered by asynchronous reset
   if (~Rst_RB)
      State_DP <= '0;   // shorthand for all bits zero
   // activities triggered by rising edge of clock
   else
```

---

[50] Several of these items will be clarified in section 4.3.4.

[51] "Sequential" here refers to code execution during simulation, not to the nature of the circuit being modeled.

```
    // when synchronous load is asserted
    if (Lod_S)
        State_DP <= '1;    // shorthand for all bits one
    // otherwise assume new value iff enable is asserted
    else if (Ena_S)
        State_DP <= State_DN;    // admit next state into state register
```

The `always_comb` and `always_ff` blocks are the most useful procedural blocks for circuit modeling. For brevity, we will not discuss the `always_latch` block that occupies an intermediate position between the two. The `initial` block is ill-suited for describing a piece of hardware as it executes only once and then suspends forever, and so is the `final` block. Section 4.3.4 will justify why the plain `always` block is not generally recommended either.

Local versus shared variables

During a simulation run, it is the totality of variables that together hold the current state of execution. A variable's value can be altered by a continuous assignment or by an assignment instruction that is part of a procedural block ("procedural assignment"). A variable declaration must specify the data type and the name of the variable. An optional `var` keyword may be added to clarify the intention,[52] and an initial value can also be assigned as part of the declaration.[53]

Example of a variable declaration   `var real Brd = 2.48678E5;`
Example of a procedural variable assignment   `Brd = Brd + Ddr;`
Example of a continuous variable assignment   `assign Brd = Brd + Ddr;`

As shown in fig.4.3, variables can be used in two distinct ways. Those declared within a procedural block are local and not visible from outside. Others participate in exchanging dynamic information between concurrent processes and/or between `modules` and must be declared at a higher level.[54] VHDL provides programmers with a special language element, aptly named `signal`, for this purpose. SystemVerilog knows of no such construct, yet it helps to distinguish between local intra-process variables and shared inter-process variables mentally.

Example of a variable declaration (for local use)   `month ThisMonth;`
Second example (for use as "signals") `real Error_D = -1.7, Actual_D = 4.3, Wanted_D = 6.0;`

> *Hint: HDL code is easier to read when "signals" can be told from local variables by their visual appearance. As part of an elaborate naming convention to be presented in section 6.7, we make it a habit to append an underscore followed by an informative suffix of a few upper-case letters to those variables that are shared between concurrent processes and serve to convey information from one to another.[55]*

---

[52] This is a feature introduced with SystemVerilog, `var` was not a reserved word in classic Verilog.

[53] Why this is inadequate for modeling a hardware reset will be explained in observation 4.31.

[54] "Dynamic" means that the data are free to evolve over time, i.e. to change value as simulation progresses.

[55] Attentive readers may notice that certain listings and schematic diagrams in this text feature identifiers that do not adhere to this scheme, e.g. CP instead of Cp_CI. This applies to standard cell connectors where the naming gets decided by the library vendor rather than by VLSI designers and HDL code writers.

Module, behavioral view

Most modules include a collection of concurrent processes that together make up the module's overall functionality. Such models are called **behavioral** because they specify how the module is to react in response to changing inputs. Potential reactions include the updating of outputs, the updating of the module's current state, the checking of compliance with some predefined timing conditions, or simply ignoring the new input.

Listing 4.13 shows a behavioral module for the LFSR circuit of listing 4.11. Taking fig.4.3 as a pattern, make a small drawing that illustrates the processes and the variables being exchanged. Find out what hardware item each process stands for and compare the drawing to that established earlier. What liberties do you have in coming up with a schematic diagram?

LISTING 4.13

Behavioral view of a linear feedback shift register (LFSR) of length 4.

```
// external interface of module
module lfsr4 (
      input logic Clk_CI, Rst_RBI, Ena_SI,    // reset is active low
      output logic Oup_DO );

// behavioral model for module

   // declare internal variables
   logic [1:4] State_DP, State_DN; // for present and next state

   // computation of next state using concatenation of bits
   assign State_DN = {(State_DP[3]   State_DP[4]), State_DP[1:3]};

   // updating of state
   always_ff @(posedge Clk_CI, negedge Rst_RBI)
      // activities triggered by asynchronous reset
      if ( Rst_RBI)
         State_DP <= 4'b0001;
      // activities triggered by rising edge of clock
      else
         if (Ena_SI)
            State_DP <= State_DN; // admit next state into state register

   // updating of output
   assign Oup_DO = State_DP[4];

endmodule
```

Observation 4.24. *In SystemVerilog, the behavior of a digital circuit typically gets described by a collection of concurrent processes that execute simultaneously, that communicate via variables, and where each such process represents some subfunction.*

Hardware modeling styles compared

Except for physical descriptions, SystemVerilog supports various circuit modeling style.

A **procedural model** essentially describes functionality in a sequence of steps much as a piece of conventional software code. A circuit is captured in one procedural block and its behavior gets implemented with the aid of sequential statements there.[56]

A **dataflow model** describes the behavior as a collection of continuous assignments that get executed under the coordination of the variables exchanged, and where each such variable stands for a signal.

A **structural model** describes the composition of a circuit by way of `module` instantiation statements along with the interconnections in between. It is equivalent to a netlist.

Listing 4.14 juxtaposes three `modules`, each coded from a different perspective while fig.4.16 illustrates their differences and commonalities.[57] Make sure you understand the profound difference between the procedural and dataflow models in spite of their apparent similarity.

Observation 4.25. *SystemVerilog allows for procedural, dataflow, and structural modeling styles to be freely combined, see fig.4.4. Circuit models almost always take advantage of this.*

LISTING 4.14

Procedural, dataflow, and structural modeling styles compared.

```
// compute result in a series of sequential steps
module fulladd_procedural
   ( input logic Aa_DI, Bb_DI, Cc_DI,
     output logic Sum_DO, Carry_DO );

   always_comb
   begin
      logic loc1, loc3, loc4;
      loc1     =   Aa_DI ^ Bb_DI;
      Sum_DO   =   Cc_DI ^ loc1;
      loc3     = ~(Cc_DI & loc1);
      loc4     = ~(Aa_DI & Bb_DI);
      Carry_DO = ~(loc3 & loc4);
   end
endmodule
//---------------------------------------------------------------------------
// spawn a continuous assignment for each logic operation
```

---

[56] Procedural, dataflow, and structural are just user-defined terms, not reserved words of the language.

[57] A full-adder has been chosen in this example for its simplicity and obviousness. Adders are normally synthesized from algebraic expressions that include a + operator.

```
module fulladd_dataflow
   ( input logic Aa_DI, Bb_DI, Cc_DI,
     output logic Sum_DO, Carry_DO );

   logic Loc1_D, Loc3_D, Loc4_D;
   assign Loc1_D    =    Aa_DI  ^ Bb_DI;
   assign Sum_DO    =    Cc_DI  ^ Loc1_D;
   assign Loc3_D    = ~(Cc_DI & Loc1_D);
   assign Loc4_D    = ~(Aa_DI & Bb_DI);
   assign Carry_DO  = ~(Loc3_D & Loc4_D);
endmodule
//-----------------------------------------------------------------------
// describe circuit network as a bunch of interconnected std cells
// note: cells from Synopsys' generic library are used here
module fulladd_structuralgtech
   ( input logic Aa_DI, Bb_DI, Cc_DI,
     output logic Sum_DO, Carry_DO );

   logic Loc1_D, Loc3_D, Loc4_D;
   GTECH_XOR2  u1 ( .A(Bb_DI),   .B(Aa_DI),   .Z(Loc1_D) );
   GTECH_XOR2  u2 ( .A(Cc_DI),   .B(Loc1_D),  .Z(Sum_DO) );
   GTECH_NAND2 u3 ( .A(Cc_DI),   .B(Loc1_D),  .Z(Loc3_D) );
   GTECH_NAND2 u4 ( .A(Aa_DI),   .B(Bb_DI),   .Z(Loc4_D) );
   GTECH_NAND2 u5 ( .A(Loc3_D),  .B(Loc4_D),  .Z(Carry_DO) );
endmodule
```

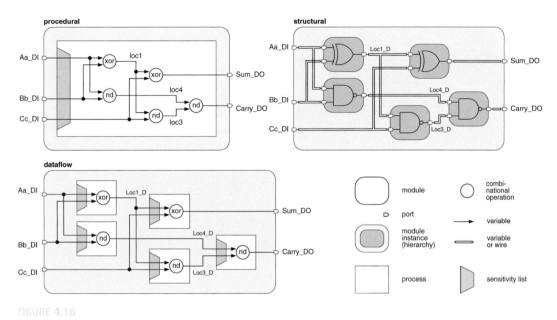

FIGURE 4.16

The modeling styles of listing 4.14 illustrated.

4.3.3 A DISCRETE REPLACEMENT FOR ELECTRICAL SIGNALS

The need for multiple logic values to describe a circuit node

An innocent approach to hardware modeling would be to use one binary digit per circuit node. In SystemVerilog, the standard 2-valued data type is `bit` which can take on value 0 or 1. Digital circuits exhibit traits that cannot be captured with a 2-valued abstraction, however. Just think of transients, indeterminate states following power-up, three-state outputs, multiple buffers driving a common node with an inherent potential for conflicts, and the like. You may want to refer to fig.4.6 for the larger picture.

Observation 4.26. *Distinguishing between logic 0 and 1 is inadequate for modeling the binary signals found in digital circuits. A more elaborate multi-valued logic system must be sought that is capable of capturing the effects of both node voltage and source impedance.*

SystemVerilog offers two more sophisticated data types named `logic` and `wire` respectively.[58] Using either of these data types, the electrical conditions of a circuit node are condensed into one **logic value** at any time as shown in table 4.3 and fig.4.17a.

---

[58] Verilog adepts will notice that this is not the full story. For one thing, `logic` was named `reg` in Verilog, a misleading choice. `reg` continues to be supported in SystemVerilog for compatibility with legacy code, but its use is discouraged. For another thing, there exist multi-valued data types other than `logic` and `wire`. As their intended purpose — switch-level simulation of transistor networks — is irrelevant to synthesis-based VLSI design, introducing them here would cause unnecessary confusion.

Table 4.3   The SystemVerilog 4-valued logic system MVL-4. Symbol x here stands for any condition of ambiguity. No distinction is made between uninitialized, unknown and don't care.

logic value → ↓	logic state		
	low	unknown	high
uninitialized		x	
strength driven	0	x	1
high-impedance	z	z	z
don't care		x	

A number of clarifications are due at this point.

- Node voltage is quantized into three **logic states**:

low	logic low, that is below U_l.
high	logic high, that is above U_h.
unknown	may be either "low", "high", or anywhere in the forbidden interval in between, e.g. as a result from a short between two conflicting drivers.

  No difference is made between a drive conflict, the outcome of which is truly unknown, and a ramping node, the voltage of which is known to assume values between thresholds U_l and U_h for a brief moment of time. Either condition is modeled as "unknown".

- The amount of current that a subcircuit can sink or source — which is inversely related to the source impedance — gets mapped onto just two discrete **drive strengths**:[59]

strong	the low impedance value commonly exhibited by a driving output.
high-impedance	the almost infinite impedance exhibited by a disabled three-state output.

  No difference is made between "charged high" and "charged low". All high-impedance conditions are merged into a single condition of undetermined state or voltage.

- One extra value has been added, namely:

don't care	whether the node is "low" or "high" is considered immaterial, used by designers to leave the choice to the logic optimization tool (applicable to synthesis only).

  The same symbol x is indeed being used as for "unknown", an oddity of the language.

- SystemVerilog knows of no particular value to describe a circuit node that has never been assigned any value following power-up. No attempt is made to distinguish between "unknown" and "uninitialized". Again, the x serves to denote either condition.

---

[59] Note the absence of drive strength "weak" as exhibited by a passive pull-up/-down resistor or a snapper.

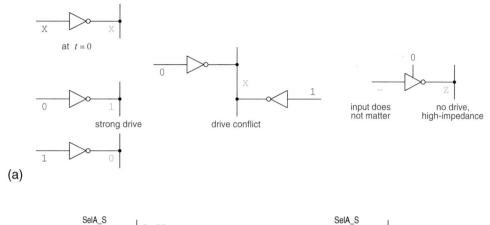

(a)

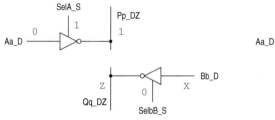

(b)

single-driver signals Pp_DZ and Qq_DZ
may assume distinct logic values,
no difference between `logic` and `wire`

if multi-driver signal Com_DZ is of type
`logic` then an error message gets issued
`wire` then the conflict is resolved to Com_DZ = 1

FIGURE 4.17

The SystemVerilog MVL-4 illustrated (a), types `logic` and `wire` compared (b).

How to model three-state outputs and busses

How do we tell we want a node to be released, i.e. reverted to an undriven condition? With the aid of the MVL-4 system, the answer is straightforward: just assign logic value z.

Example   `assign Oup_DZ = Ena_S ? Inp_D : 1'bZ;`

As illustrated in fig.4.6, many digital circuits include **multi-driver nodes** that operate under control of multiple processes. How to model them is now obvious. In the code fragment below, the common node Com_DZ is left floating, that is in a high-impedance condition, when neither of the two drivers is enabled. In the occurrence of a parallel **bus**, assigning `'{default:1'bZ}` or the shorthand form `'Z` will do so for all bits, no matter how wide the bus is.

```
wire Com_DZ;
logic Aa_D, Bb_D, SelA_S, SelB_S;
.....
assign Com_DZ = SelA_S ? ~Aa_D : 1'bZ;
.....
assign Com_DZ = SelB_S ? ~Bb_D : 1'bZ;
.....
```

A multi-driver node may give rise to drive conflicts when something goes wrong, however.[60] It is, therefore, important to know how such situations are dealt with during simulation.

If the node is modeled using a variable of type `logic`, some form of error message will be generated during compilation or simulation. On a `wire`-type node, in contrast, any conflict between diverging values is tacitly solved at simulation time by calling a **resolution function** that determines the most plausible outcome, see table 4.4. As an example, this function specifies that simulation is to continue with a 1 should an attempt be made to drive a node to z and 1 at the same time. The difference is illustrated in fig.4.17b.

Table 4.4   The SystemVerilog built-in resolution function for data type `wire`.

	X	0	1	Z
X	X	X	X	X
0	X	0	X	0
1	X	X	1	1
Z	X	0	1	Z

Observation 4.27. *Nodes modeled using data type* `wire` *can accommodate multiple drivers whereas those modeled using* `logic` *— or any other type of variable for that matter — cannot.*

Selecting adequate data types

Data type `logic` emulates the electrical behavior of digital circuits in a much more realistic way than type `bit` does and should be used whenever one wants to model an actual circuit node (as opposed to a local variable in a procedural block, for instance).[61]

---

[60] A **drive conflict** implies that two or more processes attempt to drive a signal to incompatible logic values.
[61] For efficiency considerations see footnote 20.

data type value set per binary digit	bit 2	logic 4	wire 4
for simulation purposes			
modeling of power-up phase	no	passable	passable
modeling of weakly driven nodes	no	no	no
modeling of multi-driver nodes	no	no	yes
handling of drive conflicts	n.a.	n.a.	resolved
for synthesis purposes			
three-state drivers	no	yes	yes
don't care conditions	no	yes	yes

Data types for modeling multi-bit signals

It goes by itself that scalars of the same type can be collected into a vector. The table below compares such vectors with various other types that comprise multiple data bits.

```
bit [23:0]    // 24 bit, 2-valued -> no electrical modeling capability
logic [7:0]   //  8 bit, 4-valued -> elec. modeling capab., single driver
wire [24:11]  // 14 bit, 4-valued -> elec. modeling capab., multiple drivers
```

data type(s)	byte, shortint, int, longint	integer	bit vector	logic vector	wire vector
value set per binary digit	2	4	2	4	4
word width	8/16/32/64	32	at the programmer's discretion		
arithmetic operations	yes	yes	yes	yes	yes
default signed/unsigned	signed	signed	unsign.	unsign.	unsign.
logic operations	yes	yes	yes	yes	yes
access to subwords or bits	yes	yes	yes	yes	yes
modeling of electrical effects	no	passable[a]	no	passable[a]	passable

[a] Except that multiple drivers that are not allowed with this data type.

SystemVerilog is much more relaxed on data types than VHDL. It is perfectly legal to perform arithmetic operations on bit vectors and to carry out logic operations with numerical data types, for instance. Accessing just one bit or just a few bits in an integer is also supported. What's more, data words get tacitly extended or slashed to make things fit in assignments between items of unlike widths. This latitude often comes handy as there is almost no obligation to include type conversions in the code. The down side is that it makes it impossible for EDA tools to find suspect code fragments during compilation. And type-related issues are indeed common and sometimes subtle [97].

Numerical data types can be explicitly declared as unsigned or signed with the aid of a modifier. `unsigned` indicates an unsigned integer number, and `signed` a signed integer number coded in 2's complement (2'C) format.[62] In the absence of a modifier, the number representation scheme defaults to signed for `byte`, `shortint`, `int`, `longint`, and `integer`, and to unsigned for other data types, but it is always a good idea to make the programmer's intention explicit.

LISTING 4.15

Examples of SystemVerilog multi-bit data types.

```
byte                    //  8 bit, signed,   2-valued, single driver
bit [7:0]               //  8 bit, unsigned, 2-valued, single driver
shortint signed         // 16 bit, signed,   2-valued, single driver
int unsigned            // 32 bit, unsigned, 2-valued, single driver
integer                 // 32 bit, signed,   4-valued, single driver
logic unsigned [11:0]   // 12 bit, unsigned, 4-valued, single driver
logic signed [5:0]      //  6 bit, signed,   4-valued, single driver
wire [6:1]              //  6 bit, unsigned, 4-valued, multiple drivers
```

Orientation of binary vectors

When encoding a number using a positional number system, there is a choice between spelling the data word with the MSB or the LSB first. In SystemVerilog, this primarily applies to `logic`- and `wire`-type vectors. Any misinterpretation will cause severe malfunctions.

Hint: Any vector that contains a data item coded in some positional number system should consistently be declared as $[i_{MSB} : i_{LSB}]$ where 2^i is the weight of the binary digit with index i. The MSB will so have the highest index assigned to it and will appear in the customary leftmost position because $i_{MSB} \geq i_{LSB}$.

Example   `logic [4:0] Hour_D = 5'b10111;`

All multi-bit data types introduced so far are intended to represent integer numbers with the radix point immediately following the LSB. We can thus illustrate their formats as `ii...i` and `si...i` respectively, where `s` stands for the sign bit and each `i` for one binary digit.

Let us conclude our overview on data types with a word of advice.

Hint: Rather than silently relying on defaults to keep the code as terse as possible, make your intentions reasonably explicit in the code as this will not only improve code quality but also accelerate debugging and code maintenance.

This is particularly important in SystemVerilog where defaults are irregular, where little type checking is performed, where data words get tacitly extended or slashed in width to make things fit, and where there is almost no obligation to include type conversions.

---

[62] No provisions are made to support other schemes such as 1's complement (1'C) or sign-and-magnitude (S&M). Please check section A.1.1 if not familiar with the number representation schemes used in computing.

4.3.4 AN EVENT-DRIVEN SCHEME OF EXECUTION

The need for a mechanism that schedules process execution

Recall the concurrent SystemVerilog constructs introduced in section 4.3.2.

- o Continuous assignment.
- o Procedural blocks `always_comb`, `always_ff`, and `always_latch` (for circuit modeling).
- o Procedural blocks `always`, `initial`, and `final` (for testbenches).

Simulation must, of course, yield the same result as if the many processes present in a circuit model were operating simultaneously, although no more than a few processor cores are normally available for running the simulation code. What is obviously required then is a mechanism that schedules processes for sequential execution and that combines their effects such as to perfectly mimic concurrency.

SystemVerilog and VHDL share the same model of communicating concurrent processes. The principles of event-driven simulation, explained in section 4.2.4, are also much the same. What follows here are just the SystemVerilog particularities. This, together with a general understanding of event-driven simulation as obtained from section 4.2.4 up to observation 4.9 and from fig.4.18, should suffice as a background for coding RTL circuit models.[63] Please remember that further RTL code examples are given in section 4.3.2 and on the book's companion website.

Event-driven simulation

Observation 4.28. *In SystemVerilog simulation, the continuum of time gets subdivided by events each of which occurs at a precise moment of simulation time. An* **update event** *is said to happen whenever the value of a variable (or* `wire`*) changes.*

The term "transaction" for an event scheduled for execution does not exist in SystemVerilog. Fig.4.18 is labeled accordingly. The SystemVerilog syntax also supports no distinction between signals and variables, which means that fig.4.10 does not apply. The dual role of variables has been discussed in section 4.3.2. How variables interact with the event queue depends on how exactly the assignment is coded and will be explained shortly, table 4.5 gives a summary.

Observation 4.29. *As opposed to software languages where execution strictly follows the order of statements in the source code, there is no fixed ordering for carrying out processes (including continuous assignments and assertion statements) in SystemVerilog. When to invoke a process gets determined solely by events on the variables (and* `wires`*) that run back and forth between processes.*

---

[63] The exact operation of the SystemVerilog process scheduler, advertized as "stratified event queue", is rather convoluted and beyond the scope of an introductory text on VLSI. A simulation cycle is actually organized into 17 ordered **simulation regions**, 9 reserved for executing SystemVerilog statements and 8 for programming language interface (PLI) code. Different types of statements (e.g. blocking assignments, system tasks, assertions) and even parts of such statements (e.g. right- and left-hand sides of nonblocking assignments) get executed in different regions as suggested in fig.4.18. Note that the drawing is just a first order approximation as the actual scheduler includes regions not shown and even feedback loops. The fine details matter primarily in the context of assertion-based verification and for certain testbench designs. Readers interested in full-depth discussions are referred to the specialized literature such as [98] [102] [100] [103].

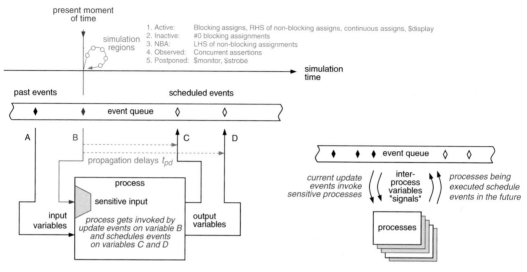

Event-driven simulation in SystemVerilog. Interactions between the event queue and a process (a), actions that repeat during every simulation cycle (b) (simplified).

Sensitivity list, process suspension and reactivation

With a background in event-driven simulation, we are now in a position to better understand the functioning of procedural blocks. As an example, reconsider the `always_ff` block of listing 4.12 and specifically the `@(posedge Clk_C, negedge Rst_RB)` parenthesis on the first line. A matching event on either signal — both actually variables in SystemVerilog — (re-)activates code execution, which is why they are said to form the sensitivity list of that process.

Constructs for temporarily suspending a process and for stating when it is to resume include:

Statement	Wake-up condition
`@(...)`	an update event (value change) on any of the signals listed here
`wait (...)`	*idem* plus the logic conditions specified here
`#...`	a predetermined lapse of time as specified here

Continuous assignments, in contrast, need and have no sensitivity list. By definition, any update event of a variable on the right-hand side of the assignment operator (re-)activates the process.

Example   `assign Oup_D = InpA_D + InpB_D;`

Delay modeling

In the context of simulation, the lapse of time between an update event at the input of a process and the ensuing event scheduled at the output reflects the delay of the piece of hardware being modeled. Circuit delays are typically conveyed by a # expression which forms an optional part of the various assignment statements. The continuous assignment below, for instance, models the propagation delay of an adder by scheduling an update event on its output t_{pd} after an event at either input.

Example   `assign #TPD Oup_D = InpA_D + InpB_D;`

The next simulation model uses a procedural block to also account for contamination delay t_{cd}.

```
always_comb
begin
    Oup_D <= #TCD '{default:1'bX};   // revert all bits to unknown after tcd
    Oup_D <= #TPD InpA_D + InpB_D;   // propagate result to output after tpd
end
```

Hint: SystemVerilog accepts time values in multiples of some time unit previously defined with a `timeunit` *statement or a* `'timescale` *compiler directive, e.g.* `#2.8`*. To avoid surprises, always specify the measurement unit, i.e. write* `#2.8ns` *instead.*

Hint: When simulating models with zero delays it becomes difficult to tell apart cause and effect in the output waveforms as the respective update events appear to coincide. A trick is to artificially postpone future events by a tiny amount of time in otherwise delayless variable assignments. To allow for quick adjustments, a constant is best declared in a package and referenced throughout a model hierarchy. Note that the largest sum of fake delays must not exceed one clock period, though.

Example   `assign #FAKEDELAY Oup_D = InpA_D + InpB_D;`

Blocking versus nonblocking assignments

Within a procedural block, a variable may be assigned a value in either of two ways, both of which may or may not carry a delay, and both of which involve the process scheduler.

A blocking assignment is identified by `=` as operator symbol. If the statement includes a non-zero delay expression, execution of the current process is suspended until that amount of simulation time has elapsed. If no delay is specified, the effect is felt immediately, that is, in the next statement exactly as with any software programming language.

Nonblocking assignments are recognized by the `<=` operator. While the right-hand side is executed instantly, the updating of the variable on the left-hand side is deferred until after the completion of all blocking assignments scheduled for the same simulation time, or even longer if a non-zero delay expression stipulates so. Meanwhile, program execution procedes with the subsequent instruction(s) in the procedural block.

Table 4.5 summarizes the effects of the various SystemVerilog assignment statements.

Always blocks

The general purpose `always` block inherited from Verilog is the equivalent of the `process` statement in VHDL: versatile but not always easy to code properly. The `always_comb`, `always_ff`, and `always_latch` blocks that came with SystemVerilog obviate numerous potential coding errors related to misadjusted sensitivity lists, wait statements, and wake-up conditions. As an example, an `always_comb` block is not only sensitive to all signals that appear on the right-hand side of assignment operators within that block, but also to those read by any `function` called.

Observation 4.30. *As a net progress over Verilog and VHDL, SystemVerilog provides specific and synthesizable constructs for combinational logic, for flip-flop-, and for latch-like behavior.*

Not only are the specialized `always_` ... blocks safer and more convenient to write, they also make HDL code easier to understand as the designer's intent is immediately apparent.

Table 4.5 The SystemVerilog constructs for updating variables compared.

Assignment Operator	Continuous `assign ... =`		Procedural			
			`=` (blocking)		`<=` (nonblocking)	
Delay term	none	non-zero	none	non-zero	none	non-zero
Execution of process	suspends until next update event on a right-hand operand		continues	suspends for delay specified	continues	
Effect on variable	immediate	deferred by delay specified	immediate	deferred by delay specified	following simultan. blocking assignm.*a*	deferred by delay specified

*a*That is, with zero delay in terms of simulation time, but only after all blocking assignments scheduled for the same moment of time have been executed.

Depending on how the code is written, a procedural block can be made to capture almost anything from a humble piece of wire up to an entire image compression circuit, for instance. This is particularly true for the general purpose `always` block.

Hint: For the sake of modularity, legibility, and smooth synthesis, do not cram too much functionality into a procedural block. As a rule, use continuous assignment statements and `always_comb` blocks to describe combinational operations while packing any kind of data storage into a separate `always_ff` (or `_latch`) block.

No binding order of execution for simultaneous events

Referring to fig.4.18, a SystemVerilog simulator is free to execute processes scheduled for the same simulation time in arbitrary order. As opposed to VHDL, SystemVerilog knows of no δ delay. To make a long story short, designers wanting to write RTL code for synthesis are advised to stick to the rules below to avoid nondeterminism, race conditions, and misinterpretation.

Hint:

1. *Prefer continuous assigns for uncomplicated combinational functions.*
2. *Do not use procedural blocks other than* `always_comb`, `always_ff` *and* `always_latch`.[64]
3. *In an* `always_comb` *block, always use blocking assignments (`=`).*
4. *In* `always_ff` *and* `always_latch` *blocks, use nonblocking assignments (`<=`) only.*
5. *Do not make* `#0` *(zero delay expression) procedural assignments.*

[64] The general purpose `always` block should be strictly limited to code not intended for synthesis. Before using that construct or the `@*` wildcard event notation, for that matter, check [102] for more detailed advice.

Initial values cannot replace a reset mechanism

As discussed earlier, the SystemVerilog syntax supports assigning an initial value to a variable as part of its declaration statement.

Observation 4.31. *The initial value given to a variable defines the objects's state at $t = 0$, just before the simulator enters the first cycle. A hardware reset, in contrast, remains ready to reconduct the circuit into a predetermined start state at any time $t \geq 0$. This asks for distributing a dedicated signal to the bistables, both in the circuit and in its HDL model.*

Make sure you understand these are two totally different things. An initialized variable does not model a hardware reset facility and will, therefore, not synthesize into one. How to code a working reset has been demonstrated in listing 4.12.

How to check timing conditions

Latches, flip-flops, RAMs, and all other sequential subcircuits impose specific timing requirements such as setup and hold times on data, and minimum pulse widths on clock inputs. Should any of these timing conditions get violated, their behavior becomes unpredictable. Checking for compliance is thus absolutely essential for meaningful simulations, see observation 4.18.

Timing checks operate by searching the event queue for past events as explained in fig.4.11. SystemVerilog provides a total of twelve specialized constructs for various timing and waveform checks, among which `$setup`, `$hold`, `$width` and `$period`. Upon detecting a violation, each of them produces a specific message without interrupting simulation. Syntax rules require that timing checks be placed in a so-called `specify` block. Listing 4.16 shows an example.

LISTING 4.16

Setup and hold time checks in a D-type flip-flop model

```
// simulation model of a single-edge-triggered flip-flop with hardcoded timing
module setff
  ( input logic Clk_CI, logic Rst_RBI, logic Dd_DI,
    output logic Qq_DO );

  logic State_DP; // state variable

  specify
    $setup ( Dd_DI, posedge Clk_CI, 1.09ns ); // data evt, clock evt, min. sep.
    $hold ( posedge Clk_CI, Dd_DI, 0.60ns ); // clock evt, data evt, min. sep.
  endspecify

  always_ff @(posedge Clk_CI, negedge Rst_RBI)
    if (~Rst_RBI)
```

```
         State_DP <= 1'b0;
     else
         State_DP <= Dd_DI;

  assign #0.92ns Qq_DO = State_DP;

endmodule
```

4.3.5 FACILITIES FOR MODEL PARAMETRIZATION

As explained in section 4.2.5, a big plus of HDL models over schematic diagrams and netlists is that they can be parametrized such as to fit multiple needs. The most salient features of SystemVerilog towards making hardware models reusable are `parameters` and the `generate` statement. Fig.4.12 shows how these fit into the general picture.

Parameters

As stated earlier, variables can carry dynamic, i.e. time-varying, information between processes and indirectly also between modules. `parameters`, in contrast, serve to disseminate static, i.e. time-invariant, details to modules. Just think of

- o Word width,
- o Active-low or -high signaling on inputs and outputs,
- o Output drive capability,
- o Functional options (e.g. details about an instruction set),
- o Timing quantities (propagation and contamination delay, setup and hold time), and
- o Capacitive load figures.

A `parameter`'s name must be included in the `module` header prior to the `ports` between a pair of parentheses identified by a hash sign #. Indicating a default value is optional. As opposed to `ports`, `parameters` do not have any direct hardware counterpart. An example follows.

```
// w-input odd parity gate
module parityoddw
   #( parameter WIDTH,    // number of inputs
      parameter real TCD = 0ns,      // contamination delay with default value
      parameter real TPD = 1.0ns )   // propagation delay with default value
   ( input logic [WIDTH-1:0] Inp_DI,
     output logic Oup_DO );
   ...
endmodule
```

The values of the `parameters` get nailed down when the `module` is instantiated, overriding any defaults specified in the `module` header. This again occurs between a pair of parentheses identified by a hash sign # with the syntax otherwise roughly following that of the port map. Then follows the customary instance identifier and next the port map.

```
parameter NUMBITS = 12;
. . . . .
// module instantiation statement
parityoddw #( .WIDTH(NUMBITS), .TCD(0.05ns), .TPD(NUMBITS * 0.1ns) )
   u173 ( .Inp_DI(DataVec_D) , .Oup_DO(Parbit_D) );
. . . . .
```

`specparam`, finally, is a reserved word for declaring a special type of parameter the value of which is intended to be overwritten during back-annotation.

The generate statement

Similarly to what has been said in section 4.2.5, SystemVerilog provides a `generate` statement that can be used to spawn processes and/or to instantiate and connect modules under control of a user-defined algorithm. The use of the `generate` and `endgenerate` keywords is optional, yet strongly recommended for the sake of clarity. A special type `genvar` helps to distinguish those variables that steer a generate procedure from ordinary simulation-time variables. The code fragment below implements the Game of Life.[65]

```
. . . . .
// spawn a process for each cell in the array
generate
for (genvar ih = 0; ih<HEIGHT; ih++)
   for (genvar iw = 0; iw<WIDTH; iw++)
      always_ff @(posedge Clk_C) begin  // sensitivity list
         integer live_neighbors;
         live_neighbors = live_neighbors_at(ih,iw);
         if (State_DP[ih][iw]=='b0 && live_neighbors==3)
            State_DP[ih][iw] <= 'b1;  // birth
         else if (State_DP[ih][iw]=='b1 && live_neighbors<=1)
            State_DP[ih][iw] <= 'b0;  // death from isolation
         else if (State_DP[ih][iw]=='b1 && live_neighbors>=4)
            State_DP[ih][iw] <= 'b0;  // death from overcrowding
      end // always_ff
endgenerate
. . . . .
```

[65] A 2-dimensional cellular automaton rather than a game; for details see footnote 35.

Note that the `generate` statement not only comes in a `for` form but also in an `if` form.

Example   `generate if (i==WIDTH-1) ... endgenerate`

Observation 4.32. `Generate` *statements get interpreted as part of the so called elaboration phase that precedes actual circuit simulation and synthesis. Variables are obviously not acceptable as conditions or as loop boundaries since they are subject to vary at run time, only* `parameters` *are.*

The need to accommodate multiple models for one circuit block

Please recall from table 4.1 that up to four HDL models may occur during a VLSI design cycle to capture a circuit-to-be at distinct levels of detail. Designers also experiment with alternative circuit architectures to compare them in terms of gate count, longest path delay, energy efficiency, and other figures of merit. SystemVerilog accommodates this need by allowing multiple `modules` for the same subcircuit, a slightly different approach than illustrated in fig.4.13.

> *Warning: Do not give two* `modules` *with functionally distinct behaviors identical names as this is extremely confusing!*

Conditional compilation of source code

In the presence of multiple `module` bodies for one circuit, there must be a way to indicate which one to include during simulation and synthesis. SystemVerilog provides roughly twenty **compiler directives**, identified by a backtick character (`` ` ``), aka grave accent. The `` `include `` directive simply inserts the entire contents of the source file named in the argument into the calling file during compilation. For selective compilation, directive `` `ifdef `` must be used.

```
// parametrized binary to Gray code converter
module binary2gray
   #( parameter ...) // parameters
   (.....); // inputs and outputs

   `ifdef usebehavioral
      // module body with behavioral model follows here
      .....
   `else
      // module body with structural model follows here
      .....
   `endif
endmodule
```

Which of the two bodies gets evaluated depends on whether the free-choice identifier immediately to the right of the `` `ifdef `` keyword, named `usebehavioral` in this example, is defined or not at compile time. Compiler directives such as `` `define `` and `` `undef `` serve to do so.

Example   `` `define usebehavioral ``
Counterexample   `` `undef usebehavioral ``

Similarly, `` `define `` can be used to specify a number for compilation as shown in listing 4.17.

LISTING 4.17

Source file `graydefs.sv` referenced in listings 4.18 and 5.1

```
// Mission: Illustrate the usage of compiler directives 'define and 'include.
//-------------------------------------------------------------------------

'define GRAYWIDTH 5    // set a word width parameter for later use
```

As a more general comment, simulation and synthesis essentially follow the same course outlined in fig.4.14 for SystemVerilog and for VHDL, except for the syntax analysis step which is most obviously language-specific. The conceptual commonalities underlying the two HDLs has enabled industry to develop EDA tools that support VHDL-SystemVerilog **co-simulation**.

4.3.6 CONCEPTS BORROWED FROM PROGRAMMING LANGUAGES

For all its procedural aspects, SystemVerilog draws heavily on the software language C. Due to the popularity of that language, we will refrain from re-iterating those concepts here and just point out a few SystemVerilog particularities.[66]

User-defined data types

Basically, a data type defines a set of values and a set of operations that can be performed on them. Users may declare their own data types or use predefined ones. Data types for modeling electrical signals have been discussed in section 4.3.3. Enumerated types — which did not exist in traditional Verilog, by the way — are by default implemented as named constants of type `int`, occupying 32 bit.

Example of a type declaration   `typedef logic signed [23:0] audiosample;`
Enumerated type declaration   `typedef enum {JANUARY, FEBRUARY, ... , DECEMBER} month;`

Subroutine, function, and task

A subroutine is either a `function` or a `task`. The primary purpose of a `function` is to return a value for use in an expression. `function`s have restrictions that make sure they return in zero simulation time without suspending the process that invokes them. Constructs that involve the event queue, including `@(...)`, `wait`, and `#...` are, therefore, not admitted. A first example of a function, named `nextmonth`, is given below, two more follow in listing 4.18.

---

[66] Incidentally, note that `source_text` serves as **start symbol** in formal definitions of SystemVerilog.

```
package calendar;

   typedef enum {JANUARY, FEBRUARY, MARCH, APRIL, MAY, JUNE, JULY,
                 AUGUST, SEPTEMBER, OCTOBER, NOVEMBER, DECEMBER} month;
   typedef logic unsigned [4:0] day;

   function month nextmonth (month given_month);
      return given_month.next; // wraps around at the end
   endfunction

   function day nextday (day given_day);
      .....
   endfunction

endpackage: calendar
```

As opposed to `functions`, `tasks` are allowed to include timing control statements that cause them to suspend, to resume, and/or to schedule events. A `task` may also call further `tasks`. Tasks are primarily used in testbenches, see listing 5.2 for examples.

Package

This instrument for sharing constants, user-defined data types, and subroutines across modules has been adopted from VHDL. Packages make it possible to sidestep the waste and perils of repeating supposedly identical declarations at multiple places. The contents of a package must be referenced in a `import` clause to make them accessible from some other source file. This is shown in listings 4.18 and 5.1, the former of which defines two functions that are being called in the latter.

LISTING 4.18

Package holding a pair of Gray code ↔ binary conversion functions.

```
// Mission: illustrate the use of packages in SystemVerilog.
// Functionality: Gray code <-> binary code conversion functions.
// Author: H.Kaeslin.
//---------------------------------------------------------------------------

`include "../sourcecode/graydefs.sv"    // the word width to be used

package grayconvPkg;

`ifndef GRAYWIDTH
   `define GRAYWIDTH 2  // default word width for input and output
`endif
```

```
//-------------------------------------------------------------------------
//
//                              |      |      |      |      |
// binary to Gray              |--.   |--.   |--.   |--.   |      # of X-ops on
// conversion                  |  `--X   `--X   `--X   `--X      longest paths
// for GRAYWIDTH=5             |      |      |      |      |      = 1
//                             v      v      v      v      v
// bit positions               4      3      2      1      0      X = XOR
//                             |      |      |      |      |
// Gray to binary             |   ,--X   ,--X   ,--X   ,--X       # of X-ops on
// conversion                 |--'   |--'   |--'   |--'   |       longest path
// for GRAYWIDTH=5            |      |      |      |      |        = GRAYWIDTH-1
//                            v      v      v      v      v
//
//-------------------------------------------------------------------------

    function automatic logic [`GRAYWIDTH-1:0] bin2gray
       (input [`GRAYWIDTH-1:0] arg);
       bin2gray=arg;                           // initial assignment
       for (int i=0; i <=$size(arg)-2; i++) // from 0 to GRAYWIDTH-2
          if (bin2gray[i+1])                    // if bit at [i+1] is 1
          bin2gray[i] = ~bin2gray[i];        // invert bit at [i]
    endfunction

    function automatic logic [`GRAYWIDTH-1:0] gray2bin
       (input [`GRAYWIDTH-1:0] arg);
       gray2bin=arg;                           // initial assignment
       for (int i=$size(arg)-2; i >=0; i--) // from GRAYWIDTH-2 downto 0
          if (gray2bin[i+1])                    // if bit at [i+1] is 1
          gray2bin[i] = ~gray2bin[i];        // invert bit at [i]
    endfunction

endpackage : grayconvPkg
```

Classes, semaphores, mailboxes, etc.

Classes are a concept from object-oriented programming that encapsulates data and binds them together with the subroutines that operate on those data. While this approach is useful for automating code generation of complex testbenches, it is currently not applicable to hardware modeling and synthesis. Much the same applies to semaphores, mailboxes, and named events, all of which are process synchronization and communication mechanisms admitted into SystemVerilog to better support system-level testbench design.

System tasks

A notorious difficulty with VHDL simulation is the formatted input and output to data files. The SystemVerilog language includes many helpful commands, collectively named "system tasks", for these and other duties. Table 4.6 gives an idea, but the multitude of commands and options means the reader will have to consult a comprehensive reference manual such as [98] for details. System tasks are commands to the simulator and, most obviously, not for synthesis.

`$write`, `$display`, `$strobe`, and `$monitor` come in four flavors, each having a different default base. `$display` assumes decimal integers, `$displayb` binary data, `$displayh` hex data, and `$displayo` octal data, and so on. The same applies to their `$f...` file-writing counterparts.

Examples

```
$display("Simulation ended after %4d checks and with %4d error(s).",
         checkcnt, errorcnt);


int simvectorfile = $fopen("../simvectors/moore6st_simvector.asc", "r");


$fclose(simvectorfile);


while( !$feof(simvectorfile)) begin
   void'($fgets(readstr, simvectorfile));
   fmatch = $sscanf(readstr, "%b %b %b",
   StimuliRec.Clr_S, StimuliRec.Inp_D, ExpRespRec.Oup_D );
   ...
end


$readmemh("../sim/vectors/stim.txt", stimuli);


$error("Expected 'b%b does not match actual 'b%b", expresp, ActResp_D);


assert (simvectorfile) else $fatal("Could not open simvector file.");


RandomSample = $random % 32768;    // interval [-32767,+32767]


RandomMonth = $urandom_range(12,1);    // interval [1,12]


RandomBit = (($urandom_range(100,0) < percentzero) ? 1'b0 : 1'b1);
```

□

Table 4.6   Selected SystemVerilog system tasks.

Command	Action
Formatted text output	
`$write`	write line to standard output immediately with no newline character
`$display`	write line to standard output immediately preceded by a newline character
`$strobe`	*idem* at the end of current time slot, i.e. before advancing simulation time
`$monitor`	*idem* when specified events occur
File operations	
`$fopen`	open a file
`$fclose`	close a file
`$fread`	read from a file
`$fscanf`	parse formatted text from a file
`$fgets`	read characters from a file and assembles them into a string
`$sscanf`	parse formatted text from a string
`$feof`	return a non-zero value when end of file found and `0` if not so
`$fwrite`	same as `$write` for writing to a file
`$fdisplay`	same as `$display` for writing to a file
`$fstrobe`	same as `$strobe` for writing to a file
`$fmonitor`	same as `$monitor` for writing to a file
Memory load and dump	
`$readmemb/h`	load memory from a text file in binary/hex format
`$writememb/h`	dump memory to a text file in binary/hex format
Simulation control	
`$time`	return current simulation time
`$reset`	reset simulation so it can restart from the beginning
`$stop`	suspend simulation
`$finish`	terminate simulation
Run time information with severity levels (standalone and for use in assertions)	
`$info`	print argument to simulator window and continue
`$warning`	print argument to simulator window, count as warning, and continue
`$error`	print argument to simulator window, count as error, and continue
`$fatal`	print argument to simulator window and terminate simulation
Random number generation (for use in stimuli preparation)	
`$random`	return a random signed integer
`$urandom`	return a random unsigned integer
`$dist_uniform`	return a uniformly distributed random number
`$dist_normal`	return a normally distributed random number
Enquiries about the event queue (for use in properties and assertions)	
`$rose`	return `1` iffargument has changed to `1`
`$fell`	return `1` iffargument has changed to `0`
`$stable`	return `1` iffargument had not changed value
`$past`	return argument's value a specified number of clock cycles earlier

4.4 AUTOMATIC CIRCUIT SYNTHESIS FROM HDL MODELS

4.4.1 SYNTHESIS OVERVIEW

Automatic synthesis aims at turning some sort of behavioral description into a gate-level netlist with as little human intervention as possible. Using the standard cells available from a target library, synthesis software attempts to come up with a gate-level circuit that meets all user-defined performance targets at the lowest possible hardware costs, see fig.4.21.

Starting from an RTL model, the synthesis process is outlined in the lower part of fig.4.14. Syntax analysis is obviously different for VHDL and for SystemVerilog models. After that, processing becomes essentially the same. Elaboration and binding have been discussed earlier in this chapter. **State reduction** eliminates redundant states, if any, while **state encoding** assigns a unique binary code to each state.[67] The subsequent **synthesis** step builds all necessary state registers and specifies the combinational subfunctions in between. The result is a preliminary network described at an intermediate level of detail, that is in terms of logic equations and generic components rather than actual logic gates.

Boolean optimization reworks the logic networks in an attempt to bring their longest signal propagation paths below the relevant user-defined timing constraint while, at the same time, minimizing hardware complexity. Finding an optimal or near-optimal circuit depends on numerous characteristics of the library cells available, so simplifying and reorganizing logic equations and networks is closely intertwined with the subsequent **technology mapping** phase [104] where the generic gates in the netlist get replaced by components that are actually available from the target library.[68] More sophisticated tools also address energy efficiency.

As stated in section 4.1.4, VHDL and Verilog were not originally intended for synthesis; and SystemVerilog is a superset of Verilog. So, while almost all VHDL simulators support the full IEEE 1076 and 1164 standards, only a subset of the legal language constructs is amenable to synthesis; and the same holds for SystemVerilog and the IEEE 1800 standard.

Observation 4.33. *As good HDL code must be portable across simulation and synthesis platform, model writers must confine themselves to safe, unambiguous, and universally accepted constructs.*

The remainder of this section is devoted to discussing the existing limitations and to presenting workable solutions for synthesis code.

---

[67] Please refer to appendix B for more details.

[68] Much of today's logic optimization software descends from programs such as ESPRESSO (two-level logic) and MIS (multi-level logic). A major challenge in coming up with adequate algorithms is to achieve a low asymptotic complexity in order to cope with complex functions and large networks.

Only a subset of the data types is amenable to hardware synthesis. Support covers the types listed next along with the pertaining array and record data types.

VHDL	SystemVerilog
+ `integer`	+ `integer, shortint, int,` and `longint`
+ `boolean` and `bit`	+ `bit`
+ `std_logic` and `std_ulogic`	+ `logic` and `wire`
+ `unsigned` and `signed`	+ `byte`
+ enumerated types	+ enumerated types
+ `ufixed, sfixed` and `float`[69]	
+ `array` and `record` of fixed size	+ `array` and `struct` of fixed size

Unlike the above items, the types below are not normally supported.

VHDL	SystemVerilog
— `real`	— `real`
— `time`	— time-related data types
— `character`	— `string, queue,` and other dynamic data types
— `file`	— file-related data types

As for the synthesizability of data operators, please refer to table 4.10.

Hardware-compatible wake-up conditions for all processes

While HDLs allow the modeling of arbitrary behavior (as long as it is causal, discrete in value, and discrete in time), automatic synthesis only supports synchronous clock-driven subcircuits and — at a higher level — conglomerates of such subcircuits. Synchronous circuit operation means that state transitions are restricted to occur exclusively at precise moments of time as defined by a clock signal.[70] Not all code that is syntactically correct and that works during simulation is thus acceptable for synthesis. More particularly:

Observation 4.34. *Any process that is supposed to model a piece of hardware must execute upon activation, return to the same instruction, and suspend there.*

The reason is that each `wait` or similar statement that may cause a process to suspend is allowed to carry its own wake-up condition. Depending on the details postulated there, the source code may imply synchronous or asynchronous behavior. While the former readily maps to a clock-driven circuit assembled from flip-flops and logic gates, the latter is likely to express a behavior that depends on a haphazard collection of events on various signals in a delicate way. Coming up with a physical circuit

---

[69] Introduced with the IEEE 1076-2008 revision, not necessarily supported by all EDA tools as yet.
[70] An in-depth discussion is to follow in section 6.2.1.

that is safe and functionally equivalent to the source code may then prove extremely difficult, if not impossible.

Before casting this into practical coding guidelines in observations 4.35 and 4.36, let us have a closer look at finite state machines.

Explicit versus implicit state models

Explicit state models conform with the above. Program execution always returns to the same line of code and suspends there after having completed one full turn. signals or variables preserve the current state from one process activation to the next. Explicit state models are intuitive to hardware designers who are accustomed to think in terms of finite state machines (FSM) along with visual formalisms such as schematics, state graphs, and the like.

Implicit state models, in contrast, are related to Nassi-Shneiderman diagrams, aka structograms, and flowcharts. They tend to come more natural to software programmers who code algorithms and are immediately recognized by the presence of multiple synchronization points per process. Upon activation, the simulator executes instructions until the next suspend statement encountered puts the process to sleep. Suspension may thus occur at distinct lines of code, and each such place represents one specific state of the model. There is no tangible state variable. Rather, it is the return address to the suspended process that assumes this role during simulation, hence the name implicit state. See fig.4.19 and table 4.7 for a comparison.

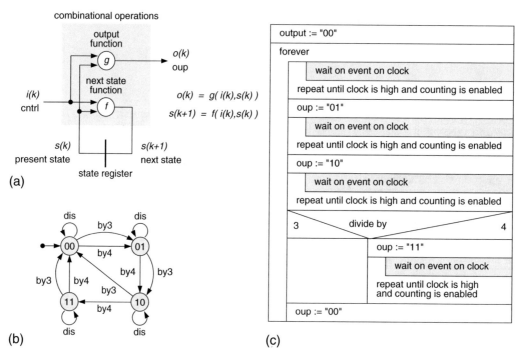

(a)

(b)

(c)

FIGURE 4.19

Three formalisms that affect the writing of software models for FSMs. Data dependency graph (a), state chart (b), Nassi-Shneiderman diagram (c). Note the programmable divide-by-3/divide-by-4 counter chosen for illustration is a Medvedev machine.

It should be clear from what has been said before that implicit state models must be translated into explicit state models before RTL synthesis can begin. Still, they occasionally find applications in the context of purely behavioral simulations in early stages of the design cycle. Code examples in this text refer to explicit state models exclusively.

How to capture a finite state machine

The restriction to explicit state models notwithstanding, one still has the choice of packing an FSM into one concurrent process (VHDL `process` statement or SystemVerilog `always` block), or of distributing it over two or more such processes.

Table 4.7   Explicit and implicit state models compared.

modeling style	explicit state		implicit state
	computed state	enumerated state	
inspired from	data dependency graph or schematic diagram	state chart, state graph, or state table	Nassi-Shneiderman diagram
synchronization mechanism	sensitivity list or single `wait` statement (semantically equivalent)		multiple `wait` statements
state variable	declared explicitly as `signal` or variable and thus of user-defined type		hidden in pointer to current statement
states captured by	(subrange of) integer or vector of bits	enumerated type	multiple `wait` statements
state transitions captured by	arithmetic and/or logic operations	one-to-one translation from state table	control flow
output function captured by	arithmetic and/or logic operations	one-to-one translation from state table	assignment statements
immediate hardware equivalent	yes		depending on `wait` conditions
synchronous	yes		*idem*
synthesizable	yes		no

Packing an entire FSM into a single process statement

The code is organized as illustrated by fig.4.20 a) through c) for Mealy, Moore, and Medvedev automata respectively.[71] Evaluation begins with the asynchronous reset and clock inputs to find out whether the machine's state must be updated. The remaining inputs are processed further down in the code. The process must be made sensitive to events on clock and reset and, in the occurrence of a Mealy machine, to events on other input signals as well.

---

[71] The three classes of automata essentially differ in the nature of their output functions. Please refer to sections B.1 and B.2 if you have doubts about the characteristics and equivalence relations among those classes.

Although this coding style is legal syntax and perfectly acceptable for simulation, its general adoption is discouraged because it is not supported by many synthesis tools and because it may result in inefficient gate-level networks.

Distributing an FSM over two (or more) concurrent processes

As depicted in fig.4.20 d) through f), a memorizing process is essentially in charge of maintaining the current state from one activation to the next. A second process of memoryless nature computes the next state and the present output value. In VHDL, both present state and next state must be modeled as signals that go back and forth between the two processes.

Capturing sequential and combinational behavior in separate processes is universally accepted and tends to lead to more economic circuit structures during synthesis. As an additional benefit, the type of automaton (Mealy, Moore, Medvedev) can be changed at any time without having to reorganize the source code too much.

In conclusion, the rules below ensure trouble-free synthesis and portability across platforms.

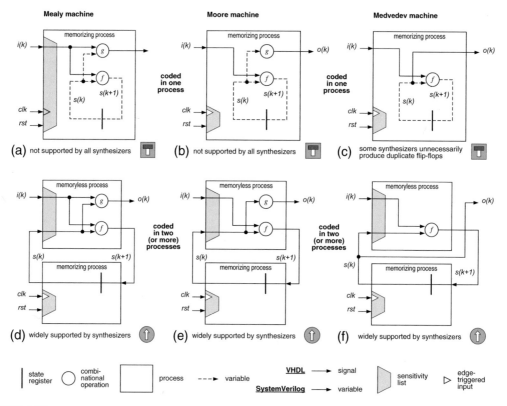

FIGURE 4.20

Coding schemes for synchronous Mealy, Moore and Medvedev machines. Note that the programmer is free to distribute the combinational operations subsumed here as "memoryless process" over two or more concurrent processes, including concurrent signal assignments (VHDL) or continuous assignments (SysVer).

Observation 4.35. *Good VHDL synthesis code shall*
a) *model circuits at the register transfer level (RTL) throughout,*
b) *collect combinational and sequential logic in separate processes,*
c) *have all memorizing* `process` *statements conform with the skeleton of listing 4.8,*
d) *prefer concurrent, conditional and selected signal assignments for combinational logic, and*
e) *have all memoryless* `process` *statements coded according to observation 4.15.*

Observation 4.36. *Good SystemVerilog synthesis code shall*
a) *model circuits at the register transfer level (RTL) throughout,*
b) *collect combinational and sequential logic in separate processes,*
c) *use* `always_ff` *blocks exclusively for memorizing behavior,[72] and*
d) *use continuous assignments or* `always_comb` *blocks for combinational logic.*

Warning: The emergence of unplanned for latches or other bistables during synthesis must alert the HDL designer that his code is badly wrong. For each subcircuit, check the number of flip-flops and latches obtained from synthesis against your expectations. As a general rule, do not ignore warnings and error messages from the synthesizer unless you understand what they mean.

Hint: In applications where state transitions depend on an enable signal, you have two options: a) include a subcondition in the memorizing process, as in listing 4.3, or b) add an extra conditional branch in the memoryless process that determines the next state. Though the two approaches are functionally identical, we recommend a) because the synthesizer may otherwise be unable to map to E-type flip-flops and, hence, to ultimately apply clock gating.

LISTING 4.19

VHDL code for a simple Mealy machine.

```
-- Mission: Illustrate how to model a Mealy machine with two processes.
--    Example designed to include an asynchronous reset, three-state outputs,
--    self loops, symbolic encoding of states and outputs, plus handling of
--    parasitic states and inputs so as to eliminate any chance for lockup.
-- Functionality: See state graph below.
-- Author: H.Kaeslin.
---------------------------------------------------------------------
library ieee;
use ieee.std_logic_1164.all;
---------------------------------------------------------------------

entity mealy5st is
   port (Clk_CI  : in  std_logic;
         Rst_RBI : in  std_logic;
         Inp_DI  : in  std_logic_vector(1 downto 0);
```

---

[72] Or, alternatively, `always_latch` blocks in the occurrence of level-sensitive clocking.

```
        Oup_DO  : out std_logic_vector(1 downto 0) );   -- ternary, 11 not used
end mealy5st;


--------------------------------------------------------------------------------
-- Inp_DI / Oup_DO                                                   (State_DP)
--
--             00 /          00 /          00 /          00 /          00 /
--            EMPTY         IBTW          IBTW          IBTW          FULL
--
--             _             _             _             _             _
--          | |  01 /     | |  01 /     | |  01 /     | |  01 /     | |
--          | |  IBTW     | |  IBTW     | |  IBTW     | |  FULL     | |
--      ---> v / ------>  v / ------>   v / ------>   v / ------>   v / ----
-- 10 / |      (0)           (1)           (2)           (3)           (4)        | 01 /
-- EMPTY -----    <--------    <--------  ^ <--------    <--------    <----  FULL
--                  10 /          10 /    |    10 /          10 /
--                 EMPTY         IBTW     |    IBTW          IBTW
--                                        o
--------------------------------------------------------------------------------

architecture enumerated_state of mealy5st is

   type state is (ST0, ST1, ST2, ST3, ST4);   -- enumerated state type
   signal State_DP, State_DN : state;         -- present state and next state
   -- symbolic encodings of output
   constant EMPTY : std_logic_vector(1 downto 0) := "10";   -- empty
   constant IBTW  : std_logic_vector(1 downto 0) := "00";   -- in between
   constant FULL  : std_logic_vector(1 downto 0) := "01";   -- full
   constant HIMP  : std_logic_vector(1 downto 0) := "ZZ";   -- high-impedance

begin

   ---- computation of next state and present outputs
   --------------------------------------------------------------------------
   p_memless : process (Inp_DI, State_DP)
   begin
      -- default assignments
      State_DN <= State_DP;   -- remain in present state
      Oup_DO <= IBTW;         -- output "in between"
      -- nondefault transitions and outputs
      case State_DP is
         when ST0 =>
            if    Inp_DI="00" then Oup_DO <= EMPTY;
            elsif Inp_DI="10" then Oup_DO <= EMPTY;
            elsif Inp_DI="01" then State_DN <= ST1;
```

```
                else Oup_DO <= EMPTY;    -- parasitic input 11, treat as 00
                end if;
            when ST1 =>
                if    Inp_DI="10" then Oup_DO <= EMPTY; State_DN <= ST0;
                elsif Inp_DI="01" then State_DN <= ST2;
                end if;
            when ST2 =>
                if    Inp_DI="10" then State_DN <= ST1;
                elsif Inp_DI="01" then State_DN <= ST3;
                end if;
            when ST3 =>
                if    Inp_DI="10" then State_DN <= ST2;
                elsif Inp_DI="01" then Oup_DO <= FULL; State_DN <= ST4;
                end if;
            when ST4 =>
                if    Inp_DI="10" then State_DN <= ST3;
                else  Oup_DO <= FULL;    -- 01, 00 or parasitic input 11
                end if;
            when others =>    -- tie up parasitic states for synthesis
                Oup_DO <= HIMP; State_DN <= ST2;
        end case;
    end process p_memless;


    ---- updating of state
    -------------------------------------------------------------------------------
    p_memzing : process (Clk_CI, Rst_RBI)
    begin
        -- activities triggered by asynchronous reset (active low)
        if Rst_RBI = '0' then
            State_DP <= ST2;
        -- activities triggered by rising edge of clock
        elsif Clk_CI'event and Clk_CI = '1' then
            State_DP <= State_DN;
        end if;
    end process p_memzing;


end enumerated_state;
```

LISTING 4.20

SystemVerilog code for a simple Mealy machine.

```
// Mission: Illustrate how to model a Mealy machine with two processes.
//    Example designed to include an asynchronous reset, three-state outputs,
//    self loops, symbolic encoding of states and outputs, plus handling of
//    parasitic states and inputs so as to eliminate any chance for lockup.
// Functionality: See state graph below.
// Author: B.Muheim.
// ---------------------------------------------------------------------------

module mealy5st
    ( input  logic Clk_CI,
      input  logic Rst_RBI,
      input  logic [1:0] Inp_DI,
      output logic [1:0] Oup_DO ); // ternary, 11 not used

//----------------------------------------------------------------------------
// Inp_DI / Oup_DO                                                 (State_DP)
//
//             00 /          00 /          00 /          00 /        00 /
//             EMPTY         IBTW          IBTW          IBTW        FULL
//
//             _             _             _             _           _
//            | | 01 /      | | 01 /      | | 01 /      | | 01 /    | |
//            | | IBTW      | | IBTW      | | IBTW      | | FULL    | |
//       ---> v / ------>   v / ------>   v / ------>   v / ------> v / ----
//  10 / |      (0)           (1)           (2)           (3)          (4)    | 01 /
//  EMPTY  -----   <--------    <--------  ^ <--------    <--------   <----  FULL
//                  10 /          10 /     |   10 /         10 /
//                  EMPTY         IBTW     |   IBTW         IBTW
//                                         o
//----------------------------------------------------------------------------

    typedef enum integer {ST0, ST1, ST2, ST3, ST4} state_type;

    // symbolic encodings of output
    const logic [1:0] EMPTY = 'b10;    // empty
    const logic [1:0] IBTW  = 'b00;    // in between
    const logic [1:0] FULL  = 'b01;    // full
    const logic [1:0] HIMP  = 'bZZ;    // high-impedance

    // present state and next state
    state_type State_DP, State_DN;
```

```
// computation of next state and present outputs
// ------------------------------------------------------------------------
always_comb  begin
   // default assignments
   State_DN = State_DP;   // remain in present state
   Oup_DO = IBTW;         // output "in between"
   // nondefault transitions and outputs
   case (State_DP)
      ST0 :
         if       (Inp_DI=='b00) Oup_DO   = EMPTY;
         else if (Inp_DI=='b10) Oup_DO   = EMPTY;
         else if (Inp_DI=='b01) State_DN = ST1;
         else Oup_DO   = EMPTY;   // parasitic input 11, treat as 00
      ST1 :
         if (Inp_DI=='b10) begin
           Oup_DO   = EMPTY;
           State_DN = ST0;
         end
         else if (Inp_DI=='b01) State_DN = ST2;
      ST2 :
         if       (Inp_DI=='b10) State_DN = ST1;
         else if (Inp_DI=='b01) State_DN = ST3;
      ST3 :
         if       (Inp_DI=='b10) State_DN = ST2;
         else if (Inp_DI=='b01) begin
           Oup_DO   = FULL;
           State_DN = ST4;
         end
      ST4 :
         if       (Inp_DI=='b10) State_DN = ST3;
         else Oup_DO   = FULL;
      default : begin
         Oup_DO   = HIMP;
         State_DN = ST2;
      end
   endcase
end

// updating of state
// ------------------------------------------------------------------------
always_ff @(posedge Clk_CI, negedge Rst_RBI)   // sensitivity list
   // activities triggered by asynchronous reset (active low)
```

```
          if (~Rst_RBI)
             State_DP <= ST2;
          // activities triggered by rising edge of clock
          else
             State_DP <= State_DN;

 endmodule
```

Listings 4.19 and 4.20 show a small Mealy machine coded for synthesis in VHDL and SystemVerilog respectively. Listings 4.21 and 5.1 do the same for a Medvedev machine. The coding of a Moore machine is left to the reader as an exercise, see problem 10. What follows next are further comments on FSM coding.

Observation 4.37. *To facilitate code readability, always try to decompose large state machines into a bunch of smaller ones that cooperate with each other. Adhere to hierarchical and modular design, consider using counters instead of long state chains, for instance.*

Observation 4.38. *The various processes that make up for an FSMs are best included into one* architecture body *(VHDL) or* module *(SysVer) along with the datapath they command. Shutting the FSM into an entity of its own just inflates the code and the effort for coding and maintenance.*

FSM optimization ignored in the language standards

As explained in section B.1.6, state machine design involves solving two optimization problems, namely state reduction and state encoding. Some synthesis tools are designed to automatically recognize FSMs in the HDL code and to carry out those optimization steps. Others must be instructed to do so using proprietary compiler directives that identify those signals or variables that act as a repository for the current state.

4.4.4 RAM AND ROM MACROCELLS

On-chip RAMs are being used to temporarily store all sorts of intermediate data whereas on-chip ROMs serve as repositories for program code, lookup tables (LUT), and other permanent information. This section will discuss how to incorporate RAM and ROM macrocells in VHDL models in a way that is acceptable for synthesis.

The most innocent approach is to declare a storage array as if the code were intended for simulation purposes and to assume the synthesizer will take care of all the rest with no human interaction. As an example, consider a 4bit-binary-to-seven-segment display decoder. The content of an adequate LUT can be captured as an array of constants as shown below. While a workable solution, this piece of code will not synthesize into a ROM, but into random logic as any other RTL model of combinational nature.

```
-- unsupported VHDL coding style
.....
-- address of array must be of type integer or natural
p_memless : process (Binary4Code_D)
   variable address : natural range 0 to 15;
   type array16by7 is array(0 to 15) of std_logic_vector(1 to 7)
   constant SEGMENT_LOOKUP_TABLE : array16by7 :=    -- segments ordered a...g
      ("1111110","0110000","1101101","1111001",     -- digits 0,1,2,3,
       "0110011","1011011","0011111","1110000",     --        4,5,6,7,
       "1111111","1110011","1110111","0011111",     --        8,9,A,b,
       "1001110","0111101","1001111","1000111");    --        C,d,E,F;
begin
   -- use binary input as index, look up in table, and assign to segment output
   address := to_integer(unsigned(Binary4Code_D));
   Segment7Code_D <= SEGMENT_LOOKUP_TABLE(address);
end process p_memless;
.....
```

Trying to do the same in the occurrence of a 64 byte RAM, for instance, would mean to include the code fragment below into the declaration section of the superordinate architecture body. Reading and writing one byte at a time would involve assignment statements and an address pointer that selects one out of the 64 storage vectors from the array of `signals`.

```
-- unsupported VHDL coding style
.....
   type array64by8 is array(0 to 63) of std_logic_vector(7 downto 0);
   signal Storage_D : array64by8;
.....
```

The idea is impractical, however, because the behavior so defined is a far cry from actual RAM macrocells and their interfaces. Worse than this, automated synthesis would hardly churn out a safe and synchronous gate-level circuit either. From a more general perspective, whether to implement a storage array in a RAM, from flip-flops, or otherwise is a decision that has far-reaching consequences both on the final circuit and on the design process.

Observation 4.39. *Spontaneous incorporation of macrocells is neither a practical nor really a desirable proposition for RTL synthesis because it would deprive designers of control over a circuit's architecture and performance.*

A more realistic approach would be to instantiate a RAM stating the macrocell generator to be used and to pass on all further specifications in a `generic` map (VHDL) or as `parameters` (SysVer). With `cmosram01` the name of some fictive generator for clocked SRAMs, for instance, this would ask for a code fragment similar to the one below.

```
-- unsupported VHDL coding style
.....
u39: cmosram01
   generic map ( NUMBER_OF_WORDS => 64, WORD_WIDTH => 8,
                 DATA_INPUT_OUTPUT_SEPARATE => false )
   port map ( CLK => Clk_C, WRENA => RamWrite_S,
              ADDR => RamAddress_D, DATIO => RamData_D );
.....
```

Regrettably, this approach is not currently feasible due to the lack of standardization and the absence of interfaces between HDL synthesis and all those proprietary macrocell generators in existence. For the time being, a macrocell must get instantiated like any other component.

```
-- supported VHDL coding style
.....
u39: myram64by8
   port map ( CLK => Clk_C, WRENA => RamWrite_S,
              ADDR => RamAddress_D, DATIO => RamData_D );
.....
```

```
// supported SystemVerilog coding style
.....
myram64by8 u39 ( .CLK(Clk_C), .WRENA(RamWrite_S),
                 .ADDR(RamAddress_D), .DATIO(RamData_D) );
.....
```

Observation 4.40. *The necessary design views of a macrocell (simulation model, schematic icon, detailed layout, etc.) must all be obtained from outside the HDL environment.*

To that end, the IC designer must either gain access to the process-specific macrocell generator software and run it with an appropriate parameter setting, or he must commission the silicon foundry do so for him. The choice is typically determined by commercial considerations.

Table 4.8 shows how a synthesis model must be organized in order to obtain various read-only and read-write storage functions.[73]

---

[73] Note that the market offers functional replacements for RAMs that are built on the basis of individual bistables. Such models are amenable to HDL synthesis in the normal way, that is they ultimately map to standard cells much as the designer's own code. Of course, such cell-based implementations cannot compete with true RAMs in terms of layout density. Also, many of them combine mix of gates, latches, and flip-flops into a circuit that does not adhere to a pure and unconditionally safe synchronous clocking discipline.

Table 4.8   **The desired circuit type determines how its synthesis model must be organized.**

Look up table (LUT) (memoryless)		
Desired hardware organization	Random logic	ROM (tiled layout)
Function must be modeled	as an array-type constant or with logic equations	by instantiating a ROM macrocell as a component
Storage array (memorizing)		
Desired hardware organization	register file built from flip-flops or latches	RAM (tiled layout)
Function must be modeled	as an array of (clocked) storage registers	by instantiating a RAM macrocell as a component
Common traits of implemented circuit		
Area-effcient when data quantity is	small	large
Technology-specific software required	no	macrocell generator
HDL code amenable to retargeting	yes	manual rework needed
Pre-synthesis simulation works from	RTL source code	extra behavioral model
Post-synthesis simulation works from	gate-level model	*idem*

4.4.5 TIMING CONSTRAINTS

A **timing constraint** is a user-defined target for some timing quantity that the final circuit must meet. Fig.4.21 illustrates a common situation where the propagation delay through a circuit has been bounded from above. The concept is very general in that a synthesis constraint can as well refer to a circuit's longest clock period, to its maximum acceptable input or output delay (all upper bounds), or even specify a minimum contamination delay (lower bound).

Any difference between the target value specified in a timing constraint and the actual delay exhibited by the circuit after synthesis is termed **slack**. For a combinational circuit this simply amounts to $t_{sl} = t_{lp\,max} - t_{lp}$ where t_{lp} stands for the propagation delay on the longest path. In the occurrence of a sequential circuit, the designer typically indicates a target clock period T_{clk} that serves as upper bound for all register-to-register paths. Slack then becomes $t_{sl} = T_{clk} - t_{ss}$. A negative slack indicates the synthesis and optimization process has failed to meet its timing target. Unless synthesizer directives suffice to correct the situation, the design will then need to be reworked at the RTL or architecture level.

Synthesis constraints are not part of the HDL standards

The timing-related constructs in the IEEE 1076 and 1800 standards have been defined exclusively with simulation in mind. Inertial effects of physical circuits are typically being modeled with `after` clauses (VHDL) or `#...` expressions (SysVer). These language constructs are meaningless in the context of synthesis. After all, it is not possible to stipulate some arbitrary timing and then to come up with a circuit that exactly meets those predefined numbers.[74] Synthesis tools thus simply ignore `after` and `reject` clauses (`#...` expressions). The same applies to `wait for` statements and to timing-related assertion statements.

Example of timing a spec ignored by VHDL synthesis   `Oup_D <= Aa_D + Bb_D after 1.7 ns;`
Example of timing a spec ignored by SysVer synthesis   `assign #1.7ns Oup_D = Aa_D + Bb_D;`

Observation 4.41. *Timing-related HDL constructs are for simulation purposes exclusively and get ignored during synthesis. They serve to model the behavior of existing circuits, not to impose target requirements for the synthesis process.*

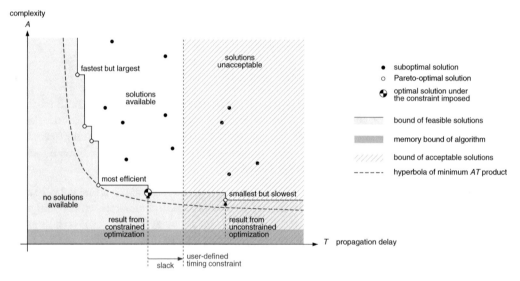

FIGURE 4.21

Trade-offs between size and performance for a hypothetical circuit.

---

[74] This is because the exact timing of a circuit not only depends on the gate-level network, but also on load capacitances, wiring parasitics, PTV and OCV, cross-coupling effects, and more. Also, the timing of library cells is not continuously adjustable. In the occurrence of a 2-input AND function, the target library would provide a few standard cells that differ in terms of transistor sizing, drive strength, and layout. Each such cell has its proper delay vs. load characteristics. Synthesis software just picks one or the other depending on the current requirements, but there is no sensible way to fashion an arbitrary delay at will.

A more sensible goal is to define bounds that could guide synthesis and logic optimization. Such timing constraints have never been adopted in the language standards, though. VHDL and SystemVerilog are, therefore, not capable of expressing any upper bound for a long path such as the one illustrated in fig.4.21.

Unsupported VHDL construct   `Oup_D <= Aa_D + Bb_D with_TPD_no_more_than 1.7 ns;`
Unsupported SystemVerilog construct   `assign #max#1.7ns OUP_D = Aa_D + Bb_D;`

As a workaround, timing and other synthesis directives must be expressed with the aid of proprietary language extensions or with scripting languages such as Tcl. Portable formulations are important as the same timing constraints are to be reused during timing verification to check whether a design indeed meets its specifications before being sent to fabrication.

How to formulate timing constraints[75]

As stated before, an upper bound for the delay from one register to the next gets imposed by the clock period in single-edge-triggered one-phase circuits.Indicating a target clock period is thus mandatory and straightforward, see fig.4.22a.

Formulating constraints for input- and output paths is more tricky because of ambiguous naming habits in commercial synthesis and timing verification tools. In fact, it is possible to define input/output timing from either of two perspectives:

o Indicate how much time is available to the circuit under construction (egocentric view).
o Quantify the amount of time that must be set aside for the surrounding circuitry
    (altruistic view).

As specifying the circuit under construction itself comes most naturally, the first perspective is normally adopted in this text. Yet, most EDA tools adopt the altruistic view, mainly because it permits to alter the target clock period without having to numerically readjust all I/O timing constraints. Unfortunately, the names typically being used, such as input and output delay, are inexpressive and give rise to confusion. The material below, including table 4.9 and fig.4.22, attempts to reconcile the two views.

---

[75] Note to the reader: For your first encounter with RTL synthesis, you may skip this section and come back later after having learned more about the operation and timing of synchronous circuits in sections 7.2 and 7.4.

Table 4.9 **Cross reference for input and output timing constraints.**

Event	Symbol	Quantity (egocentr. above, altruist. below)	Synopsys term
		relating to the interface with the upstream circuitry	
③ data-valid window begins	$t_{su\,inp}$ $\leq T_{clk} - t_{pd\,upst}$ $t_{pd\,upst} = t_{idel\,max}$	setup time of circuit under construction clock-to-output propagation delay of upstream circuitry	maximum input delay
② data-valid window ends	$t_{ho\,inp}$ $\leq t_{cd\,upst}$ $t_{cd\,upst} = t_{idel\,min}$	hold time of circuit under construction clock-to-output contamination delay of upstream circuitry	minimum input delay
		relating to the interface with the downstream circuitry	
④ data-call window begins	$t_{pd\,oup}$ $\leq T_{clk} - t_{su\,dnst}$ $t_{su\,dnst} = t_{odel\,max}$	clock-to-output propagation delay of circuit under construction setup time of downstream circuitry	maximum output delay
① data-call window ends	$t_{cd\,oup}$ $\geq t_{ho\,dnst}$ $t_{ho\,dnst} = -t_{odel\,min}$	clock-to-output contamination delay of circuit under construction hold time of downstream circuitry	<u>minus</u> minimum output delay

You will typically want to give upper bounds for the long path delays by specifying:[76]

$$t_{idel\,max} = t_{pd\,upst} = t_{pd\,ff\,upst} + t_{pd\,a} - t_{di} \ \Leftarrow\ t_{pd\,b} + t_{su\,ff} \leq T_{clk} - t_{pd\,ff\,upst} - t_{pd\,a} + t_{di} \qquad (4.1)$$

$$t_{odel\,max} = t_{su\,dnst} = t_{pd\,e} + t_{su\,ff\,dnst} + t_{di} \ \Leftarrow\ t_{pd\,ff} + t_{pd\,d} \leq T_{clk} - t_{pd\,e} - t_{su\,ff\,dnst} - t_{di} \qquad (4.2)$$

If you make no distinction between $t_{idel\,max}$ and $t_{idel\,min}$, EDA tools will assume they are the same which implies that input data get updated once per clock period at time ③ and then remain valid for one entire period. If not so, the short path delay must be constrained as well. Use a separate Tcl statement where you indicate a lower bound of

$$t_{idel\,min} = t_{cd\,upst} = t_{cd\,ff\,upst} + t_{cd\,a} - t_{di} \ \Leftarrow\ t_{cd\,b} - t_{ho\,ff} \geq -t_{cd\,ff\,upst} - t_{cd\,a} + t_{di} \qquad (4.3)$$

while observing that any physical circuit must satisfy

$$t_{idel\,min} < t_{idel\,max} \ \Leftarrow\ t_{valid\,upst} = t_{pd\,upst} - t_{cd\,upst} = t_{idel\,max} - t_{idel\,min} > 0 \qquad (4.4)$$

Similarly, if you do not distinguish between $t_{odel\,max}$ and $t_{odel\,min}$, the synthesizer will try his best to meet the setup condition of the downstream circuitry, but will do nothing particular about the hold

---

[76] While (4.1), (4.2), (4.3) and (4.5) correctly reflect the interrelations, you may have to drop the t_{di} terms there when entering numerical data into a synthesis tool. Check the documentation to find out what your tool requires. Also make sure to understand whether your tool will automatically adjust timing targets as a function of the actual t_{di} value, or whether the designer is expected to take care of that manually.

condition there. In the extreme case, a circuit that just flashes valid output data at time ④ might pass as acceptable because $t_{odel\,max} = t_{odel\,min}$ is the same as $t_{su\,dnst} = -t_{ho\,dnst}$ which indeed stands for a downstream circuit that is capable of picking up data in zero time. To prevent this from happening, you can either bank on automatic hold time fixing or explicitly constrain the short path from below by specifying

$$t_{odel\,min} = -t_{ho\,dnst} = t_{cd\,e} - t_{ho\,ff\,dnst} + t_{di} \quad \Leftarrow \quad t_{cd\,ff} + t_{cd\,d} \geq t_{ho\,ff\,dnst} - t_{cd\,e} - t_{di} \tag{4.5}$$

while observing

$$t_{odel\,min} < t_{odel\,max} \quad \Leftarrow \quad t_{call\,dnst} = t_{su\,dnst} + t_{ho\,dnst} = t_{odel\,max} - t_{odel\,min} > 0 \tag{4.6}$$

How to partition a circuit in view of synthesis and optimization

Timing constraints on propagation paths that extend across multiple circuit blocks render synthesis unnecessarily difficult and are likely to result in suboptimal circuits. This is because logic optimization and technology mapping are carried out in chunks to avoid excessive computer run times and memory requirements on large designs. Most tools accept proprietary directives for merging and segregating circuit logic into **synthesis chunks**. However, the better the initial architecture and the various design `entities` (VHDL) and `modules` (SysVer) in the RTL source code reflect a sensible hardware organization, the less effort will have to be wasted in repartitioning at synthesis time.

   Hint: Synthesis and optimization work much better if a design is organized such that

- *related or tightly connected subcircuits belong to the same design entity,*
- *all outputs from a synthesis chunk are registered, and*
- *critical paths are confined to within one synthesis chunk.*

Registered outputs further preclude the unwanted emergence of zero-latency loops and hazards.

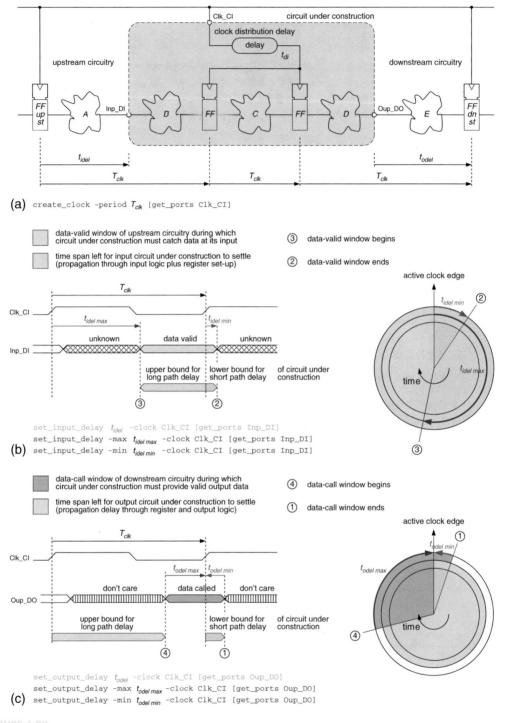

FIGURE 4.22

Timing constraints as understood by synthesis tools. Circuit overview with clock period (a), input timing (b), and output timing (c).

4.4.6 LIMITATIONS AND CAVEATS

Some circuits essentially need to be defined as gate-level netlists

Designers cannot always afford to leave decisions on a circuit's organization to the discretion of automatic synthesis, they sometimes need to exactingly control the outcome at the gate level. Arithmetic units are typical examples. Assume you had to implement a high-performance multiplier for sign-magnitude numbers, a format not really supported by HDL synthesis tools.

While schematic entry offers full control over a circuit's structure and produces highly suggestive diagrams, it is always tied to specific components and to particular circumstances in terms of word widths, pipeline depth, output format, and the like. Chances of ever reusing such a rigid circuit description are extremely low.

HDLs make it possible to write synthesis models that are structural <u>and</u> parametrized at a time. Such models make extensive use of `generics` and of conditional `component` instantiation statements, see `architecture structural` of `binary2gray` in section 4.2.5 for a simple example. The role of synthesis in the processing of HDL source code of this kind is essentially limited to elaboration, technology mapping, and timing optimization with the overall organization of the original network being preserved. The procedure as a whole can be viewed as HDL-controlled netlist generation.[77]

In addition to that, all VLSI chips include subcircuits where designers explicitly stipulate a well defined connectivity. Padframe, clock distribution network, clock gating circuitry, synchronizers, scan paths, and leakage suppression circuits are common examples. While their functionality is trivial, their structural, electrical and/or timing characteristics must conform to precise specifications. A loose collection of inverters is no valid substitute for a clock tree, for instance, nor does a simple AND operation qualify as clock gate. Similarly, scan testing implies the presence of a shift register in the actual circuit hardware as illustrated in fig.7.6, not just another way of transiting from one state to the next.

Observation 4.42. *Boolean optimization algorithms and general purpose synthesis tools are not designed to handle padfames, clock gates, synchronizers, clock distribution networks, scan paths, high-performance arithmetic circuits, and other "non-logic" portions of a design that must comply with structural rather than just with behavioral specifications.*

What are the options when tight control over a subcircuit's gate-level construction is sought?

o Use dedicated design automation tools,[78]
o fall back on schematic entry, or
o write a parametrized structural HDL model.

As for arithmetic units, take advantage of proven synthesis models (DesignWare) where possible.

---

[77] Predesigned, preverified and optimized, yet configurable and technology-independent circuit models are being marketed by Synopsys under the product name **DesignWare**. They range from fairly simple arithmetic units and register files to an entire video decoder. VHDL models of adders, multipliers, dividers, square root and trigonometric functions are available from [105] along with extraordinarily vivid explanations.

[78] Generating balanced clock trees, for instance, is postponed to the physical design phase and handled by specialized EDA software there.

Hint: Do not reoptimize subcircuits so obtained as critical properties may deteriorate. Most synthesis tools accept "don't touch" directives to prevent them from altering critical subcircuits while attempting to optimize the main body of a design.

In summary, it is important to understand that only a subset of the VHDL language defined in the IEEE 1076 standard is supported for synthesis. This is even more pronounced for SystemVerilog and the IEEE 1800 standard. The synthesizable subsets need not be exactly the same for all commercial products. Tool builders have added proprietary directives, constraints, and sometimes extra data types to fill gaps that were left open in the standards.

4.4.7 HOW TO ESTABLISH A REGISTER TRANSFER LEVEL MODEL STEP BY STEP

VHDL and SystemVerilog are perfectly suitable for coding a data processing algorithm. Yet, do not expect an EDA tool to accept a purely behavioral model and to turn that into a circuit design of acceptable performance, size, and energy efficiency. Exceptions are limited to circuits of fairly modest or fairly specific functionality. Rather, the fun and the burden of architecture design rests with the hardware developer. Only after an architecture has been worked out by human engineers does it make sense to describe the hardware organization at an intermediate level of detail, typically RTL, and to submit the HDL code so obtained to a synthesis tool.

RTL modeling is best carried out in a procedure of successive refinement:

1. Begin by drawing a fairly detailed block diagram of the architecture to be implemented.

2. Check where you can take advantage of off-the-shelf synthesis models (DesignWare).

3. Organize the circuit such as confine critical propagation paths to within circuit blocks. Make your design `entities` (`modules`) match with those circuit blocks.

4. Identify macrocells such as RAMs and ROMs and prepare for generating the necessary design views outside the HDL environment.

5. Identify <u>all</u> registers (data, I/O, pipeline, address, control, status, mode, test etc.) and loosely collect the combinational operations in between into clouds.

6. For each combinational cloud, specify the operations in mathematical terms (equations, truth tables, structograms, pseudo code, etc.) and figure out how to compute the desired outputs in an efficient and — where meaningful — also parametrizable way.

7. Establish a **schedule** that specifies what is to happen during each clock cycle. This is a table with one line per computation period and an entry for each relevant building block that expresses the following items:
   - ALU or arithmetic unit: operation being carried out, data set being processed.
   - Other major combinational block: data set being processed.
   - Finite state machine: present state, present output.
   - Register: present datum, being cleared or not, being enabled or not.
   - Important signal: present datum.
   - Output pin or connector: present datum, being driven or not.

- Input pin or connector: datum that must be available.
- Bidirectional pin or on-chip bus: present datum, being driven or not.

8. Identify all finite state machines and find out what type is most appropriate.[79]

9. Capture each register in a memoryzing `process` statement (`always_ff` block).

10. For each combinational cloud, decide on the number of processes you want to use. Prefer concurrent, selected, and conditional `signal` assignments (continuous assignments) for simpler operations. Plan to use `process` statements (`always_comb` blocks) for more convoluted computations exclusively. Give a meaningful name to each process.

11. Note that all data items that run back and forth between the various processes must be declared as `signals` (variables) and decide on the most appropriate data `type` for each.

12. Only now begin with translating your draft into actual HDL code.
- Organize finite state machines as suggested by fig.4.20 and pattern the code of registers after listing 4.3 (4.12). Take care to handle special signals such as clock, asynchronous reset, synchronous initialization, and enable properly.
- Use the schedule previously established to specify the various subfunctions in full detail.
- Fill in don't care entries wherever possible.
- Follow the recommendations of section 4.2 (4.3) and observation 4.35 (4.36).

Observation 4.43. *Writing code for HDL synthesis is not the same as writing software for a program-controlled computer. Always think in terms of circuit hierarchies and simultaneous activities (i.e. concurrent processes) rather than in terms of instruction sequences.*

> **Golden rule: Establish a block diagram of your architecture first, then code what you see!**

---

[79] Table B.5 may help in doing so.

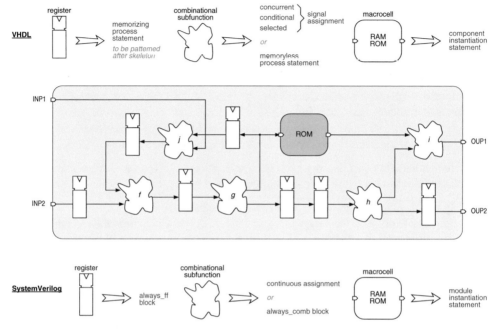

FIGURE 4.23

Translating an RTL diagram into HDL code.

4.5 CONCLUSIONS

The universal adoption of VHDL and SystemVerilog is due to their many paying benefits:

+ They support a top-down design methodology of successive refinements from behavioral simulations down to gate-level netlists using a single standard language.

+ RTL synthesis does away with all lower-level schematic drawings in a typical VLSI design hierarchy, saving significant time and effort.

+ HDLs enable sharing, reusing and porting of subfunctions and -circuits in a parametrized and therefore more useful form than schematic diagrams.

+ Automatic technology mapping makes it unnecessary to commit an HDL-based design to some specific cell library or fabrication process until late in the design cycle, even allowing for retargeting between field- and mask-programmable ASICs.

+ VHDL and SystemVerilog also support the coding of simulation testbenches, albeit not to the same degree, see fig.4.24. More on this is to follow in chapter 5.

− Learning to master VHDL or SystemVerilog may be daunting.

- While the IEEE 1076 and IEEE 1800 standards are fully supported for simulation, only a subset is amenable to synthesis because the languages have not originally been developed with synthesis in mind. This is not a problem for informed designers, however.

- Timing constraints and synthesis directives are not part of VHDL and SystemVerilog and must be captured using proprietary languages. There also is a lack of agreement between tool vendors on what constructs the synthesis subset ought to include and when to support new constructs introduced with past standard revisions.

- A gap remains between system design, which focuses on overall circuit behavior and transactions on high-level data, and actual hardware design, which involves many structural and implementation-specific issues. The necessary manual translation from a purely behavioral model to RTL synthesis code and the ensuing re-entry of design data are inefficient and lead to errors and misinterpretations.

- The impact of coding style on combinational random logic tends to be overstated. Also, do not expect timing-wise synthesis constraints to do away with architectural bottlenecks. All too often, their effects are limited to buying moderate performance gains at the expense of substantially larger circuits.

- The most important engineering decisions that set efficient designs apart from inefficient ones do not relate to HDLs, but to architectural issues. Algorithmic and architectural questions must be answered before the first line of synthesis code is written.

- HDL synthesis does not do away with architecture design!

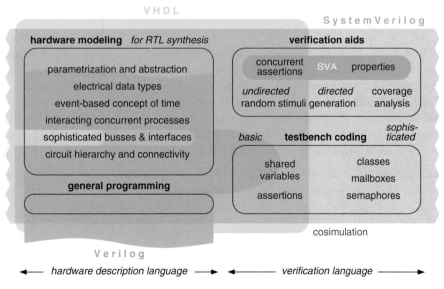

FIGURE 4.24

The capabilities of the three predominant HDLs at a glance.

4.6 PROBLEMS

1. ⁎⁎ Locate listings 4.1 and 4.4 (VHDL) or 4.11 and 4.13 (SystemVerilog). Note that each listing is accompanied in the text by a short assignment that consists in reverse engineering the code and in representing the findings graphically. If you haven't done so, catch up now.

2. ⁎ Section 4.2.2 includes examples of conditional and non-conditional signal assignments. For each such example, state the conditions that cause the code to get re-evaluated during simulation.

3. ⁎ Consider process statement `memless1` from section 4.2.2 and note that signal `Spring_D` is unconditionally set `false` before being assigned its actual value in a series of conditional statements. That value will thus evolve from `true` to `false` and back again when the process is invoked during springtime, e.g. at midnight of May 18. Do you think this trait will become visible as a brief transient during simulation? Would a circuit synthesized from this model exhibit a hazard? Explain your reasoning.

4. ⁎⁎ Listing 4.5 includes a procedural model, a dataflow model, and a structural model for a small subcircuit in VHDL (listing 4.14 does the same for SystemVerilog).
   (a) For each of the three models, find out in what way it is possible to reorder the statements without affecting the model's functionality.
   (b) Although the three models describe exactly the same functionality at exactly the same level of abstraction, they greatly differ in the total count of signals, variables, design entities, instances, processes, and statements involved. Determine the respective numbers.
   (c) Establish three schedules that list what is happening simulation cycle after simulation cycle in response of an event on any of the inputs. Think about the impact on computational efficiency when the entity gets simulated.

5. ⁎⁎ The book's companion website includes synthesis code for a so-called (4,3)-counter, a logic network that tells how many of its inputs are at logic `1`. Notice that the model assumes fixed input and output widths of 4 and 3 bits respectively. Examine how the various architecture bodies would scale if the model were to be parametrized in order to handle an arbitrary word width at the input. Add the necessary generic interface constant(s) in the entity declaration and rewrite a few architectures such as to obtain a scalable model. You may further want to synthesize and to compare the resulting networks.

6. ⁎ Explain the differences between these conditional constructs:
   `if ... then ... end if;` and `if ... generate ... end generate;` (VHDL).
   `if ... and generate if ... endgenerate` (SystemVerilog). Are there any other VHDL (SystemVerilog) statements that are related to each other in the same way?

7. ⁎⁎⁎ Using your preferred HDL, write a model for a binary-coded-decimal (BCD) counter that is amenable to simulation and synthesis. A control input of two bits is to decide between count-up (01), count-down (10), and hold (00) condition. Input-to-output latency shall not exceed one clock cycle. Under no circumstance are hazards tolerated on any of the output signals. Also, do not forget to address parasitic inputs and parasitic states.

8. ∗ Consider the VHDL code below where the output of a linear feedback state register (LFSR) and of a counter are combined into a four-bit random output. What's wrong with this design? Hint: Actually, there is one obvious and one more subtle problem. Both are related to the same clause in the process statement.

```vhdl
entity partres is
   port (
      Clk_CI : in std_logic;
      Rst_RBI : in std_logic;
      Oup_DO : out std_logic_vector(3 downto 0) );
end partres;
--------------------------------------------------------------------
architecture behavioral of partres is
   signal StateA_DP, StateA_DN : std_logic_vector(1 to 4);
   signal StateB_DP, StateB_DN : unsigned(3 downto 0);
begin

   -- computation of next state
   StateA_DN <= (StateA_DP(3) xor StateA_DP(4)) & StateA_DP(1 to 3);
   StateB_DN <= StateB_DP + "1001";

   -- updating of state
   process (Clk_CI,Rst_RBI)
   begin
      -- activities triggered by asynchronous reset
      if Rst_RBI='0' then
         StateA_DP <= "0001";
      -- activities triggered by rising edge of clock
      elsif Clk_CI'event and Clk_CI='1' then
         StateA_DP <= StateA_DN;
         StateB_DP <= StateB_DN;
      end if;
   end process;

   -- updating of output
   combine: for i in 3 downto 0 generate
     Oup_DO(i) <= StateA_DP(4-i) xor StateB_DP(i);
   end generate combine;

end behavioral;
```

9. ∗ SystemVerilog knows of no built-in construct for computing the absolute value of integers. Write a function for that.

10. ∗∗ Write a synthesis model for the finite state machine of fig.B.4. A synchronous clear shall be provided for initializing the circuit. Explain how your code reflects the fact that this is a Moore automaton.

11. ∗∗∗ Consider a shaft equipped with an angle encoder that indicates the shaft's current position using a two-bit unit-distance code, see fig.4.25. Design a state machine that accepts this code and tells whether the shaft is currently rotating clockwise or counterclockwise. The former sense of rotation shall be indicated when the shaft is at standstill. Establish a synthesis model. You may want to extend the functionality such as to indicate the angular position of the shaft and the number of turns it has made since time zero.

10   00

11   01

Unit distance encoding of shaft angle.

12. ∗∗∗ Design a synchronous first-in first-out (FIFO) queue with the features below.
   a) Separate read and write ports ("data" plus "read" or "write" respectively).
   b) A pair of outputs that flag "full" and "empty" conditions respectively.
   c) Two more outputs that indicate "almost full" and "almost empty" conditions.
   d) Parametrized queue depth, data width, and "almost full/empty" thresholds.
   e) Show-ahead capability, aka first-word-fall-through property.[80]
   Use a register file for data storage and code your RTL synthesis model in either VHDL or SystemVerilog. Chose meaningful port and signal names and observe the naming conventions of section 6.7. Note: Verifying the HDL model developed here will be the subject of problem 5 in chapter 5.

---

[80] To keep FIFO latency minimal, the oldest data item stored is made immediately available at the output with no need to activate the "read" input beforehand and/or to wait for a clock cycle or a clock edge. The situation where the output is invalid because there are no data left is flagged via the "empty" output in the normal way. Note that the meaning of the "read" input is not the same as with a standard FIFO. Whereas it means "bring me data" there, it says "I just fetched a data word, get the next one ready, if any" in a show-ahead FIFO.

4.7 APPENDIX I: VHDL AND SYSTEMVERILOG SIDE BY SIDE

Table 4.10 The most common operators in VHDL and SystemVerilog. Details may differ between the two languages and even from one EDA tool to another.

		VHDL	SysVer	amenable to synthesis
Bitwise operators				
$\overline{a}$	inversion	not a	~a	yes
$a \vee b$	or	a or b	a | b	yes
$a \wedge b$	and	a and b	a & b	yes
$a \oplus b$	xor	a xor b	a ^ b	yes
String operator				
$a \smile b$	concatenation	a & b	{ a, b }	yes
Arithmetic operators				
$\|a\|$	absolute value	abs a	n.a.	yes
$a + b$	addition	a + b	a + b	yes
$a - b$	subtraction	a - b	a - b	yes
$a \cdot b$	multiplication	a * b	a * b	for most tools
$a : b$	division	a / b	a / b	if b = const. or b = 2^n where $n \in \mathbb{Z}$
$a \bmod b$	modulo	a mod b	a % b	*idem*
a^b	exponentiation	a ** b	a ** b	if a = 2 and b $\in \mathbb{Z}$
Relation operators				
$a = b$	equal to	a = b	a == b	yes
$a \neq b$	not equal to	a /= b	a != b	yes
$a < b$	less than	a < b	a < b	yes
$a > b$	greater than	a > b	a > b	yes
$a \leq b$	less or equal	a <= b	a <= b	yes
$a \geq b$	greater or equal	a >= b	a >= b	yes
Shift operators				
shift a left logic by b positions		a sll b	a << b	yes
shift right logic (no sign extension)		a srl b	a >> b	yes
shift left arithm. (same as logic)		a sla b	a <<< b	yes
shift right arithm. (sign extension)		a sra b	a >>> b	yes

Table 4.11 **Corresponding terms and constructs in VHDL and SystemVerilog.**

Concept	VHDL	SystemVerilog
Data record	record	struct
Subroutine	subprogram	subroutine
with no side effects	function	function
with side effects	procedure	task
Circuit or subcircuit	[design] entity	module
Connector	port	port
Circuit parameter [a]	generic	parameter (#)
Circuit instance	instance	module instance
Circuit node aka net	signal (any data type allowed, retains the last value assigned)	wire (MVL-4 only, no value retention, must stay driven)
Node with multiple drivers	resolved signal	wire
Independent thread of execution	[concurrent] process	[concurrent] process
Concurrent assignment [b]	concurr. signal assignment (<=)	continous assignment (assign)
Conditional assignment	cond'al/selected signal ass. (<=)	idem plus cond'al operator (?)
Procedural descr. of ckt behav. [c]	process statement	procedural block (always)
memoryless	idem (no dedicated constr.)	always_comb block
memoryzing	idem	always_ff/_latch blocks
Procedural descr. in general	idem	always, initial, final
Wake up condition	sensitivity list	sensitivity list
Time-varying data item that retains its value until explicitly assigned a new value		
confined to within one process	variable [d] (:=)	[local] variable (var)
shared between processes	signal [e] (<=)	variable (var)
change of value	event	update event
time at which this happens	simulation time	time slot
Assignment executed		
sequentially	variable assignment in a process statement	blocking assignment (=) in an always block
concurrently	concurrent signal assignment or signal assignment in a process statement	continuous assignment (assign) or nonblocking assignment (<=) in an always block or blocking assignment (=) in a fork-join parallel block
Waiting		
on an event	wait on ...	@(...)
until a condition holds	wait until ...	wait (...)
for a lapse of time (ex.)	wait for 5 ns	#5ns (no space allowed)
a number clock cycles (ex.)	(no dedicated construct)	##17 (in SVAs only)
forever	wait	use initial block
Propagation delay	after clause	delayed assignment
example	Oup <= Inp after 3 ns;	assign #3ns Oup = Inp;
Timing conditions	via concurrent assertions	timing checks (setup, hold, etc.)

---

[a] Such as a timing quantity, a word width, or some other option.
[b] This is the simplemost kind of process.
[c] Concurrent outside, sequential inside.
[d] No interaction with the event queue.
[e] Does interact with the event queue.

Table 4.12   Key features of the predominant HDLs compared. See [106] [107] for more.

Feature	VHDL	Verilog	SystemVerilog
Background and underlying concepts			
Industry standard	IEEE 1076	IEEE 1364	IEEE 1800
Initially accepted / current revision	1987 / 2008	(1984)1995 / 2005	2005 / 2012
Originator	DoD	Gateway	Accellera
Roots	Ada	C	& VHDL, PSL, Vera
Overall character	safe, verbose	terse, precarious	powerful, bloated
Concurrent processes	yes	yes	yes
Event-based concept of time	yes	yes	yes
Circuit hierarchy and structure (netlist)	yes	yes	yes
Source code encryption mechanism	since 2008	yes	yes
Discretization of electrical signals	adjunct package	part of language	part of language
Logic system	9-valued IEEE 1164	4-valued	4-valued
Switch-level capability	no	yes (8 strengths)	yes (8 strengths)
Language features and software engineering			
Separate interface decl. and implem. module	yes	no	no
Type checking	strong	none	very limited
Type conversion functions	adjunct package	none	few required
Enumerated & other user-defined data types	supported	no	supported
Data types acceptable at block boundaries	any	binary only	any
Function arguments of variable word width	supported	no	supported
Data object with / without time attached	signal/variable	no distinction	no distinction
Signal attributes (named signal properties)	yes	n.a.	n.a.
Timing and word size parametrization	generic	supported	parameter
Conditional/repeate d process generation	generate	since 2001	generate
Conditional/repeated component instantiation	generate	since 2001	generate
Circuit block to instance binding control	configuration	since 2001	`ifdef
Simulation and testbench design			
Stringent event order in the absence of delay	yes (via δ delay)	no	no
Event queue inspection (e.g. for timing checks)	part of language	via system tasks	via system tasks
Text and file I/O	adjunct package	via system tasks	via system tasks
Back-annotation from SDF files	IEEE 1076.4	yes	yes
Acceleration of gate-level primitives	IEEE 1076.4	yes	yes
Acceptance for sign-off simulation	yes	yes	yes
Assertion-based verification	limited	no	supported
Constrained random pattern generation	no	no	supported
Coverage metering	no	no	supported
Object-oriented (not for use in circuit models)	no	no	yes
with multiple inheritance			since 2012
Queues, mailboxes, semaphores (idem)	no	no	supported
Foreign programming lang. interface (idem)	VHPI	PLI	DPI
Synthesis			
Amenable to hardware synthesis	subset only	subset only	subset only
Timing constraints	not p.o.l. (SDF)	not p.o.l. (SDF)	not p.o.l. (SDF)
Other synthesis directives	not p.o.l.	not p.o.l.	not p.o.l.
Analog and mixed-signal extension			
Designation	VHDL-AMS	Verilog-AMS	
Industry standard	IEEE 1076.1	Accellera 2.2	

Table 4.13   References on VHDL and SystemVerilog.

VHDL Reference	Year	IEEE 1076	IEEE 1164	circuit model.	test-bench	syn-thesis	Comments / special topics
		Subjects covered					
Specifications of language and syntax							
IEEE [108]	'09	-08	no	no	no	no	language reference manual
IEEE [109]	'02	-02	no	no	no	no	language reference manual
Zimmermann [110]	'02	.1	no	no	no	no	AMS syntax on www
Zimmermann [111]	'97	-93	no	no	no	no	syntax in EBNF on www
Bhasker [112]	'95	-93	no	no	no	no	syntax diagrams, +
Textbooks							
Volnei Pedroni [113]	'10	-08	yes	yes	yes	yes	many examples, FPL
Reichardt Schwarz [114]	'09	-02	yes	yes	(yes)	yes	synthesis, filter, German, +
Ashenden [115]	'08	-08	yes	yes	yes	yes	pointers, std versions, +
Ashenden Lewis [96]	'08	-08	(yes)	(yes)	(yes)	no	-08 language updates only
Chu [116]	'06	-02	yes	yes	(yes)	yes	focus on circuits, +
Molitor Ritter [117]	'04	-93	yes	yes	yes	(yes)	examples, German
Ashenden et al. [118]	'03	-93	yes	yes	yes	no	VHDL-AMS
Yalamanchili [119]	'01	-93	yes	yes	(yes)	yes	FPL, memory model
Heinkel [120]	'00	-93	yes	yes	(yes)	yes	VHDL-AMS
Zwolinski [121]	'00	-93	yes	yes	(yes)	yes	asynchronous circuits
Chang [122]	'99	-93	no	yes	no	yes	based on [?], examples
Navabi [124]	'98	-93	yes	yes	yes	(yes)	small CPU example
Chang [123]	'97	-93	yes	yes	yes	yes	testbenches, project, +
Bhasker [125]	'96	-93	yes	yes	(yes)	yes	code to circuit mappings
References with a specific focus and other resources							
Bergeron [126]	'00	-93	no	no	yes	no	functional verification, +
Hamburg Archive [127]	'06	n.a.					free models, FAQ, links

SystemVerilog Reference	Year	IEEE 1800	logic syst.	circuit model.	test-bench	syn-thesis	Comments / special topics
		\multicolumn Subjects covered					
Specifications of language and syntax							
IEEE [98]	'13	-12	yes	no	no	no	language reference manual
Textbooks							
Zwolinski [100]	'10	-05	(yes)	yes	yes	yes	digital design, +
Sutherland et al. [101]	'06	-05	(yes)	yes	(yes)	(yes)	language, syntax in EBNF
References with a specific focus and other resources							
Spear [103]	'10	-09	no	no	yes	no	object-oriented testbenches
Salemi [128]	'09	-05	yes	(yes)	(yes)	no	for VHDL converts, +
Bergeron [129]	'06	-05	no	no	yes	no	functional verification
Alabama Tutorial [130]	'13	-09	(yes)	yes	(yes)	(yes)	quick intro for designers, +
Doulos Tutorials [131]	'13	-05	(yes)	yes	yes	(yes)	upgrades over Verilog

( ) = light coverage only, + = personal preference.

4.8 APPENDIX II: VHDL EXTENSIONS AND STANDARDS

4.8.1 PROTECTED SHARED VARIABLES IEEE 1076a

The normal way of exchanging time-varying data between processes in VHDL is via `signals`. `Signals` essentially stand for electrical wires running between subcircuits and the non-zero delays of those subcircuits are expressed as part of signal assignment statements.

As opposed to this, shared `variables` are intended to support inter-process communication for bookkeeping and supervision tasks during simulation runs, e.g. counting the number of process invocations, keeping track of exceptions or other special events, coordinating activities among the different processes in a testbench, or collecting statistical data. Zero-delay communication is fine in this context. Were it not for shared `variables`, programmers would be forced to employ `signals` thereby obscuring their original intention and unnecessarily inflating execution time.

To stay clear of problems that might result from simultaneous read or write operations to a global variable by distinct processes, the access must be controlled. Protected shared `variables` provide a means for synchronization. Access to a protected shared `variable` is exclusive and must always occur by calling one of the `functions` or `procedures` written for that purpose. That is, when a first process gains access to a protected `variable` by calling one such subprogram to work on it, any further process attempting to read or modify the same `variable` by invoking the same or another subprogram must wait until the current access has terminated.

Declaring a protected shared `variable` of an existing data `type` could hardly be simpler.

Example   `shared variable event_counter : shared_counter;`

Obviously, the `variable`'s `type` is to be declared beforehand along with all `functions` and/or `procedures` necessary for access. Consistent with VHDL's guiding principles, the declaration of a protected `type` is separated from its implementation as shown in the subsequent example [132].

```
-- protected type declaration
type shared_counter is protected
   procedure : reset;
   procedure : increment ( by : integer := 1 );
   impure function value return integer;
end protected;
```

```
-- protected type body
type shared_counter is protected body

    variable count : integer := 0;

    procedure reset is
    begin
        count := 0;
    end procedure reset;

    procedure increment ( by : integer := 1 ) is
    begin
        count := count+by;
    end procedure increment;

    impure function value return integer is
    begin
        return count;
    end function value;

end protected body shared_counter;
```

Another code fragment shows that subprograms for reading or modifying a protected shared variable get invoked by prefixing their names with that of the variable meant to be accessed.

```
...
event_counter.reset;
event_counter.increment (3);
assert event_counter.value > 0;
...
```

Protected variables are part of the 2002 revision of the IEEE 1076 standard. Shared but unprotected variables had been introduced in VHDL'93 as a result from controversial debates in the standard committee, using them is discouraged as no access control mechanism was provided.

4.8.2 THE ANALOG AND MIXED-SIGNAL EXTENSION IEEE 1076.1

This standard is informally known as **VHDL-AMS** and extends the capabilities of the original language towards describing and simulating lumped analog and mixed-signal circuits. It has been a guiding principle to augment the existing VHDL constructs and to add new ones such as to make the new IEEE 1076.1 language a proper superset of VHDL which has obvious benefits.

To capture continuous quantities such as voltages, currents and charges, a new kind of object has been introduced that complements the constants, variables and signals defined in the IEEE 1076 standard. This class is termed **quantity** and any object that belongs to it takes on floating point values exclusively.

The supplemented language supports the modeling of time-continuous behavior by accommodating (possibly nonlinear) differential and algebraic equations in the time domain such as $F(\dot{x}(t), x(t), t) = 0$. So-called implicit quantities have been included to denote derivatives and integrals over time. If x has been declared as a quantity, for instance, then x'dot automatically refers to $\frac{d}{dt}$x.

Elaboration of a VHDL-AMS model yields a digital part (made up of signals and processes) and an analog part (consisting of quantities and differential algebraic equations). Simulation begins with determining the model's initial condition at time zero. The standard further defines how to synchronize the traditional event-driven simulation cycle with a solver for a system of simultaneous differential and algebraic equations.

Observation 4.44. *In a nutshell, VHDL's analog and mixed-signal extensions are as follows:*

			standard
VHDL-AMS	=	*VHDL*	*IEEE 1076*
	+	*continuous-value objects*	*IEEE 1076.1*
	+	*simultaneous differential algebraic equations*	*idem*
	+	*coupled continuous and discrete models of time*	*idem*
	+	*transistor compact models*	*(open)*

It is worthwhile to note that VHDL-AMS extends the modeling capabilities in many ways. The significance of nonlinear and/or differential and algebraic equations in stating the static and continuous-time characteristics of electrical components and circuits is immediately evident. Entire subsystems from data transmission, signal processing, control systems, etc. can be condensed into abstract high-level mathematical models.

What further sets VHDL-AMS apart from SPICE is the absence of built-in transistor models in the simulator kernel. Model writers are no longer confined to a structural view that describes how more complex (sub)circuits are pieced together from a few built-in primitives (such as resistors, capacitors, and transistors). Rather, they are put in a position to describe opamps, active filters, phase locked loops (PLL), etc. from a purely behavioral perspective using mathematical equations as building blocks and combining them with event-driven submodels where appropriate. This also enables them to create their own primitive models and to include them in circuit simulation at any time with no need for assistance from the software vendor.

Lastly, there is nothing that would confine quantities to be of electrical nature which opens the door for modeling thermal, micromechanical, optoelectronic, magnetic and other effects. Yet, we will not elaborate on VHDL-AMS as analog, mixed-signal and multi-domain models are beyond the scope of this text. Please refer to [133] [118] [134].

While VHDL provides the floating point data type `real`, it does not support operations other than basic arithmetic operators.[81] To overcome this limitation, two new `packages` have been defined and accepted as IEEE standard 1076.2.

Package `math_real` includes

- Definitions of constants including e and π.
- Sign, floor, ceiling, round, truncate, min and max functions.
- Square root, cubic root, power (x^y), exponential (e^x) and logarithm ($\ln(x)$) functions.
- Trigonometric functions (sin, cos, tan, arcsin, etc.).
- Hyperbolic functions (sinh, cosh, tanh, arcsinh, etc.).
- A pseudo-random number generator for reals uniformly distributed in the interval [0,1].

Package `math_complex` includes

- Definitions of complex number types (in Cartesian and polar form).
- Absolute value, angle, (argument), negate, and conjugate functions.
- Square and exponential (e^z) functions.
- Overloaded versions of basic arithmetic operators (for Cartesian and polar operands).
- Type conversion functions.

Both packages have been added to the existing design library `ieee`. Clearly, they are intended for modeling system behavior at higher levels of abstraction and for auxiliary functions in testbenches, not for synthesis.

As no single IEEE 1076 or IEEE 1164 data type supports computer arithmetics with adequate precision and convenience, two new packages `numeric_bit` and `numeric_std` have been developed and accepted as IEEE standard 1076.3. They include

- The definition of data types `unsigned` and `signed` (discussed in section 4.2.3) for unsigned and two's-complement arithmetic respectively,
- Overloaded versions of IEEE 1076 arithmetic, logical and relational operators.
- Arithmetic shift and rotate functions.
- Resizing functions with sign extension and reduction.
- Type conversion functions (see appendix 4.8.6).

In addition, the IEEE 1076.3 standard indicates how to interpret for the purpose of VHDL synthesis logical values, such as "L", "X" and "U", that have a physical meaning as outcomes from simulation but not as specifications for a circuit to be. This is also why the two packages are sometimes referred to as synthesis packages.

---

[81] Addition +, subtraction −, sign inversion −, multiplication $\ast$, division /, integer power ($x^n, n \in \mathbb{N}$) $\ast\ast$, and absolute value abs.

4.8.5 THE STANDARD DELAY FORMAT (SDF) IEEE 1497

The SDF was originally developed by Open Verilog International (OVI) and later modified to become SDF version 4.0 which has been accepted as IEEE 1497 standard in 2001. SDF files are written in ASCII-readable form and store timing data in a non-proprietary format for later use during the VLSI design and verification process.

SDF makes it possible to share gate delays, timing conditions, and interconnect delays between cell libraries, delay calculators, HDL simulators, and static timing analysis software. More particularly, SDF supports back-annotating existing netlists with numerical timing data obtained from layout extraction as illustrated in figs.4.14.

The language also includes constructs for forward-annotation, that is for specifying timing constraints that are to guide the synthesis process of prospective circuits shown in figs.4.14 and 4.21. Further provisions allow for documenting the PTV conditions for which the timing data stored in an SDF file apply.

4.8.6 A HANDY COMPILATION OF TYPE CONVERSION FUNCTIONS

with contributions by Reto Zimmermann and Jürgen Wassner

The table below and fig.4.26 summarize type conversion functions between the most important VHDL data `types`. Note that proprietary `types`, `functions`, and `packages` render source code awkward to port from one EDA platform to another. They should, therefore, be dismissed in favor of vendor-independent international standards.

Table 4.14 Type conversion functions.

according to as defined in package	IEEE 1076 `std.standard`
Conversion	
real ▷ integer	integer(arg)
integer ▷ real	real(arg)

according to as defined in package	IEEE 1076.3 `ieee.numeric_std`	Synopsys proprietary `ieee.std_logic_arith`
Conversion		
std_logic_vector ▷ unsigned	unsigned(arg)	unsigned(arg)
std_logic_vector ▷ signed	signed(arg)	signed(arg)
unsigned ▷ std_logic_vector	std_logic_vector(arg)	std_logic_vector(arg)
signed ▷ std_logic_vector	std_logic_vector(arg)	std_logic_vector(arg)
integer ▷ unsigned	to_unsigned(arg,size)	conv_unsigned(arg,size)
integer ▷ signed	to_signed(arg,size)	conv_signed(arg,size)
unsigned ▷ integer	to_integer(arg)	conv_integer(arg)
signed ▷ integer	to_integer(arg)	conv_integer(arg)
integer ▷ std_logic_vector std_logic_vector ▷ integer	integer ▷ unsigned|signed ▷ std_logic_vector std_logic_vector ▷ unsigned|signed ▷ integer	
Resizing		
unsigned	resize(arg,size)	conv_unsigned(arg,size)
signed	resize(arg,size)	conv_signed(arg,size)

LISTING 4.21

A Gray counter that makes use of type conversions. See listing 4.10 for the package referenced here.

```
-- Mission: Illustrate how to model a Medvedev machine in two processes.
--    Example designed to demonstrate the usage of functions from a package,
--    of IEEE numeric_std, type conversions, and a parametrized word width.
-- Functionality: w-bit Gray counter with enable and asynchronous reset.
--    Operates with double conversion: Gray->binary, increment, binary->Gray,
--    which is not necessarily the most economic nor the fastest solution.
-- Author: H.Kaeslin.
-- ----------------------------------------------------------------------------
```

```vhdl
library ieee;
use ieee.std_logic_1164.all;
use ieee.numeric_std.all;
use work.grayconv.all;            -- my own set of Gray code converter functions
-------------------------------------------------------------------------------

entity graycnt is
   generic (
      GRAYWIDTH : integer := 5 );    -- default value for number of state bits
   port (
      Clk_CI   : in  std_logic;
      Rst_RBI  : in  std_logic;
      Ena_SI   : in  std_logic;
      Count_DO : out std_logic_vector((GRAYWIDTH-1) downto 0) );
end graycnt;


-------------------------------------------------------------------------------

architecture computed_state of graycnt is

   -- present state and next state
   signal Count_DP, Count_DN : std_logic_vector(GRAYWIDTH-1 downto 0);

begin

   ---- computation of next state
   Count_DN <= bintogray(std_logic_vector(
      unsigned(graytobin(Count_DP)) + 1 ));

   ---- updating of state
   p_memzing : process (Clk_CI,Rst_RBI)
   begin
      -- activities triggered by asynchronous reset
      if Rst_RBI='0' then
         Count_DP <= (others => '0');   -- width-independent shorthand
      -- activities triggered by rising edge of clock
      elsif Clk_CI'event and Clk_CI='1' then
         -- proceed to next state only if enable is asserted
         if Ena_SI='1' then
            Count_DP <= Count_DN;
         end if;
      end if;
   end process p_memzing;
```

```
    ---- assignment of state to output only signal
    Count_DO <= Count_DP;

end computed_state;
```

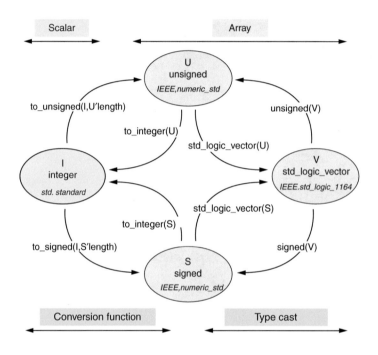

FIGURE 4.26

VHDL type conversion paths (chart courtesy of Dr. Jürgen Wassner).

4.8.7 CODING GUIDELINES

Third-party HDL code can be very painful to adopt and adapt if its style significantly differs from that of other HDL components and from the practice of the system integrators. An early book with advice on how to write reusable HDL code was [135], but most of it is part of good design practice anyway. While many companies and organisations have developed coding and reuse guidelines for internal usage, ESA was among the first to make theirs widely accessible [136]. A naming convention for VHDL signals developed by the author and his colleagues is to be introduced in section 6.7. In 2008, NXP has published their reuse standards under the label CoReUse [137]. Unfortunately, an industry-wide consensus is nowhere in sight.

FUNCTIONAL VERIFICATION

5.1 GOALS OF DESIGN VERIFICATION

The ultimate goal of design verification is to avoid the manufacturing and deployment of flawed designs. Large sums of money are wasted and precious time to market is lost when a microchip does not perform as expected. Any design is, therefore, subject to detailed verification long before manufacturing begins and to thorough testing following fabrication. One can distinguish three motivations (after the late A. Richard Newton):

1. During specification: "Is what I am asking for what is really needed?"
2. During design: "Have I indeed designed what I have asked for?"
3. During testing: "Can I tell intact circuits from malfunctioning ones?"

Any of these questions can refer to two distinct circuit qualities:

Functionality describes what responses a system produces at the output when presented with given stimuli at the input. In the context of digital ICs, we tend to think of logic networks and of package pins but the concept of input-to-output mapping applies to information processing systems in general. Functionality gets expressed in terms of mathematical concepts such as algorithms, equations, impulse responses, tolerance bands for numerical inaccuracies, finite state machines (FSM), and the like, but often also informally.

Parametric characteristics, in contrast, relate to physical quantities measured in units such as Mbit/s, ns, V, μA, mW, pF, etc. that serve to express electrical and timing-related characteristics of an electronic circuit.

Observation 5.1. *Experience has shown that a design's functionality and its parametric characteristics are best checked separately as goals, methods and tools are quite different.*

5.1.1 AGENDA

Our presentation is organized accordingly with section 5.2 presenting the options for specifying a design's functionality. The bulk of the material then is about developing a simulation strategy that maximizes the likelihood of uncovering design flaws. After having exposed the puzzling limitations of functional verification in the first part of section 5.3, we will discuss how to prepare test data sets and how to make use of assertions to render circuit models "self-checking". How to organize simulation data and simulation runs is the subject of section 5.4, while section 5.5 gives practical advice on how to code testbenches using HDLs. Neither parametric issues nor the testing of physical parts will be addressed here.

5.2 HOW TO ESTABLISH VALID FUNCTIONAL SPECIFICATIONS

Specifications available at the outset of a project are almost always inaccurate and incomplete. While parametric characteristics are relatively easy to state, expressing complex functionalities in precise yet concise terms is much more difficult. Functional specifications are, therefore, often stated verbally or graphically. There is a serious risk with doing so, however.

Warning example

An ASIC had to interface with an industry-standard microprocessor bus. Specifications made reference to official documents released by the CPU manufacturer where bus read and write cycles were described in great detail along with precise timing diagrams. Although the ASIC was designed and tested with these requirements in mind, systems immediately crashed because of bus contentions when first prototypes were plugged into the target board.

What had gone wrong? It was found that the ASIC worked fine as long as its chip select line was active. When deselected, however, the pad drivers failed to release the bus, i.e. they did not revert to a high-impedance state. This obvious necessity had been omitted in the specifications and, as a consequence, been ignored throughout the subsequent design and test phases.
□

In more general terms, the subsequent quote from [138] nicely summarizes the situation.

> Many computer systems fail in practice, not because they don't meet their specifications, but because the specifications left out some unanticipated circumstances or some unusual combination of events, so that when the unexpected occurred, the system was not able to deal with it. This is not necessarily due to sloppiness or stupidity on the part of the designer or to inadequate design methodology; it is a fundamental characteristic of the design process.[1]

This leaves us with three important issues:

"How to ascertain specifications are precise, correct and complete?"
"How to make sure specifications describe the functionality that is really wanted and needed?"
"How to have customers, marketing and engineers share the same understanding?"

As natural language and informal sketches have been found to be inadequate, let us next discuss two approaches for arriving at more dependable specifications.

---

[1] As testified by numerous tragedies, this applies to any kind of technical system, not just computers. Just study the conditions and failure mechanisms that have led to the sinking of RMS Titanic, the Challenger space shuttle accident, the crash of Airbus flight AF 447, or the nuclear disaster of the Fukushima power plant.

Ideally, all requirements for a circuit or system could be cast into a set of formal specifications which then would serve as a starting point for a rigorous mathematical proof of correctness. Over the years, a broad variety of formalisms has been devised for capturing behavioral aspects of numerous subsystems from many different fields including truth tables, signal flow graphs, equations, state graphs, statecharts, Petri nets, and signal transition graphs (STG).

A difficulty is the limited scope of each such formalism. Signal flow graphs, for instance, were developed for describing transformatorial systems, but are inadequate for modeling reactive behavior. Although Petri nets and finite state machines can, in theory, describe any kind of computation, they become unmanageable when applied to numerical computations or to highly complex situations. Most VLSI circuits, on the other hand, include diverse subsystems some of which are more of transformatorial nature (datapaths, lookup tables) and others more of reactive nature (controllers, interfaces). Relying on a single formal method for specifying the desired functionality of an entire chip or system is not normally practical.

A more mundane difficulty is that mathematical formalisms are unsuitable for communicating with customers and management. Also from a practical perspective, there must exist a straightforward and foolproof way to break down a system's specifications into specs for its various components in order to support collaborative development in a team, and to support products that comprise both hardware and software.

Prototyping often is the only viable compromise between strictly formal and totally informal specification. By rapid or **virtual prototype** we understand an algorithmic model that emulates the functionality of the target circuit but not necessarily its architectural, electrical and timing characteristics. A virtual prototype can be implemented

- as software code that runs on a general-purpose computer, microprocessor, or DSP,
- with the aid of generic software tools for system-level simulations,[2] or
- by configuring FPGAs or other FPL devices.

---

[2] Such as MATLAB/Simulink or SystemC.

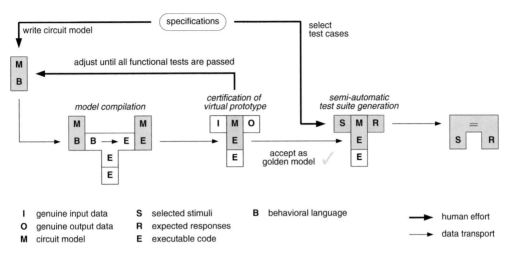

I	genuine input data	**S**	selected stimuli	**B** behavioral language
O	genuine output data	**R**	expected responses	
M	circuit model	**E**	executable code	

➤ human effort

→ data transport

FIGURE 5.1

Successive refinement of a rapid prototype and its eventual contribution toward preparing a test suite.[3]

The typical procedure goes:

1. Apply formal methods (e.g. equations and statecharts) to capture specifications.
2. Use them as a starting point for developing a virtual prototype.
3. Make the prototype as widely available as possible for a thorough evaluation.
4. Refine specifications and prototype until satisfied, see fig.5.1.

Once a virtual prototype has been thoroughly certified in a multitude of test runs, it gets elevated to a **golden model** which is assumed to be free of functional errors for the purpose of the subsequent design steps. Such a model not only defines the target functionality of the circuit-to-be but also helps to prepare the expected responses from a selection of stimuli. The pros and cons of rapid prototyping are as follows.

---

[3] Figs.5.1, 5.8 and 5.9 come as **T-diagrams**, a notation from compiler engineering that serves to plan the porting of programs from one machine to another [139]. As an extension, two extra symbols have been added. One stands for a test suite and the other for a piece of information-processing hardware, i.e. for a digital ASIC or an FPL device. Synthesis tools are viewed here as compilers that turn behavioral models into gate-level netlists, and gate-level simulators as interpreters that translate between netlists and executable code.

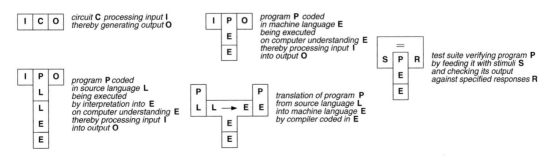

+ Demonstrations of the prospective functionality can be arranged at an early stage.
+ Functionality can be verified on the basis of real world data.
+ Shortcomings of the initial specifications are likely to get exposed in the process.
+ No time and money gets wasted for hardware design before prototype is approved by all, including management and customers.
+ Design iterations and fine tuning are not penalized by the long turnaround times associated with IC design and manufacturing.
+ Prototype is amenable to peer review and code inspection.
+ Computes expected responses for any selection of stimuli.
− There is no guarantee that all critical cases get covered during prototype testing.
− Slips that are related to timing or electrical problems are unlikely to be found because they are not rendered correctly by a purely functional model.
− Performance of the prototype does not normally come close to that of the final VLSI chip.

Observation 5.2. *Rapid prototyping gives developers the opportunity to make and uncover more mistakes earlier, and so to save precious cost and time.*

5.2.3 HARDWARE-ASSISTED VERIFICATION

This is a particular case of rapid prototyping that leverages FPGA technology on a larger scale. Special circuit boards that carry multiple reprogrammable devices are used to extensively emulate the circuit-to-be gate by gate before committing to mask-programmed VLSI. The emulator boards connect to a host computer for control, storage of stimuli and responses, tracing of internal nodes, visualization, etc. While **hardware accelerators** are available commercially,[4] developers of high-end microprocessors and ASICs tend to come up with proprietary systems optimized for their specific needs.

Example
The Bluegene/Q ASIC, a multi-processor SoC implemented in IBM's 45 nm SOI CMOS technology and targeted to run with a 1.6 GHz clock has been emulated at a speed equivalent to a 4 MHz clock [140].[5] The hardware accelerator for this $1.47 \cdot 10^9$ transistor circuit consists of 28 circuit boards designed around Virtex-5 LX330 FPGAs and standard RAMs. The design team felt that building custom hardware specifically for functional verification of Bluegene/Q was perfectly worth the effort as hardware emulation runs over 100 000 times faster than the logic-level software simulation of the same design.
□

Observation 5.3. *It is just not possible to functionally verify complex high-performance designs without recurring to complex high-performance emulation tools.*

A hardware prototype may even be tested within the target environment, provided all surrounding equipment can be made to operate at a reduced clock rate consistent with the prototype's execution speed. This approach is particularly helpful for locating interface problems.

---

[4] The Cadence Palladium box and Mentor's Veloce products are two examples.
[5] Several factors make emulation considerably slower than the real target circuit: i) FPGAs are inherently slower than hardwired ICs. ii) Complex designs need to be distributed over many FPGAs or even boards. iii) External RAM may be required to emulate on-chip memory. iv) Serialization and deserialization may be required for communicating between distinct ICs and boards whereas fast on-chip busses will ultimately do the job.

5.3 PREPARING EFFECTIVE SIMULATION AND TEST VECTORS

5.3.1 A FIRST GLIMPSE AT VLSI TESTING

Following fabrication, every single chip is subject to thorough **tests**. Most of those tests have **automated test equipment** (ATE) monitor the signal waveforms that result when predefined electrical waveforms are being applied to the **circuit under test** (CUT) over many, many clock cycles. Simulation essentially does the same, albeit with a virtual circuit model commonly referred to as **model under test** (MUT). Table 5.1 juxtaposes the respective terms.

Simulation and testing are said to be **dynamic verification** techniques. This contrasts with code inspection, formal verification, equivalence checking, timing analysis, and other **static verification** techniques that do not depend on signals, clocks, waveforms, or test data in any way. Simulation prevails when it comes to check a design's functional behavior. This is mainly because of the limited capabilities of today's formal verification methods. However, simulation and testing both raise a variety couple of fundamental and practical difficulties that we are going to address in the remainder of this chapter. Let us begin by studying the question

"What does it take to uncover a design flaw via I/O signals?"[6]

Example
Consider a multiplexer buried within a large circuit. Assume that its two control inputs have been permuted by accident, see fig.5.2. Writing (0 to 1) rather than (1 downto 0) in the VHDL code, or [0:1] instead of [1:0] in SystemVerilog, for that matter, suffices.

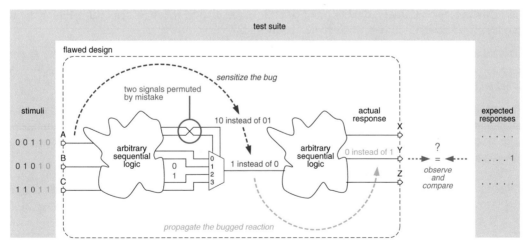

FIGURE 5.2

A typical design flaw and the preconditions necessary for uncovering it.

---

[6] By I/O signal we mean a signal that connects to a package pin or to a pad on the die so that it be driven and/or observed from externally.

Three preconditions must hold for that design flaw to become manifest.

1. **Bug sensitization.** The design error must be made to provoke a condition other than the normal one in the circuit. In the occurrence, the permuted control inputs are driven to opposite logic values while the data input to the multiplexer are adjusted such that a logic value opposed to the correct one indeed appears at its output.
2. **Bug propagation.** The stimuli must permit the erroneous condition to propagate to observable nodes by causing a cascade of intermediate nodes to assume incorrect values.
3. **Bug observation.** The corrupted value observed on one or more of the output nodes must get checked against the logic value that a correct design is expected to produce.

Unless all three conditions are met, the design flaw will have no consequence during simulation and testing, although circuits fabricated on the basis of the flawed HDL source code are almost certain to fail when put into service.

□

The same reasoning essentially applies to any bug that affects functional behavior.

5.3.2 FULLY AUTOMATED RESPONSE CHECKING IS A MUST

In the context of simulation, bug observation implies checking the MUT output. How to do so is dictated by the volume of data. Even a fairly modest subcircuit asks for keeping track of hundreds of waveforms over thousands of clock cycles. Digging through waveform plots, event lists, tabular printouts of logic values, and similar records from simulation runs is not practical for efficiency reasons. It is also unacceptable from a quality point of view because some incorrect data value hiding within myriads of correct items is very likely to be overlooked.

Observation 5.4. *Purely visual inspection of simulation data is not acceptable in VLSI design. Rather, designers must arrange for the simulator software to automatically check the actual responses from the model under test against the correct ones and to report any differences.*

The answer is to collect the **expected responses** along with the pertaining **stimuli** in a sequence of data patterns that we call a **test suite**. In its most simple expression, a test suite is a set of binary vectors listed cycle by cycle that specifies what kind of responses a correct part is supposed to output when fed with certain stimuli.[7] Test suite, test cases, and test vectors are often used as synonyms.

The word **testbench** is also used in this context, but refers to a somewhat different concept. A testbench is a piece of software used to pilot a simulation run that applies stimuli, that acquires responses, and that compares the actual against the expected responses.[8] A testbench is to a MUT

---

[7] Mechanical engineers use gauges to verify the geometric conformity of manufactured parts such as to eliminate inaccurate copies before they are being put together with other components. In essence, gauges are specifications that have materialized. Similarly, a test suite serves to verify the functional correctness of some (sub)circuit. It might as well be called a functional gauge.

[8] Put in IT terminology, a testbench is comparable to a software driver for some peripheral device whereas the test suite corresponds to a particular set of data.

in the simulation world what ATE is to a fabricated circuit in the physical reality. Testbench design will be the subject of section 5.5.

Table 5.1   Terms used in the context of dynamic verification.

In the	physical reality	world of simulation
a design exists as	fabricated circuit	HDL model or netlist
and is referred to as	circuit under test (CUT).	model under test (MUT).
As part of	prototype testing	functional verification
all those	stimuli and expected responses,	
collectively called	test suite,	
get administered by	autom. test equipment (ATE)	a software testbench
in search of	design flaws.	
As part of	production testing	fault simulation
all those	stimuli and expected responses,	
collectively called	test patterns,	
get administered by	autom. test equipment (ATE)	a software testbench
in search of	fabrication defects.	

Toy example

A test suite for a rising-edge-triggered Gray counter of word width $w=4$ with enable and asynchronous reset inputs is shown in fig.5.3. Note that it includes one stimulus/response pair per clock cycle and that both stimulus and response refer to the same cycle in each pair. How states are being encoded inside the MUT and whether the counter is actually implemented as a Medvedev or as a full Moore machine is of no importance for the input-to-output mapping.[9]

---

[9] The difference is that a Medvedev machine has no output logic whereas a full Moore machine includes a non-trivial logic that translates each state into an output value. Refer to section B.1 for further explanations.

clock cycle k	stimuli $\overline{Rst}$	Ena	test suite	expected responses Count[3:0]
0	1	0	model under test	- - - -
1	0	0		0 0 0 0
2	0	1		0 0 0 0
3	1	0		0 0 0 0
4	1	1		0 0 0 0
5	1	1		0 0 0 1
6	1	1		0 0 1 1
7	1	1		0 0 1 0
8	1	1		0 1 1 0
9	1	1		0 1 1 1
.	.	.		. . . .
.	.	.		. . . .
18	1	1		1 0 0 1
19	1	1		1 0 0 0
20	1	1		0 0 0 0
21	1	1		0 0 0 1
22	1	0		0 0 1 1
23	1	1		0 0 1 1
24	1	1		0 0 1 0
25	0	1		0 0 0 0

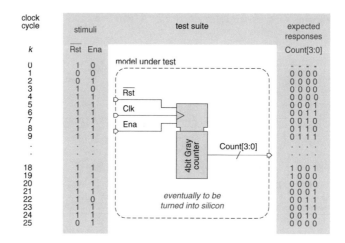

FIGURE 5.3

A test suite for a 4 bit Gray counter with enable and asynchronous reset.

□

The test suite shown in fig.5.3 is very primitive and does not extend easily to more general situations. The concepts of test suite and testbench will thus have to be refined in numerous ways in 5.4 and later sections to make them more practically useful. Before doing so, we must address more pressing problems in sections 5.3.3 and 5.3.4.

5.3.3 ASSERTION-BASED VERIFICATION CHECKS FROM WITHIN

Recall from fig.5.2 that uncovering a design flaw hidden deep in a circuit model solely via chip-level I/O signals is tedious and the outcome uncertain. Assertion-based verification comes to rescue by combining two ideas:

1. Rather than attempting to make a MUT observable via I/O signals, monitor what happens inside. After all, a simulator has access to all circuit nodes and is not limited to reading from package pins as the testing of physical parts is.
2. Instead of becoming obsessed with comparing responses, formulate functional requirements in terms of inputs, states, numerical values, outputs, or any combination thereof.

A **property** specifies one aspect of how a specific subcircuit is supposed to behave, or conversely, how it must never behave independently of circuit context, mode of operation, and the like. A concurrent **assertion** is a small piece of HDL code, that gets invoked during simulation at user-defined moments of time to check whether a given property does hold or not. The outcome may generate a text message or just contribute to the statistics of a simulation run. The MUT's functionality is not affected by the presence or absence of assertion statements.

Toy example revisited

Assume our Gray counter is just a tiny subcircuit in a much larger design. As opposed to fig.5.3, many other subcircuits stand in the way between its own I/O terminals and the top-level I/O pins accessible from externally. Yet, whatever the context, any decent Gray counter must meet the unit-distance requirement "Unless the asynchronous reset is active or the counter is disabled, the next state must differ from the current state in exactly one bit position." Listing 5.1 shows how this translates to SystemVerilog.

LISTING 5.1

A Gray counter plus three assertions that report unexpected behavioral patterns during simulation. See listing 4.18 for the package referenced here.

```systemverilog
// Mission: Illustrate how to model a Medvedev machine in two processes.
//     Example designed to demonstrate assertion-based verification, the usage
//     of functions from a package, and the absence of any type conversions.
// Functionality: w-bit Gray counter with enable and asynchronous reset.
//     Operates with double conversion: Gray->binary, increment, binary->Gray,
//     which is not necessarily the most economic nor the fastest solution.
// Author: H.Kaeslin
//--------------------------------------------------------------------------

`include "../sourcecode/graydefs.sv"    // the word width to be used
import grayconvPkg::*;    // my own set of Gray code converter functions

module graycnt
    ( input logic Clk_CI, Rst_RBI, Ena_SI,
      output logic [`GRAYWIDTH-1:0] Count_DO );

    // present state and next state
    logic [`GRAYWIDTH-1:0] Count_DN, Count_DP;

    // computation of next state
    assign Count_DN = bin2gray(gray2bin(Count_DP) + 1);

    // updating of state
    always_ff @(posedge Clk_CI, negedge Rst_RBI)
        if ( Rst_RBI)                 // if active low
          Count_DP <= 0;              // reset to all zeroes
        else                          // if rising clock
          if (Ena_SI)                 // and enable is 1
              Count_DP <= Count_DN;   // make next state the present one

    // assignment of state to output port
```

```
    assign Count_DO = Count_DP; // output assignment

//---------------------------------------------------------------------
// set of assertions for addressing the various functional mechanisms

  // check asynchronous reset mechanism
  a_async_reset: assert property (
                    @(posedge Rst_RBI)
                    Count_DO == '{default:1'b0} )
                    else
      $error("Non-zero output immediately after removal of async. reset.");
  // or alternatively:   $countones(Count_DO) == 0 );

  // check enable/disable mechanism
  a_enable_disable: assert property (
                    @(posedge Clk_CI) disable iff ( Rst_RBI)
                      Ena_SI |=> Count_DO == $past(Count_DO) )
                    else
      $error("Output has changed in spite of counting being disabled.");

  // check unit distance counting steps
  a_unit_distance_step: assert property (
      // sampling time vvvvvvvvvvvvvv    criterion vvvvvvvv of exclusion
                    @(posedge Clk_CI) disable iff ( Rst_RBI)
      //precondition vvvvvv    vvvvvvvvv property to be checked vvvvvvvvvvv
                    Ena_SI |=> $countones(Count_DO   $past(Count_DO)) == 1 )
      // optional pass action block (empty in this example)
                    else
      // optional fail action block
      $error("Consecutive output patterns differ in 0 or more than 1 bit.");

endmodule
```

The unit-distance property is captured in the last of three assertions in this example, with the `assert` keyword preceded by a user-defined name. Labeling assertions is optional, but highly recommended.

The interesting part follows on the next two lines of code. From an HDL point of view, a concurrent assertion is a passive process.[10] As with any other process, all events that are to invoke the assertion are indicated in a sensitivity list. The subsequent `disable iff (~Rst_RBI)` clause precludes any action while the reset is active. The line below carries a precondition, the implication operator `|=>`, and the property to be checked. Operator `|=>` says that when the precondition to its left is met, then the property

---

[10] That is, a process capable of producing a message and/or adding to the simulation report, but not of altering any variable's (VHDL: `signal`'s) value.

to its right is to be evaluated on the next active clock event, which is perfect for a counter.[11] $countones and $past are two system tasks with obvious meanings.

Next come two optional action blocks separated by an else. The first is the pass statement, to be executed when the property is found to hold, the second is the fail statement, to be executed when the property is violated. A message string preceded by a severity level is standard. The severity level — one of $info, $warning, $error and $fatal — defines where to send the message and whether to proceed or to abort the current simulation run. In this example, the simulator is instructed to print the argument to the simulator window and to continue in case of a fail, while a pass will cause no further action.
☐

It is a strong point of SystemVerilog that the language includes a comprehensive set of constructs, collectively known as **SystemVerilog assertions** (SVA), for expressing behavioral properties in a succinct, if somewhat cryptic way. The SVA syntax is so comprehensive, however, that the reader is referred to specialized texts such as [98] [103] for details.

VHDL provides little more than the concurrent assertion statement. Luckily for VHDL aficionados, co-simulation makes it possible to combine VHDL circuit models with SystemVerilog properties, assertions, and testbenches.

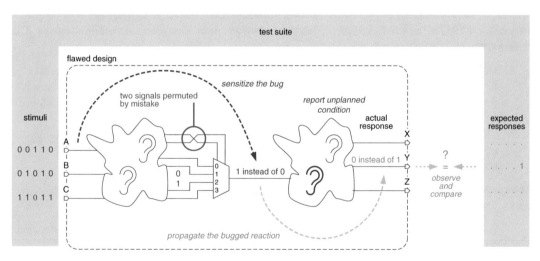

FIGURE 5.4

Assertions resemble spies placed right into the MUT itself.

In real life, a designer would hardly use an assertion to verify a behavior as trivial as an asynchronous reset. Listing 5.1 is just a toy example for illustrating the ideas behind assertion-based verification on a few lines of code. More realistic use cases follow.

---

[11] An alternative operator | - > is available for situations where evaluation has to be immediate.

- All data items entering a FIFO must eventually emanate one by one, that is in their initial order, with neither data lost nor with duplicates or bogus data added.
- In a traffic light controller, no two conflicting lanes must ever see a green light at the same time, and each light must change colors according to the standard cycle.
- State machines must not assume parasitic, illegal or otherwise suspect states.
- Memory addresses, iteration counts, and most other variables are confined to a specific legal range and must never assume values outside that range.
- Not all circuits are designed to handle all physically possible input patterns correctly. Just think of illegal instruction codes or unexpected status codes. Unforeseen inputs, symbols, or other out-of-the-ordinary conditions must be flagged should they ever occur.
- Datapath operations must not result in numeric over-/underflows, out-of-range values, and other data scaling problems without the designer being alerted.
- In a handshake protocol, data lines and the request and acknowledge signals have to switch in a well-defined order.

Many properties relate to circuit behavior over time, i.e. to sequences of events or waveforms, if you prefer. Indeed, SVAs provide several constructs for expressing and manipulating sequences. Many assertions refer to the interface between a subcircuit and the embedding circuitry, and between a subprogram and the calling code, but they are not limited to this. Assertions can equally well be used to ascertain that the result of a numerical calculation is correct, or at least plausible and inoffensive. This can be achieved in either of two ways:

a) Data produced by the MUT are cross checked against data computed by an equivalent but independently generated piece of code that is made part of the assertion. As an example, a first person may write a square root function for RTL synthesis, while a second person does the same for an assertion that double checks the output.[12]
b) The computation is reversed in the assertion. In our example, the procedure for computing the square root is fairly complex and thus prone to error, while checking the result is straightforward.

Assertions included in simulation models are effective at providing protection against design and coding errors. They nicely complement response checking because

- Feedback is immediate, see fig.5.4. There is no need for an abnormal condition to propagate to some distant node placed under constant monitoring.
- The link from the manifestation of a problem to its cause is short. There is no need to trace back a mismatching output over thousands of cycles, HDL statements, or gates in an attempt to locate its place of origin.
- Assertions are a lasting investment. There is no need to repeatedly adjust them when submodels are being assembled to form larger design entities.

Especially the last feature is in sharp and welcome contrast to the stimulus/response pairs of a test suite.

Observation 5.5. *Understand assertions as executable comments that help making sure the MUT indeed operates as you intend and believe. Include an assertion into the HDL code wherever you explicitly or implicitly assume that a certain condition shall hold, or must never hold.*

---

[12] In software terminology, this is an example of *N*-version programming.

Properties — embedded in assertions — help to make sure synthesis code does what it is supposed to do during simulation. Properties can also contribute to formal verification as they provide extra details about how a design is thought to work and what properties are supposed to hold. Properties get ignored during synthesis, however, exactly as `assert` statements.[13]

For all enthusiasm, note that the effectiveness of assertions depends entirely on the sequence of stimuli applied to the MUT that carries them. Back to the FIFO example, the queue must be completely filled and emptied during simulation as the assertion statements otherwise get no chance to check the pointer arithmetics under diverse usage conditions. It is important to check how a full queue reacts to a write request, and an empty queue to a read. This observation brings us to an even more critical topic.

5.3.4 EXHAUSTIVE VERIFICATION REMAINS AN ELUSIVE GOAL

An unfailing way — in fact the sole one — to safeguard against any possible design flaw is to verify a designs's functional behavior in perfect detail.[14] Let us see whether this is practical.

Exhaustive verification calls for traversing all edges in the design's state graph by exercising it with every possible input condition $i \in I$ in every possible state $s \in S$.[15] One might be tempted to think that the product $|I||S|$ indicates the number of clock cycles c_{exh} necessary for exhaustive verification. In almost all practical applications, traversing every edge once will necessitate traversing others twice or more, however,[16] so that we must accept

$$c_{exh} \geq |I||S| \tag{5.1}$$

as a lower bound. Let w_i, w_s and w_o denote the number of bits in the input, state, and output vector respectively. The maximum number of possible input symbols $|I|$ then is 2^{w_i} which figure must be discounted by the parasitic — i.e. unused — input codes. An analogous reasoning holds for $|S|$. Although it is not possible to accurately state $|I||S|$ in the general case, an upper bound can always be given as

$$c_{exh} \geq |I||S| \leq 2^{w_i + w_s} \tag{5.2}$$

where the "less or equal" operator holds with equality in the absence of parasitic states and input symbols, i.e. when any combination of bits is being used for encoding some legal state and input symbol respectively.

---

[13] As an alternative, a designer is free to implement part of the sanity checks in his HDL code in RTL style such as to make them synthesize into surveillance circuitry. By having that extra circuitry activate an alarm upon detection of an out-of-the-ordinary condition, he can add **self-checking** capabilities to physical circuits.

[14] Incidentally, note the same argument also applies to the testing of physical parts for fabrication defects.

[15] Parallel edges are likely to exist, yet exhaustiveness indeed calls for checking the circuit's behavior for every single edge, that is for every state/condition pair, unless the presence of Mealy-type outputs can be ruled out.

[16] Fortunate exceptions are those cases where the state graph includes an Euler line. An (open) **Euler line** is a walk through a graph that runs through every edge exactly once.

Let us plug in real figures to understand the practical significance. Consider the Intel 8080, an early microprocessor released in 1974 with an 8 bit datapath and almost trivial by today's standards. Abstracting from further details we find the following word widths:

input ports	8 bit data, 3 bit control, 1 bit reset	$w_i = 12$
registers	8 bit: A,B,C,D,E,H,L,IR; 16 bit: PC,SP; flags: 5	$w_s = 101$
output ports	8 bit data, 16 bit address, 6 bit status/control	$w_o = 30$

Assume there are no parasitic states and input symbols. The minimum number of clock cycles required for exhaustive simulation then is $2^{113} \approx 10^{34}$. Using test hardware running at 100 MHz, the process would run for more than $3 \cdot 10^{18}$ years. Software simulation would take orders of magnitude longer. To our regret, we must conclude that

Observation 5.6. *Exhaustive verification is not practical, even for relatively modest functions. Dynamic verification, therefore, must almost always do with a partial set of test cases. The problem is to come up with a test suite of practical size and sufficient coverage.*

There is no cheap answer. We are thus going to discuss a number of more and less useful approaches to this problem that plagues both circuit simulation and IC testing.

5.3.5 DIRECTED VERIFICATION IS INDISPENSABLE BUT HAS ITS LIMITATIONS

Exhaustive verification combines all data registers, controllers, counters, state machines, etc. into a single composite state. Also, each possible input gets aimlessly applied in each possible state, even if this contributes close to nothing towards uncovering more potential problems. The Cartesian product so obtained describes all situations the circuit might conceivably encounter but, at the same time, causes the number of test cases to explode.

Testing distinct functional mechanisms separately

As a way out, designers direct verification towards situations they deem critical. A pragmatic idea is to identify distinct subcircuits with fewer internal states and functional mechanisms, and to check each of them individually. In the occurrence of a CPU, this would cover circuit initialization, clock generation and distribution, the bus system including the switches involved in data and address routing, control registers, incrementing the program counter, instruction fetch and decoding, unconditional and conditional branching, the ALU in all modes of operation, status flags, accumulator and data registers, interrupt handling, stack operations, and more.

Toy example revisited

For simplicity, reconsider the Gray counter example. The test suite presented in fig.5.3 offers only partial coverage because it includes 26 cycles whereas (5.2) tells us 64 cycles is a lower bound for exhaustive verification.[17] Why is this nevertheless a reasonable compromise between functional coverage and verification costs?

---

[17] Finding the exact minimum is left to the reader as an exercise, see problem 1.

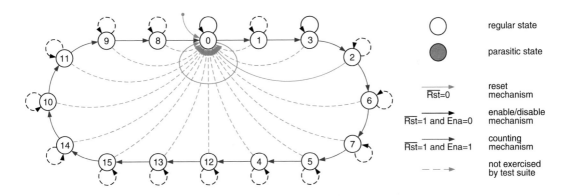

○	regular state
●	parasitic state
$\overline{Rst}=0$	reset mechanism
$\overline{Rst}=1$ and $Ena=0$	enable/disable mechanism
$\overline{Rst}=1$ and $Ena=1$	counting mechanism
- - →	not exercised by test suite

FIGURE 5.5

The state graph of a 4 bit Gray counter with edges colored according to the functional mechanism they implement.

As observed in section 5.3.3, the desired functionality requires the working of three mechanisms, and our test suite addresses each of them separately. More specifically, the reset mechanism is being verified in cycles 0, 1 and 2, and the enable/disable mechanism in cycles 3, 4 and 5. The succession of output values is being checked against the 4 bit Gray code in cycles 4 through 20. Provided the mechanisms involved operate independently from each other, one can generalize from these partial checks and conclude that the circuit should indeed work as intended. All nodes (states) of fig.5.5 get visited, and five more vectors have been added for extra confidence.
□

Still, the separate testing of functional mechanisms suffers from several limitations.

- The problem of combinatorial explosion persists as visiting all states rapidly becomes impractical on more substantial circuits.
- One could easily come up with a modified design that allows for counting to get disabled in some of the states but not in others. Although this would clearly contravene the original specifications, the test suite of fig.5.3 would fail to uncover such a flaw. And postulating to traverse all edges (state transitions) brings us back to exhaustive verification.
- Checking each mechanism and subcircuit separately holds the risk of missing problems that relate to the interaction of two (or more) of them.
- Identifying subcircuits and functional mechanisms for verification requires insight into the MUT's inner organization and operation.[18]

Observation 5.7. *Any non-exhaustive simulation or test run is tantamount to spot checking and implies compromising between functional coverage, run time, and engineering effort.*

---

[18] A situation of limited knowledge is termed **grey box probing** as opposed to the **black box** approach of exhaustive verification that makes no assumptions about the MUT whatsoever. A situation that assumes perfect knowledge of a circuit's inner details is referred to as **clear box probing**. The dilemma is this: Black box probing takes many vectors for a low probability of finding a problem. Clear box probing enables a test engineer to select test cases such as to address specific and likely problems, but may obstruct his view on other potential issues by contaminating his understanding with preconceptions from the circuit's design phase.

Ideally, an engineer would begin by enumerating all slips that might possibly occur during the design process before writing a test suite capable of sensitizing, propagating, and observing each of them. This is not possible in practice, though, because the number of potential design flaws is virtually unlimited and our imagination insufficient to list them all.

Warning example

A tiny portion from an incorrect ASIC design is shown in fig.5.6. The designer's intention was to detect the zero state of a down counter by way of a 12-input NOR function.[19] Since no 12-input gate was available, he decided to compose the function from an 8-input and a 4-input NOR gate, but mistakenly instantiated a NAND gate during schematic entry. A simulation involving these four bits would have exposed the problem, but no such check was undertaken because the test suite never had the counter assume a state in excess of 18.

Why did the designer refrain from exercising the upper bits? Firstly, he wanted to keep simulation runs short, and exhausting a 12 bit counter with enable and reset would have required 16 384 cycles. More importantly, however, the designer was convinced that all input bits to a zero detector are interchangeable. He concluded it was sufficient to check the subcircuit's functioning by initializing the counter to a small number, such as 18, followed by counting backwards to zero. He was just not prepared for a problem that would challenge his preconceptions.

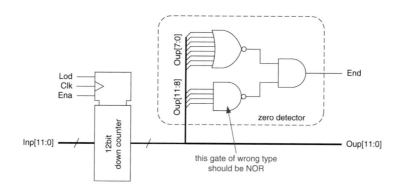

FIGURE 5.6

A silly little oversight that managed to slip through simulation unnoticed because of poor coverage.

□

What we learn from this example is that a critical difficulty of verification is to protect oneself against the unthinkable. Most examples of circuits and systems that have failed when put to service indeed confirm this.

---

[19] Using the counter's carry/borrow bit instead would probably have been a more economic choice anyway.

A simple precautionary measure consists in collecting the toggle counts of all circuit nodes during simulation. Any node that never changes its logic value points to a weakness in the test suites chosen. In the example of fig.5.6, insisting on non-zero toggle counts would definitely have helped to recognize the stimuli as being inadequate. Yet, in spite of its utility and popularity, monitoring of node activities is far from solving all problems. A test suite may well toggle all nodes of a circuit netlist back and forth and still be insufficient, see problem 3. Also, the concept of a circuit node is meaningless before a gate-level netlist has been established.

Observation 5.8. *The toggling of all circuit nodes must be considered a desirable rather than a sufficient requirement for a good test suite.*

Automatic test pattern generation (ATPG) is a technique used in the context of sorting out defective ICs following manufacturing. It is important to understand that ATPG does not normally help to uncover design flaws in nonmaterial circuit models such as HDL code or gate-level netlists. This is because ATPG starts from a presumably correct netlist and produces a set of test patterns for checking for the presence of predefined faults. Fabrication defects are almost universally assumed to follow the so-called "single stuck-at fault" model whereby one circuit node at a time is assumed to be shorted to either logic 1 or 0 for that purpose.

Functional verification, in contrast, questions the correctness of a circuit model by setting its logic behavior against some kind of functional specification or reference model.

All decent HDL simulators can calculate code coverage figures by keeping track of how many times the individual statements in a MUT's source code are being exercised during a simulation run. 100% code coverage implies all executable statements have been executed once or more.

Observation 5.9. *However useful code coverage figures are, executing all statements in an RTL or behavioral circuit model does neither imply that all states and transitions have been traversed nor that all conditions and subconditions for doing so have been checked.*

Also, code coverage relates to bug sensitization but neither to bug propagation nor to bug observation. Executing a flawed statement does not imply the bug must necessarily become manifest at an observable output.

The subsequent case study shows that functional coverage problems can take on much more subtle forms.

Warning example

Electronic dimmers for incandescent lamps work by varying the duty cycle of the load current. For every half wave of the 50 or 60 Hz mains voltage, a Triac connected in series with the lamp is turned on ("fired") at a phase angle adjustable between $0°$ and $180°$ and stays in the conducting state until the next zero crossing. A digital implementation is shown in fig.5.7.

The controller accepts commands from a touch key, converts the desired luminosity into a target phase angle, and fires the Triac via an optical coupler. The trigger impulse is initiated by a comparator when the actual phase angle matches the target value. The actual angle counter is clocked at 64 times the mains frequency so that a total of 32 intensity levels are available. Synchronization with the mains is obtained through a zero-crossing detector that resets all 5 bits of the counter whenever a new half wave begins. Post-layout simulations and testing of fabricated samples on ATE confirmed circuit operation. Yet, the design of fig.5.7 is flawed.

When the first prototype was plugged into the target board, the dimmer was found to function o.k. except for a slight but disturbing oscillation of luminous intensity. The problem was quickly located in the synchronization mechanism. Since only the actual angle counter is reset, the clock divider proceeds from its current but otherwise indeterminate state whenever a new half wave begins. As a consequence, the next increment impulse for the actual angle counter can arrive anytime between a zero-crossing and $\frac{1}{32}$ half waves later. This, together with the fact that a free-running clock oscillator is being used, leads to a beat in firing angle and luminosity.

Why had this flaw passed unnoticed during circuit simulation and testing? The answer is that all simulations were carried out with the clock frequency an integer multiple of the mains frequency. It just never had occurred to the designers that non-integer frequency ratios might give rise to specific behavioral phenomena.

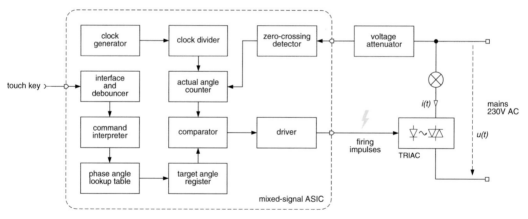

FIGURE 5.7

Block diagram of flawed digital dimmer.

The above examples have demonstrated the perils of unconscious, unspoken or unjustified assumptions and misconceptions that underlie directed verification. Independently of whether the desired functionality is embodied in a piece of hardware, of software, or both, the inconvenient truth is always the same, and we are now in a position to state it.

Observation 5.10. *The innate difficulty with selecting critical test cases for dynamic verification is that human beings can prepare only for events they can foresee.*

Routine is the dark side of experience

On this background, it is sometimes suggested that doing the functional verification right in the target environment will solve all problems. The examples below show this is not so.

Warning example

On its maiden flight on June 4, 1996, the Ariane 5 rocket had to be neutralized by its built-in self-destruct system at an altitude of 3500 m because excessive aerodynamic loads had ripped the solid boosters off the rocket shell after more than 30 s of seemingly normal flight [141]. Analysis of telemetry data revealed that there had been no structural failure but that the on-board computer had commanded the booster nozzles to maximum deflection two seconds before self-destruction occurred, and had so steered the rocket into an abnormal angle of attack.

Why did this happen? The flight control system of Ariane 5 depends on two computerized inertial reference platforms, one active and one for backup, that provide the on-board computer with velocity and attitude information. This flight control system was a proven design that had flown with Ariane 4 for years. On that fatal morning, however, both inertial reference platforms simultaneously ceased to deliver meaningful flight data and presented the on-board computer with diagnostic bit patterns instead. Misinterpreted as they got, these garbage data caused the on-board computer to initiate a sharp change in trajectory. The underlying reason was a numeric overflow that occurred when a parameter of minor importance was converted from a 64 bit floating point number to a 16 bit signed integer combined with poor exception handling.

How was it possible for such a disastrous design flaw to remain undiscovered for so long? No numeric overflow had ever occurred in an Ariane 4 flight. Yet, the Ariane 5 trajectory implied considerably higher horizontal velocity values which caused the critical parameter to accumulate beyond its habitual range. Tests for making sure that the navigational system would operate as intended in the new context had not been conducted. It was precisely the system's excellent reliability record that led the Ariane 5 design team to believe that everything worked fine and that no extra qualification steps were necessary.
□

> How a system responds to the things going wrong shows how good it really is.
> (Hans Stork, CTO of Texas Instruments)

The fact that a subsystem has performed as expected when subject to certain data sets, operating conditions, parameter configurations, and the like is no guarantee for its correct functioning in similar situations.

Even real-world data sometimes prove too forgiving

It is often argued that the functional coverage problem is best dealt with by using genuine data collected from real-world service instead of limiting dynamic verification to a small number of artificially prepared test suites. A test with data from the anticipated flight time sequence of Ariane 5 injected into the data processing section of the inertial reference system would indeed have disclosed its fatal limitation. Similarly, a soft- or a hardware prototype of the digital dimmer ASIC embedded within the remainder of the circuitry and operated with real-world waveforms would have led designers to recognize the oversight in their design.

Using actual data material is no panacea, however, because it may take an excessively large number of cycles before genuine stimuli activate some rare but critical set of circumstances where a potential misbehavior of a MUT actually becomes apparent.

Warning example

A case that was given world-wide publicity in the fall of 1994 was the flaw in the floating point division unit of early Pentium microprocessors [142] [143]. Due to a software problem, 5 out of 1066 table entries had been omitted from a PLA lookup-table employed in the radix-4 SRT division algorithm.[20] Whether results from floating point division came out wrong or not depended on the mantissa values involved. Intel scientists estimated the fraction of the total input number space that is prone to failure to be $1.14 \cdot 10^{-10}$ [144] which explains why it took several months before the user community eventually became aware of the problem.

□

We are now fully aware of the difficulties behind simulation and testing. The trouble is indeed to compose a compact set of test cases that makes a flawed design behave differently from a functionally correct one for any situation that might be relevant to the final circuit's operation. The remainder of this section will introduce helpful techniques for that.

5.3.6 DIRECTED RANDOM VERIFICATION GUARDS AGAINST HUMAN OMISSIONS

Random verification attempts to get rid of any human pre- and misconceptions that might compromise directed verification by having a random process select the test cases.

Example

The classical checkerboard test for memory circuits writes alternating bit patterns into adjacent memory locations and checks them upon readout. Due to the pattern's simple and periodic nature, this does not protect against all possible design flaws and fabrication defects, however. As a consequence, it makes sense to introduce an irregular element into the procedure.

□

Purely random patterns are not normally the best choice, however. In the memory example, you want to make sure every location gets tested with a 0 and a 1 without having to simulate for an undetermined period of time. And in order to efficiently exercise a microprocessor, it is much more effective to randomly pick from legal opcodes and to combine them with relatively few selected data sets rather than to indiscriminately load its memory with arbitrary bit patterns. Generally speaking, you want to cover as much of the circuit's state space as possible with just a limited number of test vectors.

Observation 5.11. *While random testing can be more effective at uncovering design flaws than human-selected test cases — particularly when combined with assertion-based verification and statistical coverage analysis — pattern generation needs to be biased for efficiency.*

---

[20] SRT stands for Sweeney, Robertson and Tocher who independently had invented the method in the 1950s.

Toy example revisited

The idea behind directed random testing is shown in the SystemVerilog testbench of listing 5.2.

LISTING 5.2

A testbench that applies directed random testing to a Gray counter.

```
// Mission: Provide a basic example of directed random verification.
// Functionality: w-bit Gray counter with enable and asynchronous reset.
// Author: M. Schaffner, H. Kaeslin, B. Muheim
//-------------------------------------------------------------------------

'include "../sourcecode/graydefs.sv"    // the word width to be used

module graycnt_tb;
//-------------------------------------------------------------------
    // declarations

    // set time unit and precision (could be overridden by the simulator)
    timeunit 1ns;
    timeprecision 1ps;

    // set parameters related to clocking and timing
    parameter NCLKS = 500;              // number of clock cycles
    parameter CLK_PHASE_HI = 50ns;      // clock high time
    parameter CLK_PHASE_LO = 50ns;      // clock low time
    parameter STIM_APP_DEL = 10ns;      // stimuli application delay
    parameter RESP_ACQ_DEL = 90ns;      // response aquisition delay

    // declare signals
    logic Clk_C, Rst_RB, Ena_S;         // connecting to MUT inputs
    logic ['GRAYWIDTH-1:0] Count_D;     // connecting to MUT output
    logic EndOfSim_S;                   // for coordination within testbench

    // the two tasks below ensure proper timing at the DUT's interface
    // they first wait for a number of active clock edges given as argument
    // and then wait for application or aquisition delay respectively
    task appl_wcycles(int unsigned n);
        repeat (n) @(posedge(Clk_C));
        #(STIM_APP_DEL);
    endtask

    task acq_wcycles(int unsigned n);
        repeat (n) @(posedge(Clk_C));
        #(RESP_ACQ_DEL);
    endtask
```

```
//-------------------------------------------------------------------------
   // generate clock signal

   always @*
      begin
         do begin
            Clk_C = 1; #(CLK_PHASE_HI);
            Clk_C = 0; #(CLK_PHASE_LO);
         end while (EndOfSim_S == 1'b0);

      end

//-------------------------------------------------------------------------
   // instantiate MUT

   graycnt MUT (.Clk_CI(Clk_C),          // note: the IO suffixes
                .Rst_RBI(Rst_RB),        // stand in the way of
                .Ena_SI(Ena_S),          // dot name or dot star
                .Count_DO(Count_D));     // port connections here

//-------------------------------------------------------------------------
   // stimuli application process

   initial   // runs just once
     begin : stimuli_application_p

        // declare local variables
        int randVal;
        bit ok;   // auxiliary variable for randomize with method

        // initialize testbench signal
        EndOfSim_S = 1'b0;

        // initialize signals connecting to MUT
        appl_wcycles(1);   // proceed to next application time
        Ena_S     = 1'b0;
        Rst_RB    = 1'b0;

        // initialize MUT by applying and removing asynchronous reset
        appl_wcycles(1);   // proceed to next application time
        Rst_RB <= 1'b1;

        $display("circuit initialized, stimuli application starts");
```

```
      repeat(NCLKS)
        begin
            appl_wcycles(1);    // proceed to next application time

            //random number vvvvvvv         vvvvvvv constraints vvvvvvv
            ok = randomize (randVal) with { randVal >= 0; randVal < 10; };
            // alert designer should constraints turn out to be infeasible
            assert (ok) else $error("randomization failed");
            // disable counting in 1 out of 10 cases
            Ena_S <= (randVal == 0) ? 1'b0 : 1'b1;

            // reset count to zero in 1 out of 50 cases
            ok = randomize (randVal) with { randVal >= 0; randVal < 50; };
            assert (ok) else $error("randomization failed");
            Rst_RB <= (randVal == 0) ? 1'b0 : 1'b1;

        end

      $display("stimuli application finished");

      EndOfSim_S = 1'b1;
    end

//----------------------------------------------------------------------
   // response acquisition process

   initial   // runs just once
     begin : response_acquisition_p

        // declare local variables
        int tstCnt;              // test cycles carried out
        int errCnt;              // mismatching responses found
        logic ['GRAYWIDTH-1:0] binCnt;  // enabled steps since last reset
        logic ['GRAYWIDTH-1:0] expResp; // expected response

        // initialize local variables
        tstCnt  = 0;
        errCnt  = 0;
        binCnt  = '{default:1'b0};
        expResp = '{default:1'b0};

        $display("response acquisition started");
```

```
        do begin
            acq_wcycles(1);    // proceed to next acquisition time

            if (Rst_RB == 1'b0)    // reset may have been asserted
              binCnt = 0;

            // calculate the expected response by converting the number of steps
            // to Gray code using a piece of code independent of the MUT
            expResp = (binCnt >> 1)   binCnt;

            $display("[test %04d]> checking whether expected 'b%b = actual 'b%b",
                tstCnt,expResp,Count_D);

            // check response and do the reporting and accounting
            if ( Count_D !== expResp )
              begin
                  $error("expected 'b%b != actual 'b%b",expResp,Count_D);
                  errCnt ++;    // increment incorrect responses counter
              end

            // update the number of enabled steps since last reset
            if (Rst_RB == 1'b0)
                binCnt = 0;
            else if (Ena_S == 1'b1)
                binCnt++;

            tstCnt++;    // increment test cycle counter

        end while (EndOfSim_S==0);

        $display("response acquisition stopped (%4d test cycles, %4d failed)",
            tstCnt,errCnt);

    end

endmodule
```

Instead of applying a predetermined sequence of stimuli to the MUT, the two inputs Rst_RB and Ena_S are fed with random data in a statistically controlled way. The key construct is randomize (...) with {...} used here to generate random integers constrained to an interval as indicated between the curly brackets. These numbers subsequently decide on whether to assign 0 or 1 to Rst_RB and Ena_S. A reset is specified to occur with probability of 2%, a disable with 10%.
□

In the code example of listing 5.2, simulation ends after a predetermined number of clock cycles. To what extent the MUT's specifications have been exercised at that point remains an open question. The answer lies in collecting statistical data during simulation. SystemVerilog users benefit from built-in constructs for doing so.

Example

When added to the RTL code of listing 4.20, the lines below will count how often each state gets visited, how often each input gets applied, and — most importantly — how many times each state transition gets traversed. The numbers will be reported at the end of each simulation run, they can also be combined with predefined goals to decide when to put and end to simulation.

```
.....
  // coverage checks
  // --------------------------------------------------------------------
  covergroup mealy5stCg @(posedge Clk_CI);    // name and when to sample
    state : coverpoint State_DP;              // monitor the FSM state
    inp   : coverpoint Inp_DI {               // monitor the FSM input
       bins inp[] = { 'b00,'b01,'b10 };       // declare all its legal values
    }
    stateXinp : cross state, inp;    // monitor cross coverage of the two
  endgroup

  // create an instance of the covergroup, allocating the memory required
  mealy5stCg mealy5stCgInst = new;
.....
```

Note: Multiple instances of a `covergroup` would be required if we were to monitor multiple instances of the same (sub)circuit.
□

Much as assertions, statistical coverage analysis monitors what happens inside a circuit model without having to propagate any data to an observable output. And as opposed to code coverage, it enables designers to specify what they consider relevant functional features and meaningful coverage metrics. One postulate when designing a CPU, for instance, will be that each opcode involving a memory transfer shall be combined with each addressing mode, and this several times. You would further want to make sure that various ranges of memory addresses get exercised during simulation. The higher quality and confidence in the end product justifies the one-time effort for selecting and coding the various coverage points.

As `covergroup`, `coverpoint`, `bins`, `cross`, and related SystemVerilog constructs come with so many options, syntax details must be obtained from specialized texts such as [98] [103].

Similarly to what has been found for assertion-based verification, VHDL per se does not offer as much support for coverage analysis and directed random stimuli generation. The IEEE 1076.2 VHDL extension includes a pseudo random number generator procedure named `ieee.math_real.Uniform` from which test patterns with desirable probability distributions must be obtained by way of biasing and/or filtering operations. VHDL packages and advice are available from an open source initiative named OS-VVM [145]. And again, co-simulation makes it possible to combine the exactingness of VHDL for circuit modeling with the sophisticated capabilities of SystemVerilog for verification.

5.3.8 COLLECTING TEST CASES FROM MULTIPLE SOURCES HELPS

We have found that it is absolutely essential to organize functional verification with a clear and open mind. The fact that today's VLSI circuits are so complex means that no single verification technique will suffice to uncover almost all potential design flaws, let alone to do so with a test suite of acceptable size. The conclusion is most obvious.

Observation 5.12. *Except for the simplest subsystems where exhaustive testing is feasible, an acceptable selection of test cases shall comprise:*

1. *A vast set of data that makes the MUT work in all regular regimes of operation, that exercises every functional mechanism, and that strains array and memory bounds.*
2. *Particular numeric data that are likely to cause uncommon arithmetic conditions including over-/underflow, division by small numbers incl. zero, sign reversal, carry, borrow, and NaN.[21]*
3. *Pathological cases that ask for exception handling and out-of-the-normal control flows,*
4. *Genuine data sequences collected from real-world service.*
5. *Properly biased random test cases in conjunction with coverage analysis.*

---

[21] NaN stands for **not a number**, i.e. for a binary code that does not map to any numerical value.

5.3.9 SEPARATING TEST DEVELOPMENT FROM CIRCUIT DESIGN HELPS

The idea is to safeguard a design against oversights, misconceptions, and poor functional coverage by organizing manpower into two independent teams. A first team or person works towards designing the circuit while a second one prepares the properties, the assertions, and the test suite. Their respective interpretations are then crosschecked by simulating early behavioral models of the circuit to be, see fig.5.8. The goal of the circuit designers is to come up with a model that is functionally correct whereas the test engineers essentially try to prove them wrong. The pros and cons of this healthy adversarial relationship are as follows.

  + Mistakes made by either team are likely to be uncovered in the crosschecking phase.
  + The same holds for ambiguities in the initial specifications.
  + Having the design and the test teams work concurrently helps to cut down design time.
  − A chance always remains that misleading specifications get interpreted in identical but mistaken ways by both teams.
  − The difficulty of finding test cases of adequate coverage persists.

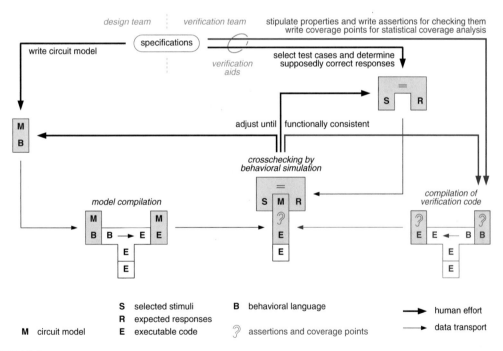

FIGURE 5.8

HDL synthesis models and verification aids being prepared by separate teams.

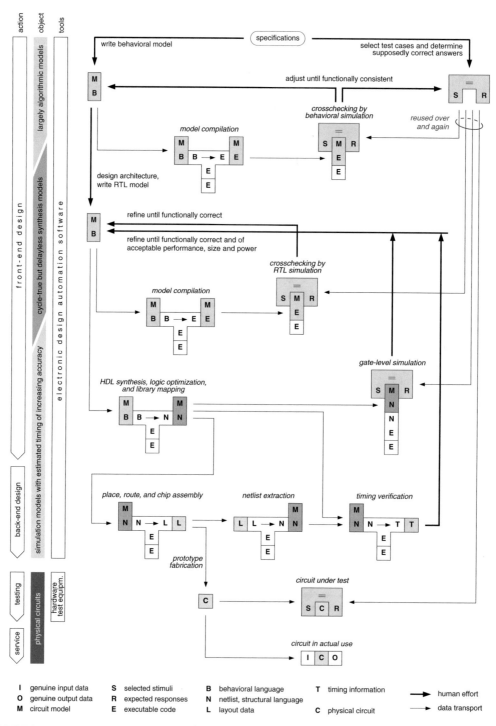

FIGURE 5.9

Life cycle of a test suite during VLSI design and test in T-diagram notation.

5.4 CONSISTENCY AND EFFICIENCY CONSIDERATIONS

Over the last decades, telling correct designs and circuits from imperfect ones has become ever more onerous with the exploding VLSI and ULSI circuit complexities. The concern is best expressed in a quote (by Walden Rhines):

> The question is whether the percentage for verification time tops out at 70% (of the total engineering effort in VLSI design) or it goes to 95% in the future.

Industry cannot afford to rewrite a test suite gathered with so much effort over and over again as a design matures from a virtual prototype into synthesis code, a gate-level netlist, and — finally — into a physical part. Moreover, it is absolutely essential that a MUT be checked against the same specifications throughout the entire design cycle. Fig. 5.9 emphasizes the reuse of stimuli and expected responses during a VLSI design and test cycle. A problem is that the various design views and tools greatly differ in their underlying assumptions, see table 5.2.

Table 5.2   Data, signal, and timing abstractions encountered during VLSI design and test.

Level of abstraction	Relevant data types and structures	Numerical precision	Modeling of electrical phenomena	Relevant time scale for latency	Timewise resolution
Immaterial circuit models					
Algorithmic model	abstract and fairly free	essentially unlimited	not an issue	system-level transaction	not an issue
Automata theory	discr. symbols	not an issue	not an issue	abstr. cycle	clock cycle
Synthesis model (HDL @ RTL)	numbers, bits, enumer. types	finite	optional	clock cycle	event sequence
Gate-level netlist Post-layout netlist	bits	finite	logic values	clock cycle	circuit delays
Physical hardware					
Autom. test equip.	bits	finite	discr. volt.	clock cycle	discr. strobes
Physical circuit			cont. quant.		continuous

The difficulty is to make sure the same stimuli and expected responses can be reused across the entire design cycle from purely algorithmic specifications to the testing of fabricated parts in spite of the many pieces of hardware and software being involved in the process. Note this is more than just a matter of format conversion. Ideally, a single test suite is reused with only minimal modifications to account for unavoidable differences in timewise and numerical resolution. Having to rewrite or to re-schedule test patterns at each development step must be avoided for reasons of quality, trustworthiness, and engineering productivity.

From a hardware engineering point of view, a good simulation set-up

- Is compatible with all formalisms and tools used during VLSI specification, design and test (such as automata theory, MATLAB, HDLs, logic simulators, and ATE).
- Translates stimuli and responses from bit-level manipulations to higher-level operations,
- Consolidates simulation results such as to facilitate interpretation by humans.
- Is capable of handling situations where the timewise relationship between circuit input and output is unknown or difficult to predict.
- Manages with reasonable run times.
- Obeys good software engineering practices (modular design, data abstraction, reuse, etc.).

We are now going to discuss a couple of measures that greatly contribute towards these goals.

5.4.1 A COHERENT SCHEDULE FOR SIMULATION AND TEST

A choice of utmost importance refers to the relative timing of a few key events that repeat in every stimulus/response cycle. Poor timing may cause a gate-level model to report hundreds of hold time violations per clock cycle during a simulation run, for instance, whereas a purely algorithmic model is simply not concerned with physical time. To complicate things further, engineers are often required to co-simulate an extracted gate-level netlist for one circuit block with a delayless model for some other part of the same design.

Observation 5.13. *To be useful for comparing circuit models across multiple levels of abstraction, a testbench must schedule all major events such as to respect the limitations imposed by all formalisms and tools involved in circuit specification, design, simulation, and test combined.*

Formalisms and tools are meant to include automata theory, HDLs, RTL models, gate-level netlists (whether delayless or backannotated with timing data), simulation software, and automated test equipment (ATE). In their choice of a schedule, many circuit designers and test engineers tend to be misled by the specific idiosyncrasies of one such instrument.

Key events

Consider a synchronous digital design[22] and note that a few **events** repeat in every clock cycle. These key events are the same for both simulation and test, namely:

- The application of a new stimulus denoted as $\triangle$ (for Application),
- The acquisition and evaluation of the response denoted as $\top$ (for Test) ,
- The recording of a stimulus/response pair for further use denoted as $\square$ (for storage),
- The active clock edge symbolically denoted as $\uparrow$, and,
- The passive clock edge denoted as $\downarrow$ (mandatory but of subordinate significance).

When these events are ordered in an ill-advised way, the resulting schedules most often turn out to be incompatible. Exchanging test suites between software simulation and hardware testing then becomes

---

[22] We assume the popular single-phase edge-triggered clocking discipline where the is no difference between clock cycle and computation period, see section 7.2.2 for details.

very painful, if not impossible. The existence of a problem is most evident when supposedly identical simulation runs must be repeated many times over, fiddling around with the order and timing of these events just to make the schedule compatible with the automated test equipment (ATE) at hand. A suspicion always remains that such belated manipulations of test vectors cannot be trusted because any change to the sequence of events or to their timing raises the question of whether the original and the modified simulation runs are equivalent, that is, whether they are indeed capable of uncovering exactly the same set of functional flaws.

A coherent stimulus/response schedule

The schedule of fig.5.10 has been found to be portable across the entire VLSI design and test cycle. Its formal derivation is postponed to section 5.9.

Observation 5.14. *At the RTL and lower levels, any simulation set-up shall*

- *provide a clock signal even if the MUT is of purely combinational nature,*
- *log one stimulus/response pair per clock cycle, and*
- *have all clock edges, all stimulus applications, and all response acquisitions occur in a strictly periodic fashion, symbolically denoted as $\triangle \downarrow (\top = \square) \uparrow$.*

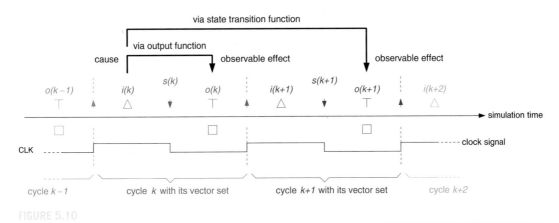

FIGURE 5.10
A coherent schedule for simulation and test events with cause/effect relationships and single-phase edge-triggered clock signal waveform superimposed.

As shown in fig.5.10, each computation period gets subdivided into four phases. Sample settings for a symmetric clock of a conservative 10 MHz are given below as an example. Of course, the numerical figures must be adapted to the situation at hand.

cycle	event with time of occurrence [ns]			
k	$\triangle$	$\downarrow$	$T = \square$	$\uparrow$
0	10	50	90	100
1	110	150	190	200
2	210	250	290	300
...	...	...	...	...

Observe that we have elected to assign the active clock edge $\uparrow$ to the end, rather than to the beginning, of a cycle because a finite state machine takes up operation from some state — typically a well defined start state s_0 — without any intervention of a clock. Further make sure to understand that a clock is needed to drive a simulation even if the circuit being modeled is of purely combinational nature.

Cause/effect relationships

It is important to know when and how the effects of a specific stimulus $i(k)$ become observable. Note from fig.5.10 that $o(k) = g(i(k), s(k))$ is visible in the response acquired after applying $i(k)$, while its effect on the state $s(k)$ can be observed only in the response acquired one clock cycle later and only indirectly as $o(k + 1) = g(i(k + 1), f(i(k), s(k)))$.

All simulation set-ups that we are going to develop next adhere to this schedule, but they differ in how the bit patterns observed at the I/O pins of the MUT relate to the stimuli and expected responses that make up for the test suite. In the most simple case, illustrated in fig.5.11, both are the same and they are firmly locked to predetermined clock cycles. This is fully adequate for subcircuits of modest complexity, such as the one of fig.5.3, for instance, and also the only way to go when it comes to the testing of physical parts using automated test equipment (ATE).

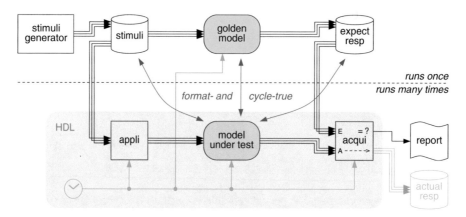

FIGURE 5.11

Elementary simulation set-up (simplified).[23]

---

[23] Note that stimuli and expected responses are prepared beforehand and stored on disk from where they are retrieved at run time under control of the testbench, hence the name **file-based simulation**. This is generally preferred over the alternative, shown in fig.5.16b, of (re-)calculating those responses at run time with the aid of a golden model because of the computing resources wasted when many iterations are required to debug a MUT.

As MUTs grow more complex, limitations of bitwise cycle-true simulations become apparent.

- Humans are easily overwhelmed by the volume of bits and bytes when confronted with raw simulation data.
- I/O operations that extend over multiple clock cycles and other sophisticated transfer protocols tend to get in the way of transparency.
- MUTs undergo profound changes during the development process and most architectural decisions have a dramatic impact on latency.[24] What begins as a purely behavioral model is later refined into an RTL model, and ultimately becomes a gate-level netlist. Only the final model will match the physical circuit it emulates in terms of bits and clock cycles.
- A golden model — and hence also the expected responses prepared using it — might not be absolutely bit-true. Many functional models do in fact ignore circuit-level details such as clock, reset, scan mode, and other test-related features.

Put in other words, a semantic gap opens between the needs of overall functional verification and the reality of circuit operation. Rather than drowning engineers with tons of 0s and 1s ascribed to specific clock cycles, advanced test suites are made to express things at more meaningful levels of detail. In addition, any decent simulation set-up must support a process of incremental design where latency is subject to change several times.

Example

JPEG image compression in essence accepts an image frame, subdivides it into square blocks, and uses a 2D Discrete Cosine Transform (DCT) to calculate a set of spectral coefficients for each block. Those coefficients are then quantized or outright replaced by zero when their impact on the perceived image quality is only minor.

For one thing, you would prefer to talk of larger data items such as image frames, blocks, and coefficient sets in the context of functional verification. A relevant operation would be the compression of one block or, alternatively, of an entire frame. Low-level details such as the reading in of pixels or the toggling of individual data bits would just distract your attention. You would not want to work at those lower levels unless forced to do so by adverse conditions.

For the other thing, JPEG decoding is a combinational function that can, in theory, be computed by a delayless model, that is, with latency zero. In practice, however, typical image decoding hardware ingests one set of quantized DCT coefficients at a time and takes many clock cycles before spitting out the pixels of a block or of the assembled image, with the exact latency figure being a function of architectural decisions.

□

---

[24] Just think of iterative decomposition, pipelining, loop unfolding, etc. The same applies to interfacing with RAMs, parallel ↔ serial conversions, specific input/output protocols, and the like.

Observation 5.15. *A testbench not only serves to drive the MUT, its more noble duties are to translate stimuli and responses across levels of abstraction, and to consolidate simulation results such as to render interpretation by humans as convenient as possible.*

Functional verification is all about data items being absorbed, digested and delivered. The chores of reformatting them and of applying/accepting them in a timely fashion must be carried out with a minimum of human attention. Test engineers feel little inclination to re-adjust their models and test suites when one component gets re-architected or clocked in a different way.

Protocol adapters aggregate bits into more meaningful data bodies

Stimuli and responses are lumped into composite data entries such as records, data packets, audio fragments, or whatever is most appropriate for the application at hand. Protocol adapters, aka bus-functional models (BFM), are inserted between test suite and MUT, see fig.5.13.

An input protocol adapter accepts a high-level stimulus (an image frame in the JPEG example), breaks it down into smaller data items (e.g. blocks and pixels), and feeds those to the MUT word by word or bit by bit over a time span that may cover hundreds of clock cycles, see fig.5.12. Another adapter located downstream of the MUT does the opposite to consolidate output bits into a higher-level response (e.g. collecting bits into a JPEG coefficient set for one image).

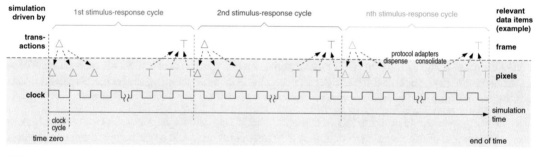

FIGURE 5.12

Protocol adapters help bridge the gap between bit-level models (RTL or gate-level netlist) and a higher level view of the same.

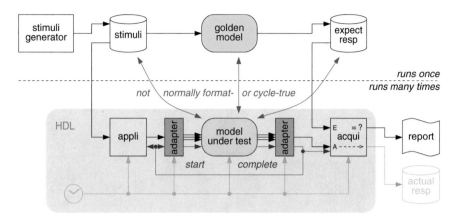

runs once
runs many times

not normally format- or cycle-true

FIGURE 5.13

Simulation set-up with protocol adapters for hiding low-level details (simplified).

Protocol adapters absorb latency changes across multiple circuit models

The upstream adapter is designed such as to feed data to the MUT just in time and, analogously, the downstream adapter such as to wait for the next valid data item to emerge at the MUT's output. The details depend on the MUT's I/O interface (which functionally follows the specifications for the target circuit).

○ The chip-to-be supports handshaking on its I/O ports, see the shaded part in fig.5.14. Each protocol adapter features the necessary control signals on the side of the MUT and participates in the handshake transfers there.

○ The MUT's interface includes some form of "start" signal and a status flag such as "ready/$\overline{busy}$" that the protocol adapters drive and interpret respectively. This situation is depicted in the shaded part of fig.5.13.

○ In the absence of any of the above provisions the protocol adapters must tacitly count clock cycles concurrently to the MUT's internal operation and/or emulate state machines that are part of the MUT to find out when to act.[25]

In conclusion, the adoption of protocol adapters minimizes the need to rework stimulus/response pairs each time a modification is made to the MUT. Any change just affects the MUT itself and one or two of the adapters but neither the remainder of the testbench nor the test suite itself thereby ensuring consistency, simplifying maintenance, and improving designer productivity. Full handshaking is the cleanest and safest way to relegate clocking details and latency figures to secondary issues as simulation proceeds in an entirely data-driven manner.

---

[25] While this may be ok for small subcircuits, we do not recommend to design larger entities in this way because the resulting circuits prove difficult to embed in a larger system. Besides, any substantial change to the MUT's architecture is likely to necessitate adjustments to the latency parameters coded into the protocol adapters. Things become really awkward when the number of clock cycles required to complete a computation depend on numerical data values. Target specifications should be reworked in such cases.

Protocol adapters can either be implemented as HDL processes of their own or merged as subfunctions into the testbench processes that handle stimulus application and response acquisition respectively. The choice is essentially a matter of software engineering.

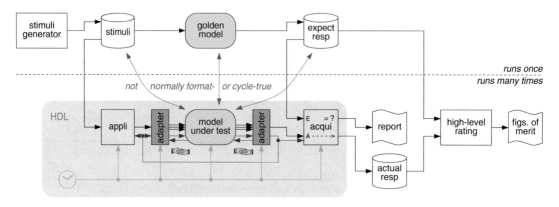

FIGURE 5.14

Simulation set-up that calculates high-level figures of merit (simplified).

5.4.3 CALCULATING HIGH-LEVEL FIGURES OF MERIT

Systems engineers do not think in terms of individual bits, they are more interested in assessing the overall correctness and performance obtained from a circuit of some given architecture. So they tend to ask: What data chunks do emanate from a MUT in response to some given set of data at the input? What do those data sets really mean? Is the MUT functionally sound? Does it meet the specifications in its current stage, or where does it need further refinements?

Example

In lossy image compression, the original and the reconstructed image will necessarily differ. Departures are also to be expected when a golden model calculates with floating point numbers of almost arbitrary precision whereas the MUT operates with finite precision arithmetics. System designers will be most interested to learn about peak signal-to-noise ratio (PSNR), maximum perceived color deviation, throughput, and similar overall ratings. Conversely, they do not care much about when which operation takes place or about occasional idle cycles.
□

Rather than just report the number of bit- or word-level inaccuracies found, a truly high-level simulation shall look back over a full simulation run and distill the essence into a couple of relevant figures of merit. A simulation set-up that does so is shown in fig.5.14. It is a good idea to carry out such calculations outside the HDL testbench and to take advantage of standard mathematics tool such as MATLAB instead. Math packages not only provide high-level functions for data and signal processing and for statistics, but also offer superior means of visualization. Assertions can also be put to service to collect relevant data.

In practice, most simulation set-ups for large circuits follow the organization of the target system itself as this encourages reuse of HDL and software modules, facilitates development in successive refinement steps, and so helps improve productivity.

Example
Fig. 5.15 shows an example from wireless telecommunication where multiple antennas are being used at the transmitter and at the receiver end to improve data rate and robustness. In the simulation set-up, a behavioral model is substituted for each RAM, IF (de)modulator, PCI interface and other subfunction that collaborates with the MUT. The various HDL models for one design entity share the same interface so that they can serve as drop-in replacements for each other during the development process. The preparation of stimuli and the evaluation of responses is implemented in MATLAB so that the HDL testbench code remains essentially limited to configuring, controlling and monitoring the MUT via the PCI interface. Protocol adapters take care of translating where necessary.
□

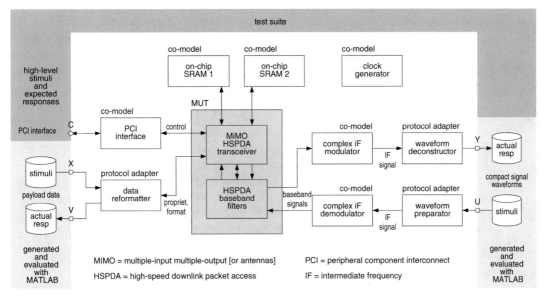

FIGURE 5.15

Example of a sophisticated simulation set-up patterned after a target system from wireless telecommunication (simplified).

5.4.5 INITIALIZATION

Almost all circuits need to be initialized prior to regular operation, and the same holds for the pertaining MUTs. Activating an asynchronous reset is just the most simple case as initialization may be much more complex. A program may need to be loaded into the instruction memory, configurations need to be set, a phase-locked loop (PLL) needs to reach operational status, the outcome from a built-in self test (BIST) needs to be waited for, etc. There are two options.

- o If done by applying stimuli vectors from the MUT's regular output, initialization may take thousands of clock cycles, but the test suite so obtained has the benefit of being portable to automatic test equipment (ATE) with no alterations whatsoever.
- o Doing the same through debugging aids provided by the simulator (forcing logic values upon nodes, setting initial values to registers and memories, and the like) may achieve the same circuit condition in almost no time, but is not reproducible with physical parts.

5.4.6 TRIMMING RUN TIMES BY SKIPPING REDUNDANT SIMULATION SEQUENCES

Memories, counters, and state machines tend to inflate the number of stimulus/response pairs necessary to verify functionality. This is because stereotypical activities eat a lot of computation time without contributing much towards uncovering further design flaws. Much of a simulation run just reiterates the same state transitions many times over without moving on to fresh states and functional mechanisms for a long time. Examples are quite common in timers, large filters, data acquisition equipment, and data transfer protocols, but the situation is notorious in image processing and man machine interfaces.

For productivity reasons, designers seek to cut back cycles that feature little or highly recurrent computational activities in a design. What follows are suggestions of what they can do.

- Monitor the evolution of the MUT's various state variables by way of properties and assertions, rather than limiting yourself to checking their effects on overall output data.
- Take advantage of the scan facility to skip uninteresting portions of a simulation run.
- Do the same using simulator commands to manipulate the MUT's state variables.
- Include auxiliary logic in the MUT that trims lengthy counting or waiting sequences and unacceptably large data quantities while in simulation mode.
- Do the same to the synthesis model to speed up the testing of physical circuits too.
- Model circuit operation on two different time scales (fine and coarse).

Example
Imagine you are designing a graphics accelerator chip. Instead of simulating the processing of full-screen frames of 1280 pixels x 1024 pixels exclusively, make your MUT code capable of handling smaller graphics of, say, 40 pixels x 32 pixels as well. Use this thinned-out model for most of your simulation runs, but do not forget to run a couple of full-size simulations before proceeding to back-end design and prior to tape out.
□

5.5 TESTBENCH CODING AND HDL SIMULATION

A testbench provides the following services during a simulation run:

a) Generate a periodic clock signal for driving simulation and clocked circuit models.
b) Obtain stimuli vectors and apply them to the MUT at well-defined moments of time.
c) Acquire the signal waveforms that emanate from the MUT as actual response vectors.
d) Obtain expected response vectors and use them as a reference against which to compare.
e) Establish a **simulation report** that lists functional discrepancies and timing violations.

VHDL and SystemVerilog not only support the modeling of digital circuits, they also provide the necessary instruments for implementing simulation testbenches.[26] This section gives guidelines for doing so based on the general principles established earlier in this chapter.

5.5.1 MODULARITY AND REUSE ARE THE KEYS TO TESTBENCH DESIGN

File-based and golden-model-based simulation are not the only meaningful ways to process test vectors, see fig.5.16 for more options. The coding effort can be kept more reasonable by recognizing that all sorts of simulation set-ups can be readily assembled from a very small number of versatile and reusable software modules. Adapting to a new design or a new simulation set-up can so largely be confined to a couple of minor adjustments to existing code.[27]

Observation 5.16. *With testbenches being major pieces of software, it pays to have a look at them from a software engineering perspective.*

Preparing verification aids is not the same as circuit design. VLSI architects are limited to a synthesizable subset of their HDL and primarily think in block diagram and RTL categories. Verification engineers, in contrast, must think in terms of functionality and behavioral properties, but are not restricted to any particular subset of the language as testbenches and assertions are not for synthesis. They are free to use other language constructs or to use the same constructs in totally different ways. VHDL users can resort to shared variables at this point, while adopters of SystemVerilog will likely take advantage of classes, inheritance, and various high-level synchronization mechanisms offered by that language.

Observation 5.17. *Verification requires a different mindset than coding synthesis models.*

---

[26] Strictly speaking, instantiating a copy of the MUT and connecting it to the corresponding testbench signals prior to simulation are the sole services that absolutely need to be realized using an HDL. All other parts of a simulation set-up could be implemented using a programming language, MATLAB, or some other software tool. In practice, however, only VHDL or SystemVerilog provide the detailed timing control necessary for generating precise clock waveforms, for correctly applying stimuli to the MUT, and for acquiring the actual responses from there just in time.

[27] In an earlier edition of this text [146], the expected response pickup facility and the golden model were indeed designed as drop-in replacements for each other. The idea was to simplify the transition between file-based and golden-model-based simulation as much as possible in a VHDL design environment. However, as golden models are typically established outside the HDL environment and as subsequent HDL simulations are mostly file-based for efficiency reasons, the necessary provisions that cluttered the code examples and their graphical illustrations have been dropped for the sake of simplicity.

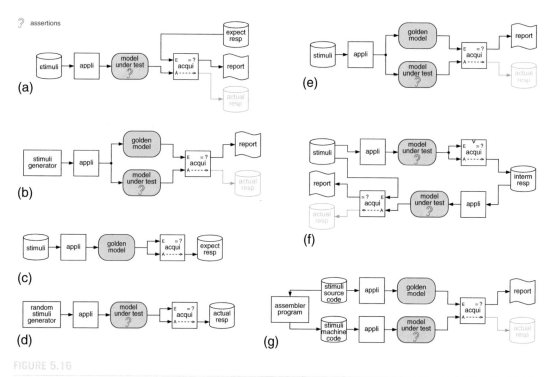

FIGURE 5.16

Software modules from which simulation set-ups can be assembled to serve a variety of needs. Basic simulation set-up that operates with a test suite previously stored on disk (a). Alternative set-up that generates stimuli and expected responses at run time (b). Preparing stimulus/response pairs with the aid of a golden model (c). Fully assertion-based verification with no evaluation of responses (d). A hybrid arrangement that combines traits from (a and b) (e). A special set-up for involutory ciphers (f). Another special set-up for a situation where the stimuli exist as source code for a program-controlled processor (g).[28]

5.5.2 ANATOMY OF A FILE-BASED TESTBENCH

Consistent with what has been said in section 5.4.1, our focus here is on the set-up for a file-based format- and cycle-true simulation. Source code organized along the lines shown in fig.5.17 is available from the book's companion website. As a circuit example, we have chosen the Gray counter of fig.5.3. Obtaining a good understanding requires working through that code. The comments below are intended as a kind of travel guide that points to major attractions.

---

[28] Incidentally, note that Figure 5.16c implements the procedure introduced in Figure 5.1.

All disk files stored in ASCII format

File-based simulation involves at least three files, namely the stimuli, the expected responses, and a simulation report. ASCII text files are definitely preferred over binary data because the former are human-readable and platform-independent, whereas the latter are not. What's more, the usage of ASCII characters makes it possible to take advantage of the full MVL-9 (SysVer: MVL-4) set of values for specifying stimuli and expected responses. Designers are so put in a position to check for a high-impedance condition by entering a z as expected response, for instance, or to neutralize a response by entering a don't care - (SysVer: x).

Separate processes for stimulus application and for response acquisition

Recall from section 5.4.1 that the correct scheduling of simulation events is crucial:

- The application of a new stimulus denoted as △ (for Application),
- The acquisition and evaluation of the response denoted as ⊤ (for Test),
- The two clock edges symbolically denoted as ↑ and ↓.

It would be naive to include the time of occurrence of those key events hardcoded into a multitude of `wait for` statements or `after` clauses (SysVer: # terms) dispersed throughout the testbench code. A much better solution is to assign stimulus application and response acquisition to separate processes that get periodically activated at time △ and ⊤ respectively. All relevant timing parameters are expressed as constants or as generics, thereby making it possible to adjust them from a single place in the testbench code. [29]

The stimulus application process is in charge of opening, reading, and closing the stimuli file. The response acquisition process does the same with the expected responses file. In addition, it handles the simulation report file. When the stimuli file has been exhausted, the stimulus application process notifies its counterpart via an auxiliary two-valued signal named `EndOfSim_S`.

Stimuli and responses collected in records

Two measures contribute towards rendering the two processes that apply the stimuli and that acquire the responses respectively independent of the MUT and, hence, highly reusable.

a) All input signals are collected into one stimulus record and, analogously, all output signals into one response record. [30]

b) The subsequent operations are delegated to specialized subprograms:
  - All file read and write operations,
  - Unpacking of stimuli records (where applicable),
  - Unpacking of expected response records (where applicable),
  - Packing of actual response records (if written to file),
  - Response checking (actual against expected), and
  - Compiling a simulation report.

---

[29] The SystemVerilog testbench in listing 5.2 achieved the same objective with a pair of `tasks`.
[30] Bidirectional input/output signals must appear both as stimuli and as expected responses.

The main processes that make part of the testbench are so put in a position to handle stimulus and response records as wholesale quantities without having to know about their detailed composition. The writing of custom code is confined to a handful of subprograms. [31]

Simulation to proceed even after expected responses have been exhausted

Occasionally, a designer may want to run a simulation before having prepared a complete set of expected responses. There will be more stimuli vectors than expected responses in this situation. To support this policy, the testbench has been designed such as to continue until the end of the stimuli file, no matter how many expected responses are actually available. The numbers of responses that went unchecked gets reflected in the simulation report.

Stoppable clock generator

A simulation run draws to a close when the processing of the last entry in the stimuli file has been completed and the pertaining response has been acquired. A mundane difficulty is to halt the simulator. There exist three alternatives for doing so in VHDL, namely

a) Have the simulator stop after a predetermined amount of time,
b) Cause a failure in an assert or report statement to abort the run, or
c) Starve the event queue in which case simulation comes to a natural end.

Alternative a) is as restrictive as b) is ugly, so c) is the choice to retain. A clock generator that can be shut down is implemented as concurrent procedure call, essentially a shorthand notation for a procedure call embedded in a VHDL process. [32]

While starving the event queue also works in SystemVerilog, the customary way to end a simulation run is to use either the `$stop` or `$finish` system task listed in table 4.6.

Reset treated as an ordinary stimulus bit

Timingwise, the reset signal, irrespective of whether synchronous or asynchronous, gets updated at time $\triangle$ like any other stimulus bit. It is, therefore, made part of the stimuli record.

---

[31] As an example, the subprograms that check the responses against each other and that handle the reporting need to be reworked when a MUT connector gets added, dropped, or renamed. Ideally, one would prefer to do away with this dependency by having those subprograms find out themselves how to handle the response records. This is, unfortunately, not possible as VHDL lacks a mechanism for inquiring about a record's composition at run time in analogy to the array attributes `'left`, `'right`, `'range` and `'length`.

[32] The reason for using this construct rather than a regular process statement is that the clock generator is reusable, and that we want to make it available in a package. This is not otherwise possible, as the VHDL syntax does not allow a package to include process statements.

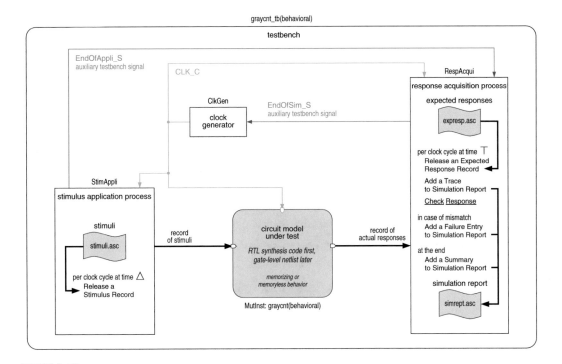

FIGURE 5.17

Organization of HDL testbench code for simulation set-up of fig. 5.11.

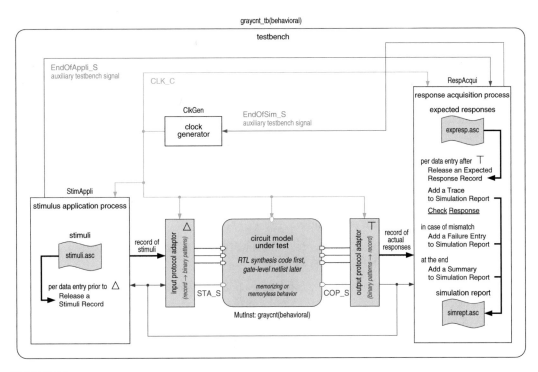

FIGURE 5.18

Organization of HDL testbench code for simulation set-up of fig. 5.13.

5.6 CONCLUSIONS

- More often than not, what is available at the onset of a VLSI project are intentions rather than specifications. Identifying the real needs and casting them into workable instructions always is the first step in the design process. Rapid prototyping often is the only practical way to condense initial conceptions into detailed and unambiguous specs.

- Simulation remains the prevalent method for functional design verification. Being a dynamic technique, it offers only limited protection against design flaws.

- Properties and assertions are great for intercepting suspect behavioral patterns right where they originate. Generously include in-code sanity checks into simulation models.

- Statistical coverage analysis allows designers to explicitly specify the features they want to be verified during simulation and to obtain relevant quality metrics. Much as with assertions, the detection of bugs so becomes less dependent on the propagation of flawed data to output pins.

- In terms of verification support — properties, assertions, random stimuli generation, coverage analysis — SystemVerilog has clearly more to offer than VHDL, see fig.4.24.

- The challenge of functional verification is to safeguard oneself against all plausible design slips without attempting exhaustive simulation. Coming up with a comprehensive collection of test cases requires foresight, care, precision, and a lot of work at the detail level. Following the guidance given in observation 5.12 should increase the likelihood of discovering problems in reasonable time.

- Make sure to understand what potential design flaws might pass undetected whenever a shortcut is taken.

- While establishing a verification plan, beware of preconceptions from the design process as to what situations and issues are to be considered uncritical. Insist on having persons other than the synthesis code writers select — or at least review — the verification strategy, the test cases, the assertions, the coverage analysis code, and the resulting data.

- Making the same test suite work across the entire VLSI design and test cycle is a necessity as functional consistency is otherwise lost. Doing so typically implies coherent event scheduling, data abstraction, and latency absorption. The event sequence $\triangle \downarrow (\top = \square) \uparrow$ has been found to work across all levels of abstractions.

- Keep in mind that assertions and shortcuts that involve simulator commands are not portable to the testing of physical parts using ATE.

- Be prepared to spend more time on verifying functionality than on designing the circuit that implements it.

There are virtually no limits to making simulation set-ups more sophisticated. The reader is referred to the specialized literature including [147] [103] [148] [129] [126] [149]. Also very helpful are industrial standards such as UVM [150] and OS-VVM [145].

5.7 PROBLEMS

1. ∗ Design a test suite for exhaustive verification of the Gray counter specified below.

Rst	Clk	Ena	Count	
0	-	-	00...0	reset
1	↓	-	Count	keep output unchanged
1	↑	0	Count	*idem*
1	↑	1	$\mathrm{graycode}((\mathrm{bincode}(\texttt{Count}) + 1)\,\mathrm{mod}\,2^w)$	increm. Gray-coded output

With how many clock cycles can you do? Indicate the formula for w-bit counters and the actual value for $w = 4$. How does this figure relate to the lower bound given in (5.2)?

2. ∗∗ Devise a test suite for a logic comparator function that tells whether two 6-bit vectors are the same or not. Go for a set of vectors that you consider a reasonable compromise between simulation time and functional coverage. Verify the VHDL architecture correct given below, or the gate-level circuit obtained after synthesis, against that test suite. No inconsistency must occur. Now check how your test suite performs on the flawed architectures given below. Note that the mistaken circuits named flawedy and flawedu are authentic outcomes from efforts by human designers that were using schematic entry tools. The deficiency of the fourth example flawedz, in contrast, has been built in on purpose to demonstrate the impact of an oversight during the editing of HDL code.

```
entity compara6 is
   port (
      ArgA_DI: in std_logic_vector(5 downto 0);
      ArgB_DI: in std_logic_vector(5 downto 0);
      Equ_DO: out std_logic );
end compara6;


-- -------------------------------------------------------------------


-- correct description of 6-bit logic comparator function
architecture correct of compara6 is
begin
   Equ_DO <= '1' when ArgA_DI=ArgB_DI else '0';
end correct;


-- -------------------------------------------------------------------


-- flawed as one of the two arguments has its bits misordered
-- note: a wrong ordering of ArgB_DI in the port list has the same effect
architecture flawedy of compara6 is
   signal Bm_DI : std_logic_vector(5 downto 0);
begin
   each_bit : for i in 5 downto 0 generate
```

```
      Bm_DI(i) <= ArgB_DI(5-i);
   end generate;
   Equ_DO <= '1' when ArgA_DI=Bm_DI else '0';
end flawedy;
```

```
-- mistaken translation of desired function into boolean operations
architecture flawedu of compara6 is
   signal C1_D : std_logic_vector(5 downto 0);
   signal C2_D : std_logic_vector(2 downto 0);
begin
   first_level : for i in 5 downto 0 generate
      C1_D(i) <= not (ArgA_DI(i) xor ArgB_DI(i));
   end generate;
   second_level : for i in 2 downto 0 generate
      C2_D(i) <= not (C1_D(i) xor C1_D(i+3));
   end generate;
   Equ_DO <= C2_D(2) xor C2_D(1) xor C2_D(0);
end flawedu;
```

```
-- corrupt due to a useless statement forgotten in the code
architecture flawedz of compara6 is
begin
   process (ArgA_DI,ArgB_DI)
   begin
      if ArgA_DI=ArgB_DI then Equ_DO <= '1';
      else Equ_DO <= '0';
      end if;
      if ArgA_DI="110011" then Equ_DO <= ArgA_DI(0);
      end if;
   end process;
end flawedz;
```

3. ∗ The purpose of this problem is to show that a test suite may ensure full toggling of all nodes in a gate-level circuit and still be inadequate for functional verification. To that end, find two combinational networks together with a non-exhaustive test suite such that
   • all nodes get toggled back and forth,
   • both networks comply with the test suite, and
   • the two networks are functionally different.

What are the simplest two such circuits you can think of? Generalizing to combinational n-input single-output functions, how does the number of test patterns necessary for full toggling relate to that required for exhaustive verification?

4. ∗ Consider a digital IC that communicates via a bidirectional data bus. A state machine inside the chip generates the enable signal for the pad drivers from its state and from `Write/Read` or some similar signal available at one of the chip's control pins. In order to stay clear of transient drive conflicts, the bus must not be driven from externally before the on-chip drivers have actually released the bus in reaction to the control pin asking them to do so. As a consequence, both testbench and physical test equipment must observe a brief delay between updating the control signal and imposing data on the bus. Extend the precedence graph and the schedule of figs.5.19 and 5.10 accordingly.

5. ∗∗∗ Refer to the specifications for a FIFO in problem 12 of chapter 4 and assume you are the person in charge of verifying the RTL model. Begin by writing a simulation testbench that generates and applies directed random stimuli. Then develop a small set of assertions. As a minimum requirement, make sure that
   • the FIFO cannot get instantiated with a depth of less than two,
   • any attempt to read from the queue when empty causes an error message, and
   • the same applies for any attempt to write when the FIFO is full.

5.8 APPENDIX I: FORMAL APPROACHES TO FUNCTIONAL VERIFICATION

Formal verification attempts to prove or disprove the correctness of some circuit representation by purely analytical means, i.e. without simulating the circuit's behavior over time. A successful proof gives the designer the ultimate confidence that his design will indeed function as previously specified at some higher level of abstraction, and this irrespective of the input as there is no need to apply stimuli. Most formal verification algorithms work by converting a given design representation into a state graph, an ordered binary decision diagram (OBDD), or some other graph-type data structure before analyzing that structure and/or comparing it against similar design representations. There are different degrees of ambition, though.

Equivalence checking

Verifying the functional equivalence between two gate-level netlists or between a netlist and a piece of HDL code is not that difficult. Logic equations are extracted from the gate-level netlist are compared against the reference set of logic equations using theorems from switching algebra. Software tools capable of doing so are typically used to check the consistency — in regular operation mode — of a gate-level netlist with the original RTL synthesis model after test structures have been added. Other relatively minor modifications such as clock tree insertion, logic reoptimization, and conditional clocking are covered as well.

While combinational subfunctions make up much of an RTL model, there are severe limitations when the checking shall be extended to sequential behavior. Automatic conformity checking of circuit models that are supposed to have equivalent external behavior but that differ in the number and/or location of registers, e.g. as a consequence from state reduction or from architectural optimizations, remains a challenging research topic [151].

Last but not least, equivalence checking always presupposes the availability of a golden model.

Model checking

As opposed to the above, model checking does not need any reference model, but aims at finding out whether a circuit model under all circumstances meets a set of specified criteria or properties that any meaningful implementation must satisfy.

A welcome quality of model checking is that it provides a counterexample when a design violates some specification. That is, the checker indicates the specific circumstances under which the maloperation becomes manifest. A serious difficulty is the combinatorial explosion that confines the approach to subsystems with a fairly limited number of states. A detailed discussion is given in [152].

Deductive verification or model proving

Deductive verification is closely related to theorem proving. The goal is a mathematical proof that a given circuit model or protocol indeed conforms with its formal specifications. The answer essentially is of type "true" or "false" and thus provides little clues to developers as to what is wrong with their designs. Deductive verification further suffers from the problems mentioned in section 5.2.1, but remains an active research area. The reader is referred to [153] [154] for accounts on formal verification technology.

5.9 APPENDIX II: DERIVING A COHERENT SCHEDULE FOR SIMULATION AND TEST

This section serves to confirm that the simulation schedule presented in section 5.4.1 is indeed a well-founded one that conforms to the fundamental timing requirements of synchronous circuits without being unnecessarily constrained further. In order to do so, we approximate timing to a degree that makes it possible to describe how a circuit behaves when viewed from outside. [33]

External timing requirements imposed by a model under test (MUT)

Four sets of data propagation paths can be identified in any synchronous design that adheres to single-phase edge-triggered clocking. [34] These paths go

- from inputs to outputs with no intervening registers ($i \rightarrow o$),
- from state-holding registers to outputs ($s \rightarrow o$),
- from inputs to state-holding registers ($i \rightarrow s$), and
- from state registers to state registers ($s \rightarrow s$).

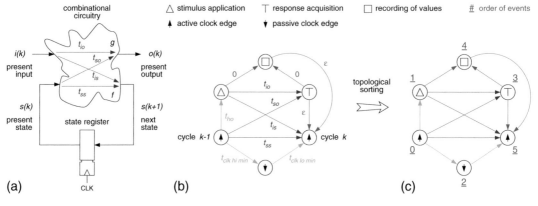

FIGURE 5.19

Data propagation paths through a synchronous single-phase edge-triggered circuit (a) along with the pertaining precedence graph (b), and the recommended order of events (c).

Not surprisingly, we find the same four sets of paths in a Mealy automaton. This is simply because the functionality of any synchronous circuit can be modeled as a Mealy type state machine. In response to a new input or state, a physical circuit finishes re-evaluating g and f after the fresh data have propagated along all paths through the combinational logic. As input and state do not, in general, switch

---

[33] We do not aim at modeling or even at understanding what exactly happens inside the circuit yet. A more accurate analysis will become feasible on the basis of a detailed timing model to be introduced in section 7.2.2.
[34] Any asynchronous reset input can be handled like an ordinary input in this context.

simultaneously, four delay parameters need to be introduced, namely t_{io}, t_{so}, t_{is} and t_{ss}. [35] Parameter t_{io}, for instance, denotes the time required to compute a new output o after input i has changed.

The necessary precedence relations for the circuit to settle to a stationary state are [36]

$$t_\top(k) \geq \max(t_\triangle(k) + t_{io}, t_\uparrow(k-1) + t_{so}) \tag{5.3}$$

$$t_\uparrow(k) \geq \max(t_\triangle(k) + t_{is}, t_\uparrow(k-1) + t_{ss}) \tag{5.4}$$

Precedence relations captured in a constraint graph

The above precedence relations can be expressed by the constraint graph of fig.5.19b where each node stands for a major event associated with clock period k. Note that there are two active clock nodes, namely one for the clock event immediately before the clock period under consideration and a second one for the clock event at its end. Each precedence relation is represented by a directed edge that runs from the earlier event to the later one. The minimum time span called for by the associated condition is indicated by the weight of that edge:

Circuit delays. The aforementioned data propagation paths t_{io}, t_{so}, t_{is} and t_{ss} map to a first set of four edges.

State register hold time requirement. A fifth non-zero weight stipulates that new stimuli must not be applied earlier than t_{ho} after the previous active clock event. Ignoring this constraint is likely to cause hold time violations at some bistables or might otherwise interfere with the precedent state transition. [37]

Clock minimum pulse widths. Two more edges of small but non-zero weight are labeled $t_{clk\,hi\,min}$ and $t_{clk\,lo\,min}$. They indicate the minimum time spans during which the driving clock signal must remain stable.

Securing a coherent vector set. There are also four edges of weight zero or close to zero. One of them leads from $\top$ to $\uparrow$ and has an infinitesimally small weight ε. It reflects the requirement that response acquisition must occur before the circuit gets any chance to change its state in reaction to the next active clock event. Three more edges define the preconditions for recording a consistent stimulus/response pair per clock cycle.

---

[35] Most practical circuits have their set of input bits grouped into a number of vectors, each of which has its own delay parameters, and similarly for outputs. Extending our approach to cover such situations as well is left to the reader as an exercise, see problem 4.

[36] Of course, precedence relations may be simpler in a given particular case, say for a combinational circuit (automaton with no state where t_{so}, t_{is}, and t_{ss} are not defined) or for a counter (Medvedev machine where t_{io} is not defined and $t_{so} = t_{ss}$). However, by consistently sticking to a scheme that is suitable for the most general case, we can avoid having to reorder events whenever we must move from one circuit type to another.

[37] Note that t_{is} and t_{ss} are meant to include the setup times of the registers. This explains why t_{su} does not appear in the constraint graph as opposed to t_{ho} which cannot be subsumed anywhere else.

Solving the constraint graph

Any desirable sequence must order the events such as to satisfy the above precedence relations for positive but otherwise arbitrary values of the seven timing parameters involved. Solutions are obtained from topological sorting of the precedence graph. [38] One such ordering is indicated by the underlined numbers in fig.5.19c. It corresponds to a periodic repetition of stimulus application, response acquisition, and clocking, and is symbolically denoted as △ ⊤ ↑ .

The precedence graph also points to minor liberties. While it is true that the recording of a stimulus/response pair may take place at any time between response acquisition and the subsequent active clock edge, there is nothing to be gained from defining an extra point in time for doing so. The events of response acquisition ⊤ and recording ☐ may as well be tied together.

The passive clock edge, on the other hand, is free to float between two consecutive active edges as long as two the constraints $t_{clk\,hi\,min}$ and $t_{clk\,lo\,min}$ are respected. As a final result, the event order △ ↓ (⊤ = ☐) ↑ will almost always represent a workable solution. The recommended simulation schedule is depicted in fig.5.10.

Anceau diagrams help visualize periodic events and timing

The Anceau diagram is very convenient for visualizing events and timewise relationships that repeat periodically. Each round trip corresponds to one clock cycle (or to one computation period). The example of fig.5.20a illustrates the simulation schedule just found.

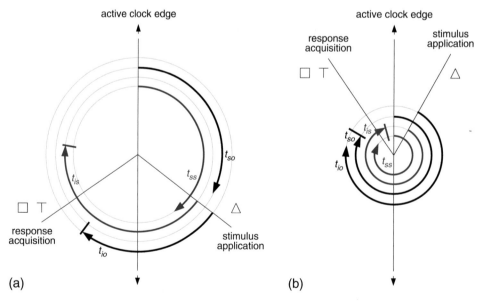

FIGURE 5.20
Anceau diagram for a Mealy type circuit operated at moderate speed (a) and close to maximum speed (b).

---

[38] The nodes of a graph are said to be in **topological order** if they are assigned integer numbers such that every edge leads from a smaller numbered node to a larger numbered one.

Each arrow stands for the delay along one of the four signal propagation paths in a Mealy machine. The two short radial bars are just graphical representations of the max operators in (5.3) and (5.4) respectively. The outer one indicates when the output has settled to a new value and becomes available for acquisition. Similarly, the inner bar tells when the computation of the next state has come to an end and so determines the earliest point in time the circuit may be safely clocked.

If the operating speed is to be increased, the clock period grows shorter while delays and timing conditions remain the same. Beyond a certain point, stimuli must get applied earlier and/or responses will have to be acquired later in the clock cycle. This is shown in fig.5.20b which corresponds to a circuit operating close to its maximum speed.

Anceau diagrams come in handy for visualizing periodic event sequences and timing conditions. We will make use of them in later chapters in the context of clocking and input/output timing.

6

THE CASE FOR SYNCHRONOUS
DESIGN

6.1 INTRODUCTION

Experience tells us that many malfunctioning digital circuits and systems suffer from **timing problems**. Symptoms include

- Bogus output data,
- Erratic operation, typically combined with a
- Pronounced sensitivity to all sorts of variabilities such as PTV and OCV.

Erratic operation often indicates the circuit operates at the borderline of a timing violation. Searching for the underlying causes not only is a nightmare to engineers but also causes delays in delivery and undermines the manufacturer's credibility.[1]

Observation 6.1. *To warrant correct and strictly deterministic circuit operation, it is absolutely essential that all signals have settled to a valid state before they are being admitted into a storage element (such as a flip-flop, latch or RAM).*

This truism implies that all combinational operations and all propagation phenomena involved in computing and transporting some data item must have come to an end before that data item is being locked in a memory element. Data that are free to change their value at any time are dangerous because they may give rise to bogus results and/or may violate timing requirements imposed by the electronic components involved.[2] Hence the need for regulating all state changes and data storage operations.

Many schemes for doing so have been devised over the years, see fig.6.1. From a conceptual perspective, we must distinguish between two diametrically opposed alternatives, namely synchronous clocking and self-timed operation. A third category that includes all unstructured ad hoc clocking styles — occasionally referred to as "clock-as-clock-can" in this text — is not practical except for the smallest subcircuits perhaps.

---

[1] Malfunctioning that occurs intermittently or that depends on minor variations of temperature, voltage, signal waveforms, and similar circumstances make debugging extremely painful. The fact that one can never be sure whether simulation accuracy suffices to predict transient waveforms, a subcircuit's reactions to a marginal triggering condition, and other details with sufficient precision does not help either.

[2] Timing requirements are meant to include set-up and hold conditions, minimum clock high and low times, and maximum clock rise and fall times. All these quantities are explained in section A.6.

The present chapter first introduces and compares these approaches before commenting on why synchronous clocking is considered to be the best choice for staying clear of timing problems in the context of digital VLSI. Several general design rules are explained. Further details such as what bistables to use, how to clock them, and how to distribute the clock signal(s) over a circuit — or a clock domain — will be the subject of chapter 7.

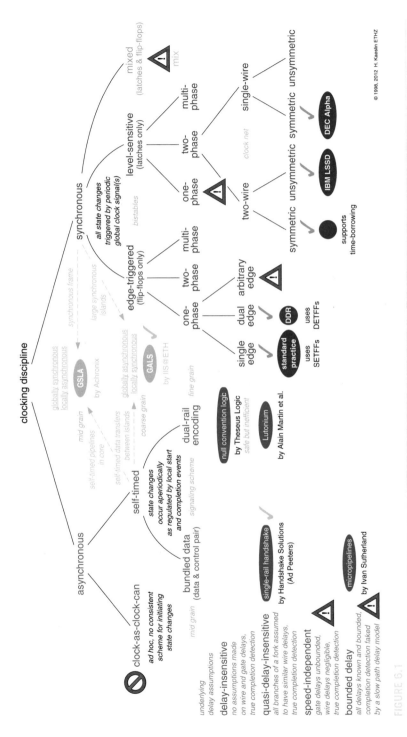

FIGURE 6.1

Family tree of clocking disciplines (simplified). Ticked are those options that this author considers safe and viable in the context of VLSI provided they are implemented correctly.

6.2.1 SYNCHRONOUS CLOCKING

Definition 1. A circuit or subcircuit is said to operate synchronously iff <u>all</u> data storage operations and, hence, <u>all</u> state transitions, are restricted to occur periodically at precise moments of time that are determined by a special signal referred to as the **clock**.

A single clock that drives a set of flip-flops is the most straightforward pattern, but the above definition is meant to include circuits where bistables are being driven by a pair of complementary clock signals, e.g. Clk and $\overline{\text{Clk}}$. The same also applies to multi-phase clocks and even to multiple clocks of distinct frequencies provided those subclocks are all locked to one primary clock signal. State changes are effectively restricted to occur synchronously to the primary clock as long as all driving clocks maintain fixed frequency and phase relationships.[3]

Definition 2. A **clock domain** or, which is the same, a synchronous island is a (sub)circuit where all clock signals maintain fixed frequency and phase relationships because they are derived from a common source.

A clock domain may be confined to one subblock on a chip or may extend over several chips or even printed circuit boards. Any line that separates two distinct clock domains is referred to as a **clock boundary**. As stated before, most designs do with a single clock signal per domain, but there are exceptions to this rule.

6.2.2 ASYNCHRONOUS CLOCKING

Definition 3. A circuit or subcircuit works asynchronously when some or all of the memory elements therein are permitted to change their states independently from a global reference.

Asynchronous circuits are easily identified by the presence of

- Unclocked bistables (SR-seesaw, Muller-C, MUTEX, etc.),
- Zero-latency feedback loops, e.g. as part of
- Asynchronous state machines (ASM),
- Logic gates other than buffers and inverters in clock nets,[4]
- Gated asynchronous (re)set signals,
- Logically redundant circuitry for hazard suppression,
- Imposed delays, ring oscillators,
- One-shots, pulse shapers and similar subcircuits.

Examples of such constructs that never appear in synchronous designs are given in fig.6.2b and discussed in more detail in section 6.4.3. A multitude of clock signals, clock domains, and clock boundaries is another indication of asynchronous circuit operation.

---

[3] This is the case when some primary clock is being subdivided by way of some counter circuitry in order to derive one or more slower clocks. The converse, that is clock frequency multiplication with the aid of a phase locked loop (PLL), also results in multiple signals with fixed frequency and phase relationships.

[4] Gated clocks are an exception, please refer to section 7.5 for advice.

6.2.3 SELF-TIMED CLOCKING

A more recent addition to the category of asynchronous circuits relies on self-timing throughout.

The output of a traditional logic circuit is valid only at certain moments of time, and it is impossible to tell from the output alone when that is the case. While a transiting voltage implies data are invalid, the inverse does not necessarily mean they have settled to a steady state. Rather, the window of validity (data-valid window) must be found by referring to a time scale, and the continuum of time gets structured by a periodic global clock.

Self-timing, in contrast, introduces a signaling mechanism that conveys when data are valid and when not. This either requires the addition of flow control signals as shown in fig.6.2c or dual-rail encoding in combination with special data patterns reserved for the purpose. The guiding principle is that no storage register is allowed to accept any new data from its predecessor before its successor has safely acquired the data item presently stored in the register.[5]

---

[5] Whether data items are subject to combinational operations while travelling from one register to the next or not, is immaterial.

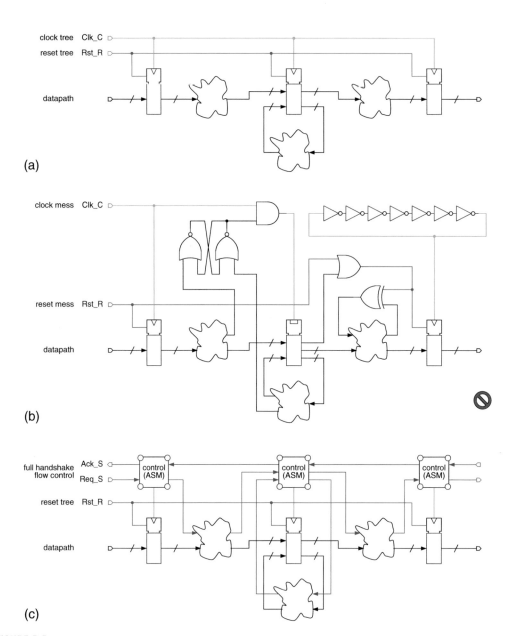

FIGURE 6.2

Toy examples intended to give a flavor of the three grand alternatives for clocking. Synchronous clocking (a), clock-as-clock-can (b), and self-timed clocking (c).

Definition 4. An asynchronous circuit or subcircuit is said to operate in a self-timed manner when all state transitions get regulated by local start and completion events on a per case basis.

As opposed to synchronous clocking, there is no global signal that would trigger operations which explains why this scheme is also known as "clockless logic". Instead, the various building blocks in the circuit coordinate their activities by way of handshake protocols which necessitates an on-going mutual exchange of status information. There is no way for an external observer to tell when a given computational step or data transfer is to happen because everything occurs in an entirely data-driven and, therefore, aperiodic way. The operation of a self-timed circuit proceeds at its own speed and automatically adjusts to PTV conditions.

Self-timed schemes come in numerous varieties [155] [156] [157] [158] that essentially differ in how status information are actually being conveyed and in the assumptions on circuit delays that underlie protocol and circuit design, see fig.6.1.

6.3 WHY A RIGOROUS APPROACH TO CLOCKING IS ESSENTIAL IN VLSI

The propagation of new data values through a digital network gives rise to transient phenomena within the network itself and at the output. Let us briefly review the impact of spurious events on the functioning of digital circuits to prepare the ground for a subsequent discussion of clocking schemes and circuit design styles.

6.3.1 THE PERILS OF HAZARDS

Hazard and glitch are two names for unwanted transients on binary signals. How they originate is examined in appendix A.5, key findings are as follows:

- Be prepared to observe glitching at the output of any combinational circuit with the sole exception of fanout trees. This is because hazards can originate
  (a) if two or more inputs change at a time — or almost so —, or
  (b) if the combinational logic includes reconvergent fanout paths.

- Glitches are difficult to predict as their manifestation — in terms of waveform, amplitude and duration — depends on many low-level details unknown at synthesis time including placement, wiring, detailed layout, gate and interconnect delays, crosstalk, PTV conditions, and on-chip variations (OCV). A minute departure may decide on whether a hazard manifests itself as a rail-to-rail excursion, becomes visible as a runt pulse, or results in no tangible appearance at all.

It is prudent to assume that any combinational network gives rise to hazards unless one has proof to the contrary. To determine whether a network actually develops glitches or not requires a detailed analysis of its transistor-level circuitry, timing parameters, layout parasitics, input waveforms, and the like. Manual elimination of hazards is a painful and unprofitable exercise as only small networks are amenable to analysis. Imposing narrow delay bounds on signal propagation paths is not compatible with automated synthesis and place-and-route (P&R) of VLSI circuits where interconnect delays often dominate over gate delays. We conclude:

Observation 6.2. *All clocks, all asynchronous reset and presets, all write lines of asynchronous RAMs plus any other signal that might trigger a state transition in a memorizing subcircuit must be kept free of hazards under all circumstances.*[6]

There is an abundance of catastrophic failures that could result if this rule were violated:

- Unwanted state transitions in any kind of finite state machine.
- Multiple registration of a single event by a counter.
- Erroneous activation of an edge-triggered interrupt request line of a microprocessor.
- Unplanned for return of a sequential (sub)circuit to its start state.
- Storage of bogus data in a register or a RAM.

---

[6] Be warned that state-change-triggering inputs are not always identified as such in the documentation of commercially available components or library elements. They sometimes hide under inconspicuous names such as "chip enable" (CEB), "write enable" (WEB), "interrupt request" (IRQ) and "strobe", just to name a few.

- Data losses or duplications during data transfer operations.
- Deadlocks in asynchronous communication protocols (such as handshaking).
- Marginal triggering and, hence, metastable behavior of bistables.[7]

Data, address, status, control, and other signals not capable of sparking off a state change without the intervention of a clock are not concerned. Similarly, combinational subcircuits are not normally sensitive to hazards because all transient effects are reversible and eventually die out.[8] Glitches also cause no problem when driving sluggish peripheral equipment such as indicator lamps, electromechanical relays, teletypes and the like.

6.3.2 THE PROS AND CONS OF SYNCHRONOUS CLOCKING

There are ten essential benefits that are shared by all synchronous clocking disciplines.

+ Hazards do not compromise functionality. Clock and asynchronous reset are the only two signals that must be kept free of hazards under all circumstances. Doing so is easy, strictly limiting distribution networks to fanout trees suffices.

+ As no timing violations ever occur within a properly designed synchronous circuit, there is no chance for inconsistent data, marginal triggering, and metastability to develop.

+ Immunity to noise and coupling effects is maximum because all nodes are allowed to settle before any storage operations and state changes occur.

+ All timing constraints are one-sided. For a circuit to function correctly, any timing quantity is either bounded from above (such as the longest propagation delay, for instance) or from below (such as the contamination delays). Two-sided constraints do not exist.[9]

+ Together, the above four properties warrant deterministic behavior of circuits independently from low-level details.[10] Synchronous designs do not rely on delay tuning in any way. What matters for functional correctness are the data operations at the RTL level exclusively. This argument cannot be overestimated in view of
  - Automatic placement, routing, and physical design verification,
  - Automatic HDL synthesis, logic optimization, clock tree generation, and rebuffering,

---

[7] The term "metastability" refers to an unpredictable behavior of a memory element that may or may not result from violating its timing conditions, see chapter 8 for details.

[8] There are few exceptions, however, where hazards are unacceptable in spite of the memoryless nature of the subsystems involved. These include:
- Digital modulators and other circuits where signals are required to follow a well-defined waveform or spectrum.
- Output enable signals where hazards could occasion transient drive conflicts thereby leading to exaggerated crossover currents, needless power dissipation, and excessive ground bounce.
- Electronically controlled power drives and power converters where unforeseen current spikes are likely to cause permanent damage.

[9] With the exception of one-phase level-sensitive operation which is considered impractical, see section 7.2.7.

[10] This is to say that buffer sizing, library changes, parameter variations (PTV and OCV), physical arrangement, layout parasitics, etc. are likely to impact maximum clock rate, I/O timing, power dissipation, and other quantitative figures of merit, but not a circuit's functionality.

- Automatic insertion of test structures,
- Reusing a HDL model or a netlist in multiple designs, and of
- Retargeting a design from one cell library and/or fabrication process to another (e.g. from FPL to a mask-programmed IC, or vice versa).

+ Synchronous operation makes it possible to separate functional verification from timing analysis and to take advantage of automata theory and related concepts.

+ There is no need for any redundant circuitry to suppress hazards, a task not supported by standard synthesis tools.

+ The compute operations that are to be carried out in each clock cycle can be stated and collected at compile time, thereby opening a door for cycle-based simulation techniques that are more efficient when circuits grow large. Asynchronous circuits, in contrast, are entirely dependent on event-driven simulation.

+ Established methods for circuit testing (such as fault grading, test vector generation, and the insertion of test structures) start from the assumption of synchronous operation. What's more, almost all test equipment is designed accordingly.

+ Synchronous clocking makes it possible to slow down and even to suspend circuit operation in any state and for an arbitrary lapse of time,[11] which greatly facilitates the tracing of state transitions, data transfers, protocol sequences, and computation flow when debugging a malfunctioning circuit. The capability to operate synchronous circuits in speed-limited environments is often welcome for prototyping purposes.

Undeniably, synchronous circuit operation also has its drawbacks.

− Performance is determined by the worst rather than by the average delay over all data.[12]

− Circuits may consume more power than necessary as a register dissipates energy in each clock cycle regardless of the extent of state change. Yet, clock gating and other techniques have been developed specifically to lower clock-induced power dissipation while maintaining overall synchronous circuit operation.

− Synchronous operation causes periodic surges in supply currents. This not only strains the power and ground nets but also entails electromagnetic radiation at the clock frequency and at higher harmonics.

− Synchronization problems are unavoidable at the interface between any two clock domains.[13] However, similar problems arise wherever an asynchronous subsystem interfaces with a clock-driven environment such as a sampled data source or data sink.

---

[11] Unless capacitive data storage is involved such as in DRAMs or in dynamic CMOS logic.
[12] A workaround is to be presented in footnote 18.
[13] This is the subject of chapter 8.

— Most synchronous clocking disciplines insist on tightly controlled delays within the clock distribution network. Special software tools that address this need during physical design make part of all major VLSI CAD suites.

6.3.3 CLOCK-AS-CLOCK-CAN IS NOT AN OPTION IN VLSI

Unsafe circuits often emanate from obsolete or perfunctory design methodologies. This is particularly true for asynchronous circuits, the design of which is very demanding, both in terms of profound understanding and engineering effort. Sporadic timing violations and a pronounced sensitivity to delay variations are typical consequences. Although popular with digital pioneers, ad-hoc clocking schemes have become unacceptable because VLSI technology has changed the picture in the following way.

— The operation of most asynchronous circuits critically depends on certain delay figures and, therefore, also on their layout arrangements. This makes it difficult to anticipate whether fabricated circuits will indeed behave as simulated as no simulation model is capable of rendering all effects that contribute to timing variations with perfect precision.

— Finding and correcting timing problems is difficult enough on a board in spite of the fact that designers have access to almost all circuit nodes and can add extra components for tuning delays. It is next to impossible on a monolithic chip that offers no such possibilities.

— Historically, emphasis was on doing with as few SSI/MSI packages as possible. The prime challenge today is first-time-right design, a few more logic gates do not normally matter.

— Early logic MOS and TTL devices were so slow that wiring-induced delays could be neglected altogether. As opposed to this, interconnect delays due to wiring parasitics tend to dominate over gate delays in VLSI, making it impossible to predict delays from circuit diagrams and netlists.

— VLSI designers cannot afford delay tuning of a vast collection of signals. For the sake of productivity, they must rely on design automation as much as possible for logic synthesis, optimization, placement, routing and verification. The higher productivity so obtained is at the expense of control over most implementation details.

In conclusion, an industrial circuit designer concerned with design productivity, first-time-right design, and fabrication yield is well advised to follow the recommendations below.

Observation 6.3. *Ad-hoc approaches to clocking are just unfortunate leftovers from the early days of digital design that are incompatible with the requirements of VLSI. Instead, strive to do with as few clock domains as possible and strictly adhere to one synchronous clocking discipline within each such domain.*

Let us bring this matter to an end with two quotes from experts

"Just say NO to asynchronous design!" [159]
"KISS those asynchronous-logic problems good-bye, Keep It Strictly Synchronous!"[14] [160]

---

[14] K.I.S.S. originally was a slogan to improve the success rate of complex operations "Keep It Simple, Stupid".

6.3.4 FULLY SELF-TIMED CLOCKING IS NOT NORMALLY AN OPTION EITHER

As opposed to unsophisticated asynchronous clocking schemes, self-timed clocking follows a strict discipline, holds the promise of achieving better performance, and provides valuable hooks for improving on energy efficiency [161]. When compared to synchronous clocking schemes, the notorious difficulty of domain-wide clock distribution is replaced by a multitude of local synchronization or arbitration problems which in turn asks for a specific design methodology.

In spite of its theoretical benefits,

— The hardware and energy overheads associated with implementing handshaking or related request-acknowledge protocols all the way down to the level of logic gates,
— The difficulties of interfacing with clock-driven peripheries and test equipment,
— The absence of self-timed components in commercial cell libraries,[15]
— The shortage of adequate EDA support,
— The excruciating subtleties of the design process, and
— The lack of widespread know-how, together with
— The ensuing time to market penalty

have prevented self-timed logic from becoming a practical alternative.[16]

6.3.5 HYBRID APPROACHES TO SYSTEM CLOCKING

Current efforts are attempting to combine the best of both worlds into **globally asynchronous locally synchronous** (GALS) circuits where synchronous islands communicate via self-timed data exchange protocols. The usage of arbiters and pausable clocks is typical for GALS circuits. Such **heterochronous** architectures are being investigated as an alternative to overly large synchronous systems in search of improved energy efficiency, better performance, more manageable clock distribution, and facilitated design reuse [163] [164] [165] [166].

A different strategy termed **mesochronous clocking** is to distribute a global clock signal without much concern for skew. Specially designed local synchronizer circuits are then being used to sample data at multiple points in time, to detect synchronization failures, and to retain valid data only [167]. A related idea is to insert tunable delay lines within the clock distribution network and to calibrate them automatically at startup time such as to make all blocks work synchronously together [168].

For the time being, most of this must be considered research, however, as industry is reluctant to embrace unproven concepts and design flows [169].

---

[15] Such as the Muller-C and the mutual exclusion elements (MUTEX) explained in appendix A.4. Dual-rail encoding further necessitates special circuits for logic gates and bistables not found in regular cell libraries.

[16] Industrial interest is documented by start-ups such as Theseus Logic Inc. and Handshake Solutions. The former has patented NULL Convention Logic (NCL) which, however, is penalized by an important overhead factor of 2 to 2.5 over traditional synchronous logic [162].

Table 6.1   The grand alternatives for clocking compared.

Desirable characteristics	Clocking discipline		synchronous
	asynchronous		
	ad hoc	self-timed	
Fundamentals			
Immune to hazards	no	yes except for protocol signals	yes except for clock and reset
No need for hazard-suppression logic	no	yes	yes
One-sided timing constraints only	no	yes	yes
No marginal triggering during circuit operation	maybe, maybe not	yes	yes within synchr. island
Avoids timing problems at interfaces	no	no	no
Design process			
No particular library cells needed	yes	no	yes
Does without tightly controlled delays in logic <u>and</u> interconnect	no	yes except for local subcircuits	yes except for clock[a]
Circuit to function irrespective of logic and layout details	no	yes	yes
Functionality and timing separable	no	mostly yes	yes
Systematic and modular design methodology, reuse facilitated	no	yes	yes
Matches with prevalent flows and tools	no	no	yes
Arbitrarily slow operation supported for debugging purposes (step-by-step)	no	no	yes[b]
Periodicity of circuit operation			
All signals to settle before clocking	no	no	yes
Non-periodic "random" supply current	more or less	yes	no
Supports cycle-based simulation	no	no	yes
Works with existing test equipment	no	no	yes
Figures of merit of final circuit			
Good area efficiency	sometimes yes, more often no	only if overhead remains modest	yes
Better than worst-case performance	more or less	yes	no
Good performance in practice	in particular applications	debatable	yes
Good energy efficiency	in particular applications	yes if overhead remains modest	yes if designed accordingly

[a] Skew-tolerant schemes are available, see section 7.2.
[b] Unless DRAMs or dynamic CMOS logic is being used.

Synchronous operation essentially rests on two guiding principles to be presented next.

While digital VLSI designers must devise circuits that function in a predictable and dependable way, they cannot afford to study the transient behavior of every single circuit node in detail. What is needed is a robust and well-understood clocking discipline that, when properly implemented, warrants correct circuit timing under all operating conditions.

From the background of our findings on transients in digital circuits in section 6.3, there is a fairly obvious solution. Rather than attempting to suppress hazards here and there — which is symptomatic for clock-as-clock-can design — the set of acceptable circuit structures is voluntarily and consistently restricted to those that
• do not let hazards originate in clock and in asynchronous (re)set nets, and that
• tolerate hazards on all other signals with no impact on functionality whatsoever.

The most efficient way to prevent dangerous hazards from coming into existence is to shut out all signals that might possibly prompt a state change from participating in logic operations. Table 6.2 begins by distinguishing between hazard-sensitive and hazard-tolerant signals before classifying signals further as a function of their respective roles in a circuit.

Reset signals cause a sequential circuit to fall into some predetermined start state without the intervention of a clock. Their effect is immediate, unconditional and always the same. Practically speaking, this includes all **asynchronous reset** and set inputs present on many bistables, but none of the synchronous initialization signals.

Clock signals are in charge of sparking off all regular transitions of a sequential circuit from one state to the next, but have no influence on what that next state will be.

Information signals is a collective term for all those signals that contribute to deciding what state a circuit is to assume in response to an active clock edge and/or what output that circuit shall produce.

> We further distinguish between functional signals and test signals with the first subclass largely outnumbering the second. Functional signals comprise data, address, control and status signals, which together implement the desired functionality. Test signals get added on top during the design process to improve circuit testability. From this perspective, **synchronous clear** and load inputs are nothing else than particular control signals.

We now stipulate

Observation 6.4. *Synchronous circuits boast a clear-cut separation into signals that decide on <u>when</u> state transitions and output changes are to take place, and into others that determine <u>what</u> data values to output and what state transitions to carry out, if any. As a consequence, (asynchronous) reset signals, clock signals, and information signals never mix. Combinational operations (other than unary negation) are strictly confined to information signals.*

Table 6.2   Taxonomy of signals within a synchronous island. The bottom part suggests a naming convention for signals in HDL models, schematics and netlists detailed further in section 6.7.

Class	Reset signal	Clock signal(s)	Electrical signals Information signals		
determines	when to (re)enter the start state	when to move from one state to the next	what state to enter next and/or what output to produce		
Hazards	inadmissible	inadmissible	harmless		
Role during simulation	general model wake up	general model wake up	evaluated at model activation time, no wake up of memorizing models		
Subclass	—	—	Functional signals		Test signals
serves to			implement the desired functionality		improve observability and controllability
Members	Asynchronous (re)set	Clock(s)	Status, control	Data, address	Block isolation, scan path(s)
Switching is	Examples and their identification by way of naming and color code				
synchronous to local clock	—	Clk	many a	many	Tst, Scm, Sci, Sco
Class_char Color		C green	S blue	D black	T yellow
asynchronous to local clock	Rst, Set	—	any input prior to synchronization		
Class_char Color	R red		A orange	A orange	A orange

a Including synchronous clear and load signals Clr and Lod.

6.4.2 SECOND GUIDING PRINCIPLE: ALLOW FOR CIRCUITS TO SETTLE BEFORE CLOCKING!

Another essential precondition for safe operation follows immediately from observation 6.1 which implies that any combinational network shall be allowed to settle to its steady-state condition before the emanating output signals are being clocked into some memory device. In the context of a (sub)circuit driven by a single clock, this amounts to the following requirement.

Observation 6.5. *Synchronous designs must be operated with a clock period long enough to make sure that all transient effects have died out before the next active clock edge instructs registers and other storage devices to accept new data.*[17]

---

[17] A more precise, quantitative formulation of this and other constraints will be given in section 7.2.

While this indeed prevents hazards on information signals from having any effect on the circuit's [next] state, it comes at a cost. The length of the computation period, and hence also the clock period, are bounded from below by the slowest signal that travels between any two consecutive registers (longest propagation path, most penalizing set of data, slowest operating condition). The clock rate must be chosen such as to conform with the worst-case timing, thereby denying the possibility to take advantage of more favorable situations no matter how frequently these might occur.[18]

6.4.3 SYNCHRONOUS DESIGN RULES AT A MORE DETAILED LEVEL

In a certain sense, observations 6.4 and 6.5 form the "constitution" of synchronous circuit design from which many "laws" of more specific nature can be easily derived given some particular situation. A couple of them will be explained and illustrated next, but we do not intend to explicit all such rules here as many of them have a rather narrow focus.[19]

HDL synthesis has greatly simplified things in that designers no longer need to manually assemble subfunctions from primitive gates and bistables. Yet, the responsibility of writing circuit models that properly synthesize to robust synchronous circuits is still that of the engineer. To do so, he must be capable of recognizing and correcting dangerous constructs.

Unclocked bistables prohibited

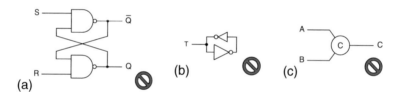

(a)   (b)   (c)

FIGURE 6.3

Examples of bistables that do not qualify for data storage in synchronous designs. Seesaw (a), snapper (b), and Muller-C element (c).

Observation 6.6. *As opposed to flip-flops and latches, unclocked bistables such as seesaws, snappers, and Muller-C elements do not qualify as storage elements in synchronous designs.*

---

[18] This is precisely the starting point for **speculative completion** that borrows from self-timed execution while preserving strictly synchronous circuit operation. The idea is based on the statistical observation that only few data vectors do indeed exercise the longest path in ripple-carry adders and related arithmetic circuits. The clock frequency is chosen such that the worst-case delay fits into two clock periods instead of one. A fast auxiliary circuit monitors carry generation and carry propagation signals in order to determine whether the current calculation involves a short or a long ripple-carry path. System operation is made to continue immediately if the adder is found to settle before the end of the first cycle. If not so, system operation is stalled for one extra clock cycle. Please refer to [170] for a more detailed account on this out-of-the-ordinary technique.

[19] [171] is a valuable reference on safe digital design. The textbook gives a list of nine detailed rules but implicitly excludes all clocking disciplines other than edge-triggered one-phase clocking which is overly restrictive.

Unclocked bistables make no difference between reset, clock and information signals which renders them vulnerable to hazards and which is against the postulates of observation 6.4. The usage of naked SR-seesaws is strongly discouraged in spite of the fact that certain latch and flip-flop designs include them as subcircuits. Snappers are not for storing data, their sole legal usage is to prevent the voltage from drifting away while a three-state node waits in high-impedance condition. The Muller-C element is a building block of self-timed circuits.

Zero-latency loops prohibited

Zero-latency loop is just another name for a circular signal propagation path in a combinational network, see fig.6.4 for examples at the gate level. The problem with such circuits is that it is not possible to determine a logic value for the output even if all input values are known as some combinational function necessarily makes reference to its own result.

Consider fig.6.4a and note that variable N can neither settle to 0 nor to 1 without causing a logic contradiction. A physical circuit may then either oscillate or assume a precarious equilibrium. Next consider fig.6.4b where the number of logic inversions along the loop is even. This circuit quickly locks into a one of two stable states of equilibrium with no contradictions, but there is no way to predict M without knowledge of its original value.

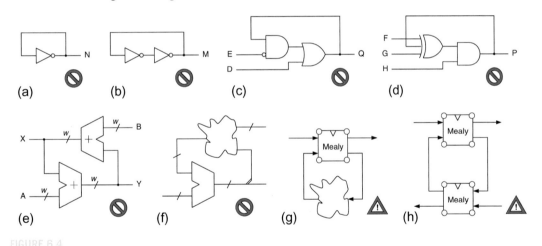

FIGURE 6.4
A few examples of undesirable zero-latency loops.

For a more substantial example consider the cross-coupled adders of fig.6.4e. Inputs A, B and outputs X, Y all have the same word width $w \geq 2$ (carry bits are left aside for simplicity).

$$X = B + Y \qquad\qquad Y = A + X \qquad\qquad (6.1)$$

Behavior depends on the input values applied. If $A + B = 0$, equations reduce to $X = X$ and $Y = Y$ which implies that X and Y preserve their values. Although built from combinational gates exclusively, a physical circuit would lock into one of many possible states and remain there as long as the inputs are kept the same. Yet, a circuit with memory but no distinction between <u>when</u> and <u>what</u> signals does not conform with the dissociation principle of synchronous design.

Conversely, if $A + B \neq 0$, one has $X = (B + A) + X$ and $Y = (A + B) + Y$. A physical circuit would be bound to add and switch in a frenzy pace forever. Failure to settle to a steady state is against the second guiding principle of synchronous operation, however. In summary, none of the circuits shown behaves safely and predictably as one would hope.

HDL code can include circular references without the author being aware of them. Examples often involve a Mealy machine plus a datapath, glue logic, or some other surrounding circuitry, see fig.6.4g. A feedback path in the logic can then combine with a through path in the finite state machine (FSM) to form a zero-latency loop. Note that not all FSM input and output bits need to participate in the loop, one each suffices. Another particularity is that such paths may open and close as a function of state and input value.[20] While this is likely to confine malfunction to just a few situations, it also renders debugging particularly difficult.

Observation 6.7. *By prohibiting zero-latency loops, gate-level circuits are effectively prevented from oscillating, from locking into unwanted states, or from otherwise behaving in an unpredictable way. In synchronous designs, each circular path must include one or more registers.[21]*

Monoflops, one-shots, edge detectors and clock chopping prohibited

Monoflop, monostable multivibrator, one-shot, and edge detector are names for a variety of subcircuits that have one thing in common: they output multiple edges or even multiple pulses for a single transition at their input. A clock chopper does the same with a clock signal. Any of these can be built from a delay line in conjunction with reconvergent fanout, though other implementations also exist. See fig.6.5 for an example and observe that the exact nature of the contraption gets determined by the logic operation at the point of reconvergence.

Unwary designers are sometimes tempted to resort to clock chopping when they have to incorporate blocks, e.g. unclocked RAMs, that impose timing constraints incompatible with normal synchronous circuit operation. Frequency multiplication is another usage. Once again, the requirements of observation 6.4 are obviously violated. Major problems are testability, ill-defined pulse width, critical place and route, timing verification, and vulnerability to delay variations.

Observation 6.8. *Monoflops, one-shots, edge detectors, and especially clock choppers are absolute no-no's in synchronous design.*

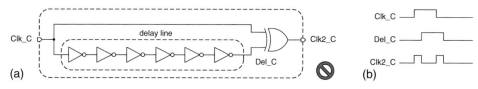

(a)   (b)

FIGURE 6.5

Clock chopper circuit (a) with waveforms (b).

---

[20] A perfidious example is given in appendix B.2.3.
[21] Exceptions exist as the absence of zero-latency loops is a sufficient but not a necessary condition for memoryless behavior. A notable example of a circuit that settles to well-defined steady states for arbitrary stimuli in spite of a zero-latency loop is the end-around carry used in 1's complement and sign-magnitude adders [172]. Proving that a feedback circuit predictably exhibits combinational behavior under all circumstances may be a major effort, though.

Clock and asynchronous (re)set signals can trigger a state transition at any time. A hazard on a signal of either of these two classes is, therefore, very likely to lead to catastrophic failure.

Observation 6.9. *No clock and asynchronous (re)set signals shall ever participate in any logic operation other than the unary operation of taking the complement. Instead, any signal capable of inducing a state change must be distributed by a fanout tree exclusively.*

This is because fanout trees are a priori known not to generate hazards.[22] Binary logic operations such as NOR, AND, XOR, are strictly prohibited. Buffers and inverters are the sole logic gates that are acceptable in clock and in reset networks.

Beware of unsafe clock gates

Clock gating implies to enable and disable state transitions by suppressing part of the clock edges depending on the present value of some control signal. Doing so with the aid of an AND or some other simple gate spliced into the clock net as shown in fig.6.6 has always been a poor technique because it is unsafe and against the principle of dissociation. However, as conditional clocking has seen a renaissance and has indeed become a necessity in the context of low-power design, safer ways of doing so will be studied in great detail in section 7.5.

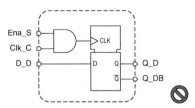

FIGURE 6.6
Unsafe D-type flip-flop with enable resulting from malformed clock gating.

No gating of reset signals

Very much like a gated clock, we speak of a **gated reset** when the asynchronous reset or preset input of a latch, flip-flop, register, counter, or some other state-preserving subcircuit participated in combinational operation with some other signal. This anachronistic practice of using a reset signal for any other purpose than for overall circuit initialization is against the principles of synchronous design and holds serious dangers.

---

[22] No hazard can arise in a single-input network that is free of reconvergence. Please refer back to section 6.3 or see section A.5 for a more complete rationale.

Warning example

Specifications had asked for a modulo-15 counter, i.e. a counter that steps from state 0 to 14 before returning to 0. Counter slices were available from the standard cell library. The designer elected to start with a modulo-16 counter and to skip the unutilized state 15 by activating the asynchronous reset mechanism whenever the counter would enter that state. He devised the circuit of fig.6.7a which behaved as expected during logic simulation. When the IC was put into service, however, the circuit missed out state 14 and went from state 13 to 0 directly. Why?

What the designer had overlooked was the transient behavior of the decode logic he had created. He implicitly started from the assumption that all inputs to the 4-input NAND would switch at the same time. In reality, all sorts of imbalances contributed to make them arrive staggered in time. In the occurrence, the LSB was slightly delayed with respect to the other bits which made the 4-input NAND temporarily see a 1111 during the transition from 1101 to 1110, see fig.6.7b.[23] Further note that other delay patterns could as well have caused the counter to return to state 0 from states 7 or 11.

The design might have developed yet another failure because the intended clearing of the counter slices in state 15 is by way of a feedback loop with latency zero. The reset condition comes to an end whenever the output of the fastest slice begins to flip. There is no guarantee that the impulse so generated is sufficiently long to clear all other bistables too. Some of the flip-flops might as well be subject to marginal triggering and eventually fall back to their previous states.

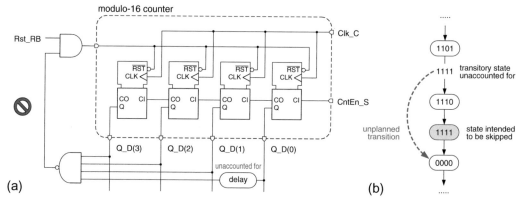

(a)   (b)

FIGURE 6.7

Unsafe modulo-15 counter subcircuit. Schematic (a) and state graph (b).

□

---

[23] The phenomenon can be recognized as a function hazard.

Observation 6.10. *The sole purpose of asynchronous (re)set inputs is to bring an entire circuit into its predefined start state. Do not gate them with information signals.*

It is interesting to study the motivations that lead people to expose themselves to the hazards of unsafe circuits such as the ones of figs 6.6 and 6.7. Two rationales are often heard:

- Some desired functionality was unavailable in a synchronous implementation from the target cell library. In the occurrence of the above modulo-15 counter, the designer flatly preferred to misuse the asynchronous reset instead of figuring out how to add a synchronous clear/load to an elementary D-type flip-flop. The need for an enable/disable mechanism, for conditional clocking, and for data transfers across clock boundaries are further situations that tend to expose designers to the temptations of asynchronous design. HDL synthesis certainly has helped designers to stay away from such practices.

- The asynchronous implementation was believed to result in a smaller circuit, faster speed, and/or better energy efficiency than a synchronous alternative. While this may be true in some cases, the contrary has been demonstrated in many others. Keep in mind that any redundant circuitry necessary to generate multiple auxiliary clocks, to stretch impulses, to suppress unwanted glitches, or to make sure local delay constraints are satisfied does not come for free either. In view of all the limitations cited in section 6.3.3, overall cost-effectiveness remains questionable to say the least.[24]

Bistables with both asynchronous reset and preset inputs prohibited

Some component and cell libraries include flip-flops that feature both an asynchronous set and an asynchronous reset. Synchronous design knows of no useful application for such subcircuits. Also, behavior is unpredictable and resembles that of a seesaw when both s and r are deasserted simultaneously.

Reset signals to be properly conditioned

Note, to begin with, that the total load controlled by the global reset is in the same order of magnitude as that driven by the circuit's clock signal. The same further applies to the scan mode signal. It typically takes a large buffer or a buffer tree to drive such nets with acceptable ramp times.

---

[24] You may want to consult section A.10 where more details on how to construct safe flip-flops with enable, with synchronous clear, with scan facility, and the like are given.

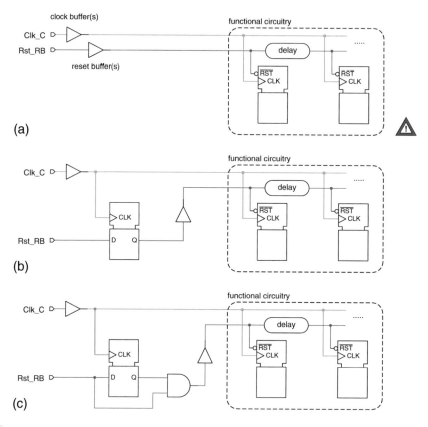

FIGURE 6.8

Conditioning of global asynchronous reset signal. Fully asynchronous (a), fully synchronized (b), and uncondi-tional application combined with synchronized removal (c).

A more specific peril associated with the usage of asynchronous (re)sets exists in applications with a **free-running clock**, that is where a reset impulse may occur at any time with respect to the driving clock, see fig.6.8a. If the circuit's asynchronous reset input gets deactivated near the active clock edge, then part of the bistables may be allowed to take up normal operation before the end of one clock cycle while others would remain locked until early in the next cycle. Going through the same efforts as with the clock distribution network to minimize **reset skew**, i.e. local delays differences within the reset distribution network, is a rather costly option.

Incorporating the auxiliary subcircuit depicted in fig.6.8b solves the problem by synchronizing the reset signal to the on-chip clock before it gets distributed to the functional bistables. All bistables are so guaranteed to get properly (re)set during the same clock cycle. Any transitions on the internal reset occur shortly after the active clock edge which leaves the designer with ample time, i.e. almost one clock period, for distributing the signal across the chip. As a side-effect, the circuit responds to the external reset signal with a latency of one clock cycle.

Yet, a new difficulty has been introduced because the reset operation now depends on the functioning of the clock subsystem. In the absence of a clock or in the presence of specific hardware faults, the circuit would fail to settle to a defined start state. Drive conflicts, static power dissipation, inconsistent output data, destructive overwrite of nonvolatile memories and other operating conditions unsafe for the surrounding system may be the undesirable consequences. In the design of fig.6.8c, a simple combinational bypass makes sure that a reset is immediately and unconditionally brought to bear. Deactivation, in contrast, takes place at well defined moments of time. An equivalent circuit alternative does away with the AND-gate by connecting the external reset input to a synchronizer flip-flop with an asynchronous reset terminal.

Keep in mind that all three designs impose conditions on the relative timing of Rst_RB and Clk_C. Violating them may cause marginal triggering in one or more of the flip-flops. Also remember that asynchronous (re)set inputs are very sensitive to glitches and noise.

Pay attention to portable design

IC designers time and again face the problem of porting a design from one implementation platform to another, e.g. when they

- Migrate a design from an FPGA to a cell-based ASIC, or vice versa,
- Upgrade to a more up-to-date target process or library,
- Incorporate a third-party circuit block into a larger design, or
- Accept a heritage design for integration.

While it is possible to fine-tune almost any design for some given technology, employing asynchronous techniques can be disastrous when it comes to porting a design to some other platform. Think ahead and design for portability in the first place.

Observation 6.11. *Accept that almost all VLSI designs are subject to porting during their lifetime and stick to the established rules of safe design and synchronous operation to render the porting smooth and cost-effective.*

Conversely, when being proposed an existing design, do not accept it unseen. Be prepared to distillate the necessary functionality and reimplement it using a clocking discipline that is compatible with VLSI or your target PLD. Heritage designs on the basis of standard parts such as microprocessors and LSI/MSI circuits notoriously include the worst examples of asynchronous design tricks. Also note that on-chip memories often are at the origin of portability problems because of the many varieties being offered.

6.5 CONCLUSIONS

- Among the three grand alternatives for clocking digital circuits, ad-hoc asynchronous operation has been found to be unsafe and inefficient in VLSI.

- While safe if implemented correctly, self-timed operation at the gate level entails an unacceptable overhead in terms of hardware and energy, and necessitates out-of-the normal design methodologies, software tools, and library cells.

- Synchronous operation of large system chunks does away with almost all timing problems, results in efficient circuits, and is compatible with today's design automation flows and cell libraries. There hardly is a better choice for VLSI, especially when there is high pressure on tight schedules and high design productivity.

- Synchronous circuits exhibit a strict dissociation of signals into
    - One clock signal (or possibly more if of fixed frequency and phase relationship),
    - One asynchronous reset signal (optional),
    - An arbitrary number of information signals.

- HDL synthesis does not dispense designers from deciding about clocking disciplines and clock domains as it is possible to express any clocking discipline in an RTL circuit model.

6.6 PROBLEMS

1. ✱✱ Fig.6.4c and d depict two feedback circuits each built from a few logic gates. Establish their respective truth tables and discuss your findings.

2. ✱ The circuit of fig.6.4c behaves much like a latch where D acts as data input and E as enable input. Explain why this construct does not qualify as a latch in synchronous designs.

3. ✱✱✱ Come up with a synchronous implementation for the modulo-15 counter of fig.6.7. Compare the relative sizes of the two alternatives based on the assumption that both counters are built from D-type flip-flops.

4. ✱✱ FireWire is the name of a serial bus for interconnecting computer and multimedia equipment. To facilitate the delimiting and recovering of individual bits from the data stream at the receiver end, data get conveyed one bit after the other coded using two peer signals (the fact that two differential signal pairs are actually being used does not matter in this context). The first signal termed "data" simply corresponds to the incoming data whereas its "companion" is to feature a transition at the boundary between any two adjacent bits iff the first signal does not. Thus, at either end of a FireWire link, a modulator circuit in the transmitter converts the incoming serial data stream into a data-plus-companion pair, while a demodulator on the receiving side is in charge of recovering the initial data from that signal pair. Design both a modulating and a demodulating circuit. What class is the automation you must use?

   ✱✱✱ Assuming that receiver and transmitter run from separate clocks is definitely more realistic but requires prior exposure to material from section 8.2 and also makes solving the problem more demanding.

6.7 APPENDIX: ON IDENTIFYING SIGNALS

> *with contributions by too many of the author's colleagues to be named here*

Many mistakes in hardware design can be traced down to minor oversights in specifications, datasheets, truth tables, HDL code, schematics, and the like. Misinterpretations are particularly likely to occur at the interface between different subsystems, subcircuits and clock domains because correct operation implies mutual agreement on the meaning behind data formats, signal waveforms, and transmission protocols. What is needed is a clear and unambiguous, yet simple, scheme for naming signals and for drawing diagrams, especially when working in a team. A helpful notational convention must keep track of

- Signal class,
- Active level,
- Signaling waveform,
- Three-state capability,
- Input, output or bidirectional,
- Present state vs. next state, and
- Clock domain.

6.7.1 SIGNAL CLASS

A total of six signal subclasses have been identified in section 6.4 and catalogued in table 6.2. The same table suggests to append one of the characters below to make a signals's role evident from its name. Ultimately, this will lead to a syntax detailed in section 6.7.7.

- R asynchronous (re)set,
- A any other signal subject to asynchronous switching,
- C clock,
- S status or control,
- D data, address or the like,
- T test.

Colors in schematic diagrams and HDL source code

Wires colored in a meaningful way expedite the understanding of large schematic diagrams and render many potential problems immediately visible.[25] Let us cite five examples on the basis of the coloring scheme of table 6.2:

- Any clock signal (green) routed through a logic gate or other combinational subcircuit with two or more inputs draws the attention to a potentially unsafe clock gating practice.
- A register with neither an asynchronous (re)set wire (red) nor a synchronous clear (blue) attached indicates the subcircuit is in need of a special homing sequence for initialization.

---

[25] The coloring scheme proposed in fig.6.2 is intended for usage in schematic diagrams that are drawn on light backgrounds. On dark backgrounds, white must be substituted for black. Also, swapping blue and yellow does re-establish the original idea of fading all test-related signals as they tend to distract from a circuit's functionality.

- Any combinational logic that drives an asynchronous (re)set line (red) indicates (re)sets are exposed to hazards as a consequence from being misused for functional purposes.
- Any signal emanating from a foreign clock domain (orange) that drives combinational logic points to a lack of synchronization.
- The absence of any test signals (yellow) makes it obvious that no test structures have been incorporated in a functional block.

Coloring signal names in HDL source code facilitates the interpretation of hardware models in much the same way as colored wires used to do in schematics. The Emacs editor has been extended by a special VHDL mode that checks and supports syntactical correctness [173]. Automatic coloring of signal names further assists authors and readers of code.

Clock symbols and clock domains in schematic diagrams

It is good engineering practice to identify the clock inputs by way of graphical symbols attached to icons, see fig.6.9a,b,c. Standard single-edge-triggering clocks are marked by a small triangle, the more unusual double-edge-triggering clocks by two dovetailed triangles, and level-sensitive clocks by a small rectangle. This rule is by no means restricted to gate-level diagrams as its application is beneficial to the clarity of schematics at any level of hierarchy.

Any signal that crosses over from one clock domain to another deserves particular attention as it may switch at any time irrespective of the receiver's clock. Making clock domains explicit in block diagrams and schematics is very important. All subcircuits that belong to one clock domain shall be placed closely together and all clock boundaries shall be made evident.

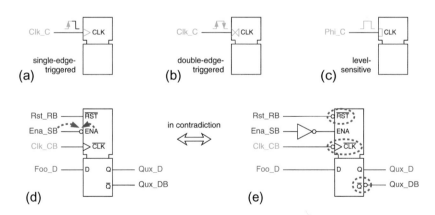

FIGURE 6.9

Notational conventions in schematic diagrams. Identification of clock inputs (a,b,c) and two competing mindsets for reflecting signal polarities (d,e).

6.7.2 ACTIVE LEVEL

Naming of complementary signals

As active-high and active-low signals coexist within the same circuit, it is important to make it clear in the identifier whether a given signal is to be understood in positive or in negative logic. Overlining the name is how negated terms are identified in mathematics, e.g. if Qux stands for an active-high signal, then its active-low counterpart is named $\overline{\text{Qux}}$ (pronounced "qux bar").

Since this notation does not yield character strings acceptable for processing with computers, many engineers prefix or postfix the basic identifier and resort to something like ~Qux, /Qux, or Qux$. However, the problem with most non-alphanumerical characters is that today's EDA tools, HDLs, and standards disagree on which such symbols they support within node names and on where they are allowed to occur in the name's character string.

For reasons of cross-platform compatibility, we prefer to use the attribute B (for "bar") in conjunction with the conventions to be given in section 6.7.7. In the occurrence, we do write Qux_B as shown in fig.6.9d and e. The popular practice of appending a simple B or N is not recommended because it may give rise to confusion in names such as Selb or OupEn.

The same argument also holds for conditions and actions related to edges rather than to logic levels. By default, the rising edge is considered to be the active one for a non-inverted signal.

Inversion symbols in schematic diagrams

Signal polarities must also become clear from schematics and icons. In fact, cell icons often include small circles that are supposed to stand as a shorthand notation for regular gate-type inverter symbols. One would thus naturally expect to find such an inversion circle wherever a signal gets translated from active-high to active-low or vice versa, and wherever polarity changes from rising-edge- to falling-edge-triggering or back. This is the mindset behind fig.6.9d.

Most industrial cell libraries follow the notation depicted in fig.6.9e, though. Please observe that signals Rst_RB and Clk_CB are <u>not</u> subject to any negation upon entering the flip-flop here, the small circles just reconfirm that the cell's inputs $\overline{\text{RST}}$ and $\overline{\text{CLK}}$ are active-low and falling-edge-triggered respectively. Much the same observation applies to output $\overline{\text{Q}}$. The reason for this departure from mathematical rigor is that a cell can so keep the same graphical icon independently from the context, a quality not shared by the notational convention of fig.6.9d.

> *Hint: As there are no generally-adhered-to standards, designers are well advised to double-check the exact meaning of signal identifiers, terminal names, and inversion circles when working with third-party components, cell libraries, or HDL code.*

6.7.3 SIGNALING WAVEFORMS

In distinguishing between active-high and active-low signals, we have tacitly assumed that it was a signal's level that conveyed information from a transmitter to the receiver(s). While this is often true, this is not the only way to code information into a signal waveform. Fig.6.10 shows a data signal along with three different interpretations of its waveform.[26]

Clock-qualified signaling means that the logic level, 1 or 0, of the signal at the instant of each active clock edge is what matters. Key characteristics of this scheme are as follows.

- Qualifier-type signals are always meant relative to a specific clock.
- They must remain stable throughout the data-call window (i.e. setup-hold interval) of the receiving circuit.
- Each logic 1 gets counted as a relevant event of its own even if the signal does not change in between (and vice versa for an active-low signal).
- Hazards between active clock edges have no effect.

Examples: The enable input of a synchronous counter, the write line of a synchronous RAM, the zero flag from an ALU when followed by a flip-flop.

Impulse signaling relies on the presence of an impulse (mark or pause) which implies that

- Each rising edge <u>or</u>, alternatively, each falling edge is counted as a relevant event.
- The signal must return to its initial, passive level before it can possibly become active for a second time.
- The durations of marks and pauses are immaterial.
- Static and dynamic hazards are unacceptable as they would cause erroneous registration of events where there are none.

Examples: Write line of an asynchronous RAM, request and acknowledge signals in a RZ (four-phase) handshake protocol.

Transition signaling does the signaling with the presence of an edge which means that

- Each rising <u>and</u> falling edge is counted as a relevant event.
- The signal must not restore its initial value between two consecutive events.
- The durations of marks and pauses are immaterial.
- Static and dynamic hazards are unacceptable as they would cause erroneous registration of events where there are none.

Example: Request and acknowledge signals in an NRZ (two-phase) handshake protocol.

---

[26] Three is by no means exhaustive. Calling upon pairs of complementary signals to improve on noise immunity (differential signaling) comes as a natural extension. Some schemes map data onto waveforms such as to allow for safe recovery of individual bits from a serial data stream at the receiving end (self-clocking), whereas others transmit two peer signals — none of which can be interpreted as a clock — for the same purpose (companion signaling). The FireWire bus, for instance, uses four wires and a waveform referred to as **non-return to zero with data strobe** (NRZ-DS) which is a combination of differential with companion signaling. We refrain from discussing such schemes here as they are mainly intended for communicating between more distant subsystems.

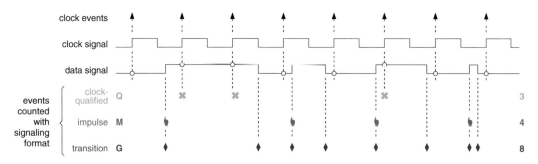

FIGURE 6.10

Three signaling waveforms in common use.

Note from fig.6.10 that the event counts and the instants of registration differ as a function of the signaling scheme. It is thus absolutely essential that transmitter and receiver agree on this if a system is to function as intended. This is not normally a problem within a clock domain where information signals are clock-qualified by nature. For simplicity, we can omit an attribute in this case. However, care must be exercised where a signal traverses the boundary of a subsystem or clock domain. The signaling waveform shall thus be stated as follows:

- o Q clock-qualified signaling,
- o M impulse signaling (for "mark"),
- o G transition signaling (for "grade").

6.7.4 THREE-STATE CAPABILITY

Signals with high-impedance capability shall be identified by a z character. This is to remind designers of the various issues associated with a three-state circuit node such as resolution function, drive conflicts, floating potential, and testability.

6.7.5 INPUTS, OUTPUTS AND BIDIRECTIONAL PORTS

HDL code is easier to understand when inputs and outputs are immediately discernible from an appended I or O.[27] IO then obviously stands for a bidirectional signal. The same information is often conveyed by way of graphical symbols in schematic diagrams, see fig.6.11.

---

[27] As an added benefit, this also circumvents a quirk of VHDL in that the language disallows reading back from an output terminal. That is, an interface signal declared as out in the port clause of some entity declaration is neither permitted on the right-hand side of a signal assignment in the pertaining architecture body nor in a relation operation (such as =, /=, <, etc.). Distinguishing between two signals does away with the problem: while Foo_D can be freely used within the architecture body, a copy named Foo_DO is to drive the output port.

Some programmers prefer to declare the port to be of mode inout instead, yet, this is nothing else than a bad habit. After all, what one wants to model from an electrical point of view is just an output and not a bidirectional. No HDL language restriction should make us obscure our honest intentions.

FIGURE 6.11

Ports and other terminals in schematic diagrams.

Observation 6.12. *Hint to SystemVerilog users:* I *and* O *suffixes are incompatible with useful features of the language such as dot name and dot star port connections,* interface, *and* modport. *Designers may elect to use* io_mode *suffixes only where unproblematic or to omit them altogether.*

6.7.6 PRESENT STATE VS. NEXT STATE

Most state machines models operate with two signals for the state variable, namely a first one that holds its present value and a second one that predicts the value it is going to assume in response to the next active clock edge. It is imperative not only to identify such signals as being closely related, but also to make it clear which is which. While this is immediately obvious in schematics, identifiers are the sole means of conveying this kind of information in HDL models. We, therefore, suggest to tell present and next state apart with the aid of suffixes P and N.

6.7.7 SIGNAL NAMING CONVENTION SYNTAX

We have now compiled a number of attributes for making a signal's nature evident. A practical problem is how to permanently attach those attributes to a signal. About the only way that can be expected to work across different EDA platforms consists in blending them right into the signal's identifier. To make a long story short, the underscore _ is the only non-alphanumerical character that universally qualifies for affixing a signal's attribute to its name.[28] This results in identifiers such as Foo_DB for an active-low data signal, for instance. In Backus Naur form:

```
signal_identifier    ::= signal_name "_" signal_attributes
signal_name          ::= letter { letter | digit }
signal_attributes    ::= class_char [ signal_waveform ] state_char
                         three-state_mode active-low_char io_mode
class_char           ::= "R" | "A" | "C" | "S" | "D" | "T"
signal_waveform      ::= "Q" | "M" | "G"
```

---

[28] The IEEE 1076.4 VITAL standard reserves the underscore as a delimiter in generics, e.g. in an identifier such as tpd_Foo_Bar for the propagation delay from an input Foo to some output Bar. As a consequence, VITAL models must not contain any underscore in a port name. While this seems to prevent the usage of the underscore as a separator between signal name and signal attribute, this is actually not so for the majority of people who are just simulating with VITAL models rather than writing their own ones. This is because any underscore in a user-defined port name always appears in the actual part of the port maps of VHDL component instantiation statements, and never in the formal part subject to VITAL ruling. The aforementioned reservations relating to the underscore character do not apply in this case. Together with the modest adoption of VITAL, this lead us to relax our naming conventions over those published in an earlier version of this text [146].

```
state_char          ::= "N" | "P" | ""
three-state_mode    ::= "Z" | ""
active-low_char     ::= "B" | ""
io_mode             ::= "I" | "O" | "IO" | ""
```

Although no attribute character has been assigned more than one interpretation, we recommend to order attribute characters as stated above and, more particularly, to always make the mandatory `class_char` come first in the `signal_attributes` substring.

Examples

`Foo_C` identifies a clock signal.

`Bar_D` indicates this is a regular data signal within some clock domain. As opposed to this, the name `Bar_A` refers to the same signal before being synchronized to the local clock and, hence, subject to toggle at any time.

`Qux_RB` refers to some asynchronous active-low reset signal.

`AddrCnt_SN` and `addrcnt_SN` are legal names for the next state of an address counter as long as the signal remains within the clock domain where it originates.

`Irq_AMI` identifies an interrupt request input that emanates from a foreign clock domain and that is meant to be stored in a positive-edge triggered flip-flop until it is being serviced.

`Carry_DB` denotes an active-low carry signal. Where a duplicate signal is required for the sole purpose of driving an output port, it shall be named `Carry_DBO`.

`GeigerCnt_DZO` reflects an output-only data signal with high-impedance capability.

`ScanMode_T` stands for an active-high scan enable signal within some clock domain.

□

For any programmer, it comes most naturally that each variable in a software module must be given its own unique identifier, and the same holds true for HDL code. It is important that the same principle also be enforced in schematic drawings and netlists because any two nodes that carry identical names in the same circuit block are considered by EDA tools as **connected by name**, i.e. as one and the same circuit node. Connections by name often help to render complex schematic diagrams easier to read.

> *Hint: While you are encouraged to take advantage of connections by name to avoid cluttering your schematic diagrams with clocks, resets, and other signals that connect to a multitude of subcircuits, beware of shorts that will arise whenever a signal name gets reused for two electrically distinct nodes. Inadvertently or consciously doing so ("... but logically they are the same ...") is a typical beginner's mistake.*

6.7.8 USAGE OF UPPER AND LOWER CASE LETTERS IN HDL SOURCE CODE

Basically, VHDL is a case-insensitive language, whereas SystemVerilog is case-sensitive. While `FooBar_DB`, `foobar_DB`, `FOOBAR_DB` and `foobar_db` are all the same in VHDL, they are read as distinct identifiers by a SystemVerilog compiler. As some EDA programs have the side effect of mapping all lower case letters to upper case ones — or vice versa — it is not a good idea to use casing to distinguish between otherwise identical names.

For the sake of legibility, we nevertheless encourage HDL programmers to write their source code in the following way:

- Reserved words: lower case throughout, e.g. `port`, `case`, `signed`.
- Subprograms, i.e. functions, procedures, tasks: lower case throughout, e.g. `foobar`.
- Constants, generics/parameters: upper case throughout, e.g. `FOOBAR` or `FOO_BAR`.
- VHDL variables: PascalCaps or camelCaps, e.g. `FooBar` or `fooBar`.
- VHDL signals: as above with suffix added, e.g. `FooBar_DB` or `fooBar_DB`.
- SystemVerilog variables: PascalCaps or camelCaps, e.g. `FooBar` or `fooBar`, optional suffix for variables that convey information between concurrent processes.[29]

Warning: Do not use case variations of any reserved word of your HDL for user-defined identifiers as this will give rise to confusion at some point, if not with EDA tools, then with human readers. E.g. neither use `WAIT` *nor* `wait` *for a state or signal.*

6.7.9 A NOTE ON THE PORTABILITY OF NAMES ACROSS EDA PLATFORMS

Porting netlists and schematics from one EDA platform to another is often painful as each has his own rules for the naming of circuit nodes. Translation steps are inevitable unless one finds a common subset accepted by all tools involved. Here are a few items to watch out for.

Non-alphanumerical characters. Most EDA data interchange formats have specific rules as to what special characters they accept as part of a node name and in what position. The underscore is about the only non-alphanumerical character tolerated in node names across many formats and EDA tools. The naming conventions presented above have been devised with this in mind and, therefore, produce widely accepted signal and node names.

Hierarchy delimiters. Slash /, period mark ., and dollar sign $ are commonly being used. For the sake of unlimited cross-platform portability, we recommend to spell node names with no special characters other than the underscore.

Busses. One finds brackets [ ], parentheses ( ), and curly brackets { } in combination with a colon :, a `to`, or a `downto` for indicating the index range of a vector.

Aliases. Although typically legal, multiple names for the same node should be avoided because many software tools are not able to handle them correctly and because of the confusion this practice tends to create with humans.

---

[29] Take note of hint in section 6.7.5.

CLOCKING OF SYNCHRONOUS CIRCUITS

7.1 WHAT IS THE DIFFICULTY WITH CLOCK DISTRIBUTION?

Up to this point, we have ignored the difficulties of distributing a clock signal over a chip or a major portion thereof. We were in good company as systems engineering, automata theory, and other theoretical underpinnings of digital design assume simultaneous updating of state throughout a circuit. Physical reality is different from such abstractions, though.

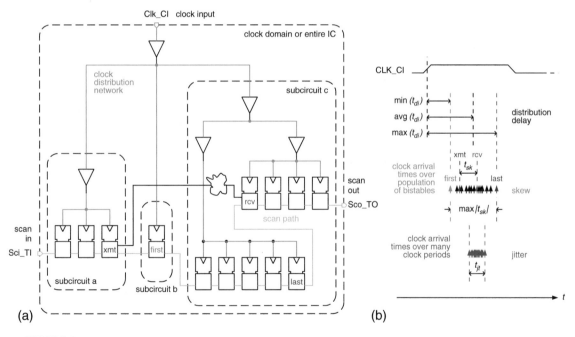

FIGURE 7.1

Clock distribution. Clock domain with clock distribution network, scan path, and just one combinational propagation path shown (a), relevant timing quantities (b).

Consider a population of flip-flops or other clocked subcircuits that make part of one clock domain in a synchronous design as shown in fig.7.1. A common clock tells them when to transit to the next state. Ideally, all such bistables are supposed to react to the clock instantly and all at exactly the same moment of time.

In practice, however, switching will be retarded due to many small delays inflicted by drivers and wires in the **clock distribution network**. As most clock signals connect to a multitude of storage elements spread out over an entire clock domain, individual switching times will differ because delays along the various clock propagation paths are not quite the same. This scattering over time is loosely referred to as **clock skew**. To make things worse, those delays will slightly vary from one clock cycle to the next thereby giving rise to **clock jitter**.

Many causes contribute to the timewise scattering of clocks:

- Unevenly distributed fanouts and load capacitances.
- Unequal numbers of buffers and/or inverters along different branches.
- Unlike drive strengths and timing characteristics of the clock buffers instantiated.
- Unbalanced interconnect delays due to dissimilar layout parasitics
  (R: wire length and thickness, via count; C: plate, fringe and lateral capacitance).
- Unequal switching thresholds of bistables (translate clock ramps into staggered switching).
- Process, temperature, voltage (PTV) and — more so — on-chip variations (OCV).
- Supply noise as caused by ground bounce and supply droop.
- Crosstalk from switching activities in the surrounding circuitry.

Excessive scattering of switching events must obviously compromise the correct functioning of a digital circuit. Design engineers thus try hard to eliminate any systematic disparities among clock arrival times.[1] However, depending on fabrication depth and design level (board, field-programmable logic, semi-custom IC, full-custom IC, hand layout), they are never able to control all of the underlying phenomena, so that a certain amount of unevenness remains.

7.1.1 AGENDA

Designing dependable circuits in spite of clock skew and jitter involves two issues, namely
- **(a)** knowing and lowering the vulnerability of a design, and
- **(b)** minimizing scattering by distributing clock signals over a domain in an adequate way.

The two issues will be discussed in sections 7.2 and 7.3 respectively. For simplicity, we will focus on signals that circulate within one clock domain, thereby dropping any input and output signals from our analysis in section 7.2. Synchronous I/O is addressed in section 7.4, whereas the problems associated with assimilating data that arrive asynchronously are postponed to chapter 8. How to safely implement clock gating is the subject of section 7.5.

---

[1] Experienced designers sometimes introduce clock skew on purpose either to accommodate RAMs and other subcircuits with larger-than-normal setup/hold times or to allow for faster clocking by adapting to uneven path delays. A better tolerance with respect to delay variations may also be sought in this way, see problem 2. The process of tuning a clock distribution network to local timing requirements is termed **clock skew scheduling**, aka useful skew. Please refer to problem 8 and to the specialized literature [174] [175] [176] for more details on this optimization technique.

7.1.2 TIMING QUANTITIES RELATED TO CLOCK DISTRIBUTION

In order to study the clocking of synchronous circuits, we need to introduce three timing parameters that specifically relate to clock distribution networks, see fig.7.1 for an illustration.

> t_{di} **Clock distribution delay**. The time lag measured from when a clock edge appears at the clock source until a state transition actually takes place in response to that edge. When referring to an IC, some package pin is normally meant to act as clock source.

> t_{sk} **Clock skew**. The inaccuracy of the same clock edge arriving at different locations within a given clock domain. There is a local and a more global view:

> In a narrow sense, skew refers to the clock terminals of two subcircuits connected by some signal propagation path. Skew is considered positive if the receiver is clocked after the transmitter, and negative if it is clocked before, i.e. $t_{sk} = t_{di\,rcv} - t_{di\,xmt}$.[2]
> From a wider perspective, one is interested in knowing the largest difference in clock arrival times between any two clock terminals within a clock domain. The term clock skew then takes on the meaning of overall skew and is defined as $\max|t_{sk}| = \max(t_{di}) - \min(t_{di})$.

> t_{jt} **Clock jitter**. The variability of consecutive clock edges arriving at the same location.

---

[2] The sign of clock skew is controversial, some authors have elected to define clock skew the other way round as $t_{sk} = t_{di\,xmt} - t_{di\,rcv}$. We prefer to have data and clock delays share the same sense of counting.

7.2 HOW MUCH SKEW AND JITTER DOES A CIRCUIT TOLERATE?

7.2.1 BASICS

Numerous schemes for driving synchronous digital circuits have been devised over the years. Some of them are more vulnerable to clock skew and jitter than others, each asks for different hardware resources, some have an impact on performance, and this section aims at comparing them. Note that while selecting a clocking discipline typically amounts to finding an optimum choice between conflicting goals, it is not concerned with functionality as any decent functionality can be combined with any decent clocking scheme.

Before entering the discussion of individual clocking schemes, let us recall from observation 6.1 that any digital signal must be allowed to settle to a valid state before it is accepted into a memorizing subcircuit. Correct and dependable circuit operation is otherwise not possible.

The time interval during which data at the input of a flip-flop, latch, RAM or other sequential subcircuit must remain valid is referred to as **data-call window**, aka aperture time, and is defined by the setup and hold times there. The term **data-valid window** designates the time interval during which data at the receiver actually do remain valid and is dependent on a circuit's propagation and contamination delays. As becomes immediately clear when comparing figs.7.2 and 7.3, the fundamental requirement simply is

Observation 7.1. *Transferring data between two subcircuits requires that the data-call window of the receiving subcircuit be fully encompassed by the transmitter's data-valid window.*

In subsections 7.2.2 through 7.2.7, each clocking discipline will be introduced by outlining its operation and elementary characteristics. Then follows a more detailed analysis which establishes conditions for how much skew can be tolerated without causing the circuit to malfunction or to exhibit undeterministic operation. The resulting inequalities will be termed **skew margins** and are visualized in the Anceau diagram of fig.7.2. While noise margins delimit the safe operating range of a digital circuit with respect to uncertainties in amplitude, skew margins do the same with respect to timewise uncertainties.

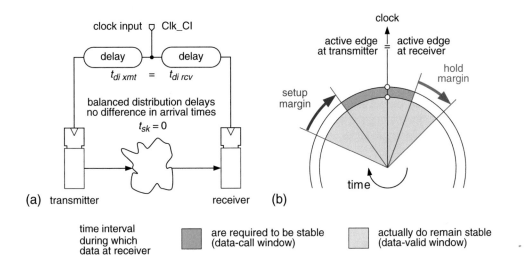

FIGURE 7.2
Clocking in the absence of skew and jitter. Circuit (a) and Anceau diagram (b).

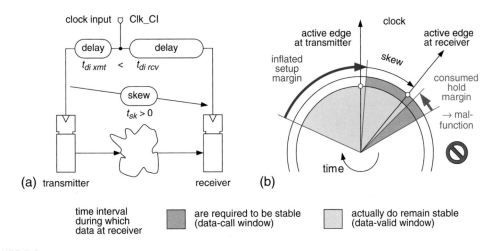

FIGURE 7.3
Impact of excessive positive clock skew. Circuit (a) and Anceau diagram (b).

Observation 7.2. *A certain amount of clock scattering is unavoidable. What really counts is that clocking discipline and clock distribution network are chosen such that the setup and hold margins they provide can absorb the combined effects of skew and jitter*
- *everywhere within a clock domain, and*
- *under any operating condition.*[3]

Although there are more clocking disciplines than those discussed here, it should be noted that any other scheme can be analyzed along the same lines. Also, the study is easily extended to include interconnect delays by adding appropriate terms to $t_{cd\,c}$ and $t_{pd\,c}$.

7.2.2 SINGLE-EDGE-TRIGGERED ONE-PHASE CLOCKING

Hardware resources and operation principle

Single-edge-triggered one-phase clocking is the most natural approach from a background in automata theory or abstract systems design. Registers are implemented from flip-flops and <u>all of them</u> get triggered by the same clock edge, henceforth termed the **active edge**. No bistables other than ordinary **single-edge-triggered flip-flops** (SETFF) are being used.

Each computation cycle starts immediately after an active clock edge and ends with the subsequent one so that $T_{cp} = T_{clk}$. All transient phenomena must die out before the active clock edge, that is within one clock period. The exact moment of occurrence of the passive clock edge is immaterial as long as the clock waveform meets the minimum pulse widths requirements $t_{clk\,hi\,min}$ and $t_{clk\,lo\,min}$ imposed by the flip-flops.

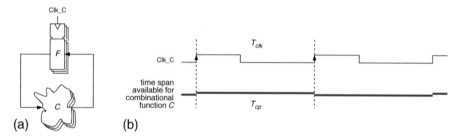

FIGURE 7.4

Edge-triggered one-phase clocking (with the rising edge being the active one). Basic hardware organization (a) and simplified timing diagram (b).

---

[3] Operating conditions are meant to include data patterns, loads, clock frequency, parameter variations (PTV and OCV), ground bounce/supply droop, and crosstalk.

Detailed analysis

Starting from setup and hold requirements of flip-flops, let us now find out how much clock skew and jitter can be tolerated without exposing a circuit to timing problems. The corresponding Anceau diagram is shown in fig.7.5.

Setup condition

Consider a pair of flip-flops with a unidirectional data propagation path in between as shown in fig.7.2. The setup condition of the receiving flip-flop is expressed as

$$t_{di\,xmt} + t_{pd\,ff\,xmt} + t_{pd\,c} \leq T_{clk} + t_{di\,rcv} - t_{su\,ff\,rcv} \tag{7.1}$$

which makes reference to timewise uncertainty of the clock after being recast into

$$t_{pd\,ff\,xmt} + t_{pd\,c} + t_{su\,ff\,rcv} \leq T_{clk} + (t_{di\,rcv} - t_{di\,xmt}) = T_{clk} + t_{sk} \tag{7.2}$$

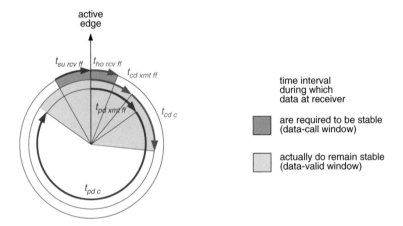

FIGURE 7.5

Anceau diagram of edge-triggered one-phase system. Note that arcs of different colors are being used to distinguish between timing quantities that relate to the setup condition and others that matter for the hold condition.

In the absence of skew and jitter, the inequality stipulates a minimum clock period. It also appears — in this limited context — that positive skew has a beneficial effect and negative skew a detrimental one. This is because a lag of the receiver clock with respect to that of the transmitter facilitates meeting the setup condition, while the opposite is true for a lead. Relation (7.2) thus effectively imposes a lower — and typically negative — bound on clock skew, which becomes even more obvious when the inequality is transformed into

$$t_{sk} \geq (t_{pd\,ff\,xmt} + t_{pd\,c} + t_{su\,ff\,rcv}) - T_{clk} \underset{typ.}{<} 0 \tag{7.3}$$

For the continuation of our analysis, we will work with the equivalent form

$$-t_{sk} \leq T_{clk} - (t_{pd\,ff\,xmt} + t_{pd\,c} + t_{su\,ff\,rcv}) \tag{7.4}$$

From a broader perspective, one has to consider the ensemble of all flip-flops in a circuit which asks for a separate inequality for any two communicating bistables. A static **timing analyzer** essentially is a software tool that takes a gate-level netlist, that checks whether inequalities (7.4) and (7.8) are met, and that flags violations, if any. To that end, the software begins by calculating the numeric delay figures along each signal propagation path in a circuit.

Rather than checking thousands of mathematical relations, however, humans prefer to come up with one simple worst-case condition which, when satisfied, guarantees that setup times are respected throughout an entire clock domain. Skew plus jitter here appear as the maximum over all signal propagation paths and as a magnitude $\max|t_{sk}|$. After all, a practical clocking scheme must accommodate data transfers between any two flip-flops, explicitly allowing for reciprocal data exchange which renders the distinction between positive and negative skew meaningless.

$$\max|t_{sk}| \leq T_{clk} - \max(t_{pdff} + t_{pdc} + t_{suff}) = T_{clk} - t_{lp} \tag{7.5}$$

The expression $\max(t_{pdff} + t_{pdc} + t_{suff})$ reflects the delay along the longest signal propagation path between any two adjacent flip-flops in the circuit. For the sake of brevity, we will refer to this path together with its delay t_{lp} as the **longest path**.[4]

Inequality (7.5) indicates that longest path, clock skew and jitter together bound the minimum admissible clock period from below and so define how fast the circuit can be safely clocked. This finding puts us in a position to estimate the performance of a given circuit more accurately than in chapter 3 and gives rise to a first observation.

$$T_{clk} \geq \max(t_{pdff} + t_{pdc} + t_{suff}) + \max|t_{sk}| = t_{lp} + \max|t_{sk}| =$$
$$\approx t_{idff} + \max(t_c) + \max|t_{sk}| \tag{7.6}$$

Observation 7.3. *Clock skew is at the expense of maximum performance in circuits that operate with edge-triggered one-phase clocking.*

In a pipeline, for instance, $t_{ff} + \max|t_{sk}|$ is nothing else but time unavailable for payload computations which is why this quantity is often referred to as **timing overhead**. Any positive difference between the left-hand and the right-hand side of (7.6) implies that the clock period is not fully utilized by computations and timing overhead. The surplus amount of time is termed **slack** and routinely calculated during static timing analysis. While slack must always remain positive, designers strive to minimize it when in search of maximum performance.

Example

A Sun UltraSPARC-III CPU implemented in 250 nm 6M1P CMOS runs from a 600 MHz clock which amounts to $T_{clk} = 1.67$ ns. Some 70% of this time is available for combinational data processing and suffices to accommodate approximately eight consecutive levels of logic, assuming that a 3-input NAND with a fanout of three is representative for a typical gate delay. The rest is taken up by the registers and the necessary allowance for clock skew and jitter [177].
□

---

[4] The term **critical path** is often used as a synonym for longest path. We prefer to understand critical path as a generic term for the longest <u>and</u> the shortest path, however, because meeting the timing conditions along the shortest path is as important for the correct functioning of a circuit as it is along the longest path.

Hold condition

Starting from the hold requirement for data travel between two flip-flops,

$$t_{di\,xmt} + t_{cd\,ff\,xmt} + t_{cd\,c} \geq t_{di\,rcv} + t_{ho\,ff\,rcv} \qquad (7.7)$$

a second condition is obtained that bounds the acceptable skew and jitter from above this time because it is positive clock delay that puts the hold condition at risk.

$$t_{sk} = (t_{di\,rcv} - t_{di\,xmt}) \leq t_{cd\,ff\,xmt} + t_{cd\,c} - t_{ho\,ff\,rcv} \qquad (7.8)$$

Again from a circuit perspective one finds

$$\max|t_{sk}| \leq \min(t_{cd\,ff} + t_{cd\,c} - t_{ho\,ff}) = t_{sp} \qquad (7.9)$$

Here the right-hand side $\min(t_{cd\,ff} + t_{cd\,c} - t_{ho\,ff})$ stands for the **shortest path** t_{sp} through a circuit which is sometimes being referred to as "race path". As opposed to the situation found for the setup margin (7.5), the hold margin (7.9) either holds or not, no matter how much the clock period is being stretched. As both conditions must be met, the latitude to clock skew does not depend on the speed at which the circuit is operated. Another interesting observation is that any combinational logic placed between two adjacent flip-flops facilitates meeting (7.9) because of its inherent contamination delay. The same also applies for slow interconnect lines.

Implications

The most precarious circumstances occur when no combinational logic — and thus no contamination delay — is present between two consecutive flip-flops or other edge-triggered storage elements. Consider a shift register, for instance. The skew margin then collapses to

$$\max|t_{sk}| \leq \min(t_{cd\,ff} - t_{ho\,ff}) \qquad (7.10)$$

which also indicates there is no way to improve the situation by adjusting clock waveforms. Whenever an IC suffers from insufficient hold margins, a painful and costly redesign is due.

Table 7.1   Timing problems and their remedies in edge-triggered one-phase circuits.

Remedies for timing problems	if identified during the design process	if found once prototypes have been manufactured
on long path(s)	redesign circuit such as to reduce its max. prop. delay (may be hard)	extend clock period or renegotiate PTV conditions (most likely unacceptable)
on short path(s)	insert delay buffers between consecutive flip-flops (comparatively easy)	adjusting clock waveform or PTV does not help (fatal)

In practice, there is a deplorable difficulty with giving numerical figures for the skew margins defined by (7.9) and (7.10) because most datasheets lack indications on contamination delay.[5] Yet, as the inequality $0 \leq t_{cd} < t_{pd}$ closely bounds it from both sides, it is safe to state that $t_{cd\,ff} - t_{ho\,ff}$ always is a very small quantity. Since any differences in clock arrival times are at the expense of this tiny safety margin, edge-triggered one-phase systems must be considered inherently sensitive to clock skew. Note that interconnect delay has a beneficial effect, however.

Example

Table 7.2 is an excerpt from the datasheet of a CMOS flip-flop.[6] The maximum admissible clock skew between any two such flip-flops where the Q output of one cell directly connects to the D input of the next is $124\,\mathrm{ps} - (-14\,\mathrm{ps}) = 138\,\mathrm{ps}$. Please keep in mind this is just an estimate that assumes identical MOSFETs and PTV conditions throughout. The beneficial impact of interconnect delay is also ignored, on the other hand.

Table 7.2   Timing characteristics of a standard cell flip-flop in a 130 nm CMOS technology.

D Flip-flop with reset 1x drive propagation delay for fanout 2 @ typ. process, 25 °C, 2.5V/2.5V	Timing parameter				Max. toggle frequency [GHz]
	$t_{pd\,ff}$ [ps]	$t_{cd\,ff}$ [ps]	$t_{su\,ff}$ [ps]	$t_{ho\,ff}$ [ps]	
from a high-density library	160	124	70	−14	4.35

□

Observation 7.4. *Matching of clock distribution delays and careful timing analysis are critical when designing circuits and systems with edge-triggered one-phase clocking. Shift registers and scan paths are especially vulnerable to (positive) clock skew.*

Scan-type testing requires the presence of shift registers

A scan path is a feature included in almost all VLSI designs to make the circuit testable from outside. It essentially permits to control and observe circuit nodes hidden deep inside the circuit and works by chaining the existing flip-flops into a shift register while in test mode. Fig.7.6 shows the basic arrangement. What matters here is that a scan path includes no logic between flip-flops because the necessary input multiplexers are part of the standard cells.

---

[5] The reasons for this are given in appendix A.6.

[6] As the official datasheets give no indications for contamination delay, the number included in the table is an educated guess obtained from the computer models prepared for simulation and timing analysis. The so-called **timing library format** (TLF) describes a cell's overall delay as a sum of an initial inertia followed by output ramping. Both inertial delay and output ramp time are modeled as functions of switching direction (rise, fall), capacitive load, and input ramp time. The contamination delay figure of table 7.2 is the minimum inertial delay for a cell characterized with no external load attached and for clock ramp time zero, or almost so.

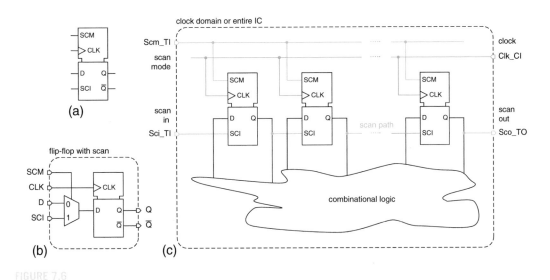

Scan test structures. Scan flip-flop (b), icon (a), and overall circuit structure (c).

Scan paths typically traverse many subcircuits and are, therefore, particularly exposed to clock skew at the boundaries in between. In the example of fig.7.1a, hold time violations are most likely to occur in the first flip-flop of subcircuit c.[7] When suffering from timing violations in scan mode, a chip will have to be scrapped even if otherwise fully functional because of the inability to complete the mandatory testing procedures.

Watch out when mixing cells from different libraries

Our previous estimates were based on numerical data from table 7.2. Yet, real circuits are being assembled from flip-flops of many different types which further puts hold margins at risk.

A key idea behind the concept of a **logic family** — whether available as a cell library or as a set of physical SSI/MSI components — is the capability to liberally combine components with others from the same family in spite of fabrication tolerances. Timingwise, this implies that the shortest contamination delay must be greater or equal to the longest hold time over all edge-triggered storage devices.

$$\max(t_{hoff}) \leq \min(t_{cdff}) \tag{7.11}$$

Note that no margin whatsoever is left when this relation holds with equality. Only flip-flops with hold time zero warrant free interchange, **negative hold times** are even safer. High-quality standard cell libraries are usually designed along this line, yet beware of exceptions that do not meet the requirement of (7.11).

---

[7] Scan flip-flops sometimes feature a separate scan out terminal SCO that differs from the ordinary Q output by passing through two extra inverters. We now understand why the dilated contamination delay t_{cdff} so obtained is very welcome. Similarly, automatic scan insertion tools can splice in so-called **lookup latches** at the boundaries between major subcircuits to bolster up hold margin. We consider this option as a workaround, however.

Observation 7.5. *Timing is particularly at risk when different logic families are mixed so that signals travel from fast storage elements to slower ones.*

Mixing logic families, a practice with a long tradition at the board level, is finding its way into VLSI design when high-speed and low-power cells are being combined on a single chip in search of the optimum performance-energy tradeoff. Also note that the timing parameters of macrocells, such as on-chip RAMs, tend to differ significantly from those of simple and, hence, much faster flip-flops and latches.

Hold time fixing

As edge-triggered one-phase-clocking is so vulnerable to clock skew and jitter, modern EDA tools support automatic hold time fixing, a technique whereby buffers get inserted into all signal propagation paths found to provide insufficient hold margins until the extra contamination delay suffices to compensate for the deficit. Extensive buffer chains not only inflate circuit size, however, but also waste energy without contributing to computation.

> *Hint: Wait with hold time fixing until load capacitances and interconnect delays can be estimated with good precision during the routing phase. Have the P&R tool or the timing verifier list the shortest paths as a function of their respective hold margins. You are now in a good position to find out where delay buffers must be inserted to ensure a reasonable minimum hold margin in your design.*

For a concluding remark of more fundamental nature, reinspect inequalities (7.1) through (7.11). Notice that there is a law that will be confirmed throughout our timing analyses.

Observation 7.6. *Within timing conditions, quantities always combine as follows:*

inequality	critical path	relevant timing parameters
setup condition	longest	propagation delays and setup time
hold condition	shortest	contamination delays and hold time

7.2.3 DUAL-EDGE-TRIGGERED ONE-PHASE CLOCKING

Hardware resources and operation principle

Dual-edge-triggered clocking has long been left aside before gaining acceptance along with the buzzword **double data rate** (DDR) in the context of computer memory and mainboard design where it serves to increase memory bandwith while avoiding excessive clock frequencies. Conceptually, this technique is very similar to single-edge-triggered one-phase clocking but rests on special storage elements that operate on both clock edges. As suggested in fig.7.7c, it is possible to construct a **dual-edge-triggered flip-flop** (DETFF) from a pair of latches and a multiplexer.

Exactly as in single-edge clocking, each computation period gets bounded by two consecutive active clock edges. Yet, the fact that either edge causes the state to get updated cuts the clock frequency in half for the same computation or memory access rate. The fundamental timing requirement thus becomes $T_{cp} = \frac{1}{2}T_{clk} \geq t_{lp}$, see fig.7.7b.

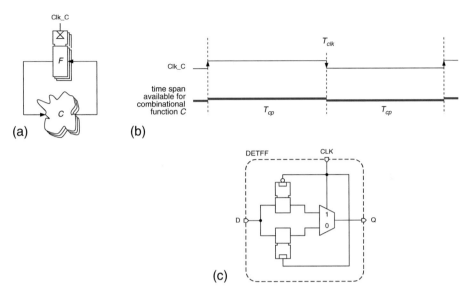

FIGURE 7.7

Dual-edge-triggered one-phase clocking. Basic hardware organization (a) and simplified timing diagram (b). Logically equivalent circuit for a dual-edge-triggered flip-flop (c).

Implications

As a DETFF is described by the same set of timing parameters as a SETFF, analysis yields much the same findings, only the numerical values are bound to differ. The differences relate to clock frequency and energy efficiency. In a single-edge-triggered circuit, the clock net toggles twice per computation cycle. A DDR circuit carries out the same computation with a single clock edge which cuts the energy spent for charging and discharging that net in half, at least in theory. This explains the name **half-frequency clocking** sometimes being used.

Unfortunately, the higher capacitive loads of a DETFF on clock and data inputs partially offset this economy. The slightly more complex circuit does not help either. The economy may even turn negative if the data input is subject to intense glitching [178]. Still, reductions on the order of 10 to 20% of overall power over an equivalent single-edge-triggered design seem more typical.

While dual edge clocking has gained acceptance with DDR RAM interfaces, it has yet to make it into cell libraries and EDA tools, notably for synthesis and timing verification. As HDL synthesizers do not currently accept dual-edge-triggered circuit models, one must model for single edge clocking and replace every SETFF by an analogous DETFF following synthesis.[8]

---

[8] Where qualified DETFF cells are unavailable, one may consider using soft macros. Internal wires shall then be given so much weight that the cells involved always get placed next to each other. As opposed to SETFFs, the output of a DETFF is not connected to a bistable's output, however, but passes through a multiplexer, that is through a combinational network. Careless design or routing might thus engender hazards. It is imperative that a highly consistent timing be guaranteed in spite of minor variations in the soft macro's interconnect routing. Also, any decent collection of DETFFs must be made to satisfy (7.11) to ensure interoperability.

Another particularity is that dual-edge-triggered clocking assumes a strictly symmetrical clock wave-form that satisfies $t_{clk\,lo} = t_{clk\,hi} = \frac{1}{2}T_{clk}$. Put differently, the clock must be made to maintain a duty cycle δ_{clk} of exactly $\frac{1}{2}$ under all operating conditions.

7.2.4 SYMMETRIC LEVEL-SENSITIVE TWO-PHASE CLOCKING

Hardware resources and operation principle

Circuits are **latch-based** which is to say that no bistables other than level-sensitive latches are being used. A subset of them is controlled by clock signal clk1, the complementary set by clk2. Data flow occurs exclusively <u>between</u> the two subsets, there is no exchange <u>within</u> a subset. Put differently, the latches together with the data propagation paths in between always form a bipartite graph.[9] Violating this rule might lead to unpredictable behavior as a consequence from zero-latency loops that would inevitably form when the latches at either end of a combinational logic become transparent at the same time.

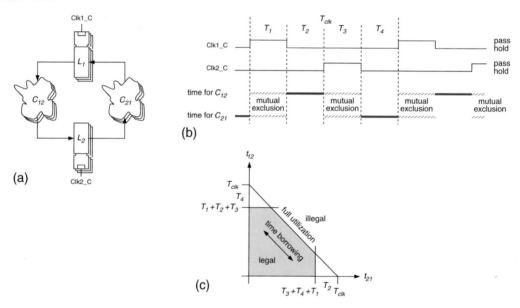

(a)
(b)
(c)

Symmetric level-sensitive two-phase clocking. Basic hardware organization (a), simplified timing diagram (b), and range of operation (c).

Ignoring the setup times and propagation delays of the latches, for a moment, one finds the situation shown in fig.7.8. The two clock signals subdivide the clock period T_{clk} into four intervals labeled T_1 through T_4. During each computation period, data complete a full circle from latch set L_1 through logic

---

[9] A graph is said to be **bipartite** iff it is possible to decompose the set of its vertices into two disjoint subsets such that every edge connects a vertex from one subset with a vertex from the other.

C_{12} to latch set L_2 and from there through logic C_{21} back to latch set L_1 Computation period and clock period are the same $T_{cp} = T_{clk}$.

The earliest moment combinational logic C_{21} can take up evaluating a new input is when latch set L_2 becomes transparent, provided logic C_{12} has completed its evaluation at that time. Otherwise this point in time is delayed into clock interval T_3 until C_{12} completes.

The last possible moment for C_{21} to complete its evaluation is right before latch set L_1 stores the result when switching to hold mode, i.e. at the end of T_1. One thus obtains

$$\max(t_{21}) \leq T_3 + T_4 + T_1 \tag{7.12}$$

and analogously for combinational logic C_{12}

$$\max(t_{12}) \leq T_1 + T_2 + T_3 \tag{7.13}$$

The two inequalities seem to suggest that more than one clock period is available to accommodate the cumulated evaluation times of the two combinational blocks. Of course, this is not possible as C_{21} and C_{12} must work strictly one after the other to make sure the circuit operates properly. This condition of mutual exclusion is captured in a third constraint that requires

$$\max(t_{12} + t_{21}) = \max(t_{pd\,full-circle}) \leq T_{clk} = T_1 + T_2 + T_3 + T_4 \tag{7.14}$$

Note there exists no dead clock interval during which neither C_{21} nor C_{12} is allowed to evaluate new input data. Depending on the actual timing figures, any of the intervals T_1 through T_4 can actually become productive. Also note that one pipeline stage must include two latch sets, L_1 and L_2, as C_{21} and C_{12} do not operate concurrently. As a rule, two latches are needed in a level-sensitive two-phase system where an edge-triggered system has one flip-flop.

Together (7.12), (7.13) and (7.14) confine the legal operating range with respect to timing as depicted in fig.7.8. Full utilization of the clock cycle is achieved, when (7.14) holds with equality, i.e. when $\max(t_{pd\,full-circle}) = T_{clk}$. Further observe from fig.7.8 that symmetric level-sensitive two-phase clocking offers the potential for trading evaluation time left unused by C_{12} against time for C_{21}, and vice versa, without any modification to the clock phases. This somewhat surprising characteristic is referred to as **time borrowing**.

With the above traits, namely mutual exclusion, alternation with no dead time, and time borrowing, symmetric level-sensitive two-phase system resembles a relay race, during which two runners alternate but can decide themselves — within certain bounds — where exactly to hand over the baton.

Detailed analysis

Incorrect timing is likely to lead to corrupt circuit states when input data are not properly stored by the target latches. This phenomenon, from which there is no recovery other than resetting the entire circuit, is termed **latch fall-through** or latch race-through. The skew margins necessary to stay clear of this problem are obtained from the setup and hold requirements of the two latch sets L_1 and L_2, also see figs.7.8 and 7.9.

Setup condition

More precise formulations of equations (7.13) through (7.14) are obtained when the various delays introduced by the latches themselves are taken into consideration.

$$\max|t_{sk}| \leq T_{clk} - T_2 - \max(t_{pd\,lc\,2} + t_{pd\,21} + t_{su\,la\,1}) \qquad (7.15)$$

$$\max|t_{sk}| \leq T_{clk} - T_4 - \max(t_{pd\,lc\,1} + t_{pd\,12} + t_{su\,la\,2}) \qquad (7.16)$$

$$0 \leq T_{clk} - \max(t_{pd\,ld\,1} + t_{pd\,12} + t_{pd\,ld\,2} + t_{pd\,21}) \qquad (7.17)$$

All three inequalities must hold in order to ensure correct operation. Whatever values the various timing parameters may have, it is always possible to find a minimum for the clock period T_{clk} that satisfies them all.

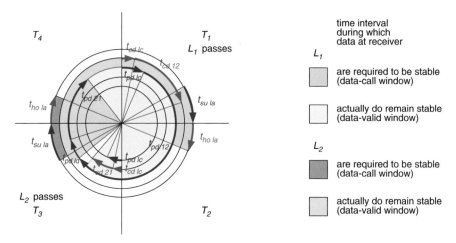

Anceau diagram of symmetric level-sensitive two-phase system shown for a time-borrowing situation where C_{12} occupies approximately $\frac{5}{8}$ and C_{21} $\frac{2}{8}$ of a clock period.

Hold condition

Correct timing at the inputs of L_1 and L_2 requires that

$$\max|t_{sk}| \leq T_2 + \min(t_{cd\,lc\,2} + t_{cd\,21} - t_{ho\,la\,1}) \qquad (7.18)$$

$$\max|t_{sk}| \leq T_4 + \min(t_{cd\,lc\,1} + t_{cd\,12} - t_{ho\,la\,2}) \qquad (7.19)$$

respectively. As in edge-triggered circuits, the most critical situation occurs when no logic is present. In contrast to the situation there, however, the margin against hold violations is largely determined by the duration of the two non-overlap intervals T_4 and T_2. This quality helps to absorb skew and jitter across a single clock distribution net as well as between the two nets.

Implications

The latitude for skew and jitter is largely determined by the waveforms of the clock signals. Room to accommodate unknown or unpredictable timing variations can be designed into level-sensitive two-phase systems just by sizing the non-overlap intervals accordingly. In principle, the most excessive skew can be coped with even after circuits have been fabricated, provided it is possible define the two clock waveforms from outside the chip, e.g. by driving clk1 and clk2 from two separate pins. As both non-overlap intervals are productive, this entails no loss of speed, only the time borrowing capability gets somewhat restricted.

This picture sharply contrasts with edge-triggered one-phase clocking, where the hold margin is imposed by flip-flop parameters that cell-based designers are unable to control.

The price paid for this advantage lies in the routing overhead and in the extra design effort for distributing two clock signals. In order to keep variations between the two clocks within reasonable limits, the two nets should be made similar both geometrically and electrically, i.e. they should run close together with similar loads attached at corresponding points. Area is another cost factor because two latches occupy more die area than one flip-flop. Probably the most important handicap, however, is the fact that it is not currently possible to obtain level-sensitive two-phase circuits from HDL synthesis without rewriting the RTL source code.

7.2.5 UNSYMMETRIC LEVEL-SENSITIVE TWO-PHASE CLOCKING

Hardware resources and operation principle

Unsymmetric level-sensitive two-phase clocking is obtained from the symmetric scheme by transferring combinational logic C_{12} to the opposite side of latch set L_2 and merging it into C_{21}, thereby making intervals T_3, T_4 and T_1 available for evaluating the combined function. Clock interval T_2 becomes unproductive due to the absence of combinational logic between L_1 and L_2. Most often, T_2 is shortened at the benefit of the other intervals to utilize the clock period as much as possible.

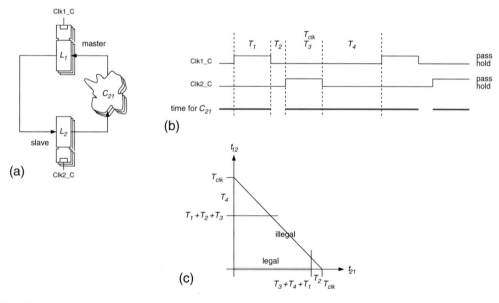

FIGURE 7.10

Unsymmetric level-sensitive two-phase clocking. Basic hardware organization (a), simplified timing diagram (b), and range of operation (c).

Detailed analysis

Setup condition

For latch set L_1 and L_2 respectively is

$$\max|t_{sk}| \le T_{clk} - T_2 - \max(t_{pd\,lc2} + t_{pd\,21} + t_{su\,la1}) \qquad (7.20)$$

$$\max|t_{sk}| \le T_2 + T_3 - \max(t_{pd\,ld1} + t_{su\,la2}) \qquad (7.21)$$

Observe that setup time for latch set L_2 is minimal when L_1 accepts fresh data at the very end of T_1.[10] Inequality relation (7.21) is uncritical for any case of practical interest and (7.20) can always be met by selecting an adequate clock period.

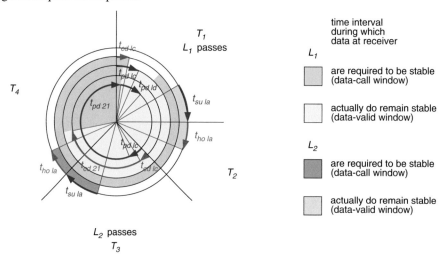

FIGURE 7.11

Anceau diagram of unsymmetric level-sensitive two-phase system shown for near maximum utilization of clock period.

Hold condition

For latch set L_1 and L_2 respectively we find

$$\max|t_{sk}| \le T_2 + \min(t_{cd\,lc2} + t_{cd\,21} - t_{ho\,la1}) \qquad (7.22)$$

$$\max|t_{sk}| \le T_4 + \min(t_{cd\,lc1} - t_{ho\,la2}) \qquad (7.23)$$

---

[10] The finding that (7.21) does not follow from (7.16) whereas (7.20) is identical to (7.15) is irritating at first sight. The underlying reason is as follows.

The starting point for formulating setup conditions in the symmetric case was that both C_{12} and C_{21} would take up as much time as possible. i.e. they would start as soon as their input latches become transparent and end just before their output latches switch to hold. As a consequence, (7.17) had to be added as a third condition to make sure the clock period suffices to accommodate the cumulated delays of C_{12} and C_{21}.

As there is no C_{12} in the unsymmetric case, the above way of looking at the setup condition for L_2 is ill-guided. Instead, (7.21) is obtained from the assumption that C_{21} eats as much time as it can thereby leaving the bare minimum for the propagation path from L_1 to L_2. As a side effect, a third condition is dispensed with.

Either condition provides a non-overlap interval that offers protection in case of excessive skew and jitter. Shortening the unproductive non-overlap interval T_2 in search of maximum performance is limited only by the quest for an adequate skew margin.

Implications

In summary, benefits and costs of unsymmetric level-sensitive two-phase clocking are almost identical to those of its symmetric counterpart. A difference is that wide skew margins are bought at the detriment of operating speed because non-overlap interval T_2 is unproductive.

Observation 7.7. *What makes level-sensitive two-phase clocking disciplines more tolerant to clock skew and jitter are their non-overlap intervals. Liberal skew margins can be designed into such circuits by adjusting the waveforms of their clock signals.*

Example

IBM's patented **level-sensitive scan design** (LSSD) technique [180] [181] is a clever combination of unsymmetric level-sensitive two-phase clocking with a scan-type test facility. Fig.7.12 shows the key ideas behind it. The output Q of each LSSD storage element connects to an input labelled SCI on a subsequent LSSD cell much as in an edge-triggered scan path design.

The three clock nets notwithstanding, LSSD follows a two-phase clocking scheme. During normal operation, Clk1 and Clk2 exhibit non-overlapping pulses just as in any other unsymmetric level-sensitive two-phase clocking system. Terminal D of each LSSD storage element then acts as data input and Q as data output. Clock Clk0 is made to remain inactive at logic 0 by the clock generator, thereby disabling all scan inputs SCI.

In scan mode, the waveforms of Clk0 and Clk1 are swapped which causes all LSSD elements to read data from their respective SCI terminals. Data at the D inputs are being ignored because all CLK1s rest at 0. The LSSD cells together act as a long shift register, which makes it possible to serially write out their state to pin Sco_TO and to read in a new state from pin Sci_TI.

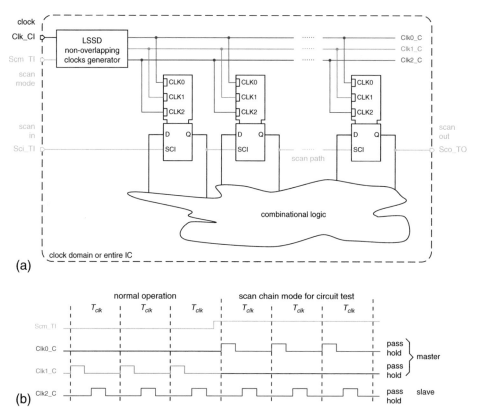

FIGURE 7.12

Level-sensitive scan design. Overall circuit structure (a) and clock waveforms (b).

Incidentally, note that the power dissipated in driving the clock nets is roughly the same as for a two-phase scheme because only two out of the three clock nets are active at any time.
□

Practically speaking, the relative robustness of the two level-sensitive clocking schemes discussed in sections 7.2.4 and 7.2.5 implies that their clock distribution networks need not necessarily be balanced to the same degree of perfection as in the occurrence of edge-triggered circuits. Clock skew even becomes close to harmless when circumstances permit one to drive a circuit from a slow clock that affords ample non-overlap phases. Relaxed timing constraints are very welcome to experienced designers that take advantage of them for doing with less and lighter clock buffers in order to improve on overall energy efficiency and on switching noise.

Distributing two or three clock signals is not always popular, however, because of the extra wiring resources required, the inferior EDA support, the less intuitive circuit operation, and the more complicated timing. Also note that considerable skew can build up between the various clock nets if they are driven from distant buffers or if they significantly differ in their electrical or geometric characteristics. Thus, one can't help to ask

"Is it possible to design latch-based circuits that are driven by a single clock signal, and what characteristics would such circuits have?"

The answer is positive and two very different solutions are going to be examined next.

7.2.6 SINGLE-WIRE LEVEL-SENSITIVE TWO-PHASE CLOCKING

Hardware resources and operation principle

Single-wire level-sensitive two-phase clocking comes in a symmetric and in an unsymmetric variation. Either variation differs from its two-wire counterpart in that all latches are being driven from one common clock signal. The usage of latches that pass and hold on opposite clock polarities does away with the need for a second clock net.

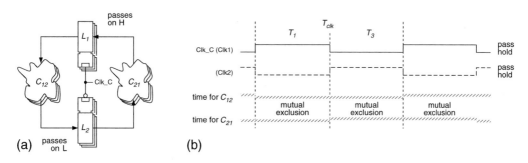

FIGURE 7.13

Symmetric single-wire level-sensitive two-phase clocking. Hardware organization (a) and simplified timing diagram (b).

As far as timing is concerned, this approach is equivalent to driving the two subsets of latches in the original configuration of fig.7.8 with complementary signals or, which is the same, with clock signals whose non-overlap intervals have been removed.

Detailed analysis

The critical hold margins are those of (7.18) and (7.19) with zero substituted for T_2 and T_4. Assuming identical timing figures for either latch bank, these two inequalities reduce to

$$\max|t_{sk}| \leq \min(t_{cd\,lc} + t_{cd\,c} - t_{ho\,la})$$ (7.24)

where $\min(t_{cd\,c})$ is a shorthand for $\min(t_{cd\,21}, t_{cd\,12})$.

$$\max|t_{sk}| \leq \min(t_{cd\,lc} - t_{ho\,la})$$ (7.25)

then holds in the absence of combinational logic between the latches. This constraint can only be relaxed if the circuitry is organized such as to never connect two latches directly with no logic in between, inserting dummy buffers for their contamination delay where necessary.

With all non-overlap intervals that could protect against timewise variability gone, single-wire level-sensitive two-phase clocking is essentially as exposed to skew and jitter as edge-triggered one-phase clocking is. On the positive side, there is an unrestricted capability for time borrowing between C_{12} and C_{21}.

Yet, level-sensitive two-phase clocking offers another and less obvious opportunity for improving circuit performance. Assume the data input of a latch is being driven from some non-inverting gate, e.g. a 3-input AND. We then have a cascade that consists of a 3-input NAND, a NOT, and the bistable's own inverting input buffer. The circuit can be improved by merging the AND operation into the bistable and by collapsing the cascaded inverters. The usage of a "3-input AND function latch", as the combined cell is called, does away with two inverter delays, or almost so.[11] Energy efficiency also benefits from the circuit modification.

What sets symmetric level-sensitive two-phase clocking apart from other schemes is that it becomes possible to play this trick <u>twice</u> per computation period by embedding logic in both of the two latch subsets. Whether one or two nets are being used for clock distribution does not matter in this context, so most of what has been said here applies to standard [two-wire] level-sensitive two-phase clocking as discussed in section 7.2.4 too.

Example
Probably the most prominent VLSI circuits constructed along this line were those of the Alpha processor family introduced by the now defunct Digital Equipment Corp.[12] Please refer to section 7.3 for details on how the clock is distributed over each of those chips.
□

We will now turn our attention to a clocking scheme the properties of which sharply differ from those discussed so far. In spite of advantages in performance and energy efficiency, level-sensitive one-phase clocking can not be recommended for ASICs where limiting the design effort and the design risk are paramount. The subsequent analysis will show why.

Level-sensitive one-phase clocking stores state in latches exclusively and uses a single clock to drive them. The circuit arrangement corresponds to edge-triggered one-phase design with all flip-flops replaced by latches, see fig.7.14a. The same arrangement can be obtained by bypassing all slave latches in an unsymmetric level-sensitive two-phase design.

---

[11] You may want to see section 3.4.3 for more information on function latches.

[12] Depending on the authors, the clocking discipline used in the 21064 is referred to as "level-sensitive single phase with no dead time" or as "single-wire two-phase clocking".

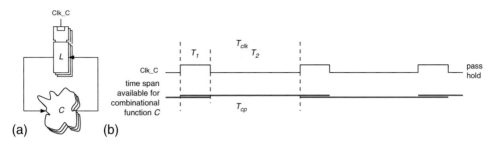

FIGURE 7.14

Level-sensitive one-phase clocking. Basic hardware organization (a) and simplified timing diagram (b).

The fact that a latch must become transparent before it can accept a new data item for storage, together with the simultaneous clocking of all bistables, implies that the combinational logic is fed with new data (at the beginning of T_1) <u>before</u> the previous results have been latched (which takes place at the end of T_1). As a consequence, the combinational network must assure that data at its output remain unchanged while another wave of data has begun to propagate from the input through that very network. Correct circuit operation rests on inertial effects as transient phenomena are no longer allowed to die out before clocking occurs.

Spinning the idea further, one may want to arrange for two or more data waves to propagate through the combinational logic at all times, a bit like a juggler who keeps several objects in the air simultaneously. This bold approach is known as **wave pipelining** [182] [183] and accepts path delays in excess of one clock period in an attempt to break the barrier that normally limits throughput.[13]

$$T_{cp} = T_{clk} < t_{lp} = t_{id\,la} + \max(t_c) \qquad\qquad \Theta = \frac{1}{T_{cp}} = \frac{1}{T_{clk}} > \frac{1}{t_{lp}} \qquad (7.26)$$

---

[13] While it is indeed possible to design feedforward wave pipelines that satisfy (7.26), recall from section 3.7 that first order feedback loops are not amenable to pipelining (but must be tackled with loop unfolding instead). Since that finding does not depend on how the necessary latency is obtained, it applies to wave pipelining too. Also note that the clock frequencies and throughputs made possible by wave pipelining will not exceed those obtained from a register-based fine grained pipeline because some minimum separation must always be maintained between any two consecutive waves. Wave pipelining manages with far less registers, though.

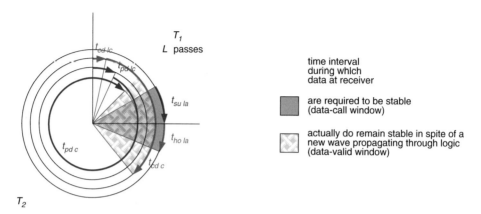

FIGURE 7.15

Anceau diagram of level-sensitive one-phase system. The formation of multiple waves becomes apparent by $t_{pd\,c}$ covering more than a full circle.

Detailed analysis

Setup condition

Combinational logic is fed with new data at the beginning of transparent interval T_1 and must yield a stable result no later than at the end of the subsequent transparent interval.

$$\max|t_{sk}| \leq T_{clk} + T_1 - \max(t_{pd\,lc} + t_{pd\,c} + t_{su\,la}) \tag{7.27}$$

Once more, the setup condition is easily satisfied by acting on T_{clk}. Relation (7.27) exhibits a more intriguing characteristic, however, especially when compared to (7.5) or (7.20). More than a full clock period becomes available for the combinational logic, unless the cumulated latch delay, clock skew and jitter eat up more time than T_1 provides. In other words, the circuit can be made to operate in a wave pipelined regime as shown in fig.7.15.

Hold condition

Data applied to latches at the beginning of transparent interval T_1 must not change before the end of this very transparent interval.

$$\max|t_{sk}| \leq -T_1 + \min(t_{cd\,lc} + t_{cd\,c} - t_{ho\,la}) \tag{7.28}$$

When compared to edge-triggered one-phase clocking, the meagre skew margin has been reduced further by T_1 and is bound to become negative in the absence of combinational logic. It is now evident at what expense the extra leeway in the setup condition has been bought.

Implications

Equation (7.28) imposes an upper bound for clock high phase T_1 which contrasts with all other clocking schemes analyzed so far. As any bistable comes with minimum clock widths requirements, the high phase gets bounded from below too. We thus end up with a two-sided constraint for T_1.

$$\max(t_{clk\,hi\,min\,la}) \leq T_1 \leq \min(t_{cd\,lc} + t_{cd\,c} - t_{ho\,la}) - |t_{sk}| \tag{7.29}$$

Whether there is a solution or not depends on the detailed timing figures of the cells being used. Yet, in the absence of combinational logic and relevant interconnect delay, the remaining upper bound $\min(t_{cd\,lc} - t_{ho\,la}) - |t_{sk}|$ leaves very little room indeed for a comfortable clock high time. The problem is further aggravated by the necessity to find a valid timing for a variety of latch types and under all possible operating conditions.

As a consequence, it is impossible to build a shift register — or a scan path — in the normal way, even if skew and jitter are optimistically assumed to be zero. For the circuit to work properly, delay elements must be inserted into each overly short propagation path until (7.28) is safely satisfied. Redundant gates have to be added for their contamination delay alone.

There is another peculiarity worthwhile to note. Slowing down a circuit's operation is often helpful for locating problems. For all clocking disciplines discussed earlier, this could be achieved simply by stretching the clock's waveform. With (7.28) specifying an upper bound, this is no longer possible. Slowing down a level-sensitive one-phase clocked circuit asks for keeping one clock phase constant (T_1 in the occurrence of fig.7.14) while extending the other (T_2).

From our analysis of level-sensitive one-phase clocking, we conclude:

+ When compared to two-phase designs, the number of latches and clock nets is cut in half.
+ In theory, and with the exception of computations organized as first order feedback loops, there is a potential for shortening the clock period to below the logic's propagation delay.
− As a consequence from the feedback loop being closed during the transparent intervals, correct operation critically depends on clock pulse width, gate delays, and interconnect delays.
− All library cells must be bindingly characterized for their contamination delays.
− Interconnect delays, and hence also layout parasitics, must be precisely controlled.
− Unchecked timing variations may prove deadly.
− Signals that arrive too early must be delayed artificially at extra costs
  in terms of area and energy.
− Sufficient skew margins must also be bought with additional contamination delay.

Delay tuning is a nice word for the process of adjusting circuit delays until a design appears to work for a given set of timing parameters. This practice is not compatible with regular high-productivity EDA design flows. While some HDL synthesizers and automatic place and route (P&R) tools are capable of hold time fixing, most EDA software is designed to meet long path constraints at minimum hardware and energy costs. Also, contamination delays are neither guaranteed by manufacturers nor normally indicated in library datasheets. Most simulation models do not accurately reproduce contamination delay either.

Observation 7.8. *Level-sensitive one-phase clocking critically depends on fine tuning propagation and contamination delays through both combinational logic and interconnect. This practice is incompatible with existing EDA design flows, with industrial cell libraries, with reliable circuit operation in the presence of timing variabilities, and with the drive towards ever higher productivity in VLSI design and test.*

However, note that level-sensitive one-phase clocking has been applied successfully to supercomputers such as the Amdahl 580 and the Cray-1 [184] at a time when circuits were assembled at the board level from SSI/MSI parts. A milder form is also being discussed in problem 6.

The main problem of **clock distribution** is to drive many thousands of clocked subcircuits, such as latches, flip-flops and RAMs, spread over a die, a board or a system while keeping the unavoidable skew within narrow limits. Slow clock ramps are problematic for several reasons:

- As illustrated in fig.7.16, unavoidable disparities of the switching thresholds across clocked cells translate ramp times into skew.
- The timing figures published in datasheets and stored in simulation models are obtained from stimulating cells with clock ramps of 50 ps and less during library characterization.
- When clocked with slow ramps, setup and hold times tend to grow, putting correct behavior and timing accuracy at risk.

As a consequence, clock signals must literally snap from 0 to 1 and back again.

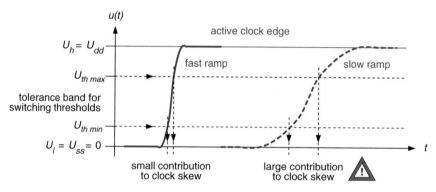

FIGURE 7.16
Sluggish clocks translate into extra skew.

Observation 7.9. *Among all signals on a chip or in a system, clocks typically feature the largest fanout, travel over the longest distance, and operate at the highest speed. Yet, their waveforms must be kept particularly clean and sharp.*

The most innocent idea for distributing a clock is to treat it like any other signal, i.e. to use a standard minimum-width line to connect all clocked subcircuits to a common source. That such an approach is not viable becomes immediately clear from a numerical example.

Warning example
Consider a clock domain in a CMOS IC that includes a modest 500 flip-flops. Clock distribution is via a metal line 130 nm wide that meanders through the chip's core area much as in fig.7.17a. Sheet resistance is 70 mΩ/□, a typical value for a second or third level metal. For simplicity, assume that the flip-flops are connected to that wire at regular intervals thereby forming an RC network of ladder type

with $\#_{sct} = 500$ identical sections of length $100 \,\square$ and with a capacitance of 12 fF each. The delay and the ramp time at the end of such a net in response to a step at the input then roughly are [185]

$$t_{pd\,wire} \approx 0.4\,R_{sct}\,C_{sct}\,\#_{sct}^2 = 0.4\,R_{wire}\,C_{wire} = 0.4 \cdot 3500\ \Omega \cdot 6\ \text{pF} = 8.4\ \text{ns} \tag{7.30}$$

$$t_{ra\,wire} \approx R_{sct}\,C_{sct}\,\#_{sct}^2 = R_{wire}\,C_{wire} = 3500\ \Omega \cdot 6\ \text{pF} = 21\ \text{ns} \tag{7.31}$$

where R_{sct} and C_{sct} refer to the lumped resistance and capacitance of one section respectively as illustrated in fig.7.18a. Obviously, such figures are totally inadmissible for a clock.
$\square$

The above example suggests two starting points. Firstly, the clock net must be reshaped and resized such as to cut down, control and balance interconnect delays in a better way. Secondly, interconnect lengths, capacitances and resistances must be lowered by subdividing a chip-wide clock net into many smaller nets each with a driver of its own. These are indeed the basic ideas behind two alternative approaches that we are going to discuss next.

7.3.2 COLLECTIVE CLOCK BUFFERS

The collective approach has one single buffer that connects to all clocked cells directly via metal lines. To handle the huge fanout, the buffer consist of multiple stages of increasing drive strengths. The idea failed in the above example because the clock net was shaped like a long, narrow and winding alpine road. The picture improves if length, width and shape are chosen more carefully. As the ensemble of clocked cells sets a lower bound for the load capacitance, it is the resistance of the distribution network that must be kept low, while, at the same time, attempting to make all interconnect delays about the same.

The subsequent countermeasures all help to improve the situation:

- Keep clock wires as short as possible, place the driver close to the center of the circuit.
- Make clock distribution wires reasonably wide. Narrow wires yield a high resistance whereas plate capacitance dominates in excessively wide wires.
- Use the upper metal layers as only they combine low resistance with low capacitance.
- Avoid unnecessary layer changes as contacts and vias contribute to resistance.
- Equalize delays by making all clock paths electrically and geometrically similar.

Three layout arrangements developed with these rules in mind are shown in figs.7.17b to d. Although the recursive H-tree immediately strikes as the optimal solution, it is difficult to implement in practice unless clock loads are uniformly distributed and unless one is prepared to set aside one metal layer for the sole purpose of clock distribution. Spine, comb and grid topologies similar to those adopted for power distribution purposes are or — rather — were more realistic. Yet, wide clock lines unnecessarily inflate parasitic capacitance and, hence, power dissipation. Another difficulty is that the important switching currents associated with driving large loads concentrate at a few points in a chip's floorplan.

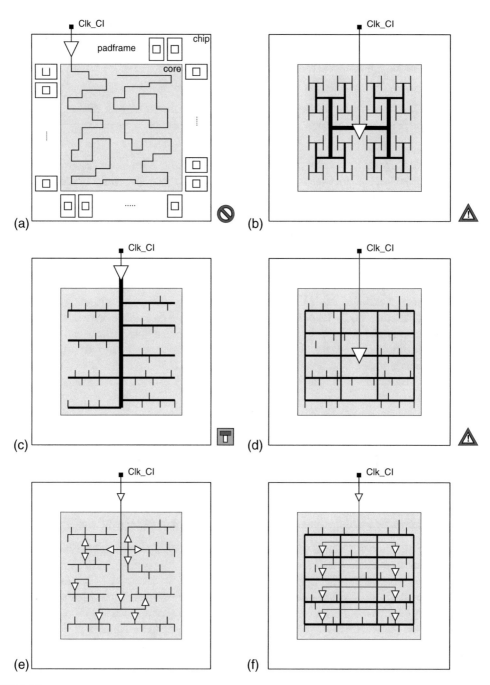

FIGURE 7.17

On-chip clock distribution schemes (simplified). Narrow meander (a), central buffer combined with a balanced H-tree layout (b), collective buffer combined with spine-like wiring (c), central buffer combined with grid-like wiring (d), distributed buffer tree (e), distributed buffer tree combined with a grid (f).

Example

Consider a circuit with 10 000 bistables. Clocking is from a collective buffer with a voltage swing of 1.2 V. Let the aggregate load capacitance per bistable be the same as in the previous example since the clock lines are shorter but also much wider this time. Further assume that the drive current follows a triangular waveform during the signal's ramp time of 100 ps. It then peaks out at

$$\hat{I}_{clk} \approx 2 \, \frac{C_{clk} \, U_{dd}}{t_{ra \, clk}} = 2 \, \frac{10\,000 \cdot 12 \text{ fF} \cdot 1.2 \text{ V}}{100 \text{ ps}} \approx 2.9 \text{ A} \tag{7.32}$$

At a clock frequency of 500 MHz, the final inverter stage driving the clock net dissipates

$$P_{clk} > \frac{\alpha_{clk}}{2} \, C_{clk} \, U_{dd}^2 f_{clk} = 1 \cdot 10\,000 \cdot 12 \text{ fF} \, (1.2 \text{ V})^2 \, 500 \text{ MHz} \approx 86 \text{ mW} \tag{7.33}$$

☐

Current spikes as strong as this necessitate special precautions. Older cell libraries used to provide special **clock drivers** designed to fit into a chip's padframe where they were fed from dedicated power and ground pads in order to avoid ground bounce problems.

Example

The original Alpha processor, the 21064, followed a collective driver approach in its purest form. Total capacitive load of the clock node was 3.25 nF, requiring final driver transistors with $W_n = 100$ mm and $W_p = 250$ mm [186]. Power dissipation was 30 W from a 3.3 V supply at a clock frequency of 200 MHz. Clock rise and fall times were 500 ps and a peak switching current of 43 A had been measured. Clock distribution roughly followed fig.7.17d with the central driver extending along more than half of the chip's centerline. Transmitting a clock edge from the chip's center to a corner was found to take less than 300 ps, which compared favorably to the 5 ns clock period and paved the way for single-wire level-sensitive two-phase clocking where there are no non-overlap intervals that could provide extra room for skew.

☐

7.3.3 DISTRIBUTED CLOCK BUFFER TREES

The concern for energy efficiency subsequently mandated the introduction of gated clocks and drove designers away from the flat distribution schemes that were popular at the time when the Alpha 21064 was developed.

What's more, observe from (7.30) that interconnect delay grows quadratically with line length l because $R_{wire} \propto l$ and $C_{wire} \propto l$ (and $\#_{sct} \propto l$ too). Let us examine what happens when a long line gets turned into a **repeated wire** by subdividing it into $\#_{seg}$ shorter segments separated by $\#_{seg} - 1$ inverting or non-inverting buffers in between as shown in fig.7.18b. To first order, overall propagation delay then becomes

$$t_{pd\,rept} \approx \#_{seg} \, (t_{pd\,buf} + 0.4 \, R_{sct} \, C_{sct} \, (\frac{\#_{sct}}{\#_{seg}})^2) = \#_{seg} \, t_{pd\,buf} + \frac{0.4}{\#_{seg}} \, R_{wire} \, C_{wire} \tag{7.34}$$

which means that the overall delay can be made a <u>linear</u> function of line length just by choosing $\#_{seg} \propto l$. To minimize the overall delay, the propagation delay of each wire segment should be made to match that of one repeater, in which case one speaks of an optimally repeated wire. In practice, repeaters are typically spaced further apart such as not to unnecessarily inflate the overhead in terms of area and power.

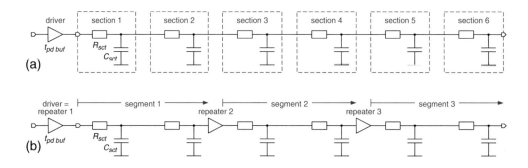

FIGURE 7.18

Lumped RC model of a bare wire (a) vs. that of a repeated wire (b). $\#_{sct} = 6$ and $\#_{seg} = 3$ in this example.

A distributed clock tree thus consists of a multitude of repeaters of moderate size physically located close to the loads they drive and inserted into every major branch, see fig.7.17e. As opposed to the collective approach, no large currents need to flow over large distances. Also, the power dissipated for driving the clock net is distributed over the chip's area. Conversely, a tree-like structure requires careful delay matching. The goal is to equalize the delays along all branches in spite of unevenly distributed loads and distances. This implies:

- Hierarchically partition the design such as to balance clock loads to a reasonable degree.
- Plan for local subtrees and size clock buffers as a function of the load they must drive.
- Prefer the upper metal layers for their lower resistance and capacitance.
- Retard overly fast clock distribution paths by inserting extra buffers.
- For fine tuning, consider adding dummy loads and detours to early branches.

While both collective clock buffers and distributed clock trees have been made to work in commercial circuits, distributed trees won because they support clock gating, help to keep distribution delay low, and tend to be less demanding in terms of routing resources, overall power, and current crowding.

One difficulty in implementing them is that accurate data on which to base detailed delay calculations do not become available until late in the design cycle. Another problem is that the clock tree must be readjusted whenever a design modification entails a minor change in the loads to drive or in their geometric locations.

The EDA industry has come to the rescue by developing automatic software tools that are capable of inserting balanced clock trees into circuits of substantial size, largely doing away with the need for manual delay budgeting and iterative tuning. As a matter of fact, distributed trees prevail since the introduction of automatic **clock tree generators** that are run as part of physical design, e.g. between place and route. At that point, calculations can be based on fairly accurate estimates for layout parasitics and interconnect delays.

Hint: Restricting repeaters in a clock tree to non-inverting buffers (no inverters) eliminates the risk of ending up with contradictory branches that differ in their numbers of negations.

Example

An unprecedented degree of sophistication has been achieved with Intel's Itanium 2 9000, a dual core design clocked at 1.6 GHz with $1.72 \cdot 10^9$ transistors manufactured in a 90 nm bulk CMOS process with seven layers of Cu interconnect. The sub 100 nm technology, the huge die size of 27.7 mm by 21.5 mm, the maximum power dissipation of 104 W in conjunction with local hot spots all contribute to important and unpredictable on-chip variations (OCV). While an oversize clock grid would have helped to contain skew, designers were afraid of the large equalizing currents and of the power overhead this would have entailed [187]. Instead of contenting themselves with statically balancing clock distribution delays at design time, their regional active deskew scheme adaptively fine-tunes delays in the different clock tree branches at run time with the aid of multiple phase comparators and adjustable delay lines.
□

7.3.4 HYBRID CLOCK DISTRIBUTION NETWORKS

Hybrid schemes have been devised for high-performance VLSI in an attempt to combine the efficiency of a distributed clock tree with the robustness of the grid approach. Instead of having the leaf cells of a distributed buffer tree drive the clocked cells directly, their outputs are shorted together by a domain-wide metal grid as sketched in fig.7.17f. The grid minimizes local skew by providing low-resistance current paths between nearby points. The fairly stiff and uniform structure so obtained facilitates circuit design as timing does not depend too much on details of cell placement. This approach results in a small distribution delay and acceptable power without eating too much wiring resources [188] [181].

Example

The clock distribution concept pursued by Sun Microsystems in their Niagara processor that includes $279 \cdot 10^6$ transistors manufactured on a die of 378 mm^2 in 90 nm 9LM (Cu) bulk CMOS technology and rated with a maximum power of 63 W at 1.2 GHz and 1.2 V is comparatively straightforward. By cleverly combining a global H-tree with multiple regional clock-gated grids and with local subtrees, its designers were able to keep clock skew below 50 ps without recurring to complex active deskewing circuitry [189]. While overall power density is comparable to the above example, the smaller die and the better spreading of heat afforded by the concurrent operation of eight relatively modest processor cores was certainly helpful in avoiding inordinate delay variabilities. Another particularity is post-layout hold time fixing with the aid of metal-programmable delay buffers. Following static timing analysis, delays get statically adjusted using a local metal with no impact on the previously frozen detailed signal routing.
□

7.3.5 CLOCK SKEW ANALYSIS

Functional simulation is inadequate for uncovering clock skew problems. As exhaustive simulation is not practical, it is very likely that not all critical patterns get applied and that some skew-related timing problems pass unnoticed. The good news is that there is no need to simulate a circuit functionally to find out whether timing conditions are met or not in synchronous designs. Timing verification, aka static timing analysis, is much more appropriate as it

- Circumvents all coverage problems,
- Quantifies the timing margins on all signal propagation paths, both short and long,
- Locates slack, if any, and
- Does with a more reasonable computational effort.

Do not forget to include layout parasitics, to use adequate interconnect models, and to account for crosstalk, PTV and OCV variations during static timing verification.

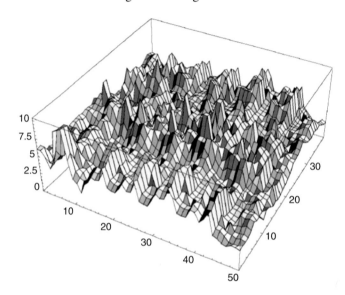

FIGURE 7.19

Clock skew map for a Cell processor chip measuring roughly 17.8 mm by 12.3 mm. Horizontal dimensions are given in arbitrary units, skew is in [ps] (photo copyright IEEE, reprinted from [190] with permission).

Observation 7.10. *Whatever the clocking discipline and clock distribution scheme, careful design and timing verification are essential.*

7.4 HOW TO ACHIEVE FRIENDLY INPUT/OUTPUT TIMING

7.4.1 FRIENDLY AS OPPOSED TO UNFRIENDLY I/O TIMING

We now extend our discussion to the board-level interface of entire chips that are part of the same clock domain, see fig.7.20a.[14] The basic requirement of observation 7.1 that any data-call window must be fully enclosed by the pertaining data-valid window remains exactly the same as for data transfers between simple flip-flops within a chip.

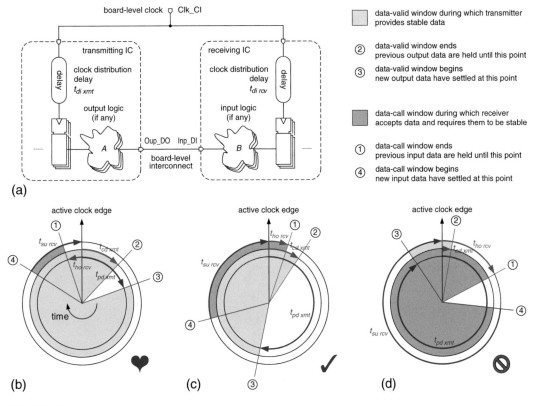

(a)

(b)   (c)   (d)

FIGURE 7.20

Two VLSI chips with a data transfer path (a). Anceau diagrams for friendly (b), average (c), and incompatible I/O timing characteristics (d). Any counterclockwise arrow implies a negative sign.

---

[14] Note that much the same argument also applies to virtual components and other major circuit blocks that contain a clock distribution network of their own. How to synchronize data at the boundaries between distinct clock domains is to be addressed in chapter 8.

Interfacing with outputs that provide valid data for most of a clock period is straightforward as there is plenty of time for the receiving circuit to evaluate and store the acquired data. Similarly, most of the clock period is available for the transmitting circuit to settle if the receiver calls for stable data during a brief time slot just prior to the active clock edge. Note the ample setup and hold margins in fig.7.20b.

As opposed to this, interfacing with a circuit that just flashes output data for a brief moment of time is a real nuisance. Data are likely to settle too late or to vanish too early for the receiver to get hold of them. Similarly, inputs that ask for stable data for an extended period leave little time for output logic and interconnect delays. A long hold time proves especially awkward since it obliges the transmitter to maintain its former output long after the active clock edge has sparked off a transition to a new state. The two I/O timings shown in fig.7.20d are in effect incompatible: it is impossible to transfer data if transmitter and receiver are to be driven from a common clock. Engineers are then forced to recur to delay lines, stopover registers, adventurous clocking schemes, and other makeshift improvisations.

The desiderata for friendly input-output timing are as follows.

I/O timing	data-call window	inputs $t_{su\,inp}$	$t_{ho\,inp}$	data-valid window	outputs $t_{pd\,oup}$	$t_{cd\,oup}$
friendly	narrow	small	close to zero or, better, negative	wide	small	large, as close to $t_{pd\,oup}$ as possible
unfriendly	wide	large	large (positive)	narrow	large	close to zero

How to express timing constraints for commercial synthesis and timing verification tools has been the subject of section 4.4.5.

7.4.2 IMPACT OF CLOCK DISTRIBUTION DELAY ON I/O TIMING

Driving the huge capacitances associated with clock nets necessitates large multi-stage drivers which bring about important clock distribution delays. So far, we have not cared much about this effect as only skew matters within a circuit. Yet, the impact of clock distribution delay on I/O timing is twofold, just compare figs.7.21 and 7.22.

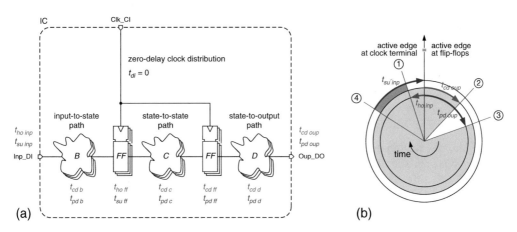

FIGURE 7.21

I/O timing of a VLSI chip in the absence of clock distribution delay. Circuit (a) and Anceau diagram (b). Note: Events refer to input and output terminals of a single circuit here, which view contrasts with all Anceau diagrams shown so far in this chapter where events related to the interface between two circuits.

On outgoing signals, distribution delay simply adds to both propagation and contamination delays. The resulting data lag remains largely uncritical unless a system is to operate at high clock frequencies in which case the prolonged propagation delay will be felt.

On incoming signals, distribution delay shortens setup and extends hold time. Even a moderate distribution delay is likely to render setup time negative on those inputs that directly connect to a register. More importantly, however, the same amount of delay will inflate hold time way beyond any realistic value for a transmitter's contamination delay.

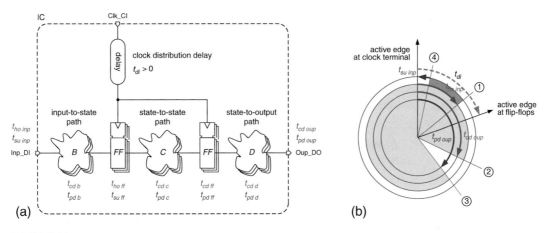

FIGURE 7.22

Impact of excessive clock distribution delay. Circuit (a) and Anceau diagram (b).

For a quantitative analysis, assume the circuit operates with edge-triggered one-phase clocking. The circuit's input and output terminals then exhibit these timing parameters

$$t_{su\,inp} = \max(t_{pd\,b} + t_{su\,ff}) - t_{di} \tag{7.35}$$

$$t_{ho\,inp} = \max(-t_{cd\,b} + t_{ho\,ff}) + t_{di} \tag{7.36}$$

$$t_{pd\,oup} = \max(t_{pd\,ff} + t_{pd\,d}) + t_{di} \tag{7.37}$$

$$t_{cd\,oup} = \min(t_{cd\,ff} + t_{cd\,d}) + t_{di} \tag{7.38}$$

In conclusion, distribution delay adds to input hold time and may so easily turn a circuit with perfect I/O timing characteristics into an unacceptable one. What's more, within a population of digital components, any differences between clock distribution delays translate into clock skew unless they get counterbalanced within the board- or system-level clock distribution network.

Observation 7.11. *While it is imperative to minimize clock skew, it is also important to keep on-chip clock distribution delay within tight bounds to avoid the awkward timing characteristics at the chip's I/O interface that otherwise result.*

Example

FPGAs of the Actel SX-A family include three dedicated low-skew clock networks. One of them is referred to as hardwired which is to say that there are no more than two programmable links between the clock input and any on-chip flip-flop. This is done not only to minimize clock skew but also distribution delay. The resulting skew is said to be as low as 0.1ns and the pin-to-pin delay from the clock input to any register output is specified as 5.3ns (A54SX72A-3 under worst case commercial conditions).
□

7.4.3 IMPACT OF PTV VARIATIONS ON I/O TIMING

Keeping a receiver's data-call window within the transmitter's data-valid window under all circumstances is particularly difficult at the interface between two components because

- I/O timing is subject to vary with the off-chips loads attached,
- I/O timing is affected by board-level interconnect delays,
- What matters is the <u>difference</u> between the distribution delays of receiver and transmitter,
- The PTV conditions of two chips are bound to differ, and because
- PTV variations notoriously have a large impact on timing figures.[15]

While clock distribution delay is of little importance as long as all clocked subcircuits are affected alike, it becomes highly critical at the interface between two components and at the interface between two supply domains. It is not uncommon to find that maintaining all components properly synchronized within a clock domain is impossible when those components are subject to moderate but non-uniform PTV variations.

Observation 7.12. *PTV variations put data transfers between ICs at risk. A healthy latitude to timing variations is thus even more important at the I/O interfaces than within a chip itself.*

---

[15] It is not exceptional to see timing quantities vary by a factor of three from the slowest to the fastest operating conditions acceptable for some given CMOS technology.

Before discussing a more radical solution for all the above problems in section 7.4.8, let us suggest a few countermeasures that alleviate the undesirable effects of clock distribution delay.

7.4.4 REGISTERED INPUTS AND OUTPUTS

Reconsider fig.7.22. An obvious idea for simplifying I/O timing is to insert data registers between the input circuitry from Inp_DI and combinational logic *B* and/or between combinational logic *D* and the pad drivers for Oup_DO. Let us see whether that helps.

On the output side, signals should not be made to propagate through deep combinational networks. Including an **output register** right before the pad drivers maximizes the data-valid window and — as a welcome side effect — also yields hazard-free signals. A price to pay is the extra latency. If needed, the circuit's contamination delay can be extended by driving the output register from a local clock designed with a slight lag.

   Input registers, on the other hand, must not be allowed to suffer from substantial and/or unpredictable clock distribution delay as input hold time otherwise becomes unmanageable. The remedies to be presented next are, therefore, particularly important on the input side.

7.4.5 ADDING ARTIFICIAL CONTAMINATION DELAY TO DATA INPUTS

As data lag compensates for clock distribution delay, any combinational network inserted between the input pads and the first bistable helps because of its contamination delay. After all, it is the timing of the clock relative to the data signals that decides when exactly data voltages get interpreted, also see problem 8.

Example
Various Xilinx FPGA families include a user-configurable multiplexer in each I/O block that selects or bypasses a delay line in the data input propagation path, see fig.7.23. Note that the data-call window is being shifted and observe the impact on the FPGA's hold time requirement.

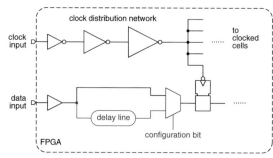

Data input configured	Input timing if			
	straight		delayed	
Xilinx FPGA	$t_{su\,inp}$ [ns]	$t_{ho\,inp}$ [ns]	$t_{su\,inp}$ [ns]	$t_{ho\,inp}$ [ns]
XCS05-4	1.2	1.7	4.3	0.0
XCS40-4	0.4	3.5	5.3	0.0
XC2V40-6	0.84	-0.36	3.24	-2.04

FIGURE 7.23

Compensating clock distribution delay with configurable delay lines on FPGA data inputs. Principle (top) and excerpts from Xilinx data sheets (bottom).

□

7.4.6 DRIVING INPUT REGISTERS FROM AN EARLY CLOCK

The interfacing of ICs or virtual components (VC) that share a common clock would be easier if their input registers — and possibly their output registers too — could be driven from a separate clock that does not suffer from a distribution delay as important as the rest of the circuit. This is indeed possible based on the observation that I/O registers account only for a small fraction of a circuit's total clock load. An early clock is tapped from the main tree close to its root, see fig.7.24. The resulting lead of the input registers must be taken into account when designing the remainder of the circuit, however.

7.4.7 CLOCK TAPPED FROM SLOWEST COMPONENT IN CLOCK DOMAIN

Complex circuits often feature comparatively long hold times, making it difficult to interface them with smaller and faster components. This is mainly because clock distribution delay tends to grow with circuit size and clock load. The practice of mixing high-density CMOS core functions with fast SSI/MSI components — such as high-speed CMOS, bipolar or BiCMOS — for glue logic functions on a circuit board exacerbates the problem.

A workaround exists provided that the component that exhibits the longest clock distribution delay within the clock domain can be arranged to have a special output tapped from the its clock distribution network close to its leafs and brought to an extra package pin, see fig.7.24. The original clock is then first fed into that slower part and redistributed from its clock tap to all faster components within the same clock domain.

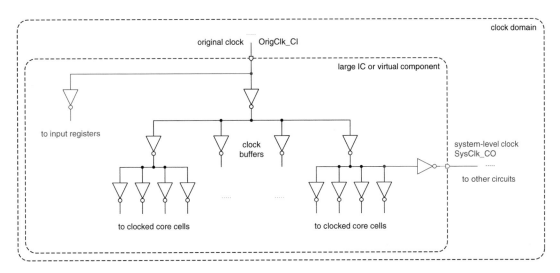

FIGURE 7.24

Two more options for compensating clock distribution delay (data I/O not shown.)

7.4.8 "ZERO-DELAY" CLOCK DISTRIBUTION BY WAY OF A DLL OR PLL

All problems of I/O timing would be gone if clock distribution delay could be made zero as the events that trigger state changes and the external clock would then perfectly align. This is indeed possible with an on-chip **delay locked loop** (DLL) or **phase locked loop** (PLL). Once locked, both DLLs and PLLs act as a servo loop that keeps a local signal in phase with a reference signal supplied from externally. The difference essentially relates to the actuator in the servo loop: PLLs use a controlled oscillator whereas DLLs include an adjustable delay line or — which is the same — a controlled phase shifter. In either case, the servo loop is made to compensate for the cumulated propagation delays of clock buffers and interconnects, see fig.7.25. The overall circuit then behaves like a distribution network with zero or close to zero insertion delay. Better still, PTV variations are automatically compensated for.

As every overclocker knows, most microprocessors operate at rates that are an integer multiple of the clock frequency distributed over the motherboard. This is achieved by inserting a frequency divider into the PLL feedback path as shown in fig.7.25a. Both DLLs and PLLs take a number of cycles to lock and have a limited lock range which makes them intolerant to abrupt frequency changes and vetoes single-cycle operation. Unless frequency multiplication is truly required, DLLs are preferred for their unconditional stability, faster lock times, and low jitter.

DLL and PLL subcircuits are not always part of ASIC design libraries. They can be designed either from voltage-controlled delay lines and oscillators or around numerically adjustable delay lines. [191] [192] [193] [194] include material on the design of integrated PLLs and DLLs. A particular difficulty with adjustable delay lines in the context of clocking is that the respective propagation delays for rising and for falling edges must closely match.

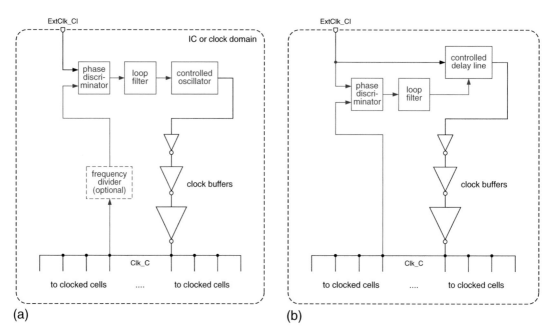

(a)   (b)

Using a PLL (a) or a DLL (b) to minimize and stabilize clock distribution delay.

Fast lock-in and stable phase alignment in spite of PTV variations heavily depend on the phase discriminator. So-called **bang-bang** PLLs use a simple D-type flip-flop where the reference signal is applied to data input D and the local clock to flip-flop terminal CLK so that the Q output essentially reflects the sign of the phase difference. The servo loop is then designed such as to drive the local clock into alignment with the external reference signal. Bang-bang PLLs do not obey the standard linear PLL theory, though. Please refer to [195] [196] [197] for details.

The **injection-locked oscillator** (ILO) is another special case of a PLL where the reference clock is injected directly into an LC- or ring-type oscillator (ILRO) with no special phase discriminator. Rather than continuing to oscillate at its free-running frequency, the oscillator will lock to the injected frequency provided the difference between the two is sufficiently small. Because of the simple circuit, ILOs dissipate less power than full-fledged PLLs and lend themselves for use in clock distribution networks [198].

Examples

An analog and a mostly digital DLL circuit designed for clock alignment in the context of a memory system have been compared by [199] and the results are summarized below. Both circuits have been fabricated on the same chip with a 400 nm 3.3 V standard CMOS process. Nominal clock frequency is 400 MHz, yet data are transferred on both edges of the clock so that the time interval is a mere 1.25 ns per bit. Also consider that the digital circuit takes less effort to design and is more easily ported to a new process than the analog alternative.

DLL architecture	analog	digital	
area	0.68	0.96	mm^2
power dissipation	175	340	mW @ 3.3 V
maximum clock frequency	435	> 667	MHz
minimum supply voltage	2.1	1.7	V
lock time	2.0	2.9	μs
clock jitter	195	245	ps peak-to-peak

Actel's ProASIC$^{\underline{PLUS}}$ FPGAs include two clock conditioning blocks each of which consists of a PLL, four programmable clock dividers, and a few delay lines. The output frequency range is 6 to 240 MHz with a maximum acquisition time of 20 μs. A lock signal indicates that the PLL has locked onto the incoming clock signal. Delay lines are programmable in increments of 250 ps. Supply voltage must be within 2.3 V and 2.7 V, nominal power dissipation is 10 mW.

The clock distribution subsystem in Intel's Itanium CPU is by itself very complex and organized into global, regional, and local clock distribution. It makes use of one PLL at the global and of 30 (sic!) DLLs at the regional level. Implemented in 180 nm CMOS technology, the sophisticated distribution scheme limits clock skew to 28 ps for a 1.25 ns (800 MHz) clock [200].

Intel's Nehalem multi-core architecture includes multiple clock domains and implements a dynamic voltage and frequency scaling (DVFS) scheme whereby supply voltages and clock frequencies get adjusted as a function of the momentary workload on a per-core basis. The PLLs have been designed such that clock rate tracks supply droop, i.e. transient voltage variations caused by changing current demands of the load circuitry. Those PLLs fabricated in 45 nm CMOS HKMG technology exhibit 1 μs lock time and 3.7 ps RMS jitter for a 375 ps (2.67 GHz) clock [201]. A duty cycle correction mechanism is made necessary by the fact that the circuit includes double data rate (DDR) subcircuits.[16]
□

In conclusion, the adoption of sophisticated closed-loop phase control schemes not only improves I/O timing but must be understood as a yield enhancement measure without which it would not have been possible to run chips as complex as modern microprocessors at extreme clock rates in spite of unpredictable process variations and aging.

---

[16] Duty cycle correction and DLLs are also part of Intel's so-called Quickpath on-chip interconnects that transmit multiple data bits in a differential format along with the pertaining clock.

7.5 HOW TO IMPLEMENT CLOCK GATING PROPERLY

7.5.1 TRADITIONAL FEEDBACK-TYPE REGISTERS WITH ENABLE

In most designs, part of the flip-flops operate at a lower rate than others. Just compare a pipeline register against a mode register that preserves its state for millions of clock cycles or even until power down. Other registers occupy positions in between. This situation suggests the usage of enable signals to control register operation as shown in the code fragments below.

```
.....
process (Clk_C)
begin
    -- activities triggered by rising edge of clock
    if Clk_C'event and Clk_C='1' then
        if Ena_S='1' then        -- control signal that governs
            State_P <= State_N;    -- register update
        end if;
    end if;
end process;
.....
```

```
.....
always_ff @(posedge Clk_C)
    // activities triggered by rising edge of clock
    if (Ena_S)                 // control signal that governs
        State_P <= State_N;    // register update
.....
```

When presented with such code, synthesizers call upon a multiplexer to close and open a feedback loop around a basic D-type flip-flop under control of the enable signal as shown in the example of fig.7.26b. As the resulting circuit is simple, robust, and compliant with the rules of synchronous design,[17] this is a safe and often also a reasonable choice. Some libraries indeed offer an E-type flip-flop that combines a flip-flop and a MUX in a single cell.

---

[17] Notably with the dissociation principle presented in section 6.4.

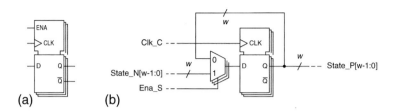

FIGURE 7.26

Register with enable. Icon (a) and safe E-type flip-flop circuit (b).

On the negative side, this approach takes one fairly expensive multiplexer per bit and does not use energy efficiently. This is because any toggling of the clock input of a disabled flip-flop wastes energy in discharging and recharging the associated node capacitances for nothing. Note that the capacitance of the clk_c input is not the only contribution as any clock edge causes further nodes to toggle within the flip-flop itself. Also, the wider a register, the larger the overhead in terms of energy and transistor count.

Most of the energy could be conserved by selectively turning off the clock within those circuit regions that are currently disabled.[18] Any such conditional clocking or **clock gating** scheme must be implemented very carefully, however, as safe circuit operation is otherwise compromised.

7.5.2 A CRUDE AND UNSAFE APPROACH TO CLOCK GATING

Using a simple AND gate to shut off the clock of positive-edge-triggered D-type flip-flops is particularly tempting, but also particularly dangerous, see fig.7.27a. This is because any glitch of the enable signal Ena_s is passed on to the gated clock Ckg_c while the clock input Clk_c is 1. All downstream bistables are exposed to hazards during the clock's first phase where hazards are particularly likely, jeopardizing their correct functioning. To make the circuit work, designers would have recourse to extensive hazard control with all its undesirable implications.[19]

---

[18] Refer to [202] for a conceptually different approach to conditional clocking that does not depend on the presence of a special enable signal but compares the data value at the input of a flip-flop against that stored.

[19] Such as complicated timing analysis, redundant logic for hazard suppression, finicky delay tuning, layout dependencies, difficult design verification, low design productivity, and all the other side-effects of ad hoc (i.e. non-self-timed) asynchronous operation.

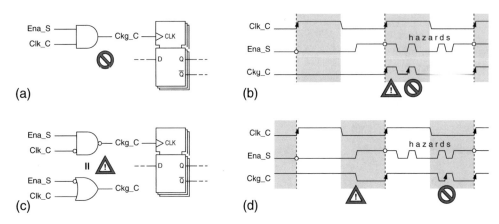

(a)   (b)   (c)   (d)

FIGURE 7.27

Unsafe clock gating circuits (a,c) along with the waveforms they produce (b,d).

7.5.3 A SIMPLE CLOCK GATING SCHEME THAT MAY WORK UNDER CERTAIN CONDITIONS

The circuit of fig.7.27c fares better because the NOT-NAND gate in the clock net is transparent for glitches during the second clock phase rather than during the first phase. This approach can be made to operate safely provided

- The timewise position of the passive clock edge on clock Clk_C is always well defined,
- All transients on the enable signal Ena_S are guaranteed to die out before the end of the first clock phase under any circumstance,
- The gated clock Ckg_C is rechecked for hazards following layout, the enable signal Ena_S is checked for hazard conditions during timing analysis, and
- The extra delay introduced by the gate is accounted for in the clock distribution network.

Making sure all of these requirements are consistently met complicates the design process, however. On the positive side, this approach minimizes circuit overhead, clock load, and energy dissipation during those periods when the clock remains disabled.

7.5.4 SAFE CLOCK GATING SCHEMES

More robust schemes are based on a special **clock gate**, a three-port cell that accepts the enable signal at terminal ENA plus the main clock at CLK and that outputs at CKG an intermittent clock signal which is to toggle only when the enable asks it to do so. Here come the specifications for such a contraption in the context of edge-triggered clocking:

- No latency, that is the enable input must affect the next active clock edge.
- The gated clock output CKG must be free of hazards under all circumstances.
- The enable input ENA must be immune to hazards
  (the absence of glitches from the input clock applied at CLK is taken for granted).
- The only timing constraints imposed on the enable input ENA must be the observation of reasonable setup and hold times (exactly as for an ordinary E-type flip-flop), any signal that emerges from combinational logic must qualify as enable.
- The gated clock must have the same duty cycle as the input clock
  (applies to single-edge-triggering exclusively).
- The propagation delay from terminal CLK to CKG must be small and fixed.
- Overall power dissipation must be low, especially while the clock is disabled.
- The gated clock should come up with a predictable polarity after circuit reset.

Table 7.3 Truth tables for safe conditional clocking in single-edge (left) and dual-edge triggered circuits (right).

ENA	CLK	CKG	
1	↑	↑	forward active clock edge
-	0	0	clamp output to zero
↑	1	CKG	maintain output
↓	1	CKG	*idem*

ENA	CLK	CKG	
1	↑	CKG	toggle output
1	↓	$\overline{\text{CKG}}$	*idem*
0	-	CKG	maintain output
↑	-	CKG	*idem*
↓	-	CKG	*idem*

The corresponding subcircuits are shown in fig.7.28 b) and e).[20] Ideally, clock gates are available from the standard cell library. If not so, such components should be added before HDL synthesis begins.[21] Compliance with the specifications must be checked and numeric timing figures must be obtained by way of transistor-level simulations exactly as for any other library cell.

---

[20] Though unnecessary from a purely functional point of view, a reset facility serves to bring the subcircuit into a predictable start state following power up.

[21] A soft macro that pieces together a clock gate from regular standard cells shall only be considered as a workaround because uncontrollable interconnect delays between the bistable and the combinational gate may put correct operation at risk. The internal wires must then be given so much weight that the cells involved get placed next to each other in an attempt to tightly control interconnect delays and their variations. Also, soft macros must be prevented from being torn apart by latter logic (re)optimization steps, see observation 4.42 for practical advice.

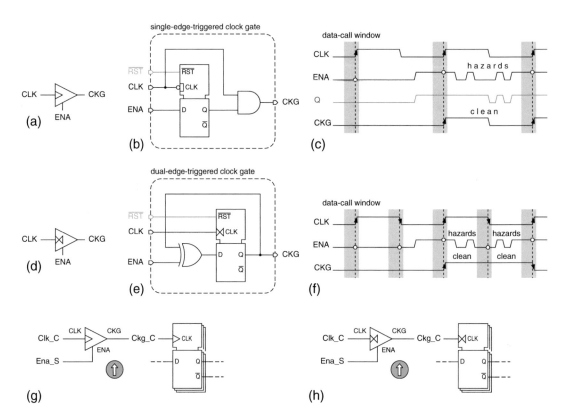

FIGURE 7.28

Safe clock gates for single-edge (b) and dual-edge-triggered circuits (e). Suggested icons (a,d). Pertaining clock waveforms (c,f). Final register circuits (g,h).

Make sure you understand that one clock gate gets shared by all flip-flops in a register as indicated in fig.7.28 g) and h). The overall power dissipated for clocking a *w*-bit register will be much lower than if *w* individual E-type flip-flops were being used as in fig.7.26. And the idea extends to larger storage arrays and even to entire circuit regions. The more flip-flops get served by a common clock gate, the more important the economy. It goes without saying that the propagation delay from CLK to CKG must be taken into account during clock tree generation much as with any regular clock buffer.

Table 7.4 compares all four options. The quest for optimum energy efficiency puts pressure on designers to resort to a multitude of local clocks with a small fanout each and, as a consequence, to prefer NOT-NAND-type gates over latch-based clock gates. Keep in mind this is dangerous unless you exactly know what you do. Further note that situations where a hierarchy of conditions is to exert control over local clocks require special attention, see problem 11. As a last word of caution, recall that scan testing mandates that all flip-flops of a scan path be enabled during scan-in and scan-out operations.

> *Hint: Before opting for a clock gating scheme, check what kind of subcircuits the EDA tool suite available can handle properly as part of clock tree generation.*

Table 7.4   Conditional clocking techniques for single-edge-triggered clocking compared.

Enable/disable mechanism	feedback via MUX	AND gate	NOT-NAND gate	latch-based clock gate
Illustration	fig.7.26b	fig.7.27a	fig.7.27c	fig.7.28b
Robustness	safe	unsafe	vulnerable	safe
Circuit overhead for a w-bit register [GE]	$1w \ldots 3w^a$	1.5	1.5	≈ 4.5
Energy balance	wasteful if disabled most of the time	optimal	optimal	good if many flip-flops get gated together
Impact on clock tree synthesis	none	must compensate for skew	must compensate for skew	must compensate for skew
Design effort	minimal, HDL only	not an option	recurring for each clock net	one-time to establish cell

^{a}The lower bound refers to an E-type flip-flop library cell, the upper to a separate MUX

Warning example

In search of energy efficiency, the developers of a low-power DSP core marketed as virtual component had come up with sophisticated clocking and clock gating schemes that synthesized properly for cell-based ASICs. Troubles began when one of their customers asked for rapid prototyping with field-programmable logic to speed up software development. The relatively few nets and buffers dedicated to clock distribution in the target FPGA did not suffice to accommodate the complex fine-grained clock gating scheme. To make things worse, reverting to standard E-type flip-flops turned out to be exceedingly difficult as a consequence from clock gating being entangled with functional operation at many places in the latch-based design.
□

Observation 7.13. *A good clocking scheme and HDL coding style must always allow for a straightforward conversion from enable-controlled bistables to clock gating and back with no impact on functional behavior. It further should be compatible with both ASICs and field-programmable logic devices.*

7.6 SUMMARY

- The prime problem in clock engineering is not the lack of solutions but the plethora of divergent alternatives, see table 7.5. Strictly <u>adhering to a consistent clocking discipline throughout</u> is paramount to successfully designing large and dependable digital circuits.
  - Edge-triggered clocking disciplines are most simple conceptually. However, their small latitude to clock skew and jitter requires careful balancing of the clock distribution network all the way to physical design. Careful timing verification is crucial, hold time fixing may help.
  - Single-edge-triggered one-phase clocking is most popular with both ASIC designers and EDA vendors.
  - All latch-based clocking disciplines with non-overlapping clocks boast a large tolerance with respect to skew and jitter. This is at the expense of performance in the unsymmetric case only.
  - Symmetric level-sensitive two-phase clocking provides room for optimizing the clock subsystem either for performance or for energy efficiency.
  - Translating an edge-triggered one-phase design into an equivalent unsymmetric level-sensitive two-phase design is straightforward which greatly helps in designing circuits of the latter category.
  - Level-sensitive one-phase clocking is hardly applicable to ICs due to its fatal dependency on stable contamination and interconnect delays.

- Macrocells, megacells and subcircuits synthesized from third-party HDL code tend to exhibit unexpected, undesirable and often undocumented timing features such as excessive hold times, massive internal clock distribution delays, absence of adequate clock buffers, a mix of rising- and falling-edge-triggered flip-flops, amalgams of level-sensitive and edge-triggered bistables, and more. Watch out for such peculiarities.

- Experience tells us that any departure from a plain one-phase-edge-triggered-flip-flops-only approach complicates the design process and asks for a more substantial engineering effort. Just about everything from interface design down to timing verification and design for test (DFT) tends to become more onerous to handle in the presence of
  - Asynchronous subcircuits and constructs,
  - Hybrid clocking schemes (e.g. mixes of edge-triggered with level-sensitive bistables),
  - Unclocked memories or other macrocells that have timing characteristics incompatible with standard flip-flops,
  - Multiple clock domains,
  - Multi-cycle paths,[22]
  - Conditional clocking, and — to a lesser extent — also

---

[22] A **multi-cycle path** is a signal propagation path that extends over more than one clock period (computation period in the occurrence of dual-edge triggering). Multicycle paths come into existence when VLSI architects decide to have parts of a circuit run at a lower rate than others, or when they accept to wait two — or more — clock cycles for the results from complex combinational operations to appear at the output of the pertaining logic in exchange for being allowed to clock the entire circuit at a faster rate than the longest path would normally permit.

- Latch-based design, and
- Dual-edge triggering.

Especially in combination, such departures tend to strain EDA tools beyond what they have been designed for.

- Ideally, one could think of a synthesis tool that accepts some functional model and that lets the designer select among the more popular clocking disciplines before generating a circuit netlist. Commercial HDL software does not work in this way, though, and designers have to adjust their RTL source code whenever they opt for a different clocking discipline.

- A good clock distribution network exhibits
  - Low clock skew and jitter,
  - Fast clock slopes,
  - Modest distribution delay, and
  - A good tolerance towards PTV and OCV as well as towards diverse circuit arrangements that emanate from automatic placement and routing (P&R).

While distributed trees and hybrid clock distribution networks prevail, careful design and post-layout timing verification are critical in any case.

- Designing clock distribution networks must be considered part of back-end design. In the occurrence of distributed clock buffering, use a trustworthy clock tree generator to synthesize, place and route the clock distribution networks. Never entrust clock tree insertion to standard logic synthesis tools as such products have not normally been designed to handle clock distribution.

- Circuit simulation does not suffice to detect skew-related problems. Rather, the circuit together with its clock distribution network must be subject to static timing verification. In synchronous designs, this basically implies the checking of setup and hold conditions along all signal propagation paths between any two bistables. Layout parasitics must be accounted for by working from post-layout netlists as obtained from layout extraction. Software tools for doing so are commonplace.

- Conditional clocking requires particular attention as it interferes with scan test, and as the simplest and most efficient clock gates are highly exposed to malfunction. The presence of logic gates other than buffers and inverters in clock distribution networks should always alert the expert during a design review.

- Most cell libraries are written with no particular clocking discipline or clock distribution scheme in mind. Thus, when evaluating a clocking strategy, do not let you be misguided by some arbitrary feature of your design environment. Put your system requirements first. That is, take into account
  - The compatibility with other subsystems to be integrated on the same chip, if any,
  - The library elements available such as flip-flops, latches, LSSD elements, memories, clock drivers, clock gates, PLLs, etc. along with the timing data that go with them,
  - The tools available for load estimation, layout extraction, circuit simulation, timing verification, clock tree generation, and circuit synthesis in general,

- - The accuracy of the timing model(s) being used,
  - The available software tools for inserting test structures,
  - The available EDA tools for generating and grading of test vectors,
  - The hardware test equipment at your disposal, and, last but not least,
  - Personal experience.

- Clock skew and jitter have become more acute with the ever shrinking feature sizes and clock periods. Luckily, today's semiconductor manufacturing processes provide many low resistance metal layers of which designers can take advantage for better clock distribution.

Table 7.5 Comparison of the most important clocking disciplines along with their compatibility and usage with clock distribution schemes.

Characteristics	Synchronous clocking discipline					
	edge-triggered		level-sensitive			
	single edge	dual edge	two-phase			one-phase
			sym-metric	unsym-metric	single wire	
Bistables per computation stage	SETFF	DETFF	two latches			latch
Clock nets, wiring resources	1	1	2	2	1	1
Relative clock power (approx.)[a]	$1 \cdot 1 \cdot 2$	$1 \cdot 1.3 \cdot 1$	$2 \cdot \frac{1}{2} \cdot 2$	$2 \cdot \frac{1}{2} \cdot 2$	$1 \cdot 1 \cdot 2$	$1 \cdot \frac{1}{2} \cdot 2$
All timing constraints one-sided	yes	yes	yes	yes	yes	no
Relatively tolerant to skew	no	no	yes	yes	no	no
Allows for time-borrowing	n.a.	n.a.	yes	no	no	n.a.
Utilization of clock period	≤ 1	≤ 1	≤ 1	< 1	≤ 1	> 1
Function latches	[0,1]	[0,1]	[0,1,2]	[0,1]	[0,1,2]	[0,1]
EDA support	full	limited	limited	limited	limited	poor
Key quality	slim, straight	slower clock	highly flexible	easily derived	high perf.	risky, onerous
Applied in conjunction with collective buffer	past ASICs				DEC Alpha	not for ASICs
Applied in conjunction with distributed tree	most ASICs	DDR circuits	some ASICs	IBM LSSD[b]		not for ASICs

[a] expressed as # clock_nets· # FF_load_caps_per_net· # edges_per_computation_cycle
[b] Three clock nets two of which are active at any time.

7.7 PROBLEMS

1. ** For all clocking disciplines discussed in this chapter visualize the skew margins in their Anceau diagrams. Use different colors for margins that (a) are a function of the operating speed of the circuit, (b) vanish unless the components involved have non-zero delays, and (c) do not depend on either one.

2. * Old hands among IC designers have long known that edge-triggered shift registers are particularly in danger of malfunctioning. As a remedy, they used to oppose the flows of clock and data during physical design, e.g. they had the clock wire run from right to left in a shift register where data travel from left to right. Explain why this helped.

3. * Reconsider the flip-flop timing data given in table 7.2. What do you think, is this a well-behaved flip-flop or does it impose timing conditions that are awkward to meet?

4. * Some designers like to combine rising-edge-triggered flip-flops with others that trigger on the falling edge in the same clock domain. Analyze the timing characteristics of this scheme and discuss its merits (wrt skew tolerance, performance, scan path insertion, energy efficiency, design verification, engineering effort, etc.). Does it offer any advantage over the disciplines discussed in this chapter?

5. * Yet another approach to edge-triggered clocking uses flip-flops that are built from three latches instead of two, see fig.7.29b. Consider a circuit where all standard flip-flops have been replaced by such bistables and carry out the same analysis as before.

 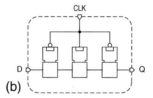

(a)   (b)

FIGURE 7.29

Master-slave-slave flip-flop. Proposed icon (a) and logically equivalent circuit (b).

6. ** The circuit of fig.7.30 occupies a position somewhere between single-edge-triggered and level-sensitive one-phase clocking. The motivation behind replacing part of the flip-flops by (pulse-clocked) latches is the search for improved energy efficiency. Savings are expected from a lighter clock load and from the reduced node count and switching activities in a latch when compared to those in the more complex master-slave flip-flop. Substitution takes place only where the combinational logic between two consecutive bistables can be demonstrated to meet the setup and hold time conditions of the receiver due to the circuitry's inherent contamination delay alone, i.e. with no extra gates or wiring detours.

Draw the Anceau diagram assuming that clock frequency remains the same as in the original circuit. Establish all relevant setup and hold margins. Formulate adequate replacement rules in a more formal way. Comment on the viability and effectiveness of this hybrid clocking scheme.

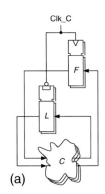

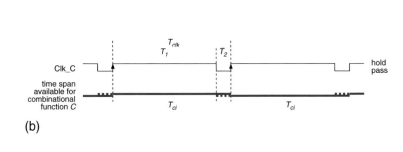

(a)

(b)

FIGURE 7.30

A hybrid clocking scheme that combines latches with flip-flops. Basic hardware organization (a) and simplified timing diagram (b).

7. ∗ Reconsider the timing data in fig.7.23 and calculate the length of the FPGA data-call windows for the various situations. What do you observe? Do you have an explanation?

8. ∗∗ Assume you want to include an on-chip RAM in a clock domain to be designed from a standard cell library. The edge-triggered synchronous RAM is to interface with a 14 bit address register and two 24 bit data registers, one for reading and one for writing. For simplicity, let's assume all flip-flops are identical. Both RAM and flip-flops trigger on the rising edge, their timing data are given next.

Cell	Timing parameters			
	t_{pd} [ns]	t_{cd} [ns]	t_{su} [ns]	t_{ho} [ns]
Flip-flop	0.5	0.2	0.3	0.1
RAM macrocell	2.5	1.5	1.0	0.8

a) Assuming zero skew, estimate the setup and hold margins if the circuit is to run at a clock frequency of 250 MHz. Where's the difficulty?

b) Make several proposals for solving the problem and evaluate them!

c) Study the options of inserting some delay τ in the various input and output lines of the RAM macrocell. Compile a table that lists the impact on the RAM's apparent timing figures if inserted (α) in address and data inputs, (β) in the data outputs, and (γ) in the clock input.

d) What is the result of deliberately designing the clock tree such as to slightly retard or advance the clock signal to the RAM? Does one of these help?

e) Can you imagine situations where it makes sense to apply such tricks? Is it possible to improve performance by designing a carefully imbalanced clock distribution network?

9. ∗∗ A commercial synthesis tool from the 1990s claimed to construct FSMs with hazard-free outputs by adding latches to all input and output lines of a Mealy type automaton. These extra latches were controlled by the same clock signal that drove the edge-triggered state register such as to always have either the input latches transparent and the output latches on hold, or vice versa. Find out when outputs are supposed to switch! As both latch sets are not allowed to be transparent simultaneously, outputs were believed to be isolated from asynchronous input changes. Show this is not always true!

Note: If you find it hard to analyze this state machine, imagine the difficulty of designing a larger circuit that mixes level-sensitive with edge-triggered clocking.

10. ∗ The impact of clock distribution delay on I/O timing has been illustrated in fig.7.22. Come up with a revised Anceau diagram for an analogous circuit with registered input and output this time.

11. ∗ Fig.7.31 shows two clock distribution trees that make use of clock gating. Whether local clocks Ckg1_C, Ckg2_C, Ckg3_C and Ckg34_C are to be active or not obviously depends on enable signals Ena1_S through Ena4_S generated from within the circuit itself. Circuit testing is supposed to happen via a full scan path that gets turned on from the Scm_TI scan mode terminal. Unfortunately, though, fig.7.31a includes four oversights. Locate them and explain what's wrong!

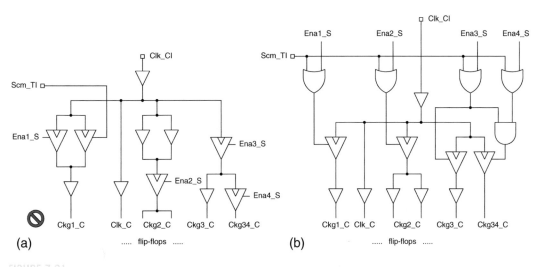

FIGURE 7.31

Hierarchical clock gating and scan test.

8

Most digital systems that interact with the external world must handle asynchronous inputs because events outside that system appear at random points in time with respect to the system's internal operation. As an example, a crankshaft angle sensor generates a pulse train regardless of the state of operation of the electronic engine management unit that processes those pulses.

Synchronization problems are not confined to electromechanical interfaces. Much the same situation occurs when electronic systems interact that are mutually independent, in spite — or precisely because — of the fact that each of them is operating in a strictly synchronous way. Just think of a workstation that exchanges data with a file server over a local area network (LAN), of audio data that are being brought to a D/A converter for output, or of a data bus that traverses the borderline from one clock domain to another. Obviously, the processing of asynchronous inputs is more frequent in digital systems than one would like.

The subsequent episode can teach us a lot about the difficulty of accommodating asynchronously changing input signals. The account is due to Charles E. Molnar who was honest enough to tell us about all misconceptions he and his colleagues went through until the problems of synchronization were fully understood.

Historical example

Back in 1963, a team of electronics engineers was designing a computer for biological researchers that was to be used for collecting data from laboratory equipment. In order to influence program execution from that apparatus, a mechanism was included to conditionally skip one instruction depending on the binary status of some external signal. The basic idea behind that design, sketched in fig.8.1a, was to increment the computer's program counter one extra time via a common enable input of its flip-flops iff the external signal was at logic 1.

When monitoring the computer's operation, the engineering team observed that the next instruction was occasionally being fetched from a bogus address after the external signal had been asserted thereby causing the computer to lose control over program execution. As an example, either of 0 1111 = 15 or 1 0000 = 16 is expected to follow after address 0 1110 = 14 depending on the external input. Instead, the program counter could be observed to enter a state such as 1 1101 = 29 or 0 0010 = 2. The team soon found out such failures developed only if the external signal happened to change just as the active

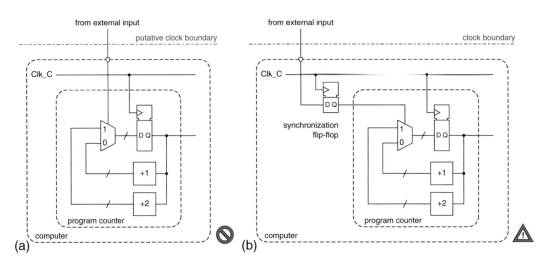

FIGURE 8.1

Control unit with conditional instruction skip mechanism (simplified). Original design without synchronization (a), improved design with external signal synchronized (b).

clock edge was about to occur, causing the program counter to end up with a wild mix of old and new bits.

The obvious solution was to insert a synchronization flip-flop such as to make a single decision as to whether the external level was 0 or 1 at the time of clock. Although the improved design, sketched in fig.8.1b, performed much better than the initial one, the team continued to observe sporadic jumps to unexpected memory locations, a failure pattern for which it had no satisfactory explanation at the time. It took almost a decade before Molnar and others who worked on high-speed interfaces[1] dare publicly report on the anomalous behavior of synchronizers and before journals would accept such reports that contrasted with general belief.

☐

Two failure mechanisms are exposed by this case, namely data inconsistency and synchronizer metastability, both of which will be discussed in this chapter together with advice for how to get them under control.

---

[1] What was considered high speed yesterday, no longer is high speed today. In the context of synchronization, "high-speed" always refers to circuits that operate at clock frequencies and data rates approaching the limits of the underlying technology.

8.2 DATA CONSISTENCY IN VECTORED ACQUISITION

We speak of vectored acquisition when a clock boundary — the hypothetical line that separates two clock domains — is being traversed by two or more electrical lines that together form a data, address, control or status word, or some other piece of information.

8.2.1 PLAIN BIT-PARALLEL SYNCHRONIZATION

Consider the situation of fig.8.2 where data words traverse a clock boundary on w parallel lines before being synchronized to the receiver clock clkQ in a register of w flip-flops. Due to various imperfections of practical nature,[2] some of the bits will switch before others do whenever the bus assumes a new value. If this happens near the active clock edge, the register is bound to store a **jumbled data** pattern that mixes old and new bits in an arbitrary way.

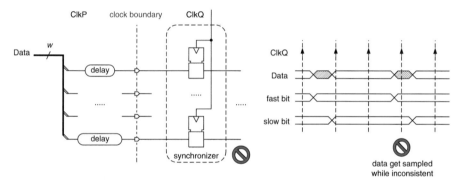

FIGURE 8.2

Non-simultaneous switching in a parallel bus may result in jumbled data.

Depending on context, the impact of occasional data jumble on a circuit's operation ranges between benign malfunction and fatal disaster:

- Data error for one cycle before being overwritten with correct value.
- Episodic disturbance lasting for several clock cycles (e.g. in a filter).
- Value outside legal range (e.g. if data is a complementary pair, follows one-hot encoding, or is coded in some other redundant format).
- Finite state machine (FSM) deflected to a state unplanned for.
- Finite state machine (FSM) trapped in a lockup situation.
- An address counter pointing to mistaken memory locations.

---

[2] Such as unmatched gate delays, unlike loads, distinct wire routes, unequal layout parasitics, clock skew, PTV variations, OCV, and crosstalk.

Warning example

A stream of digital audio data available in a 16bit parallel format is resynchronized to a 44.1 kHz output clock before being fed to a D/A converter. As audio samples are coded in 2's complement format (2'C), their range covers the interval $[-32\,768\,,\, +32\,767]$. Only one sample will be affected in the occurrence of a synchronization failure. Yet, to apprehend the impact of jumbled data, imagine two consecutive samples of low amplitude end up intermingled as follows.

sample	decimal	relative		2's complement code
$s(t)$ (correct)	+47	+0.0014	→	0000 0000 0010 1111
$s(t+1)$ (correct)	−116	−0.0035	→	1111 1111 1000 1100
maximum mix	+32 687	+0.9976	←	0111 1111 1010 1111
random outcome	+5 388	+0.1644	←	0001 0101 0000 1100
minimum mix	−32 756	−0.9997	←	1000 0000 0000 1100

□

Observation 8.1. *Data that cross a clock boundary on parallel lines cannot be synchronized with the aid of parallel registers alone. Data may otherwise get jumbled and upset downstream circuits.*

8.2.2 UNIT-DISTANCE CODING

We first observe that any pattern that results from jumbling two data words necessarily matches either of the two words iff their Hamming distance is one or less. The consistency problem can thus be solved by adopting a unit-distance code[3] provided data are known never to change by more than one step in either direction at a time, a requirement that confines unit-distance coding to applications such as the acquisition of position and angle encoder data.

clock count	unsigned binary decimal		code	Gray coding decimal		code
$c(t)$	+47	→	0010 1111	+47	→	0011 1000
$c(t+1)$	+48	→	0011 0000	+48	→	0010 1000
maximum mix	+63	←	0011 1111	+48	→	0010 1000
minimum mix	+32	←	0010 0000	+47	→	0011 1000

---

[3] **Unit-distance codes** have the particularity that any two adjacent numbers are assigned code words that differ in a single bit. They include the well-known Gray code (2^w) and the Glixon, O'Brien, Tompkins, and reflected excess-3 codes (binary coded decimal (BCD)). 4bit Gray coding, for instance, goes as follows:

decimal	Gray code	decimal	Gray code	decimal	Gray code	decimal	Gray code
15	1000	11	1110	7	0100	3	0010
14	1001	10	1111	6	0101	2	0011
13	1011	9	1101	5	0111	1	0001
12	1010	8	1100	4	0110	0	0000

Example

A sampling rate converter IC for digital audio applications included a subfunction for tracking the ratio of the two sampling frequencies in real time. The circuit was built on the basis of an all-digital phase locked loop (PLL) and asked for a counter clocked at frequency f_p the state of which had to be read out periodically with frequency f_q. The frequency ratio of the two clocks was known to be contained in the interval $[\frac{1}{2}...2]$ and to vary slowly. Experience had shown that any jumbled clock counts would impair the final audio signal in a critical way, even if increment and read-out operations happened to coincide only sporadically. An error-safe solution had thus to be sought. Fig.8.3 shows the relevant hardware portion which eliminated data jumble by having the clock count data traverse the clock boundary as Gray code numbers.
□

8.2.3 SUPPRESSION OF JUMBLED DATA PATTERNS

The idea here is to detect and ignore transients by comparing subsequent data words at the receiver end. If any mismatch is detected, data are declared corrupted and are discarded by having the synchronizer output the same value as in the cycle before. A comparison over just two words as in the circuit of fig.8.4 suffices, provided input data are guaranteed to settle within one clock period. To avoid loss of data, any data item must get sampled correctly at least two times in a row which implies that up to three clock intervals may be necessary until a new item becomes visible at the synchronizer output. A related but more onerous proposition is to use error detection coding to find out when a freshly acquired data word should be ignored.

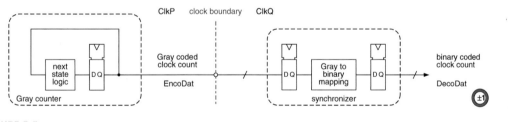

FIGURE 8.3
Frequency ratio estimator with vectored synchronizer based on a unit-distance code (partial view).

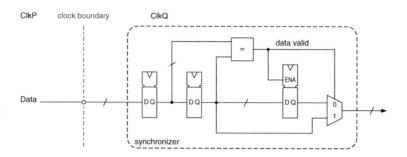

FIGURE 8.4

A vectored synchronizer that detects and ignores jumbled data.

Note that specialized hardware is not the only possible approach to reject jumbled data patterns. The same idea can also be implemented in software provided data rate, data transition time, and sampling rate can be arranged to be consistent with each other.[4]

8.2.4 HANDSHAKING

This approach contrasts with the previous one in that it prevents jumbled data from being admitted into the receiving circuit. The general idea is to avoid sampling data vectors while they might be changing. Instead, the updating and the sampling of the data get coordinated by a **handshake protocol** that involves both the producing and the consuming subsystems. As will become clear later in this section, handshake protocols also have other applications than just avoiding the emergence of jumbled data at clock boundaries. **Full handshaking** is essentially symmetrical and requires two control lines, termed **request** Req and **acknowledge** Ack respectively, that run in opposite directions, see fig.8.5a.

Observation 8.2. *The rules of full handshaking demand that any data transfer gets initiated by some specific event on the request line and that it gets concluded by an analogous event on the acknowledge line.*

Handshake protocols come in many variations, let us focus on a few important ones.[5]

---

[4] Observe the resemblance to the debouncing of mechanical contacts.

[5] What follows are comments on further variations of the basic handshake protocol. Notice from fig.8.5 that the request line has the same orientation as the data bus. This assumption, which we have made throughout our discussion, is referred to as a **push protocol** because any data transfer gets initiated by the producer. The data-valid message is transmitted over the request line, and the data secured message over the acknowledge line.

   In a **pull protocol** everything is reversed. The request line runs in the opposite direction than the data bus, transfers get initiated by the consumer, the request line carries the data secured message (that prompts the producer to deliver another data item), and the data-valid message is encoded on the acknowledge line. Yet, it is important to understand that these designations relate to naming conventions only, by no means do they imply that an active producer is driving a passive consumer, or vice versa.

   Further observe in fig.8.5d that the data bus holds valid data for only half of the time. The precise name of this scenario is "four-phase prolonged early push protocol". The freedom in deciding when to withdraw or overwrite a data vector, represented in fig.8.5d by the event labelled Val-, suggests this is not the only option. You may want to refer to [203] for an exhaustive discussion of handshake protocols and their terminology.

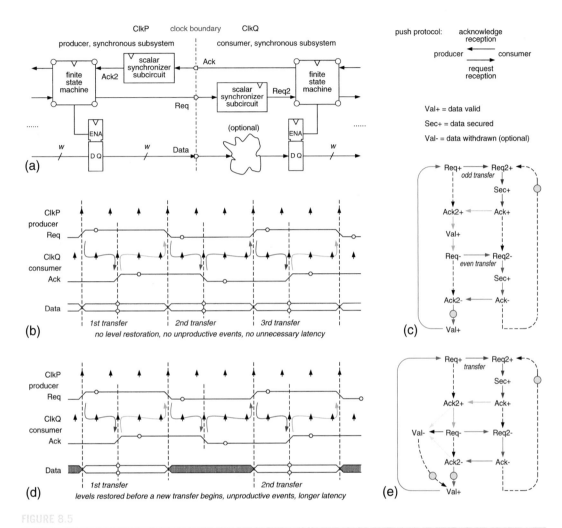

FIGURE 8.5

Full handshaking put to service for vectored synchronization. Circuit (a), waveforms with NRZ protocol (b), signal transition graph (STG) (c), same for RZ protocol (d,e).

NRZ or two-phase handshake protocol

The waveforms and event sequences of figs.8.5b and c show a possible scenario. Let both control lines be at logic 0 before the first data transfer begins. When the producer has finished preparing a new data word, he stores it in his output register. By toggling the Req line, he then informs the consumer that the Data vector has settled to a new valid state and requests it to ingest that data item. He thereby becomes liable of maintaining this state until data reception is confirmed by the consumer as the latter is free to accommodate and process the pending data item whenever he wants to do so.

When, some time later, the consumer has safely got hold of that data item and he is ready to accept another one, he toggles the Ack line running back to the producer to make this manifest.[6] Upon arrival of this confirmation, the producer does no longer need to hold the present data item but becomes free to withdraw or to overwrite it at any time. When the first data transfer comes to an end, both control lines are at 1, ready for a second transfer.

Notification occurs with transition signaling and each data transfer operation involves one event on each of the two control lines. This explains why the non-return-to-zero (NRZ) protocol is also known as two-phase, two-stroke, and transition protocol. An implication is that it always takes two consecutive data transfers before the control lines return to their initial logic values.

RZ or four-phase handshake protocol

The alternative scenario depicted in fig.8.5d and e contrasts with the above in that the control lines get restored to their initial values between any two consecutive transfer operations, hence the name return-to-zero (RZ). Four-phase, four-stroke, and level protocol are synonyms as this variation is based on level signaling and involves two extra events. Although data rate is cut to half of what is achievable with a NRZ protocol at the same clock frequency, this approach is more common because the hardware tends to be somewhat simpler and easier to understand. Observe that latency also suffers from the extra two phases that must be traversed.

With either signaling scheme, the handshake protocol makes sure that data vectors get sampled only while the electrical lines are kept in a stable and consistent state. The need for synchronization — together with the risk of metastability — is confined to two scalar control signals no matter how wide the data word is. Two-stage synchronizers are typically being used.

Full handshaking works independently of the relative operating speeds of the subsystems involved.[7] If, for instance, the producer were replaced by a much faster implementation, data transfers would continue correctly with no change to the consumer nor to the interface circuits. Although the producer would spend much less time in preparing a new data item, the mutual exclusion principle inherent in the protocol would take care that data items do not get produced at a rate that is incompatible with the consumer.

Observation 8.3. *The strict sequence of events imposed by a full handshake protocol precludes any loss of data, confines synchronization to two control lines, and makes it possible to design communicating subsystems largely independently from each other.*

The latter property greatly facilitates the modular design of complex systems as nothing prevents VLSI designers from taking advantage of full handshaking to govern data transfer operations even when all subsystems involved sit in the same clock domain. On the negative side, the symmetric protocol requires that both subsystems involved be **stallable**. That is, they must be capable of withholding their data production and consumption activities for an indeterminate amount of time if necessary. This is not always possible, though.

---

[6] Note that this does not necessarily imply that the processing of the last data item is actually completed. A pipeline, for instance, can accommodate new data as soon as the result from the processing of the previous item by its first stage has been properly stored in the subsequent pipeline register. There is no need for this data item to have finished traversing the second or any later stage. It should be obvious, however, that data must have propagated through any combinational logic placed between the producer's output register and the first data register in the consumer before the acknowledge is asserted.

[7] A questionable implementation where this is not the case is being discussed in problem 4 of section 8.6.

8.2.5 PARTIAL HANDSHAKING

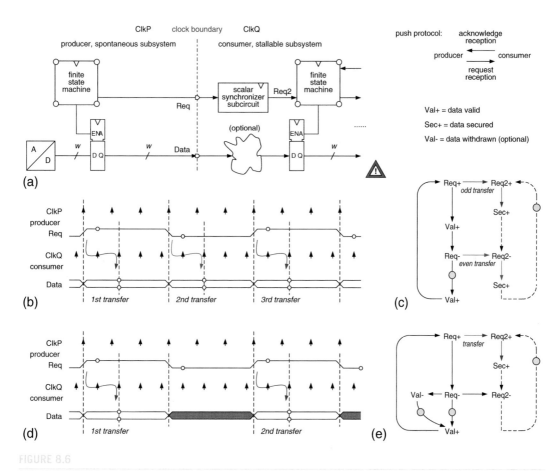

FIGURE 8.6

Partial handshaking in the occurrence of a fixed-rate producer. Circuit (a), waveforms and STG with NRZ protocol (b,c), and the same for RZ protocol (d,e).

Consider a digital data source operating with a fixed sampling rate. An example is given in fig.8.6a with an A/D converter acting as producer of audio samples. Such a data source works spontaneously and neither needs nor cares about protocols. Input terminal Ack thus becomes meaningless and must be dropped, thereby rendering the circuit and the protocol unsymmetrical.

More importantly, a fixed-rate producer is not stallable. This is to say that it does not wait, thereby forcing its counterpart to always complete the data processing in time before the next data item becomes available. Data items may otherwise get lost. By making assumptions about the response time of the partners involved, partial handshaking becomes exposed to failure when these assumptions change. In the example of fig.8.6c, Sec+ must occur before Val+ under any circumstances, which imposes restrictions on the mutual relationship of producer and consumer clock rates and latencies.

It is important to understand that porting an existing design to a new technology, reusing part of it in a different context, or applying dynamic voltage and/or frequency scaling are likely to challenge such assumptions. As opposed to this, full handshaking scales with frequency and operating speed because no assumptions are made.

Observation 8.4. *Partial handshake protocols cannot function properly unless the response times of the communicating subsystems are a priori known and can firmly be guaranteed to respect specific relationships under all operating conditions.*

For reasons of symmetry, all the above also holds for the converse situation where it is the consumer that is of non-stallable nature, see fig.8.7. Note the presence of an `Ack` line and the absence of a `Req`. The typical failure mode here is that data items will get read multiple times by the consumer should its counterpart fail to deliver new data in time.

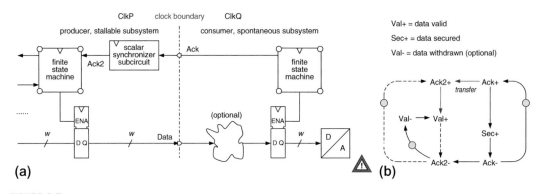

FIGURE 8.7

Partial handshaking in the occurrence of a fixed-rate consumer. Circuit (a) and STG with RZ protocol (b).

8.2.6 FIFO SYNCHRONIZERS

The idea is to insert some form of first-in first-out (FIFO) queue between two clock domains where writing in occurs from one domain and reading out from the other. Some people seem to believe this does away with all synchronization problems. Be cautioned, however, that keeping track of the fill level is more tricky than in a synchronous FIFO where read and write pointers belong to one clock domain. FIFO synchronizers clearly are about asynchronous circuit design!

Observation 8.5. *Subtle design flaws in FIFO synchronizers may become manifest under highly specific delay and timing conditions exclusively and are, therefore, unlikely to be found with RTL and other simulations due to idealizing assumptions there.*

Fig.8.8 shows a proven architecture where a dual-port RAM sits across the clock boundary and gets operated as a cyclic buffer by a read and a write pointer located in the producer and consumer domains respectively.[8] Observe that what gets actually synchronized are the memory pointers, not the data per se.

---

[8] One might as well use a register file instead of a RAM.

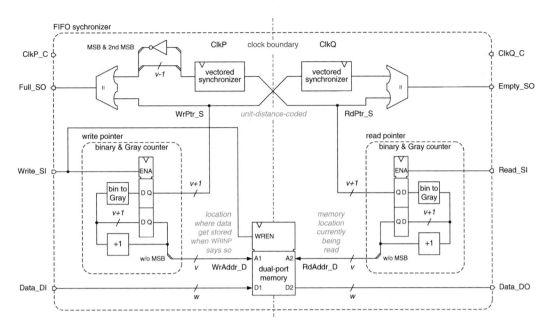

FIGURE 8.8

FIFO synchronizer built around a dual-port memory (simplified).[9]

The generation of the `Full` and `Empty` flags deserves further explanations. As both the full and the empty conditions are identified by the read and write counters pointing to the same address in a circular buffer, something more must be consulted to tell the two situations apart. A common trick is to prefix the address pointers with an extra digit. Make sure to understand that this new MSB is used in comparing the pointers exclusively, it is not part of the memory address busses. If buffer capacity is 2^v by w bit, then the pointer includes bits $(v, v-1, ..., 1, 0)$ out of which $(v-1, ..., 1, 0)$ actually serve to access the memory.

Next the two pointer vectors must be made to the traverse the clock boundary without being jumbled. Let us use a Gray code. As one can easily verify, the two conditions then become:

empty iff `WrPtr = RdPtr`

full iff `WrPtr`(i) $= \overline{\texttt{RdPtr}}(i)$ for $i = v$ and $i = v-1$ and
         `WrPtr`(i) = `RdPtr`(i) for all other bits $i = v-2, ..., 1, 0$.

There is one more difficulty. Simply using the lower bits $(v-1, ..., 1, 0)$ from the $v+1$-bit pointers to address the memory will not work with Gray codes as the two pointers would not visit memory locations

---

[9] Not shown is the reset mechanism which is straightforward to add. The two odd-looking icons near the top have been created as a more distinctive alternative for the box typically used for (arithmetic) comparators. The two dimples are meant to suggest the two pans of a kitchen scale.

in the same order, causing some data words to be duplicated and others to be lost. The reason is that truncating the MSB from the $v+1$-bit Gray code does not yield a v-bit Gray code because the lower bits are reflected rather than repeated, see footnote 3. One solution is to start from a $v+1$-bit Gray counter and to combine its MSB and 2nd MSB into the MSB of the v-bit code using an XOR operation. Instead, the approach behind fig.8.8 is to come up with a combined counter that produces binary and Gray codes in pairs, the former being used for accessing the memory and the latter for pointer comparison purposes.

The circuit shown here works by synchronizing the two pointers into the respective other clock domain. Variations of this architecture and more detailed advice can be found in [204] [205]. A somewhat leaner alternative where read and write pointers get compared asynchronously before synchronizing just the "full" and "empty" outputs is described in [206].

When compared to handshaking, FIFO synchronizers are more costly in terms of latency and circuit complexity. Their strength is that they can handle temporarily unequal rates of production and consumption with an amount of leeway proportional to buffer length v. Still, the producer must honor the "full" flag and never attempt to overstuff the queue, and vice versa for the consumer. This leaves the designer with two options.

- Producer and consumer must be arranged to converge to the same data rate before exhausting the buffer's length, e.g. by making data production and consumption dependent on extra "nearly full" and "nearly empty" flags.
- Either the producer or the consumer can be ensured to operate consistently faster than the other, in which case it must be made stallable by its slower counterpart.

What contributes to the popularity of FIFO synchronizers is that they are available as virtual and DesignWare components. As an aside, note that a synchronous FIFO queue can be useful for interfacing between major building blocks even when these are driven from a common clock.

8.3 DATA CONSISTENCY IN SCALAR ACQUISITION

As the name suggests, scalar acquisition implies that a clock boundary is being traversed by just one line. It may appear surprising that the acquisition of a single bit should give rise to any problem that is worth mentioning, yet, there are a few subtle pitfalls. In order to better understand the peculiarities, let us first examine how unsophisticated schemes fail in the presence of asynchronously changing inputs before proceeding to more adequate approaches.

8.3.1 NO SYNCHRONIZATION WHATSOEVER

In the circuit of fig.8.9a, a scalar input signal Data is being fed into two combinational subcircuits g and h that are part of a synchronous consumer circuit without any prior synchronization to the local clock ClkQ. Two deficiencies are likely to lead to system failure.

Firstly, signals G and H emanating from g and h respectively will occasionally get sampled during the time span between contamination and propagation delay when their values correspond neither to the settled values from the past interval t nor to those for the upcoming interval $t + 1$.[10] In the timing diagram of fig.8.9a such unfortunate circumstances apply to the central clock event.

Secondly, even though G and H may happen to be stable at sampling time, they may relate to distinct time intervals if $t_{cd\,g} > t_{pd\,h}$. If so, an inconsistent set of data gets stored in the two registers before being passed on to the downstream circuitry for further processing. This undesirable situation typically occurs when one of the paths includes combinational logic whereas the other does not. For an example, check the rightmost clock event in fig.8.9a.

8.3.2 SYNCHRONIZATION AT MULTIPLE PLACES

Adding synchronization flip-flops in front of all combinational subcircuits as in fig.8.9b improves the situation but does not suffice. This is because the flip-flops involved would sample the input slightly offset in time as a consequence from unbalanced delays along the paths Data $\rightarrow$ Data$_g$ and Data $\rightarrow$ Data$_h$, clock skew, unlike switching thresholds, noise, etc. Every once in a while, input data would get interpreted in contradicting ways as depicted in the timing diagram.

---

[10] Remember from section A.5 that combinational outputs do not necessarily transit from one stable value to the next in a monotonous fashion. Rather, they may temporarily assume arbitrary values due to hazards.

To stay clear of inconsistencies across multiple flip-flops, synchronization must be concentrated at a single place before any data are being distributed. This is the only way to make sure that all downstream circuitry operates on consistent data sampled at a single point in time. It is, therefore, standard practice to use synchronizers similar to those depicted in figs.8.9c and d.

Observation 8.6. *Any scalar signal that travels from one clock domain to another must be synchronized at one place by a single synchronization subcircuit driven from the receiving clock. When dealing with two complementary signals, it is best to transmit only one of them and to (re)obtain the complement after synchronization.*

A synchronizer clocked at rate f_{clkq} is bound to miss part of the input data unless all data pulses are guaranteed to last for at least one clock period $1/f_{clkq} = T_{clkq} < \min(t_{data\,valid\,i})$. Note the incongruity of data rate and sampling rate in fig.8.9c and the ensuing data loss. Actually, the sampling period of the synchronizer must provide sufficient leeway to accommodate setup and hold times as well as data transients. What if the clock available on the consumer side is simply too slow?

- Convert the data stream from its bit-serial into a bit-parallel format with the aid of a shift register, thereby making it possible to transmit data at a much lower rate. Then use one of the vectored acquisition schemes found to be safe.
- Use an analog phase locked loop (PLL) to clock the synchronizer at a faster rate.
- Recur to dual-edge-triggered one-phase clocking if clocking the synchronizer at twice its original frequency suffices.

Even with the best synchronization scheme, an active clock edge and an input transition will occasionally coincide when signals get exchanged between two independent clock domains, see fig.8.9d. For the system to function correctly, the synchronizer must then decide for either one of the two valid outcomes as the downstream circuitry cannot handle ambiguous data. This is precisely the subject of the subsequent section.

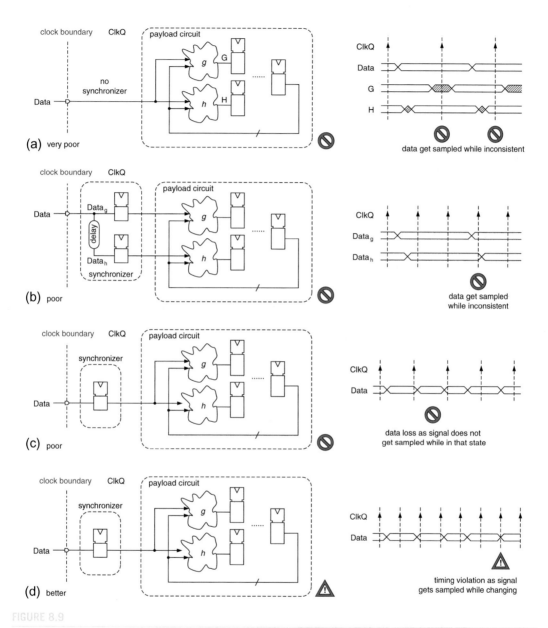

FIGURE 8.9

Acquisition of scalar inputs (irrelevant details not shown). No synchronization (a), synchronization at multiple places (b), and synchronization at a single place with inadequate (c), and with adequate sampling rate (d) (recommended).

8.4 MARGINAL TRIGGERING AND METASTABILITY

8.4.1 METASTABILITY AND HOW IT BECOMES MANIFEST

Reconsider the various synchronization schemes discussed. What they all have in common is a flip-flop — or a latch — the data input of which is connected to the incoming signal. As no fixed timing relationship can be guaranteed between data and clock, the data signal will occasionally switch in the immediate vicinity of a clock event, thereby ignoring the requirement that input data must remain stable throughout a bistable's data-call window.

"What happens when an input change violates a bistable's setup or hold condition?"

Technically, this kind of incident is referred to as **timing violation** or **marginal triggering**.[11] Until the mid 1970s, it was generally believed that the bistable would then decide for either of the possible two outcomes right away and output either a 0 or a 1. As stated in the introduction to this chapter, it took a long time until the design community began to understand this was not necessarily true.

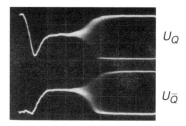

U_Q

$U_{\bar{Q}}$

FIGURE 8.10

Measured behavior of a CMOS latch in response to marginal triggering (photo reprinted from [207] with permission).

What had been observed were intermediate voltages and excessive delays before the two complementary outputs of the bistable eventually settled to a normal steady-state condition in response to marginal triggering. It was found that delays occasionally exceeded officially specified propagation delay figures by orders of magnitude, which made it clear that a better understanding of the process was imperative. We will, in the remainder of this chapter, summarize important results from empirical measurements and theoretical analysis of this phenomenon.

---

[11] Please recall that clocked (sub)circuits not only impose
- minimum setup and hold times t_{su} and t_{ho} but also
- minimum clock pulse widths $t_{clk\,lo\,min}$ and $t_{clk\,hi\,min}$,
- maximum clock rise and fall times $t_{clk\,ri\,max}$ and $t_{clk\,fa\,max}$, and
- absence of runt pulses and glitches on clock and asynchronous (re)set inputs.

Disregarding any such timing condition is likely to cause marginal triggering. The customary abstraction of flip-flops, latches and RAMs into bistable memory devices the behavior of which is entirely captured by way of truth tables, logic equations, and the like then no longer holds.

Today's digital subcircuits are inherently analog networks, and bistables are no exceptions. The data retention capability of latches, flip-flops, and SRAM cells is essentially obtained from closed feedback loops built from two inverting amplifiers. Fig.8.11b shows the transfer characteristics of two CMOS inverters where the output of either inverter drives the input of the other. Two stable points of equilibrium reflect the binary states 0 and 1 respectively. A third and unstable point of equilibrium exists in between. More precisely, the state space features two "valleys of attraction" separated by a line of unstable equilibrium. Marginal triggering implies bringing a bistable very close to that separation line before leaving it to recover. The bistable is then said to hang in an evanescent **metastable condition**.[12]

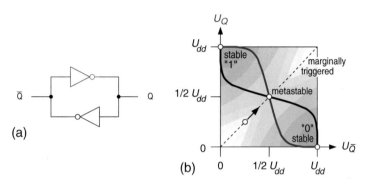

(a)

(b)

FIGURE 8.11

Points of equilibrium and marginal triggering. Cross-coupled inverting amplifiers (a) and state space with superimposed transfer curves (b).

Mathematical analyses of the metastable behavior of cross-coupled inverters can be found in the literature such as in [208] [207] [60]. While correctly modeling the behavior of the feedback loop, they do not necessarily describe the waveforms observable at the output terminals of actual latches and flip-flops. Similarly, node voltages half-way between zero and U_{dd}, such as those depicted in fig.8.10, are not normally visible at the outputs of a library cell or physical component. This is because of the various auxiliary subcircuits inserted between the metastable memory loop itself and the I/O terminals. Output buffers, for instance, tend to restore intermediate voltages to regular logic 0s and 1s. Device characteristics and physical layout also matter. Fig.8.12 shows how metastable conditions normally become manifest in various bistables.

Eventually the circuit returns to either of the two states of stable equilibrium. Yet, not only is the outcome of the decision process unpredictable, but the time it takes to recover from a metastable equilibrium condition necessarily exceeds the bistable's customary propagation delay, see fig.8.13. In fact, **metastability resolution time** has been reported to outrun propagation delay by orders of magnitude on occasion, so that we must write $t_{mr} > t_{pd}$ and sometimes even $t_{mr} \gg t_{pd}$. Most alarming are the facts that it is not possible to guarantee any upper bound for the time a bistable takes to recover

---

[12] A philosophical concept known as Buridan's Principle states that a discrete decision based upon input having a continuous range of values cannot be made within a bounded length of time. It is named after the fourteenth century philosopher Jean Buridan who claimed that a donkey placed at the same distance from two bales of hay would starve to death because it had no reasons to choose one bale over the other.

and that the behavior of a physical part is no longer consistent with the one published in the datasheet and captured in the simulation model.

Observation 8.7. *Metastability is a problem because of unpredictable delay, not because of unpredictable logic outcome.*

While it is true that actual behavior greatly varies from one type of bistable to the next, it should be understood that the phenomenon of metastability is a fundamental one observed independently from circuit and fabrication technology. There are currently no flip-flops or latches that are free of metastable behavior.

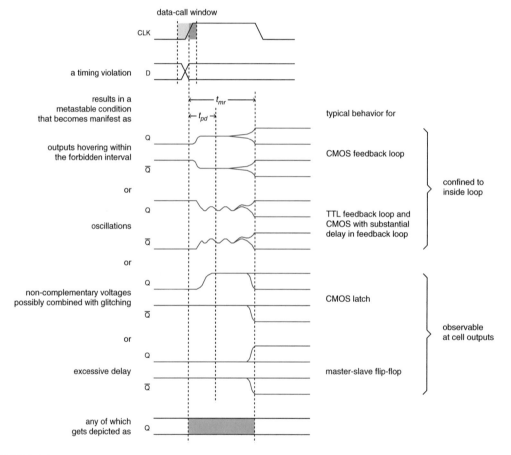

FIGURE 8.12

Patterns of metastable behavior in response to timing violations.

As a consequence from a synchronizer hanging in the metastable condition,

- o The downstream circuitry may process wrong data, or
- o May be presented with voltages within the forbidden interval, or
- o May find itself short of time for settling to a steady-state value before the next clock event arrives, see fig.8.14.

In the latter two cases, some downstream storage elements will be subject to marginal triggering themselves thereby permitting metastability to spread further into the clock domain. In any of the three cases, part of the system is likely to malfunction.

Warning example

The erratic behavior that Molnar and his colleagues had observed with their laboratory computer even after they had added a flip-flop for synchronizing the external input was in fact the result of metastability in the program counter. The problem was exacerbated by the slow germanium transistors then available and the poorly designed flip-flop circuits. Yet, it is unfair to blame the circuit designers for this as nobody was aware of metastability and of the importance of quick recovery in synchronizers at the time.

□

Observation 8.8. *Metastability together with its fatal consequences is completely banned from within a clock domain iff care is taken to meet all timing requirements under all operating conditions. As opposed to this, there is no way to truly exclude metastability from occurring at the boundaries of independently clocked domains.*

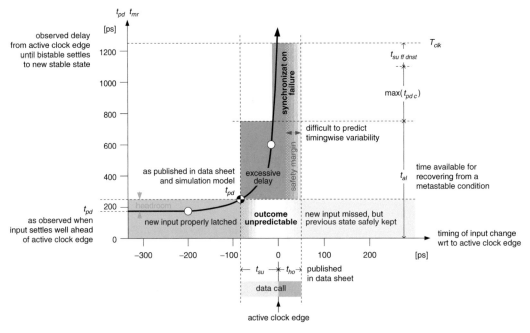

FIGURE 8.13

Measured timing characteristics of a bistable when input data change in the vicinity of the data-call window (arbitrary numerical data).

8.4.3 A STATISTICAL MODEL FOR ESTIMATING SYNCHRONIZER RELIABILITY

As metastability has been found to be unavoidable in synchronizers, it is essential to ask

"How serious is the metastability problem really?"

As illustrated by the two quotes below,[13] even experts do not always agree.

> "Having spent untold hours at analyzing and measuring metastable behavior, I can assure you that it is (today) a highly overrated problem. You can almost ignore it."

> "Having spent untold hours debugging digital designs, I can assure you that metastable behavior is a real problem, and every digital designer had better understand it."

Within their respective contexts, both statements are correct which indicates that we must develop a more precise understanding of the phenomenon. Theoretical and experimental research has come up with statistical models, that can serve as a basis for estimating the probability of system failures

---

[13] By Peter Alfke and Bruce Nepple respectively.

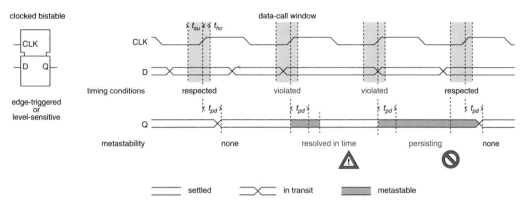

FIGURE 8.14

Recovery from marginal triggering and the impact on downstream circuitry.

due to metastability. Consider a flip-flop driven by a clock of frequency f_{clk} that is connected to an asynchronous data signal with an average edge rate of f_d. We will speak of a **synchronization failure** whenever a metastable condition persists for longer (after the synchronizer has been clocked) than the situation allows, i.e. when $t_{mr} > t_{al}$. The **mean time between errors** (MTBE) for such an arrangement has been found to obey the general law

$$t_{MTBE} = \frac{e^{K_2 t_{al}}}{K_1 f_{clk} f_d} \tag{8.1}$$

where parameters K_1 and K_2 together quantify the characteristics of the synchronizer flip-flop with respect to marginal triggering and metastability resolution [208] [209]. K_2 is an indication of the gain-bandwidth product around the memory loop of its master latch. K_1 represents a timewise window of susceptibility for going metastable.

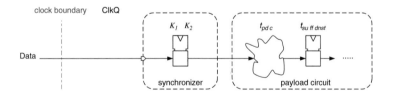

FIGURE 8.15

Circuit diagram of single-stage synchronizer.

Now consider a single-stage synchronizer like the one of fig.8.15 and ask yourself for how much time the first flip-flop is allowed to rave in an out-of-the-normal condition for the downstream logic to stay clear of trouble? The timewise allowance for resolving metastability t_{al} is simply the clock period T_{clk} diminished by the setup time of the downstream flip-flop(s) $t_{su\,ff\,dnst}$ and the propagation delay

$\max(t_{pdc})$ along the longest path through any combinational circuitry inserted between the flip-flops. You want to refer back to fig.8.13 for an illustration.

$$t_{al} = T_{clk} - \max(t_{pdc}) - t_{su\,ff\,dnst} \quad \text{with} \quad T_{clk} = \frac{1}{f_{clk}} \tag{8.2}$$

Example

The metastability characteristics of a flip-flop in a XC2VPro4 FPGA from the Xilinx Virtex II Pro family are listed in table 8.1 for typical operating conditions. For $f_{clk} = 100$ MHz, $f_d = 10$ MHz, $t_{su\,ff\,dnst} = 0.5$ ns and $t_{pdc} = 4$ ns, the calculated MTBE exceeds the age of the universe. Just doubling the clock frequency reduces the MTBE to a mere 4 s, which demonstrates that the impact of metastability is extremely dependent on a circuit's operating speed. Whether the single-stage synchronizer of fig.8.15 is acceptable or not, thus turns out to be a quantitative question.

□

8.4.4 PLESIOCHRONOUS INTERFACES

Being a statistical model, (8.1) assumes that there is no predictable relationship between the frequencies and phases of the data and clock signals. This does not always apply, though. A notable exception are **plesiochronous** systems where data producer and consumer are clocked from separate oscillators that operate at the same nominal frequency. Think of a local area network (LAN), for instance. Once clock and data have aligned in an unfortunate way at the receiver end, successive data updates are subject to occur during the critical data-call window many times in a row, thereby rendering the notion of MTBE meaningless.

Observation 8.9. *Plesiochronous interfaces are exposed to burst-like error patterns and are not amenable to analysis by simple statistical models that assume uncorrelated clocks.*

Plesiochronous interfaces require some sort of self-regulating mechanism that avoids consecutive timing violations and repeated misinterpretation of data. A tapped delay line may be used from which an adaptive circuit selects a tap such as not to sample a data signal in the immediate vicinity of a transient [210]. Two related ideas are adaptively shifting the data or the clock signal via a digitally adjustable delay line [211], and oversampling the input data before discarding unsettled and duplicate data samples. None of these approaches is free of occasional coincidences of clock events and data changes, however. The problem is just transferred from the data acquisition flip-flop to the subcircuit that adaptively selects a tap or controls sampling time.

8.4.5 CONTAINMENT OF METASTABLE BEHAVIOR

Limiting the harmful effects of metastability is based on insight that directly follows from (8.1) and (8.2). Keep the number of synchronization operations as small as possible and allow as much time as practical for any metastable condition to resolve. More specific suggestions follow.

Estimate reliability at the system level

Having a fairly accurate idea of the expected system reliability is always a good starting point. It makes no sense to try to improve synchronization reliability further when the MTBE already exceeds the expected product lifetime. However, always remember that (8.1) and all further indications in this text refer to one scalar synchronizer and that a system may include many of them. Also keep in mind that statistical models do not apply to plesiochronous operation.

In practice, a frequent problem is that only few bistables come with datasheets that specify their metastability resolution characteristics. Luckily, there exists a workaround.

Observation 8.10. *As a rule of thumb, synchronization failure is highly infrequent if a flip-flop is allowed three times its propagation delay or more to recover from a metastable condition.*

In the absence of numerical K_1 and K_2 values, refraining from detailed analysis is probably safe if the application is not overly critical and if $t_{al} \geq 3\, t_{pd}$ can be guaranteed throughout.

Select flip-flops with good metastability resolution

Flip-flops optimized for synchronizer applications feature a small K_1 and, above all, a large K_2. They shall be preferred over general-purpose flip-flops with inferior or unknown metastability recovery characteristics.[14] Unfortunately, component manufacturers and library vendors continue to be extremely reticent to disclose quantitative metastability data.[15]

Table 8.1   Metastability resolution characteristics of various CMOS flip-flops. Note: The figures below have been collected from distinct sources and do not necessarily relate to the same operating conditions. Still, a massive improvement over the years is evident.

D-type flip-flop				Metastability	
reference or vendor	cell type or name	technology	F [nm]	K_1 [ps]	K_2 [GHz]
Horstmann et al. [208]	n.a.	std cell	1500	47 600	3.23
VLSI Technology	DFNTNS	std cell	800	140 000	12.3
Ginosar [211]	n.a. "conservative"	std cell	180	50	100
Xilinx [209]	XC2VPro4 CLB	FPGA	130	≈ 100	27.2

---

[14] [212] finds that static bistables should be preferred over their dynamic counterparts.

[15] There are two reasons for this. For one thing, the issue is no longer perceived as urgent now that the speed and metastability resolution characteristics of flip-flops have improved so much when compared to older fabrication technologies. For another thing, it takes a considerable effort and degree of sophistication to properly determine the K_1 and K_2 parameters for a cell library. The burden of doing so is thus left to VLSI designers in critical high-speed applications, see [208] [209] [213] for measurement set-ups. It is important that such characterizations be carried out under operating conditions that are as identical as possible to those actually encountered by the synchronizer when put into service in the target environment. Relevant conditions include capacitive load, layout parasitics, clock slew rate, fabrication process, and, last but not least, PTV conditions.

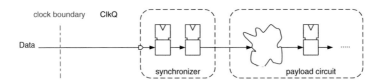

FIGURE 8.16

Two-stage synchronizers obtained from adding an extra flip-flop.

Remove combinational delays from synchronizers

Recall from (8.2) that any extra delay $t_{pd\,c}$ between two cascaded flip-flops is at the expense of recovery time t_{al} for the first bistable. As a consequence, the one-stage synchronizer circuit of fig.8.15 is far from being optimal at high clock rates. When the MTBE proves insufficient, a better solution must be sought.

Two flip-flops cascaded with no combinational logic in between extend the time available for metastability resolution to almost an entire clock period. The low error rates typically obtained in this way have contributed to the popularity of **two-stage synchronizers** shown in fig.8.16. In case the additional cycle of latency resulting from the extra flip-flop proves unacceptable, try to reshuffle the existing registers or check [214] for a proposal that operates multiple two-stage synchronizers in parallel.

Drive synchronizers with fast-switching clock

Experience has shown that clocking synchronizers with fast slew rates tends to accelerate recovery from a marginal triggering condition [208]. What's more, overly slow clock ramps tend to dilate setup and hold times beyond their nominal values as obtained from library characterization under the assumption of a sharp clock edge with zero or close-to-zero ramp time. Yet, these are the figures stored in simulation models and listed in datasheets on which design engineers necessarily base all their reasoning.

Free synchronizers from unnecessary loads

Not surprisingly, capacitive loading has been found to slow down the metastability resolution process in a bistable. It is therefore recommended to keep the loads on synchronizer outputs as small as possible by using buffers and buffer trees where necessary.

Lower clock frequency at the consumer end

As a minor change of the clock frequency has a large impact on the MTBE, it is always worthwhile to check whether it is not possible to operate the entire consumer from a somewhat slower clock.[16]

---

[16] A more exotic proposal is the concept of a **pausable clock**, where metastability is detected by way of analog circuitry designed for that purpose, and where the consumer's clock is frozen until it has been resolved [215].

Use multi-stage synchronizers

In those — extremely infrequent — situations where a two-stage synchronizer does not leave enough time for metastability to resolve, the available time span can be extended well beyond one clock period by recurring to synchronizer circuits that make use of multiple flip-flops. Please refer to [215] where the merits of cascaded and clock-divided synchronizers are evaluated.

Keep feedback path within synchronizers short

Digital VLSI designers normally work with predeveloped cell libraries. In case you must design your own synchronizers at the transistor level, consult the specialized literature on the subject [216] [217] [208] [207] [215]. For the purpose of analysis, the two cross-coupled amplifiers can be replaced by a linear model in the vicinity of the metastable point of equilibrium. As a rule, the internal feedback path should be kept as fast as possible. A fast master latch is desirable because a high gain-bandwidth product tends to improve recovery speed, K_2 and MTBE. This is also why a higher supply voltage has been found to be beneficial.

A variety of misguided approaches to synchronization, such as a "metastability blocker" or a "pulse synchronizer", for instance, are collected in [211].

8.5 SUMMARY

- Asynchronous interfaces give rise to two problems, namely jumbled data and metastability. While it is always possible to avoid data inconsistencies altogether by making use of adequate data acquisition schemes and protocols, the metastability problems that follow from marginal triggering can be tackled in a probabilistic fashion only.

- Metastability is often put forward as a welcome explanation for synchronization failures because of its unavoidable and unpredictable nature. Yet, we tend to believe that most problems actually result from data consistency problems that have been overlooked.

- Metastability becomes a threat to system reliability when synchronizers are being operated close to their maximum clock and data frequencies, and/or when synchronizers are involved in very large quantities. Circuits that can afford a few extra ns of the clock period for the synchronizers to recover should be safe, on the other hand. Two-stage synchronizers normally prove more than adequate in such situations.

- Do not expect functional simulation or testing to uncover many synchronization problems as fatal flaws may pass unnoticed. In addition to the usual coverage problem, provoking a synchronization failure may require many repeated runs that just differ in relative shifts of events in the sub-ns range. This is neither efficient nor does it inspire much confidence. A procedure for finding which clock domain crossings in a circuit may need to be improved is proposed in [218].

- To steer clear of the imponderabilia and the extra costs associated with synchronization, implement the rules below.
  - Eliminate uncontrolled asynchronous interfaces wherever possible, that is partition a system into as few clock domains as technically feasible.
  - If you must traverse a clock boundary, do so where data bandwidth is smallest.
  - Estimate error probability or mean time between errors at the system level.
  - Whenever possible, set aside some extra time for synchronizers to settle.
  - Within each clock domain, strictly adhere to a synchronous clocking discipline.
  - Avoid (sub)circuits that tend to fail in a catastrophic manner when presented with corrupted data.

8.6 PROBLEMS

1. ∗ What is wrong with the two-stage synchronizer circuit of fig.8.17?

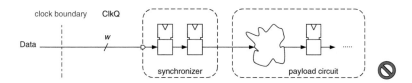

FIGURE 8.17

Bad synchronizer circuit.

2. ∗ Reconsider Molnar's original circuit of fig.8.1a and recall that the computer failed when its program counter became filled with a bogus address as a consequence from a synchronization failure. At first sight, it appears that only a scalar signal is being acquired from externally, so that there should be no chance for an inconsistent address word to develop. Find out why this is not so.

3. ∗∗ Establish detailed state diagrams for the two finite state machines in fig.8.5. Generate all necessary control signals and try to keep latency small. Depending on how the interfaces with the surrounding circuitry are defined, there may be more than one acceptable solution. Can you design the models such as to minimize the differences in the HDL codes of two- and four-phase protocols?

4. ∗∗ Fig.8.18 shows an arrangement that has a long tradition for carrying data vectors from one clock domain to another. What sets it apart from the handshake circuit of fig.8.5 is a bistable that sits right on the clock boundary. This has earned the circuit names such as "shared flip-flop" or "signaling latch" synchronizer although an unclocked data-edge-triggered seesaw is typically being used (a level-sensitive seesaw is sometimes also found). The shared bistable functions as a flag, set by the producer and reset by the consumer, that instructs one partner to carry out its duty and the other to wait. Much as in fig.8.5, each of the two control signals gets accepted into the local clock domain by a standard two-stage synchronizer. Compare the two circuits and their detailed operation.

The correct functioning of the circuit of fig.8.18 rests on an assumption that may or may not hold in real world applications, however. Find out what that assumption is. Establishing a signal transition graph (STG) may help. Hint: Consider situations where the two clock frequencies greatly differ from each other.

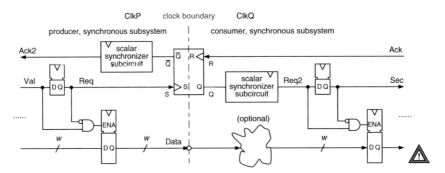

FIGURE 8.18

Vectored synchronization on the basis of a shared bistable.

5. ∗ In fig.8.13, three points are marked by empty circles or a propeller respectively. Explain what the three points have in common. What sets the propeller apart from the other two marks?

ELEMENTARY DIGITAL ELECTRONICS

A.1 INTRODUCTION

Working with electronic design automation (EDA) tools requires a good understanding of a multitude of terms and concepts from elementary digital electronics. The material in this chapter aims at explaining them, but makes no attempt to truly cover switching algebra or logic optimization as gate-level synthesis is fully automated today. Readers in search of a more formal or more comprehensive treatise are referred to specialized textbooks and tutorials such as [219] [220] [36] and the seminal but now somewhat dated [221].[1] Textbooks that address digital design more from a practical perspective include [222] [223] [224] [225] [226].

Combinational functions are discussed in sections A.2 and A.3 with a focus on fundamental properties and on circuit organization respectively before section A.4 gives an overview on common and not so common bistable memory devices. Section A.5 is concerned with transient behavior which then gets distilled into a few timing quantities in section A.6. At a much higher level of abstraction, section A.7 finally sums up the basic microprocessor data transfer protocols.

A.1.1 COMMON NUMBER REPRESENTATION SCHEMES

Our familiar decimal number system is called a **positional number system** because each digit in a number contributes to the overall value with a weight that depends on its position (which was not so with the ancient Roman numbers, for instance). In a positional number system, there is a natural number $B \geq 2$ that serves as a **base**, e.g. $B = 10$ for decimal and $B = 2$ for binary numbers. Each digit position i is assigned a weight B^i so that when a non-negative number gets expressed with a total of w digits, the value follows as a weighted sum

$$(a_{i_{MSD}}, a_{i_{MSD}-1}, a_{i_{MSD}-2}, ..., a_{i_{LSD}+1}, a_{i_{LSD}})_B = \sum_{i=i_{LSD}}^{i_{MSD}} a_i B^i \qquad (A.1)$$

where $i_{MSD} \geq i_{LSD}$ and $w = i_{MSD} - i_{LSD} + 1$. A radix point is used to separate the integer part made up of all digits with indices $i_{MSD} \geq i \geq 0$ from the fractional part that consists of those where $-1 \geq i \geq i_{LSD}$. When writing down an integer, we normally assume $i_{LSD} = 0$. As an example, 173_{10} stands for $1 \cdot 10^2 + 7 \cdot 10^1 + 3 \cdot 10^0$ ($i_{MSD} = 2$, $i_{LSD} = 0$, $w = 3$). The leftmost digit position has the

---

[1] Those with a special interest in mathematics may want to refer to appendix 3.11 where switching algebra is being put into perspective with fields and other algebraic structures.

largest weight $B^{i_{MSD}}$ while the rightmost digit has the smallest weight $B^{i_{LSD}}$. In the context of binary numbers, these two positions are referred to as the most (MSB) and as the least significant bit (LSB) respectively. 101.01_2, for instance, stands for $1 \cdot 2^2 + 0 \cdot 2^1 + 1 \cdot 2^0 + 0 \cdot 2^{-1} + 1 \cdot 2^{-2} = 5.25_{10}$ ($i_{MSB} = 2$, $i_{LSB} = -2$, $w = 5$).

Table A.1 Representations of signed and unsigned integers with four bits.

bit pattern MSB LSB $a_3\,a_2\,a_1\,a_0.$	interpreted as unsigned	interpreted as signed			
		offset-binary O-B	2's complem. 2'C	1's complem. 1'C	sign & magn. S&M
1111.	15	7	-1	$[-]\,0$	-7
1110.	14	6	-2	-1	-6
1101.	13	5	-3	-2	-5
1100.	12	4	-4	-3	-4
1011.	11	3	-5	-4	-3
1010.	10	2	-6	-5	-2
1001.	9	1	-7	-6	-1
1000.	8	0	-8^a	-7	$[-]\,0$
0111.	7	-1	7	7	7
0110.	6	-2	6	6	6
0101.	5	-3	5	5	5
0100.	4	-4	4	4	4
0011.	3	-5	3	3	3
0010.	2	-6	2	2	2
0001.	1	-7	1	1	1
0000.	0	-8^a	0	0	0
bit weights	$2^3\,2^2\,2^1\,2^0.$	idem $-\,2^3$	$-\,2^3\,2^2\,2^1\,2^0.$	$-(2^3-1)\,2^2\,2^1\,2^0.$	$\pm(2^2\,2^1\,2^0).$
sign inversion	n.a.	$(\overline{a_3\,a_2\,a_1\,a_0}) + 1$		$\overline{a_3\,a_2\,a_1\,a_0}$	$\overline{a_3}\,a_2\,a_1\,a_0$
VHDL type	unsigned	n.a.	signed	n.a.	n.a.
SysVer modif.	unsigned	n.a.	signed	n.a.	n.a.

^{a} Has no positive counterpart, sign inversion rule does not apply.

As for signed numbers, several schemes have been developed to handle them in digital circuits and computers. Table A.1 illustrates how the more common ones map between bit patterns and numbers. For conciseness, integers of merely four bits are tabulated there.

The leftmost bit always indicates whether a number is positive or negative. Except for that one bit, offset-binary and 2's complement encodings are the same. What they further have in common is that the most negative number has no positive counterpart (with the same number of bits). Conversely, two patterns for zero exist in 1's complement and in sign-and-magnitude representation which complicates the design of adders, subtractors, comparators, and other arithmetic units. What makes the 2's complement format so popular is the fact that any adder circuit can be used for subtraction if arguments and result are coded in this way.

Observation A.1. *Digital hardware just deals with* 0s *and* 1s *and attaches no semantic meaning to the symbols it manipulates. What gives a bit pattern a meaning as a character, as a signed or unsigned, as an integer or fractional number, as a fixed-point or floating point number, etc., essentially is the interpretation by humans, or by human-made software.*

A bit pattern remains absolutely devoid of meaning unless the pertaining number representation scheme is known.[2] Hardware description languages (HDL) such as VHDL and SystemVerilog provide designers with various data types and with index ranges to help them keep track of number formats.

A.1.2 FLOATING POINT NUMBER FORMATS

By definition, a floating point number a encoded using binary digits obeys

$$a = s \cdot f \cdot 2^e \tag{A.2}$$

where s is the sign, f the mantissa, and e the exponent. Word width follows as $w = 1 + \#e + \#f$. The industry standard IEEE 754 defines coding formats for $w = 16, 32, 64$ and 128 respectively. A single precision floating point number, for instance, occupies $w=32$ bit with $\#e=8$ and $\#f=23$, and is formatted `s eeee eeee . ffff ffff ffff ffff ffff fff`.

The standard further stipulates that numbers are to be normalized to the interval $[1...2)$ prior to encoding which implies that the binary digit immediately preceding the radix point is always 1 and, hence, devoid of information. Rather than storing the full mantissa `1.ff...f`, only the fractional bits to the right of the radix point `ff...f` are made part of the data word. The uncoded bit to the left is referred to as **hidden bit**.

The exponent is coded in offset-binary format with a bias of $2^{\#e-1} - 1$. In the occurrence of $\#e=8$, this means that e must be augmented by 127 before being stored in the `ee...e` field, and vice versa for decoding. In theory, the exponent can so cover a range of $-(2^{\#e-1} - 1) \le e \le 2^{\#e-1}$.

As for the sign bit, $s = 0$ for positive and $s = 1$ for negative numbers by convention.

Now for the exceptions. First observe that it is not possible to express zero as permanently tying the hidden bit to 1 does not allow for a zero mantissa, and so renders the above coding scheme utterly useless. A more viable format must accommodate at least one non-normalized number. Then there must be a way to tell that a result lies beyond the numerical range representable with the bits available, just think of $\frac{1}{0}$ and other calculations. We further wish to express that $\frac{0}{0}$, taking the square root of a negative number, and similar operations yield undefined results. **Not a number** (NaN) is the name for such arithmetic exceptions.

A practical floating point system must thus set aside a few bit patterns to accomodate particular situations. In the IEEE 754 std, those special codes are easily identified by their exponent field made of either 0s or 1s throughout. Interpretation thus departs from a strictly mathematical one for two out of the $2^{\#e}$ possible exponent patterns. Table A.2 illustrates all this with the aid of a toy format `seee.ffff`

---

[2] As an analogy, a pocket calculator handles only numbers and does not know about any physical unit involved, e.g. [m], [kg], [s], [μA], [kΩ] and [USD]. It is up to the user to enter arguments in correct units and to know how to read the results.

Table A.2 Toy example for a floating point format with eight bits $w = 8$.

bit pattern MSB LSB $s\,e_2\,e_1\,e_0.f_{-1}...$	binary code	exp.	mant.	num. value	interpreted as	mant.	end value
`1111[1].1111`	255	4	$1+\frac{15}{16}$	-31	NaN	n.a.	none
...	...	...	...	...	NaN	n.a.	none
`1111[1].0000`	240	4	$1+\frac{0}{16}$	-16	neg. infinity	n.a.	$-\infty$
`1110[1].1111`	239	3	$1+\frac{15}{16}$	-15.5			
...	...	...	...	...			
`1110[1].0000`	224	3	$1+\frac{0}{16}$	-8.0			
...	...	...	...	...			
four more binades of negative numbers							
...	...	...	...	...			
`1001[1].1111`	159	-2	$1+\frac{15}{16}$	$-0.484\,375$			
...	...	...	...	...			
`1001[1].0000`	144	-2	$1+\frac{0}{16}$	$-0.250\,000$			
`1000[1].1111`	143	-3	$1+\frac{15}{16}$	$-0.242\,187\,5$	`1000[0].1111`	$0+\frac{15}{16}$	$-0.117\,187\,5$
...	...	...	...	...	...	...	...
`1000[1].0010`	130	-3	$1+\frac{2}{16}$	$-0.140\,625\,0$	`1000[0].0010`	$0+\frac{2}{16}$	$-0.015\,625\,0$
`1000[1].0001`	129	-3	$1+\frac{1}{16}$	$-0.132\,812\,5$	`1000[0].0001`	$0+\frac{1}{16}$	$-0.007\,812\,5$
`1000[1].0000`	128	-3	$1+\frac{0}{16}$	$-0.125\,000\,0$	`1000[0].0000`	$0+\frac{0}{16}$	$[-]\,0$
`0111[1].1111`	127	4	$1+\frac{15}{16}$	31	NaN	n.a.	none
...	...	...	...	...	NaN	n.a.	none
`0111[1].0000`	112	4	$1+\frac{0}{16}$	16	pos. infinity	n.a.	$+\infty$
`0110[1].1111`	111	3	$1+\frac{15}{16}$	15.5			
...	...	...	$1+...$	...			
`0110[1].0000`	096	3	$1+\frac{0}{16}$	8.0			
`0101[1].1111`	095	2	$1+\frac{15}{16}$	7.75			
...	...	...	...	...			
`0101[1].0000`	080	2	$1+\frac{0}{16}$	4.00			
`0100[1].1111`	079	1	$1+\frac{15}{16}$	3.875			
...	...	...	...	...			
`0100[1].0000`	064	1	$1+\frac{0}{16}$	2.000			
`0011[1].1111`	063	0	$1+\frac{15}{16}$	$1.937\,5$			
...	...	...	...	...			
`0011[1].0000`	048	0	$1+\frac{0}{16}$	$1.000\,0$			
`0010[1].1111`	047	-1	$1+\frac{15}{16}$	$0.968\,75$			
...	...	...	...	...			
`0010[1].0000`	032	-1	$1+\frac{0}{16}$	$0.500\,00$			
`0001[1].1111`	031	-2	$1+\frac{15}{16}$	$0.484\,375$			
...	...	...	...	...			
`0001[1].0000`	016	-2	$1+\frac{0}{16}$	$0.250\,000$			
`0000[1].1111`	015	-3	$1+\frac{15}{16}$	$0.242\,187\,5$	`0000[0].1111`	$0+\frac{15}{16}$	$0.117\,187\,5$
...	...	...	...	...	...	...	...
`0000[1].0010`	002	-3	$1+\frac{2}{16}$	$0.140\,625\,0$	`0000[0].0010`	$0+\frac{2}{16}$	$0.015\,625\,0$
`0000[1].0001`	001	-3	$1+\frac{1}{16}$	$0.132\,812\,5$	`0000[0].0001`	$0+\frac{1}{16}$	$0.007\,812\,5$
`0000[1].0000`	000	-3	$1+\frac{0}{16}$	$0.125\,000\,0$	`0000[0].0000`	$0+\frac{0}{16}$	0

where $\#e = 3$, $\#f = 4$, and where the offset for the exponent is $+3$.[3] For clarity, the hidden bit is shown between square brackets [.] there. Last but not least, note that handy IEEE 754 converters are available on the Internet.

A.1.3 NOTATIONAL CONVENTIONS FOR TWO-VALUED LOGIC

A.1.3 NOTATIONAL CONVENTIONS FOR TWO-VALUED LOGIC

The restriction to two-valued or bivalent logic[4] seems to suggest that the two symbols o and 1 from switching algebra should suffice as a basis for mathematical analysis. This is not so, however, and two more logic values are needed so that we end up with a total of four symbols.[5]

- o stands for a logic zero.
- 1 stands for a logic one.
- x denotes a situation where a signal's logic state as o or 1 remains **unknown** after analysis.
- - implies that the logic state is left open in the specifications because it does not matter for the correct functioning of a circuit. One is thus free to substitute either a o or a 1 during circuit design which explains why this condition is known as **don't care**.

The mathematical convention for identifying the **logic inverse** of a term, aka "Boolean complement", is by overlining it. That is, if a is a variable, then its complement shall be denoted $\bar{a}$.[6] Most obviously, one has $\bar{0} = 1$, $\bar{1} = 0$, and $\bar{\bar{a}} = a$.

---

[3] While the data type is `float(3 downto -4)` in VHDL notation, such a format is too restrictive to be useful.

[4] Note that **binary** is almost always used instead of **bivalent**. This is sometimes misleading as the same term also serves to indicate that a function takes two arguments. The German language, in contrast, makes a distinction between "zweiwertig" (bivalent) and "zweistellig" (taking two arguments).

[5] This is still insufficient for actual circuit design. A more adequate set of nine logic values has been defined in the IEEE 1164 standard and is discussed in full detail in section 4.2.3; what we present here is just a subset.

[6] Unfortunately, this practice is not viable in the context of EDA software because there is no way to overline identifiers with ASCII characters. A more practical and more comprehensive naming scheme is being proposed in section 6.7. Taking the complement is expressed by appending the suffix _B to the original name so that the Boolean complement of Qux is denoted as Qux_B (pronounced "qux bar"). For the present appendix, however, we mostly stick to one-letter names, such as q and $\bar{q}$, as multi-character strings tend to obfuscate the math.

A digital circuit is said to be **combinational** if its present output gets determined by its present input exclusively when in steady-state condition. Such a circuit has no state and can, therefore, just as well be qualified as **memoryless**.

This contrasts with **sequential** logic, the output of which depends not only on present but also on past input values. Sequential circuits must necessarily keep their state in some kind of storage elements, which is why they are also referred to as state-holding or as **memorizing**.

In this section, we confine our discussion to combinational functions and begin by asking

"How can we state a combinational function and how do the various formalisms differ?"

Probably the most popular way to capture a combinational function, is to come up with a truth table, that is with a list that indicates the desired output for each input.

Table A.3   A truth table of three variables that includes don't care entries.

x	y	z	g
0	0	0	1
0	0	1	1
0	1	0	–
0	1	1	–
1	0	0	1
1	0	1	0
1	1	0	–
1	1	1	0

Let us calculate the number of possible logic functions of n variables. Observe that a truth table comprises 2^n fields each of which must be filled either with a 0 or a 1 (don't care conditions do not contribute any extra functions). So there are 2^{2^n} different ways to complete a truth table and, hence, 2^{2^n} distinct logic functions.

n	functions
1	4
2	16
3	256
4	65 536

as listed in table A.5

A.2.2 THE n-CUBE

A geometric representation is obtained by mapping a logic function of n variables onto the n-dimensional unit cube. This requires a total of 2^n nodes, one for each input value. Edges connect all node pairs that differ in a single variable. A drawing of the n-cube for truth table A.3 appears in fig.A.1. Note that the concept of n-cubes can be extended to arbitrary numbers of dimensions although representing them graphically becomes increasingly difficult.

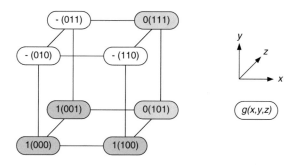

FIGURE A.1

3-cube equivalent to table A.3.

A.2.3 KARNAUGH MAP

The Karnaugh map, an example of which is shown in table A.4, is another tabular format where each field stands for one of the 2^n input values. The fields are arranged such as to preserve adjacency relations from the n-cube when the map is thought to be inscribed on a torus. Although, extensions for five and six variables have been proposed, the merit of easy visualization which that Karnaugh maps so popular tends to get lost beyond four variables.

Table A.4 Karnaugh map equivalent to table A.3.

g		00	01	11	10
		yz			
x	0	1	1	–	–
	1	1	0	0	–

A.2.4 PROGRAM CODE

Logic operations can further be captured using an HDL or some other formal language, an example is given in listing A.1. Note that, while the function described continues to be combinational, its description is procedural in the sense that the processing of the associated program code must occur step by step.

LISTING A.1

A piece of behavioral VHDL code that is equivalent to table A.3

```
entity gfunction is
   port (
       X : in Std_Logic;
       Y : in Std_Logic;
       Z : in Std_Logic;
       G : out Std_Logic );
end gfunction;

architecture procedural of gfunction is
begin
   process (X,Y,Z)
       variable temp: Std_Logic;
   begin
       temp := '-';
       if Y='0' then temp := '1'; end if;
       if X='1' and Z='1' then temp := '0'; end if;
       G <= temp;
   end process;
end procedural;
```

A.2.5 LOGIC EQUATIONS

What truth tables, Karnaugh maps, *n*-cubes, and procedural HDL code examples have in common, is that they specify — essentially by enumeration — input-to-output mappings. Put differently, they all define a logic function in purely **behavioral** terms.

Logic equations, in contrast, also imply the operations to use and in what order to apply them. Each such equation suggests a distinct gate-level circuit and, therefore, also conveys information of **structural** nature. Even in the absence of don't care conditions, a great variety of logically equivalent equations exist that implement a given truth table. Since, in addition, it is always possible to expand a logic equation into a more complex one, we note

Observation A.2. *For any given logic function, there exist infinitely many logic equations and gate-level circuits that implement it.*

Fig.A.2 illustrates the symbols used in schematic diagrams to denote the subcircuits that carry out simple combinational operations. Albeit fully exchangeable from a purely functional point of view, two equations and their associated gate-level networks may significantly differ in terms of circuit size, operating speed, energy dissipation, and manufacturing expenses. Such differences often matter from the perspectives of engineering and economy.

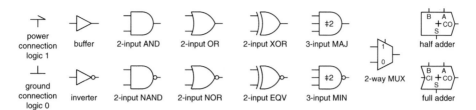

Schematic icons of common combinational functions.

Example

The Karnaugh map below defines a combinational function of three variables. Equations (A.3) through (A.12) all implement that very function. Each such equation stands for one specific gate-level circuit and three of them are depicted next. They belong to equations (A.6), (A.11) and (A.12) respectively. More circuit alternatives are shown in fig.A.4.

f		yz			
		00	01	11	10
x	0	1	1	0	0
	1	0	1	0	1

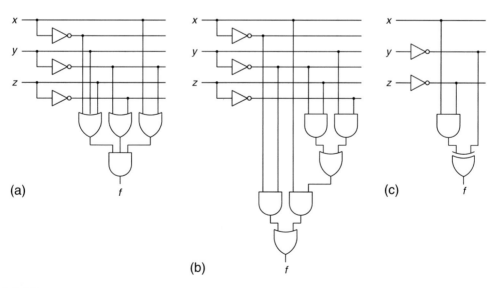

A selection of three circuit alternatives for the same logic function.

□

A.2.6 TWO-LEVEL LOGIC

Sum-of-products

Any switching function can be described as a sum-of-products (SoP) where sum and product refer to logic OR and AND operations respectively, see equations (A.3) and (A.4), for instance.[7] Disjunctive form is synonymous to sum-of-products. A product term that includes the full set of variables is called a **minterm** or a fundamental product. The name **canonical sum** stands for a sum-of-products expression that consists of minterms exclusively. The right-hand side of (A.3) is a canonical sum whereas that of (A.4) is not.

$$f = \bar{x}\bar{y}\bar{z} \vee \bar{x}\bar{y}z \vee x\bar{y}z \vee xy\bar{z} \tag{A.3}$$

$$f = \bar{x}\bar{y} \vee \bar{y}z \vee xy\bar{z} \tag{A.4}$$

Product-of-sums

As the name suggests, product-of-sums (PoS) are dual to SoP formulations. Not surprisingly, the concepts of conjunctive form, **maxterm**, fundamental sum, and **canonical product** are defined analogously to their SoP counterparts. Two PoS examples are given in (A.5) and (A.6), you may want to add the canonical product form yourself.

$$f = (\bar{x} \vee y \vee z)\,(\bar{x} \vee \bar{y} \vee \bar{z})\,(x \vee \bar{y})\,(x \vee \bar{y})\,(x \vee \bar{y} \vee z) \tag{A.5}$$

$$f = (\bar{x} \vee y \vee z)\,(\bar{y} \vee \bar{z})\,(x \vee \bar{y}) \tag{A.6}$$

Other two-level logic forms

SoP and PoS forms are subsumed as **two-level logic**, aka two-stage logic, because they both make use of two consecutive levels of OR and AND operations. Any inverters required to provide signals in their complemented form are ignored as dual-rail logic is assumed.[8] As a consequence, not only (A.3) through (A.6), but also (A.7) and (A.8) qualify as two-level logic.

$$f = \overline{\overline{(\bar{x}\bar{y})}\,\overline{(\bar{y}z)}\,\overline{(xy\bar{z})}} \tag{A.7}$$

$$f = \overline{x\bar{y}\bar{z} \vee yz \vee \bar{x}y} \tag{A.8}$$

Incidentally, observe that (A.7) describes a circuit that consists of NAND gates and inverters exclusively. As illustrated in fig.A.4, this formulation is easily obtained from (A.4) by applying De Morgan's

---

[7] We will denote the sum and product operators from switching algebra as $\vee$ and $\wedge$ respectively to minimize the risk of confusion with the conventional arithmetic operators $+$ and $\cdot$. However, for the sake of brevity, we will frequently drop the $\wedge$ symbol from product terms and write xyz when we mean $x \wedge y \wedge z$. In doing so, we imply that $\wedge$ takes precedence over $\vee$.

[8] The term **dual-rail** logic refers to logic families where each variable is being represented by a pair of signals a and $\bar{a}$ that are of opposite value at any time (e.g. in CVSL). Every logic gate has two complementary outputs and pairwise differential inputs. Taking the complement of a variable is tantamount to swapping the two signal wires and requires no extra hardware.

This situation contrasts with **single-rail** logic where every variable is being transmitted over a single wire (e.g. in standard CMOS and TTL). A complement must be obtained explicitly by means of an extra inverter.

theorem[9] followed by **bubble pushing**, that is by relocating the negation operators from all inputs of the second-level gates to the outputs of the first-level gates.

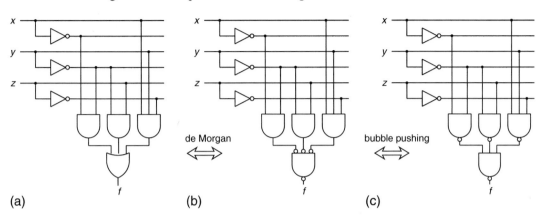

de Morgan   bubble pushing

(a)   (b)   (c)

FIGURE A.4

Translating an SoP logic (a) into a NAND-NAND circuit (c) or back.

Observation A.3. *It is always possible to implement an arbitrary logic function with no more than two consecutive levels of logic operations.*

This is why two-level logic is said to be universal. The availability of manual minimization methods, such as the Karnaugh or the Quine-McCluskey method [221], the multitude of circuit alternatives to be presented in sections A.3.1 through A.3.3, together with the — now largely obsolete — belief that propagation delay directly relates to the number of stages have further contributed to the popularity of two-level logic since the early days of digital electronics.

A.2.7 MULTI-LEVEL LOGIC

Multi-level logic, aka multi-stage logic, differs from two-level logic in that logic equations extend beyond two consecutive levels of OR and AND operations. Examples include (A.9) and (A.10) where three stages of ORs and ANDs alternate, (A.11) with the same operations nested four levels deep also belongs to this class.

$$f = (\bar{x} \vee z) \bar{y} \vee x (y \bar{z}) \tag{A.9}$$

$$f = \overline{(x \bar{z} \vee y) (\bar{x} \vee \bar{y} \vee z)} \tag{A.10}$$

$$f = \bar{x} \bar{y} \vee x (\bar{y} z \vee y \bar{z}) \tag{A.11}$$

Equation (A.12) below appears to have logic operations nested no more than two levels deep as well, yet the inclusion of an exclusive or function[10] makes it multi-level logic. This is because the XOR function

---

[9] The **De Morgan theorem** of switching algebra states $\bar{x} \vee \bar{y} = \overline{x y} \ (= \overline{x \wedge y})$ and $(\bar{x} \wedge \bar{y} =) \ \overline{x} \overline{y} = \overline{x \vee y}$.

[10] The exclusive or XOR is also known as antivalence operation, and its negated counterpart as equivalence operation EQV or XNOR. Please further note that OR and AND operations take precedence over XOR and EQV.

is more onerous to implement than an OR or an AND and because substituting those for the XOR results in a total of three consecutive levels of logic operations.

$$f = x\bar{z} \oplus \bar{y} \tag{A.12}$$

The circuits that correspond to (A.11) and (A.12) are depicted in fig.A.3b and c respectively. Drawing the remaining two schematics is left to the reader as an exercise.

Originally somewhat left aside due to the lack of systematic and affordable procedures for their minimization, multi-level logic has become popular with the advent of adequate computer tools. VLSI also destroyed the traditional preconception that fewer logic levels would automatically bring about shorter propagation delays.

A.2.8 SYMMETRIC AND MONOTONE FUNCTIONS

A logic function is said to be **totally symmetric** iff it remains unchanged for any permutation of its variables; **partial symmetry** exists when just a subset of the variables can be permuted without altering the function. A logic function is characterized as being **monotone** or unate iff it is possible to rewrite it as a sum-of-products expression where each variable appears either in true or in complemented form exclusively. If all variables are present in true form in the SoP, then the function is called **monotone increasing**, and conversely **monotone decreasing** if all variables appear in their complemented form.

Examples

$$c = xy \vee xz \vee yz \tag{A.13}$$

$$s = xyz \vee \bar{x}\bar{y}z \vee \bar{x}y\bar{z} \vee x\bar{y}\bar{z} \tag{A.14}$$

$$h = \overline{\bar{x} \vee \bar{y}\bar{z}} = xy \vee xz \tag{A.15}$$

$$m = xz \vee y\bar{z} \tag{A.16}$$

$$n = xy \vee x\bar{z} \tag{A.17}$$

$$o = \overline{wx \vee yz} = \bar{w}\bar{y} \vee \bar{w}\bar{z} \vee \bar{x}\bar{y} \vee \bar{x}\bar{z} \tag{A.18}$$

function	name	symmetric	monotone
c (A.13)	3-input majority (MAJ)	totally	increasing
s (A.14)	3-input exclusive or (XOR)	totally	no
h (A.15)	anonymous	partially	increasing
m (A.16)	2-way multiplexer (MUX)	no	no
n (A.17)	anonymous	no	yes
o (A.18)	anonymous	partially	decreasing

☐

A.2.9 THRESHOLD FUNCTIONS

A.2.9 THRESHOLD FUNCTIONS

Many combinational functions can be thought to work by counting the number of variables that are at logic 1 and by producing either a 0 or a 1 at the output depending on whether that figure exceeds some fixed number or not.[11] Per force, all such threshold functions are totally symmetric and monotone. Examples include OR and AND functions along with their inverses.

Probably more interesting are the **majority function** (MAJ) and its inverse the **minority function** (MIN) that find applications in adders and as part of the Muller-C element. MAJ and MIN gates always have an odd number of inputs of three or more. This is because majority and minority are mathematically undefined for even numbers of arguments and are of no practical interest for a single variable. In the occurrence of a 3-input MAJ gate (A.13), the condition for a logic 1 at the output is $\#\,1's \geq 2$ as reflected by its icon in fig.A.6c.

A.2.10 COMPLETE GATE SETS

A set of logic operators is termed a (functionally) **complete gate set** if it is possible to implement arbitrary combinational logic functions from an unlimited supply of its elements.

Examples and counterexamples

Complete gate sets include but are not limited to the subsequent sets of operations: $\{$AND, OR, NOT$\}$, $\{$AND, NOT$\}$, $\{$OR, NOT$\}$, $\{$NAND$\}$, $\{$NOR$\}$, $\{$XOR, AND$\}$, $\{$MAJ, NOT$\}$, $\{$MIN$\}$, $\{$MUX$\}$, and $\{$INH$\}$. As opposed to these, none of the sets $\{$AND, OR$\}$, $\{$XOR, EQV$\}$, or $\{$MAJ$\}$ is functionally complete.[10]
□

Though any complete gate set would suffice from a theoretical point of view, actual component and cell libraries include a great variety of logic gates that implement up to one hundred or more distinct logic operations to better support the quest for density, speed and energy efficiency.

Several complete gate sets have cardinality one which means that a single operator suffices to construct arbitrary combinational functions. One such gate that deserves special attention is the 4-way MUX. It is in fact possible to build any combinational operation with two arguments from a single such MUX without rewiring. Consider the circuit of fig.A.5 where the two operands are connected to the multiplexer's select inputs. For each 4-bit value that gets applied to data lines p_3 through p_0, the multiplexer then implements one out of the 16 possible switching functions listed in table A.5. The 4-way MUX so effectively acts as a 2-input gate the functionality of which is freely programmable from externally.[12]

---

[11] Incidentally, observe the relation to artificial neural networks that make use of similar threshold functions.

[12] As an extension, note that a 4-way MUX can be made to implement a 3-input function by allowing its inputs p_3 through p_0 not only to connect to 0 or 1 but to a third argument in direct z or negated $\bar{z}$ form as well.

Table A.5   The 16 truth tables and switching functions implemented by the circuit of fig.A.5.

	assignment				function implemented			assignment				function implemented	
	p_3	p_2	p_1	p_0				p_3	p_2	p_1	p_0		
x	1	1	0	0			x	1	1	0	0		
y	1	0	1	0			y	1	0	1	0		
p	setting				$q =$	name	p	setting				$q =$	name
0	0	0	0	0	0	null, never	15	1	1	1	1	1	unity, always
1	0	0	0	1	$\overline{x \vee y}$	NOR, Pierce	14	1	1	1	0	$x \vee y$	OR, sum
2	0	0	1	0	$\overline{x}\,y$	INH x, inhibit	13	1	1	0	1	$x \vee \overline{y}$	y implies x
4	0	1	0	0	$x\,\overline{y}$	INH y, inhibit	11	1	0	1	1	$\overline{x} \vee y$	x implies y
3	0	0	1	1	$\overline{x}$	NOT x	12	1	1	0	0	x	pass x
5	0	1	0	1	$\overline{y}$	NOT y	10	1	0	1	0	y	pass y
6	0	1	1	0	$x \oplus y$	XOR, antival.	9	1	0	0	1	$\overline{x \oplus y}$	EQV, equival.
7	0	1	1	1	$\overline{xy}$	NAND, Sheffer	8	1	0	0	0	xy	AND, product

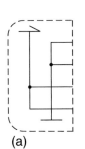
(a)

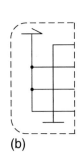
(b)

(c)

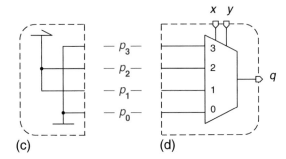
(d)

FIGURE A.5

A programmable logic gate. 4-way multiplexer (d) with the necessary settings for making it work as an inverter (a), a 2-input NAND gate (b), and as an XOR gate (c).

A.2.11 MULTI-OUTPUT FUNCTIONS

All examples presented so far were single-output functions. We speak of a multi-output function when a **vector** of several bits is produced rather than just a **scalar** signal of cardinality one.

Example

The **full adder** is a multi-output function of fundamental importance. It adds two binary digits and a carry input to obtain a sum bit along with a carry output. With x, y, and z denoting the three input bits, (A.13) and (A.14) together describe the logic functions for the carry out bit c and for the sum bit s respectively.

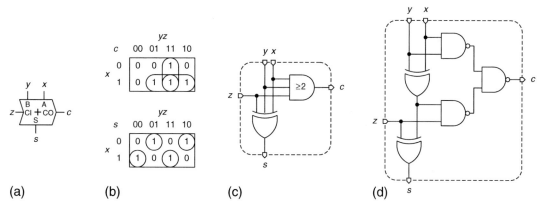

(a)   (b)   (c)   (d)

FIGURE A.6

Full adder. Icon (a), Karnaugh maps (b), and two circuit examples (c,d).
☐

A.2.12 LOGIC MINIMIZATION

Given the infinitely many solutions, we must ask ourselves

"How to select an appropriate set of logic equations for some given combinational function?"

Metrics for logic complexity and implementation costs

The goal of logic minimization is to find the most economic circuit for a given logic function under some speed and energy constraints. The criterion for economy depends on the technology targeted. Minimum package count used to be a prime objective at a time when electronics engineers were assembling digital systems from SSI/MSI components. Today, it is the silicon area occupied by gates and wiring together that counts for full-custom ICs. The number of gate equivalents (GE) is more relevant in the context of field-programmable logic (FPL) and semi-custom ICs.

From a mathematical point of view, the number of literals is typically considered as criterion for logic minimization. By **literal** we refer to an appearance of a logic variable or of its complement. As an example, the right-hand side of (A.4) consists of seven literals that make up three composite terms although just three variables are involved.

An expression is said to contain a **redundant** literal if the literal can be eliminated from the expression without altering the truth table. Equation (A.19), the Karnaugh map of which is shown in fig.A.7a, contains several redundant literals. In contrast, none of the eleven literals can be eliminated from the right-hand side of (A.20) as illustrated by the Karnaugh map of fig.A.7b. The concept of redundancy not only applies to literals but also to composite terms.

Redundant terms and literals result in redundant gates and gate inputs in the logic network. They are undesirable due to their impact on circuit size, load capacitances, performance, and energy dissipation. What's more, **redundant logic** causes severe problems with testability, essentially because there is no way to tell whether a redundant gate or gate input is working or not by observing a circuit's behavior from its connectors to the outside world.

Minimal versus unredundant expressions

Unredundant and minimal are not the same. This is illustrated by (A.21), an equivalent but more economical replacement for (A.20) which gets along with just eight literals. Its Karnaugh map is shown in fig.A.7c.

Observation A.4. *While a minimal expression is unredundant by definition, the converse is not necessarily true.*

$$e = \bar{x}\bar{y}\bar{z}\bar{t} \vee \bar{x}\bar{z}t \vee \bar{x}yzt \vee xz \vee x\bar{y}z \tag{A.19}$$

$$e = \bar{x}\bar{y}\bar{z} \vee \bar{x}\bar{z}t \vee yzt \vee xz \tag{A.20}$$

$$e = \bar{x}\bar{y}\bar{z} \vee \bar{x}yt \vee xz \tag{A.21}$$

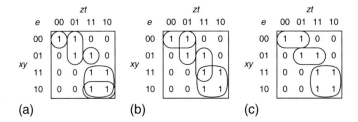

(a) (b) (c)

FIGURE A.7

Three Karnaugh maps for the same logic function. Redundant form as stated in logic equation (A.19) (a), unredundant form as in (A.20) (b), and minimal form as in (A.21) (c).

Note that for obtaining the minimal expression (A.21) from the unredundant one (A.20), a detour via the canonical expression is required during which product terms are first expanded and then regrouped and simplified in a different manner. Thus, there is more to logic minimization than eliminating redundancy.

Next consider function d tabulated in fig.A.8. There are two possible sum-of-products expressions shown in equations (A.22) and (A.23) both of which are minimal and use six literals. The minimal product-of-sums form of (A.24) also includes six literals. We conclude

Observation A.5. *A minimal expression is not always unique.*

$$d = \bar{x}\bar{y} \vee xz \vee y\bar{z} \tag{A.22}$$

$$d = xy \vee \bar{x}\bar{z} \vee \bar{y}z \tag{A.23}$$

$$d = (\bar{x} \vee y \vee z)(x \vee \bar{y} \vee \bar{z}) \tag{A.24}$$

Karnaugh maps for equations (A.22) through (A.24), all of which implement the same function with the same number of literals.

Multi-level versus two-level logic

Observation A.6. *While it is possible to rewrite any logic equation as a sum of products and as a product of sums, the number of literals required to do so grows exponentially with the number of input variables for certain logic functions.*

The 3-input XOR function (A.14), for instance, includes 12 literals. Adding one more argument t asks for 8 minterms each of which takes 4 literals to specify thereby resulting in a total of 32 literals. In general, a n-input parity function takes $2^{(n-1)} \cdot n$ literals when written in two-level logic form. Asymptotic complexity is not the only concern, however. Multi-level circuits are often faster and more energy-efficient than their two-level counterparts.

The process of converting a two-level into an equivalent multi-level logic equation is referred to as **factoring**, aka structuring, and the converse as **flattening**.

$$\overline{x}\,\overline{y}\,\overline{z} \vee \overline{x}\,\overline{y}\,z \vee x\,\overline{y}\,z \vee x\,y\,\overline{z} \quad \underset{\text{flattening}}{\overset{\text{factoring}}{\rightleftharpoons}} \quad \overline{x}\,\overline{y}\,(\overline{z} \vee z) \vee x\,(\overline{y}z \vee y\overline{z}) \tag{A.25}$$

Multi-output versus single-output minimization

Probably the most important finding on multi-output functions is

Observation A.7. *Minimizing a vectored function for each of its output variables separately does not, in general, lead to the most economical solution for the overall network.*

This is nicely illustrated by the example of fig.A.9. Solution (a) which is obtained from applying the Karnaugh method one output bit at a time requires a total of 15 literals (and 7 composite terms). By reusing conjunctive terms for two or more bits, solution (b) manages with only 9 literals (and 7 composite terms). In terms of gate equivalents, overall circuit complexity amounts to 12.5 and 9.5 GEs if all ORs and ANDs get remapped to NAND gates by way of bubble pushing.

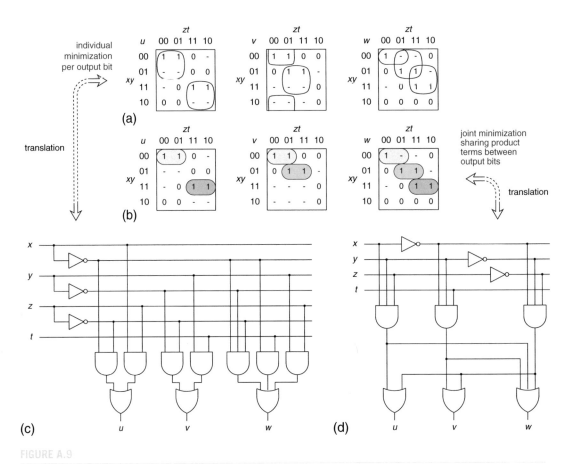

FIGURE A.9

A multi-output function minimized in two different ways.

Manual versus automated logic optimization

Observation A.8. *Manual logic optimization is not practical in VLSI design.*

For real-world multi-output multi-level networks, the solution space of this multi-objective optimization problem (area, delay, energy) is way too large to be explored by hand. Also, solutions are highly dependent on nasty details (external loads, wiring parasitics, cell characteristics, etc.) that are difficult to anticipate during logic design. Logic minimization on the basis of AND and OR gates with unit delays is a totally unacceptable oversimplification.

A.3 CIRCUIT ALTERNATIVES FOR IMPLEMENTING COMBINATIONAL LOGIC

A.3.1 RANDOM LOGIC

The term is misleading in that there is nothing undeterministic to it. Rather, **random logic** refers to networks built from logic gates the arrangement and wiring of which may appear arbitrary at first sight. Examples have been given earlier, see fig.A.9 for instance. Standard cells and gate arrays are typical vehicles for implementing random logic in VLSI.

As opposed to random logic, **tiled logic** exhibits a regularity immediately visible from the layout because subcircuits get assembled from a small number of abutting layout tiles. As tiling combines logic, circuit, and layout design in an elegant and efficient way, we will present the most popular tiled buildings blocks.

A.3.2 PROGRAMMABLE LOGIC ARRAY (PLA)

A PLA starts from two-level logic and packs all operations into two adjacent rectangular areas referred to as AND- and OR-**plane** respectively. Each input variable traverses the entire AND-plane both in its true and in its complemented form, see fig.A.10. A product term is formed by placing or by omitting transistors that act on a common perpendicular line. Each input variable participates in a product in one of three ways

true	1
complemented	0
not at all	-

Parallel product lines bring the intermediate terms to the OR-plane where they cross the output lines. The sums are then obtained very much like the products, the only difference being that products are not available in complemented form, so that any product enters a sum in either of two ways, namely

true	1
not at all	0

The general arrangement as two pairs of interacting meshs yields a very compact layout. What's more, a PLA is readily assembled from a small set of predefined layout tiles for any logic function. The criteria for logic minimization are not the same as for random logic. The PLA's width being fixed by the number of input and output variables, PLA minimization software[13] must act on the number of conjunctive terms to minimize layout height.

---

[13] Such as the seminal ESPRESSO [227], which also introduced the above notation for capturing PLA codings.

Example

Fig.A.10 shows a PLA-style circuit that implements the logic function of fig.A.9c. For the sake of simplicity, switches and resistors have been substituted for the MOSFETs of an actual circuit.[14] When comparing to the random logic of fig.A.9d, keep in mind that a PLA gets penalized on such a small function by the overhead associated with complementing all inputs, with distributing signals over both planes, and with restoring voltages to proper levels.

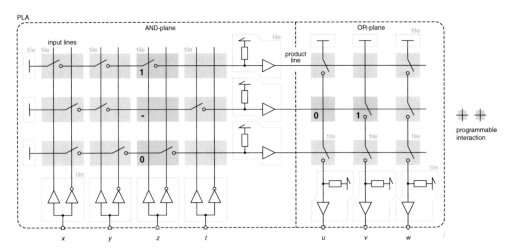

FIGURE A.10

The circuit organization of a NAND-NAND-type PLA drawn with switches and resistors instead of transistors. The function implemented is the same as in fig.A.9.

Using the notation introduced before, the PLA's configuration and programming are expressed in a concise manner as follows.

AND-plane				OR-plane		
x	y	z	t	u	v	w
1	1	1	–	1	0	1
0	1	–	1	0	1	1
0	0	0	–	1	1	1

□

Owing to their superior layout densities, PLAs used to be popular building blocks for combinational functions with many inputs and outputs. They have lost momentum when automatic synthesis of random logic and multiple metal layers became available, but the underlying concepts continue to play an important role in field-programmable logic (FPL).

---

[14] In fig.A.10, the switches in both planes work against static pull-up loads which circuit style is a departure from truly complementary CMOS style. Transistor networks trimmed down in this or a similar fashion are typical for PLAs.

While both AND- and OR-planes are configurable in a PLA, programming of a ROM is confined to the OR-plane.[15] The role of the fixed AND-plane is a assumed by a built-in address decoder that computes all possible minterms from the input variables, see fig.A.11. A ROM is an attractive option when full decoding of inputs is indeed required, but otherwise remains of limited appeal because its size doubles for each extra input or address bit.

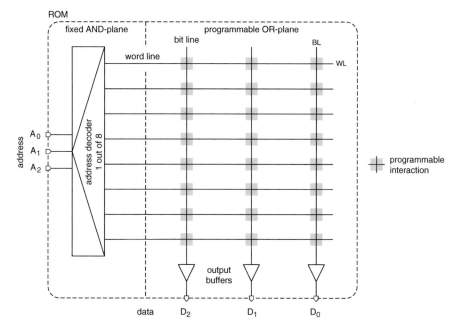

General ROM arrangement (8 words by 3 bits, grossly simplified).

Two-level logic is extremely uneconomic for addition, multiplication, and other functions where the SoP includes many minterms that cannot be merged. Early IC designers have thus extended the tiling approach to multi-level logic and specifically to various forms of multipliers.

Circuit organization is patterned after the classic procedure thought at elementary school whereby multiplier and multiplicand are placed on two orthogonal sets of parallel lines, with a 1-digit by 1-digit multiplication being carried out at every intersection. Addition is distributed over the grid by

---

[15] Make sure you understand that, in spite of its name, a ROM is a purely combinational function or — which is the same — a memoryless subcircuit unable to hold a state. Further note that the term programmable array logic (PAL) denotes a third breed of tiled two-level logic where the AND-plane is programmable and the OR-plane is predefined.

including an adder at every intersection and by having each multiply-add operation propagate its sum and carry to two out of the four adjacent tiles for further processing: the sum towards the bottom and the carry towards the left. The entire multiplier thus consists of largely identical tiles arranged as a two-dimensional array. Communication within the array remains strictly local which minimizes parasitic capacitances and interconnect delays.

This is where the commonalities between the various types of array multiplier end. Popular examples include the **Braun multiplier** for unsigned and the **Bough-Wooley multiplier** for signed numbers. **Booth-recoded multiplier**s have also been constructed along this line. While tiled multipliers have fallen behind much as PLAs have, their basic circuit organization lives on in random logic implementations.

A.3.5 DIGEST

Fig.A.12 summarizes the circuit options for implementing a combinational function. Tiled logic does not waste nearly as much resources for wiring as cell-based random logic does because signals are brought from one layout tile to the next directly. Another benefit of ROMs and PLAs is that any reprogramming is limited to minor modifications to one metal mask or two (as long as the array's overall capacity is not exceeded) whereas any modification to random logic necessitates redoing multiple mask levels and is bound to affect the subcircuit's footprint.

Conversely, the area overhead associated with auxiliary circuits (such as input and output circuitry, decoders, and pull-ups) makes ROMs and PLAs uneconomical for small subfunctions. Developing a layout generator also represents an important investment, part of which must be renewed with each process generation. Only ROMs continues to be routinely supported as combinational macrocells today (together with RAMs for memory functions).

One reason is that the availability of multiple metal layers has narrowed down the difference between cell-based and tiled logic in terms of layout density. What really leveraged random logic, however, is that automatic HDL synthesis and multi-level logic optimization have during the 1990s matured into powerful software tools that not only cover combinational but also sequential circuits.

Assembling lookup tables, arithmetic logic units, and the like from abutting layout tiles is no longer an option unless the word widths involved are extremely large and unless maximum density, performance, and/or energy efficiency must be sought with limited wiring resources.

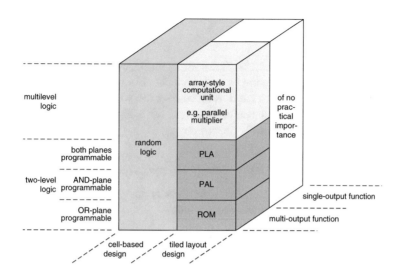

FIGURE A.12

Design space for implementing combinational functions.

A.4 BISTABLES AND OTHER MEMORY CIRCUITS

Bistable subcircuits are essential building blocks of sequential circuits. What they all have in common is the ability to store one bit of information by assuming either of two stable states. An almost infinite variety of circuit implementations has been developed over the years using various design styles and fabrication technologies. As a consequence, designations for bistables proliferate which continues to generate confusion today. Yet, this is totally unnecessary.

Observation A.9. *From the perspectives of architecture and logic design, it suffices to consider a bistable's behavior and to ignore everything about its internal structure and operation.*

We are going to present a simple taxonomy that rests on behavioral criteria exclusively before deriving a coherent and unambiguous naming convention from that.

Clocked   storage elements clearly distinguish between input terminals that determine <u>what</u> state transitions shall take place, and others which determine <u>when</u> such transitions must occur. Any terminal that belongs to the latter category is referred to as a **clock input**.[16]

Unclocked   storage elements, in contrast, do not evidence such a separation.

Table A.6   **Taxonomy of bistables as a function of their behavior.**

	Bistable		
	clocked		unclocked
Behavior	edge-triggered	level-sensitive	
Data inputs get evaluated	at every active clock edge	while clock is at active level	at any time
Name	**flip-flop**	**latch**	no single name
Examples	D-flip-flop E-flip-flop T-flip-flop JK-flip-flop	D-latch	SR-seesaw Muller-C MUTEX snapper
Clock terminal	identified by ∧	identified by ⊓	none

Clocked bistables must be subdivided further into edge-triggered and level-sensitive ones. In any **edge-triggered** bistable, it is a transition of the clock signal that causes the data present at the input terminal to be admitted into the circuit and to be stored there.

The behavior of **level-sensitive** memory circuits is slightly more complex in that such circuits may be in pass or in hold mode depending on the logic value of the clock. In **pass mode**, data are simply propagated from the input to the output which is why this mode is also termed transparent mode.

---

[16] Most clocked bistables are driven from a single clock. Although not really popular with circuit designers, some bistables require a dual-rail clock of two signals CLK and $\overline{CLK}$ driven by complementary waveforms. We are not concerned with this subordinate detail here.

In **hold mode**, output data are kept frozen and input data are being ignored which explains why the device is sometimes said to be opaque.

Throughout this text, we will consistently refer to a bistable that is both clocked and edge-triggered as a **flip-flop**. Conversely, we reserve the word **latch** for any bistable that is clocked and level-sensitive. The phrase **clocked bistable** is used as a generic term for both.[17]

Table A.6 puts the names and behaviors of all popular bistables into perspective.

A.4.1 FLIP-FLOPS OR EDGE-TRIGGERED BISTABLES

The data or D-type flip-flop

The D-type flip-flop exhibits the simplest behavior an edge-triggered bistable can have. Most engineers find it easiest to think in terms of D-flip-flops and to convert their designs into other forms, if necessary.

The basic D-flip-flop has a clock terminal CLK and a data input D. The output datum is either available in true form Q, or in complemented form $\bar{Q}$, or both. Similarly, the clock can induce a state transition either on its rising or on its falling edge, referred to as **active edge**.[18] Please see fig.A.13 for the truth table of a basic rising-edge-triggered D-type flip-flop. Icon and signal waveforms are also shown along with the causality relation.

CLK	D	Q	$\bar{Q}$	
0	–	Q	$\bar{Q}$	keep state unchanged
1	–	Q	$\bar{Q}$	*idem*
↓	–	Q	$\bar{Q}$	*idem*
↑	0	0	1	adopt D as new state
↑	1	1	0	*idem*

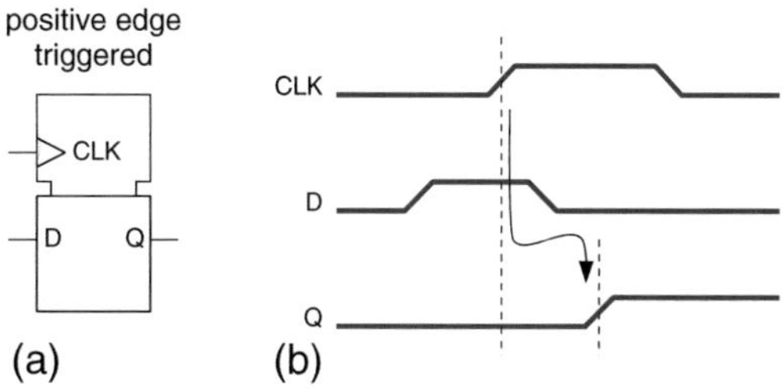

(a)   (b)

FIGURE A.13

Rising-edge-triggered D-flip-flop. Truth table (left), icon (a), and waveforms (b).

A vast collection of more elaborate flip-flop variations are obtained from the basic D-type by extending its functionality in numerous directions and by combining these new features.

Initialization facilities

An extra input found on most flip-flops makes it possible to put the circuit into some predefined start state. Such initialization mechanisms come in two flavors.

---

[17] Be warned that many sources use either word indiscriminately for any kind of bistable. The term latch, in particular, is often meant to include some forms of unclocked bistables, see footnote 24. Also, the fact that latches have a pass mode is sometimes emphasized by calling them **transparent latches** although this property is shared by all latches. For the sake of clarity and simplicity, we refrain from such practices.

[18] Flip-flops that trigger on either edge exist, but are not as widely used.

Both synchronous **clear** CLR and synchronous **load** LOD inputs affect flip-flop operation solely on the active clock edge. As the names say, the former imposes logic state 0 while the latter brings the flip-flop into state 1. Either one can be considered as just another data input that masks the regular data input D, see table A.7. Any basic D type flip-flop is easily upgraded to include a synchronous clear or load by adding one or two gates in front of it.

The asynchronous **reset** RST and the asynchronous **set** SET, in contrast, have an immediate effect on the flip-flop's operation because they directly act on the state-preserving memory loop with no intervention from the clock, see table A.7.[19] This also explains why the asynchronous (re)set mechanism must be incorporated into the elementary flip-flop circuit itself, there is no way to add it later. We therefore conclude:

Observation A.10. *A D-type flip-flop with an asynchronous reset input forms a fundamental building block from which any more sophisticated flip-flop, counter slice, or other edge-triggered bistable can be assembled with the aid of a few extra logic gates.*

Table A.7   Truth tables of rising-edge-triggered D-type flip-flops with synchronous clear (left) and with active-low asynchronous reset (right).

CLK	CLR	D	Q	$\overline{Q}$	
0	–	–	Q	$\overline{Q}$	keep state unchanged
1	–	–	Q	$\overline{Q}$	*idem*
↓	–	–	Q	$\overline{Q}$	*idem*
↑	1	–	0	1	enter state 0
↑	0	0	0	1	adopt D as new state
↑	0	1	1	0	*idem*

$\overline{RST}$	CLK	D	Q	$\overline{Q}$	
0	–	–	0	1	enter state 0
1	0	–	Q	$\overline{Q}$	keep state unchanged
1	1	–	Q	$\overline{Q}$	*idem*
1	↓	–	Q	$\overline{Q}$	*idem*
1	↑	0	0	1	adopt D as new state
1	↑	1	1	0	*idem*

In case a bistable is equipped with two conflicting initialization inputs, i.e. with CLR and LOD or with RST and SET, it must be specified which of the two takes precedence over the other.[20] Although flip-flops with both asynchronous reset and set inputs exist, there is no meaningful application for them in synchronous designs.

Scan facility

An effective and popular way to ensure the testability of sequential logic is to replace all ordinary D-type flip-flops with special scan flip-flops and to connect them such as to make them cooperate like a shift-register while in scan test mode. A **scan flip-flop** essentially includes a select function at the input. Depending on the logic value present at the scan mode control terminal SCM, the data admitted into the flip-flop is either taken from the data input D (during normal operation) or from the scan input SCI (during scan test), see table A.8.[21]

---

[19] Incidentally, note that asynchronous (re)set signals often are of **active-low** polarity as this used to offer better protection against noise and other fugitive events when combined with unsymmetric TTL levels.

[20] Simultaneous activation of asynchronous set and reset inputs of equal precedence levels is disallowed as this could lead to irregular behavior — similarly to that observed when an seesaw is forced into the forbidden state — and/or to anomalous static power dissipation.

[21] Please refer to fig.7.6 in the main text for a schematic and the very basics of scan path testing.

Enable/disable facility

Not all flip-flops in a circuit need to be updated at every active clock edge, many of them must conserve their state for many consecutive clock cycles. This requires that flip-flops be equipped with a mechanism to enable or disable state transitions via a special control input ENA, see table A.9. A data flip-flop is readily extended to become a so-called enable or **E-type flip-flop**; it suffices to add a multiplexer in front of its data input that feeds the uncomplemented output back as long as the enable signal remains inactive, see fig.7.26 for an illustration.

Table A.8   Truth table of a rising-edge-triggered scan flip-flop.

CLK	SCM	SCI	D	Q	$\overline{Q}$	
0	–	–	–	Q	$\overline{Q}$	keep state unchanged
1	–	–	–	Q	$\overline{Q}$	*idem*
↓	–	–	–	Q	$\overline{Q}$	*idem*
↑	0	–	0	0	1	adopt D as new state, "normal operation mode"
↑	0	–	1	1	0	*idem*
↑	1	0	–	0	1	adopt SCI as new state, "scan mode"
↑	1	1	–	1	0	*idem*

Table A.9   Truth table of a rising-edge-triggered E-type flip-flop.

CLK	ENA	D	Q	$\overline{Q}$	
0	–	–	Q	$\overline{Q}$	keep state unchanged
1	–	–	Q	$\overline{Q}$	*idem*
↓	–	–	Q	$\overline{Q}$	*idem*
↑	0	–	Q	$\overline{Q}$	*idem*
↑	1	0	0	1	adopt D as new state
↑	1	1	1	0	*idem*

The toggle or T-type flip-flop

A toggle flip-flop is a bistable that changes state at every active clock edge, see table A.10. It is obtained from a D-flip-flop by providing an inverting feedback from the true output back to the data input. Similarly to the D-type, the T-flip-flop is easily upgraded to include an enable input, in which case toggling takes place only if the enable is active at the active clock edge.

Table A.10   Truth table of a rising-edge-triggered T-type flip-flop.

CLK	Q	$\overline{Q}$	
0	Q	$\overline{Q}$	keep state unchanged
1	Q	$\overline{Q}$	*idem*
↓	Q	$\overline{Q}$	*idem*
↑	$\overline{Q}$	Q	change state, "toggle"

The nostalgia or JK-type flip-flop

In lieu of a single data input D, the JK-flip-flop has two inputs labeled J and K that, together, determine the state the bistable is going to enter at the next active clock edge, see table A.11. The JK-flip-flop is essentially a leftover from the days of SSI that is kept in today's cell libraries mainly for reasons of compatibility. What made it popular is its versatility. Permanently tying J and K to logic 1 results in a T-type flip-flop. A toggle flip-flop with enable is obtained when J and K are connected to form a common input. D-type behavior asks for K to be the inverse of J which then serves as data input D.

Table A.11 Truth table of a rising-edge-triggered JK-type flip-flop.

CLK	J	K	Q	$\overline{Q}$	
0	–	–	Q	$\overline{Q}$	keep state unchanged
1	–	–	Q	$\overline{Q}$	*idem*
↓	–	–	Q	$\overline{Q}$	*idem*
↑	0	0	Q	$\overline{Q}$	*idem*
↑	0	1	0	1	adopt J = $\overline{K}$ as new state
↑	1	0	1	0	*idem*
↑	1	1	$\overline{Q}$	Q	change state, "toggle"

A.4.2 LATCHES OR LEVEL-SENSITIVE BISTABLES

The data or D-type latch

Very much like a basic flip-flop, a basic latch features just a data input D and a clock input CLK. Please note the latter terminal is often referred to as "enable" E or G in datasheets and icons. However, as this input carries the only signal that defines <u>when</u> the latch is to leave its present state and <u>when</u> it is to enter a new one, it clearly must be understood and handled as a clock.

CLK	D	Q	$\overline{Q}$	
0	–	Q	$\overline{Q}$	hold output
1	0	0	1	pass D
1	1	1	0	*idem*

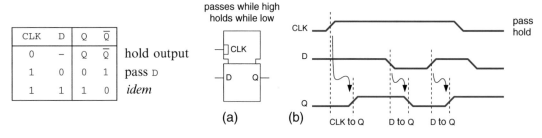

(a)   (b)

D-latch transparent on logic 1. Truth table (left), Icon (a), and waveforms (b).

D-type latches not only find applications as subcircuits of flip-flops but also as bistable memory devices of their own when used in conjunction with a level-sensitive clocking scheme.[22]

---

[22] See sections 7.2.4 and 7.2.5 which corroborate the notion of terminal CLK as a clock rather than as an enable input.

A.4.3 UNCLOCKED BISTABLES

Unclocked bistables differ from their clocked counterparts in that there is no clock terminal that might trigger a state transition without, at the same time, also contributing towards defining the next state. Put in simple words, there is no distinction between <u>when</u> and <u>what</u> inputs.

Observation A.11. *Any information stored in an unclocked bistable is vulnerable to spurious events such as glitches, runt pulses, and other noise on the inputs.*

This is because the absence of a dedicated clock input implies that a transition at some data or control terminal may spark off a state change at any time. The situation sharply contrasts with clocked bistables where there exists no input other than the clock and an optional asynchronous (re)set that can possibly cause the state to flip.

The SR-seesaw

Any two inverting gates interconnected such as to form a loop exhibit bistable behavior because there is positive feedback and zero latency. Let us begin by analyzing two cross-coupled NOR gates. The truth table of the circuit shown in fig.A.15 exposes two alarming peculiarities.

Firstly, the input vector $s = R = 1$ causes the two output terminals to assume identical values thereby violating the rule that outputs Q and $\bar{Q}$ must always assume complementary values. The situation is often referred to as the **forbidden state**.

Secondly, the outcome is unpredictable when the circuit is switched from the forbidden state to the data storage condition. In the occurrence, when inputs s and R simultaneously change from 1 back to 0, nodes Q and $\bar{Q}$ will eventually reassume complementary values, but there is no way to tell whether they will settle to 01 or to 10 after abandoning the forbidden 00 state.[23]

S	R	Q	$\bar{Q}$	
0	0	Q	$\bar{Q}$	maintain output, data storage condition
0	1	0	1	enter state 0, "reset"
1	0	1	0	enter state 1, "set"
1	1	0	0	noncomplementary output, "forbidden state"

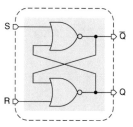

FIGURE A.15

Seesaw built from NOR gates. Truth table (left), circuit example (right).

---

[23] This is because the two stable states are separated by a thin line of metastable equilibrium. Any bistable that is brought close to that line must revert to one stable state or the other before normal operation resumes. However, as explained in section 8.3.4, the course is undeterministic and the time it takes is unbounded.

An alternative circuit is obtained from NAND gates, fig.A.16, yet analysis yields analogous results. As both circuits exhibit bistable behavior but operate with no intervention from a clock, we refer to them as level-sensitive SR-type seesaws or simply as seesaws.[24]

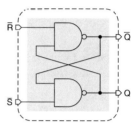

$\overline{R}$	$\overline{S}$	Q	$\overline{Q}$	
0	0	1	1	noncomplementary output, "forbidden state"
0	1	0	1	enter state 0, "reset"
1	0	1	0	enter state 1, "set"
1	1	0	$\overline{Q}$	maintain output, data storage condition

FIGURE A.16

Seesaw built from NAND gates. Truth table (left), circuit example (right).

Although seesaws are at the heart of many latch and flip-flop circuits, naked zero-latency feedback loops are unsuitable for data storage in synchronous designs because of various shortcomings they suffer from.[25] Safe and useful applications of seesaws are very limited and include the generation of non-overlapping clock signals and the debouncing of mechanical contacts.

The edge-triggered SR-seesaw

As opposed to the seesaws of figs.A.15 and A.16, the set and reset inputs are edge-triggered rather than level-sensitive. As a consequence, there is no such thing as a forbidden state with noncomplementary output values. Still, the outcome remains unpredictable when both inputs transit from 0 to 1 in overly rapid succession. Also, with two flip-flops, the circuit of fig.A.17b is more costly in terms of transistor count than any other bistable discussed in this section.

---

[24] SR is just an acronym for set-reset. We have coined the name **seesaw** to avoid the confusion that arises when the more popular terms (asynchronous) SR-flip-flop, SR-latch, $\overline{S}\,\overline{R}$-latch, and NOR|NAND-latch are being used. As shown in table A.6, we make it a habit to distinguish between latches, flip-flops, and unclocked bistables.

[25] As explained in section 6.4, synchronous circuits boast a consistent separation into signals that trigger state changes and others that determine the sequence of states. Unclocked bistables make no such distinction which makes them vulnerable to hazards and may render their behavior unpredictable.

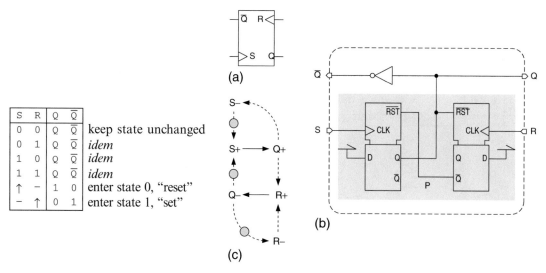

S	R	Q	Q̄	
0	0	Q	Q̄	keep state unchanged
0	1	Q	Q̄	*idem*
1	0	Q	Q̄	*idem*
1	1	Q	Q̄	*idem*
↑	−	1	0	enter state 0, "reset"
−	↑	0	1	enter state 1, "set"

FIGURE A.17

Edge-triggered seesaw. Truth table (left), icon (a), circuit example (b), and signal transition graph (STG) (c).[26]

The Muller-C element

The Muller-C element is a bistable with one output and two interchangeable inputs. The output immediately assumes whatever value the two inputs agree on, but preserves its past value when the two input values differ. The behavior can be likened to hysteresis or to a majority seesaw. The idea is easily generalized to more than two inputs. One may also observe that the Muller-C behaves like an AND-gate while the output is low, and like an OR-gate while the output is high.

Some circuit implementations combine logic gates into a zero-latency feedback loop, the most elegant solution being with a 3-input majority gate as shown in fig.A.18c. Other circuits use a memory element of four transistors reminiscent of a snapper, see fig.A.18d.[27]

---

[26] **Signal transition graphs** (STG) are extremely helpful for describing the behavior of asynchronous circuits and controllers. Each node stands for an event such as a signal transition. A rising edge is identified by an appended + and a falling edge by a −. The updating of data and the withdrawal of data — returning to a high impedance condition, for instance — are also considered to be events. Each solid edge captures a cause/effect relationship implemented within the subcircuit being modeled while a dashed edge indicates the waiting for a condition to be satisfied by the surrounding circuitry. The position of all marks reflects the present state of the circuit. STGs belong to a subclass of Petri nets known as "marked graphs" and obey the same rules: For an event to take place, each incoming edge must carry a mark. When the transition actually fires, those marks get absorbed and a new mark is placed on every outgoing edge. The number of marks does, therefore, not necessarily remain the same.

[27] [228] compares four alternative circuits and concludes that the majority gate implementation is superior to the weak feedback approach in terms of delay and, above all, energy efficiency.

Table A.12   Truth table of Muller-C element.

standard form			
A	B	C	
0	0	0	enter state 0, "reset"
0	1	C	maintain output
1	0	C	maintain output
1	1	1	enter state 1, "set"

with B inverted			
A	B	C	
0	0	C	maintain output
0	1	0	enter state 0, "reset"
1	0	1	enter state 1, "set"
1	1	C	maintain output

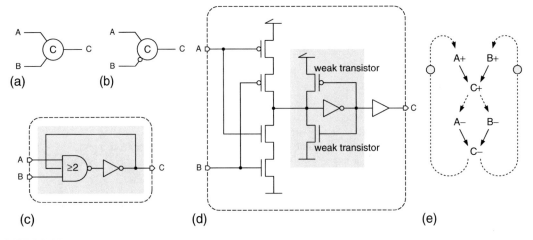

Muller-C element. Icon (a), alternative circuit examples (c,d), and signal transition graph (STG) (e). Icon for a Muller-C with one inverting input (b).

Similarly to seesaws, the Muller-C must not be used as part of synchronous designs because of the absence of a clock input. It finds useful applications for the processing of handshake signals in self-timed systems, however [229] [158].

The mutual exclusion element

Also known as interlock element, the mutual exclusion element — or MUTEX for short — features two symmetric inputs R1 and R2 that are associated with outputs G1 and G2 respectively. The two outputs are never active at the same time, see table A.13. At rest, both inputs are inactive R1 = R2 = 0 and so are the outputs. When a positive impulse arrives on either input, it gets immediately propagated to the pertaining output. Should a second impulse arrive on the other input later, it will not get passed on until the first impulse has come to an end. Impulses are thus propagated on a "first come, first serve" basis unless two of them arrive simultaneously, in which case the circuit arbitrarily selects one to pass through and withholds the other.[28]

---

[28] Note the analogy with two persons simultaneously arriving at the door of a closet or phone booth.

A look back tells us that the seesaw already had the capability of discriminating between two events on the basis of their order of arrival. A comparison of the MUTEX truth table with those of figs.A.16 and A.15 reveals that they are in fact the same (except for swapped and inverted output or input terminals respectively). Not surprisingly, one finds a level-sensitive seesaw in the mutual exclusion circuit, see fig.A.19b for an example. What, then, are the four extra transistors good for?

Note, to begin with, that the seesaw waits in the forbidden state while R1 = R2 = 0. It then enters either the set or the reset state depending on which input switches from 0 to 1 first. If both inputs go high at the same time, or nearly so, the seesaw is subject to marginal triggering. The circuit then lingers in a state of metastable equilibrium before eventually returning to a stable state and letting one input through.[29] From a practical point of view, it is important to make sure that the output signals G1 and G2 remain logically unambiguous, free of glitches, and consistent with the truth table in spite of the seesaw hovering in an irregular condition.

The four transistors must, therefore, be understood to form some kind of filter that defers the circuit's response until the seesaw has recovered. An observation made in [166], namely that the closer in time a pair of rising transitions arrives on the inputs, the longer the MUTEX takes to decide which of them to propagate, comes as no surprise from this perspective.

The mutual exclusion element plays a key role in a subcircuit known as arbiter and is instrumental in self-timed circuits.

Table A.13   Truth table of the mutual exclusion element.

R1	R2	G1	G2	
0	0	0	0	wait
0	1	0	1	let R2 pass
1	0	1	0	let R1 pass
1	1	G1	G2	let the earlier impulse pass, resolve conflict in case of simultaneous arrival

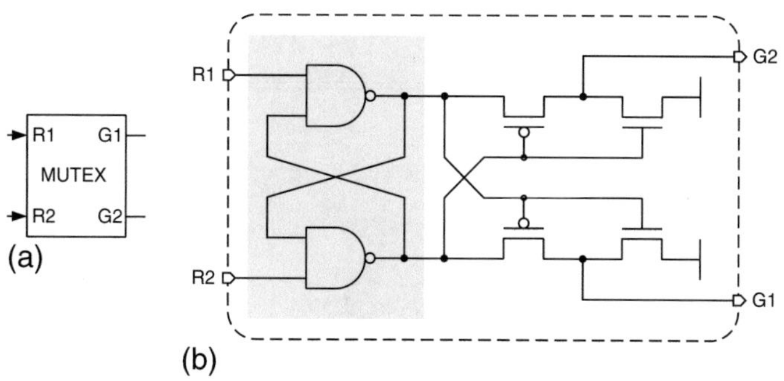

FIGURE A.19

Mutual exclusion element. Icon (a) and circuit example (b).

---

[29] See footnote 23.

As everyone knows, random access memories serve as short-term repositories for large quantities of data. A RAM essentially consists of a large array of elementary binary storage cells that share a common input/output or data port D, see fig.A.20. Any access has to occur one data word at a time and the address A serves to identify the data word currently being accessed. As an example, a 1 Mi x 4 bit RAM accepts and returns data as 4-bit quantities and requires a 20 bit address to select one out of the $2^{20} = 1\,048\,576$ memory locations available.[30]

The overall organization bears many common traits with ROMs in that both the bit cell array and the address decoder are assembled from a few layout tiles. What sets a RAM apart from a ROM are bistable storage cells the state of which can be changed from the data port in very little time. A write enable input

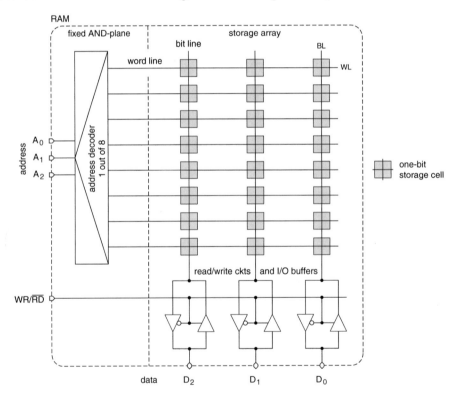

FIGURE A.20

General RAM arrangement (8 words by 3 bits, grossly simplified).

[30] Kibi- (ki), mebi- (Mi), gibi- (Gi) and tebi- (Ti) are binary prefixes recommended by various standard bodies for 2^{10}, 2^{20}, 2^{30} and 2^{40} respectively because the more common decimal SI prefixes kilo- (k), mega- (M), giga- (G) and tera- (T) give rise to ambiguity as $2^{10} \neq 10^3$. As an example, 1 Mibyte = 8 Mibit = $8 \cdot 2^{20}$ bit.

WR/$\overline{\text{RD}}$ that controls the operation of the bidirectional input/output buffers for write and read operation is another important departure.

Two techniques for implementing two-valued memory cells prevail today. In a **static RAM** (SRAM), each bit of data is being stored with the aid of two cross-coupled inverters that form a positive feedback loop. The two stable points of equilibrium so obtained are identified with logic 0 and 1 respectively.

As opposed to this, it is the presence or absence of an electrical charge on a small capacitor that reflects the binary information in a **dynamic RAM** (DRAM). The elementary bit cell is utterly simple and small thereby maximizing the memory capacity available from some given piece of silicon. The ensuing low costs per storage bit are the main reason why DRAMs dominate the mass market for computer main memory in spite of their longer access times.

In either case, data storage is **volatile** exactly as for latches and flip-flops which is to say that the information is lost upon disruption of the supply voltage.

To save on the overall pin count, the address is typically being time-multiplexed over a single address port in commodity DRAM components, e.g. 20 bit as a pair of two 10 bit chunks. Some memories handle write and read transfers over separate input and output ports. Also available are dual-port RAMs that feature two independent I/O ports and that allow for two concurrent read/write transfers.

A.5 TRANSIENT BEHAVIOR OF LOGIC CIRCUITS

Our discussion of logic circuits has, so far, been concerned with steady-state conditions and with the end points of switching processes exclusively. As a consequence from delays and inertial effects, circuits assembled from transistors, wires and other real-world components exhibit various transient phenomena, though. A key question is

"How do the outputs of digital (sub)circuits evolve from one value to the next?"

Let us first be concerned with transient waveforms as witnessed on an oscilloscope hooked to the output of a combinational circuit before turning our attention to the underlying mechanisms that cause them.

A.5.1 GLITCHES, A PHENOMENOLOGICAL PERSPECTIVE

One might naively expect that binary signals progress from one value to the next in monotonic ramps. Experiments with real circuits reveal that this is not always the case, though. Any kind of fugitive or non-monotonic event on a binary signal is termed a **glitch** or a **hazard**.[31] Transients are sometimes catalogued from a purely phenomenological point of view, i.e. as a function of the signal waveform observed.

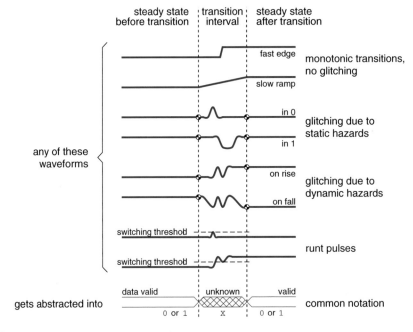

FIGURE A.21
Transients classified according to their waveforms.

A **static hazard** manifests itself as a temporary deviation from its steady-state logic value that occurs in response to some other signal change. If a signal moves back and forth before eventually settling to a logic value opposite to the initial one, we speak of a **dynamic hazard**. As illustrated in fig.A.21, both static and dynamic hazards may be classified in further detail.

The voltage excursion during a glitch does not necessarily make a full swing, stunted pulses are being observed as well. They occur when a signal has begun to change (order) just before an antagonistic effect (counterorder) sets in, thereby preventing the first transition from completing. Such spurious events, termed **runt pulses**, may render circuit operation unpredictable and irreproducible if they reverse direction in the vicinity of the logic family's switching threshold or if they cut across it for a very short lapse of time before turning back.

A.5.2 FUNCTION HAZARDS, A CIRCUIT-INDEPENDENT MECHANISM

This first mechanism can be understood from the logic function alone and does not depend on any specific circuit implementation. Consider the Karnaugh map in table A.14, for instance. When the input vector xyz changes from 001 to 011, then the output h switches from 1 to 0. Conversely, the output stays 1 when the input goes from 001 to 000.

Table A.14   A Karnaugh map that may give rise to function hazards.

h		yz 00	01	11	10
x	0	1	1	0	0
	1	0	1	1	0

What happens when two or more inputs change at a time? As before, the final output value is found in the appropriate field. The intermediate steps leading to that result will, however, depend on the actual sequence of events at the input. Assume input vector xyz is to change from 001 to 111. Depending on whether variable x or y is switching first, the intermediate input is 101 or 011. While in the first case output h remains at 1 throughout, it temporarily assumes value 0 in the latter case and so gives rise to a static hazard on 1.

Similarly, three inputs that change shortly one after the other may give rise to a dynamic hazard, e.g. when input xyz goes from 001 to 011 followed by 111 before settling on 110.

This type of transient output in response to two or more input transitions is termed **function hazard** since it depends on the logic function exclusively. Whether the resulting hazards are of static or of dynamic nature is immaterial in this context. Even the most humble binary functions, such as AND and OR, exhibit function hazards for an appropriate sequence of input transitions. n-cubes are most convenient for tracing function hazards, see fig.A.22.

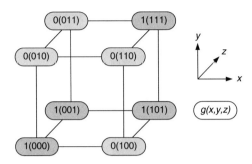

3-cube equivalent to table A.14.

One might argue that no intermediate steps and, therefore, also no function hazards would arise if the switching of the inputs were to occur simultaneously and instantaneously. However, due to dissimilar propagation delays along interconnect lines and within the logic circuitry itself, the result is very much the same as if inputs had switched with a timewise offset.

In practice, the waveforms that result from function hazards range from barely noticeable runt pulses to multiple full blown glitches, depending on how much the input signals are skewed and depending on the logic and interconnect delays involved. Please note that the existence of a hazard can also pass totally unnoticed from the output waveform, a situation which is sometimes referred to as **near hazard**.[31]

A.5.3 LOGIC HAZARDS, A CIRCUIT-DEPENDENT MECHANISM

The second mechanism differs from the first in that the switching of a single input variable suffices to generate unwanted transients. Also, the emergence of transients cannot be explained from the function alone but is related to a particular gate-level circuit structure. As an example, consider fig.A.23, a circuit implementing the Karnaugh map of table A.14.

---

[31] Strictly speaking, the term **glitch** refers to visible waveforms while **hazard** designates the underlying mechanism that may or may not cause some combinational circuit to develop non-monotonic transients. This distinction is not always maintained, though, as the term hazard is often meant to include the observable effect too. Incidentally, note that a glitch can have causes other than hazards such as ground bounce, crosstalk, or electrostatic discharge (ESD).

There is an analogy with medicine in that the clinical picture with its observable symptoms (phenotype) is what matters from a therapeutic and personal point of view. A more profound analysis, in contrast, is concerned with why some living creatures have a higher predisposition for being struck by certain diseases than others do (genotype). Whether the illness actually manifests itself in a given individual or not is of little interest from this epidemiologic perspective.

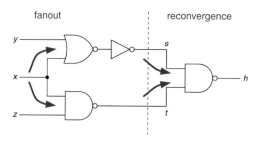

A simple combinational network exposed to logic hazards.

Let's assume that input vector xyz changes from 101 to 001. The fact that the combinational function maps either input to logic 1 may lead us to believe that the output would steadily remain at that value. Yet, unequal propagation delays will cause the switching of inner nodes to be slightly skewed. In the occurrence, the additional inverter in the upper path might be responsible for delaying the switching of s with respect to t. Vector st then changes from 10 via 11 to 01 and output h switches from 1 via 0 back to 1 which amounts to a static hazard. This behavior, which cannot be explained as a function hazard because initial and final node are adjacent in the n-cube, is referred to as a **logic hazard**.

Note, by the way, that the inverse input change, i.e. xyz from 001 to 101, produces no glitch since vector st goes from 01 via 00 to 10, three values all of which result in a 1 at the output. This is not a general characteristic of logic hazards, however.

From a more detailed point of view, logic hazards can be traced back to a combination of **reconvergent fanout** and a function hazard. Let us view the logic circuit of fig.A.23 as being composed of two subcircuits. For certain inputs yz, the subcircuit to the left will broadcast a change of variable x to both of its output nodes s and t. The subcircuit to the right where these two signals recombine is then confronted with multiple changing inputs, and output h may or may not develop a glitch depending on the circuit's detailed timing characteristics.

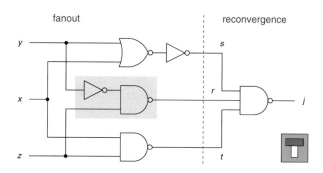

Same circuit as in fig.A.23 with logic hazards suppressed.

Reconvergent fanout is a necessary but not a sufficient condition for logic hazards. This is documented by the circuit of fig.A.24, a modification of fig.A.23 obtained from adding two gates. When xyz changes from 101 to 001 the new node r stays 0, thereby preventing output j from moving away from its steady-state value. Both the original and the modified circuits contain reconvergent fanout, but only the original one is glitching in response to a single input change. Exact conditions for the existence of logic hazards are being given in [221]. As a rule, multi-level networks tend to glitch more intensively than two-level networks because multiple paths of different lengths are more likely to coexist. Similarly to what was found for function hazards, the waveforms caused by logic hazards depend on timing and other circuit details.

The suppression of hazards in fig.A.24 has been bought at the expense of introducing redundant hardware. In fact, the added gates — represented by two extra literals in the logic equation — do not affect the logic function of the network and are, therefore, redundant. As redundant logic is almost impossible to test and entails superfluous switching activity, this approach is not recommended in the context of VLSI design.

A.5.4 DIGEST

Our findings on transient phenomena in combinational logic are best summed up as follows.

Observation A.12. *Hazards may or may not bring forth glitches, extra signal events unwanted and unaccounted for on the logic and higher levels of abstraction.*

Whether a hazard actually materializes as a rail-to-rail pulse, as a runt pulse, or not at all depends on circuit structure, gate and interconnect delays, load conditions, wiring parasitics, layout arrangement, operating conditions (PTV), on-chip variations (OCV), and other implementation details of relatively minor importance.

Observation A.13. *A combinational circuit is susceptible to develop hazards*
○ *if two or more input variables change at a time,*[32] *or*
○ *if the circuit includes reconvergent fanout and if one or more inputs change at a time.*

In conclusion, all logic networks able to carry out some form of computation may give rise to hazards and glitches. Needless to say this includes almost all digital circuits of technical interest.[33]

---

[32] "At a time" here means before all transients in response to the earlier input change have died out.
[33] How to design safe circuits in spite of hazards is a major topic of section 6.3.

A.6 TIMING QUANTITIES

We have now learned about transient events that are associated with the working of combinational logic. Next we want to abstract all relevant inertial effects of digital components and subcircuits to a small set of quantities. Essentially, we want to unambiguously state when a signal is valid, when it is subject to evolve, and when it is safe to update it without worrying about inner details of the circuits concerned.

A.6.1 DELAY PARAMETERS SERVE FOR COMBINATIONAL AND SEQUENTIAL CIRCUITS

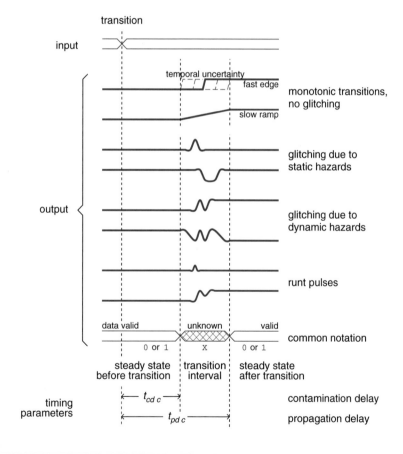

FIGURE A.25

Transients revisited.

Note from fig.A.25 that it takes a pair of delay parameters to adequately describe how long transient phenomena persist at the output of a digital circuit in response to a change at one of the circuit's inputs. One of these two timing quantities is very popular, while the other is not.

t_{pd} **Propagation delay.**[34] The time required to process new input from applying a stable logic value at a (data or clock) input terminal until the output has settled on its final value, i.e. until all transients in response to that input change have died out.

t_{cd} **Contamination delay**, aka (output) retain delay and retain time.[35] The inertial time from altering the logic value at a (data or clock) input until the output value no longer remains the same, i.e. until transients begin in response to that input change. By definition, $0 \leq t_{cd} < t_{pd}$ must hold for any physical component or (sub)circuit.

Contamination delay figures are rarely published in datasheets, though.[36] As a stopgap, safety-minded engineers often substitute the lower bound for the missing parameters by admitting $t_{cd} \equiv 0$. In doing so, causal behavior is taken for granted but any inertial effects are ignored. While this is a viable workaround in many cases, contamination delay is an essential quality of any flip-flop.[37]

Observation A.14. *Signal propagation along some path through a digital network is captured by a pair of timing parameters termed contamination and propagation delay respectively. Their numeric figures are chosen so as to enclose all transient phenomena (such as ramps, hazards, glitches, and runt pulses) that might occur at the output in response to some input change.*

What does this mean for a typical digital circuit that features multiple input and/or output terminals? If the propagation delay of such a circuit is to be characterized by a single quantity, it is obviously necessary to consider all of the terminals. Propagation delay of vectored inputs and outputs is, therefore, determined by whichever signal takes the uppermost amount of time to traverse a circuit (longest path). Conversely, it is the quickest travel time through the logic that determines the contamination delay (shortest path).

In the occurrence of a (level-sensitive) latch, the point of time when output data become valid depends on the data input in some situations and on the clock input in others. Two propagation delays are thus required to characterize the timing of a latch. $t_{pd\,ld}$ indicates the **data-to-output delay** (from input D to output Q) while $t_{pd\,lc}$ states the **clock-to-output delay** (from CLK to Q). The same argument obviously also applies to the contamination delay.

---

[34] Also referred to as "settling time" or "maximum delay" by some authors.

[35] The terms "internal delay", "output hold time", and "minimum (propagation) delay" are synonyms found in the literature. We do not support their usage as they lend themselves to confusions, however.

[36] There are various reasons for this. For one thing, the concept of contamination delay is virtually unknown to most practitioners and textbooks in digital electronics. For another thing, manufacturers are reluctant to commit themselves to minimum values for any delay parameter because they reserve the right to upgrade their fabrication processes at any time in search of better performance or lower manufacturing costs. Also, there is a common industry practice known as **down-binning** which implies shipping a faster device against an order for a slower part. Because of its superior speed, a faster device is bound to exhibit a shorter contamination delay than the original part.

[37] Equating all contamination delays with zero is an inadmissible oversimplification that makes it impossible to gain a more profound understanding of how clocked circuits operate. The functioning of a simple shift register, for instance, cannot be explained under that assumption. Refer to section 7.2.2 for an in-depth analysis.

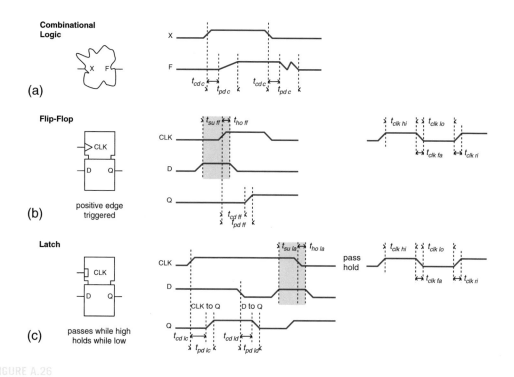

Timing characteristics of combinational subcircuits and of clocked bistables.

A.6.2 TIMING CONDITIONS GET IMPOSED BY SEQUENTIAL CIRCUITS ONLY

The orderly functioning of any memoryzing (sub)circuit — flip-flop, latch, RAM, or larger sequential circuit — requires that it be driven by a clock of clear-cut waveform. Ambiguous voltages, glitches, sluggish ramps, and stunted waveforms are unacceptable. Clocked circuits further impose dead times during which the clock must remain stable before it may be allowed to toggle again. Two pairs of timing conditions have been defined to capture the characteristics of an acceptable clock. Note that all four parameters relate to no other signal than to the clock itself.

$t_{pu\,clk\,min}$ **Clock minimum pulse width**. The time span during which the clock signal must firmly be kept either low or high before it is permitted to swing back to the opposite state. Shorter clock pulses must be avoided as a bistable is otherwise likely to behave in an unpredictable way. In practice, it is often necessary to distinguish between
$t_{lo\,clk\,min}$ clock minimum low time (MPWL) which refers to logic 0 (pause), and
$t_{hi\,clk\,min}$ clock minimum high time (MPWH) which refers to logic 1 (mark).
Both combine into $t_{pu\,clk\,min} = \max(t_{lo\,clk\,min}, t_{hi\,clk\,min})$.

$t_{ra\,clk\,max}$ **Clock maximum ramp time**. The timewise allowance for the clock to ramp from one logic state to the opposite one. Driving a bistable with overly slow waveforms may cause inner nodes to float or to be placed under control of conflicting drivers which may lead to an irrecoverable loss of data. To prevent this from happening, a pair of maximum transition or ramp times is imposed on the clock.[38]
$t_{ri\,clk\,max}$ clock maximum rise time refers to the transition from 0 to 1 while
$t_{fa\,clk\,max}$ clock maximum fall time is concerned with the inverse transition.
Both combine into $t_{ra\,clk\,max} = \min(t_{ri\,clk\,max}, t_{fa\,clk\,max})$.

Any clocked (sub)circuit further imposes requirements on the timewise relationship between any of its data inputs and the clock that is driving that (sub)circuit.

t_{su} **Setup time**. The lapse of time immediately before the active clock edge during which an input is required to assume a fixed logic value of either 0 or 1 at the input of a clocked (sub)circuit. The setup condition is here to make sure all inner nodes have settled to values determined by new input data before the (sub)circuit locks into the corresponding state in response to the subsequent active clock edge. Violating the setup requirement must be avoided under any circumstance because bistables are otherwise likely to behave in an unpredictable way.

t_{ho} **Hold time**. The lapse of time immediately after the active clock edge during which data are required to remain logically unchanged at the input of a clocked (sub)circuit. The hold condition assures that all inner nodes have properly settled so that the new state is maintained even when the stimuli that caused the transitions in the first place are being removed. Violating the hold requirement must be avoided for the reasons explained before. Although either the setup or the hold time may assume a negative value for certain components or (sub)circuits,[39] $t_{su} + t_{ho} > 0$ always holds.

---

[38] There is another reason for doing so. Timing data of clocked subcircuits vary with the waveform of the driving clock. Yet, for reasons of economy, it is common industrial practice to characterize flip-flops, latches, RAMs, and the like for one typical clock ramp time, e.g. for $t_{clk\,ra} = t_{clk\,ri} = t_{clk\,fa} = 50$ ps. Driving them with a clock waveform that exhibits much slower ramps will cause the actual parameter values of t_{pd}, t_{cd}, t_{su}, and t_{ho} to significantly deviate from the figures published in datasheets and simulation models.

[39] A negative hold time means that a data input is not required to preserve its value until <u>after</u> the active clock edge for being properly stored in the (sub)circuit, but is free to resume new transient activities <u>before</u> that time. Similarly, a negative setup time means that the input is allowed to switch until after the active clock edge.

Observation A.15. *Setup time and hold time together demarcate a brief lapse of time in the immediate vicinity of the active clock edge. Their numerical values are chosen so as to guarantee that data get stored and/or processed as intended under all circumstances.*

Any ambiguous or changing input during the aperture so defined is likely to expose any bistable or other clocked (sub)circuit to marginal triggering and to cause it to fail in an unexpected and unpredictable way.[40] To prevent this from happening, the logic value must be kept constant and logically well-defined throughout. In the occurrence of a circuit with multiple input bits, such as a register, every bit must remain stable either at 0 or 1 throughout.

Bistables with extra control inputs and/or with multiple clocks impose additional timing conditions. More specifically, asynchronous set and reset signals must not be released in the immediate vicinity of an active clock edge otherwise exposing the bistable to marginal triggering.

$t_{su\,rst}$ **Recovery time**, aka release time. Indicates by how much time the deactivation of an asynchronous (re)set input must precede the active clock edge such as to allow a bistable to unlock properly before taking up normal operation.

$t_{ho\,rst}$ **Home time** is more suggestive than the popular synonym removal time. Indicates for how much time an asynchronous (re)set input must remain activated following the active clock edge in order to safely bring a bistable home into its reset state. If the asynchronous (re)set is released too early, then chances are that the bistable falls into some undetermined state or a metastable condition.

Some datasheets explicitly specify recovery and home times while others refer to these quantities as particular cases of setup and hold times.[41]

A.6.3 SECONDARY TIMING QUANTITIES ARE DERIVED FROM PRIMARY ONES

The timing quantities introduced so far suffice for characterizing digital circuits. Yet, it is sometimes convenient to define quantities that are derived from those primary parameters.

t_{cw} **Data-call window**, aka aperture time, setup-and-hold window, and sampling time. The overall time span during which data must maintain a constant and well-defined value at the input of a memorizing (sub)circuit $t_{cw} = t_{su} + t_{ho} > 0$.

t_{id} **Insertion delay**. As the name suggests, this term denotes the extra delay that is inflicted on a signal when a given (sub)circuit is being inserted into a signal's propagation path. In the occurrence of a combinational circuit, insertion delay is the same as the propagation delay on the longest path, that is $t_{id\,c} = \max(t_{pd\,c}) = t_c$.

---

[40] You may want to consult section 8.4.1 for more information on what exactly happens with a bistable circuit in this case. How to determine the setup and hold time figures of a given bistable is also explained there.

[41] The difference is that an asynchronous control signal must subsequently retain its passive value indefinitely unless the circuit is to be reinitialized to its start state whereas any ordinary (synchronous) control signal is essentially free to switch after the hold time has expired. Using separate terms thus seems justified, but we will not insist on this.

For flip-flops, one has $t_{idff} = t_{suff} + t_{pdff} = t_{ff}$ (where t_{pdff} refers to the non-inverting output unless indicated otherwise) because this is the minimum lapse of time an edge-triggered bistable takes to store and propagate a data item provided the active clock edge is optimally timed.

$f_{toff\,max}$ **Maximum toggling rate**. No flip-flop can be made to operate faster than its insertion delay and its minimum clock pulse widths permit. The utmost clock frequency thus is $f_{toff\,max} = \frac{1}{T_{toff\,min}}$ where $T_{toff\,min} = \max(t_{idff},\, t_{hi\,clk\,min\,ff} + t_{lo\,clk\,min\,ff})$. Although this quantity is given much publicity in advertisements, it remains of little practical interest as it leaves no room for any data processing activity.

r_{sl} **Slew rate**. The average velocity of voltage change during a logic transition, that is $r_{sl} = \frac{\Delta u}{\Delta t}$. The quantity is positive for rising edges $r_{sl\,ri} = \frac{U_h - U_l}{t_{ri}}$ and negative for falling ones $r_{sl\,fa} = \frac{U_l - U_h}{t_{fa}}$.

δ **Duty cycle**. The average proportion of time during which a signal is active or a load is energized. For an active-high signal of period T, one has $\delta = \frac{t_{hi}}{T} = \frac{t_{hi}}{t_{hi} + t_{lo}}$ if the switching is so fast that ramp times can be ignored. $0 \leq \delta \leq 1$ holds by definition.

A.6.4 TIMING CONSTRAINTS ADDRESS SYNTHESIS NEEDS

Timing constraints differ from delay parameters and timing conditions in that they do not describe the timewise behavior of an existing component or design for the purposes of analysis and simulation, but serve to express target characteristics of a circuit-to-be for the purposes of design and synthesis. As illustrated in fig.4.21, most timing constraints specify an upper bound for some propagation delay. Asking for a minimum contamination delay also makes sense in certain situations.

A.7 BASIC MICROPROCESSOR INPUT/OUTPUT TRANSFER PROTOCOLS

While microcomputer architectures are beyond the scope of this text, we briefly review the three fundamentally different I/O transfer protocols because many ASICs are to interface with a microprocessor bus system.

A peripheral device that wishes to deliver data to a microcomputer has to ask for an input transfer operation. Conversely, an output transfer is solicited when the peripheral needs to obtain data. In either case, the peripheral sets a **service request** flag. Three conceptually different ways exist for notifying the microcomputer about such a condition and for handling the subsequent data transfer, see fig.A.27.

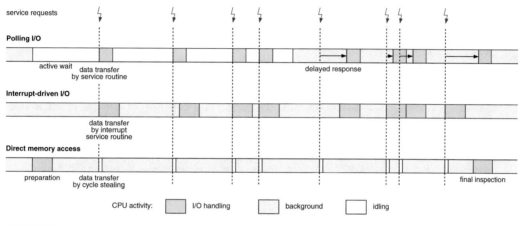

FIGURE A.27

The three basic data transfer protocols for microcomputer input/output.

1. **Polling.** In this scheme, the CPU actively waits for service requests to arrive. The peripheral is wired to some port on a peripheral interface adapter (PIA) so that the CPU can examine its status by way of read operations on the pertaining PIA address. Whenever the operating system expects a peripheral device to ask for I/O operations in the near future, it enters a **program loop** which makes it periodically read that port and check the appropriate bit position there to find out whether a service request is pending or not.

   If so, the program branches to a service routine which tells the CPU how many data words to transfer, where to get them from, how to process them, and where to deliver them to. The routine is also in charge of setting the service request flag back to its inactive state once the peripheral has been serviced.

   If not so, the CPU proceeds with executing the loop's code which makes it repeat read-and-check operations until the expected service request finally arrives.

   Designing the polling loop always is a compromise, see fig.A.27. A tight loop ensures a fast response but leaves little room for doing anything useful while waiting, whereas an ampler loop provides room to carry out computations in the background, but results in prolonged response times as the request bit per force gets examined less frequently.

2. **Interrupt-driven.** As opposed to polling, the CPU is not locked in a loop but is free to execute code in a background process. In order to make this possible, the request flag is brought out to a line that connects to a special input of the CPU termed **interrupt request**. Activation of this line diverts program execution to an interrupt service routine once execution of the current instruction has completed. The routine first causes the CPU to properly suspend the current process by saving the contents of critical registers to memory or on a special stack. The remainder of the I/O operation occurs in much the same way as for polling. After having completed the interrupt service routine, the CPU resumes the suspended process.

Processor instruction sets typically include a pair of special instructions that allow programmers to temporarily suspend the interrupt mechanism in order to bring critical code sequences to an end without having their execution delayed or broken up. Withholding interrupt requests in this way is referred to as interrupt masking.

3. **Direct Memory Access** (DMA). The CPU is freed from most of the burden associated with I/O transfers by delegating them to a special hardware unit termed **DMA controller** that is hooked to the service request line in lieu of the CPU. Before a series of I/O operations can begin, the CPU in a preparatory step instructs the DMA controller how many data words to transfer and indicates their destination and/or source addresses. The CPU itself is not involved in the subsequent transfer operations, which contrasts sharply with polling and interrupt-driven I/O. Instead, the DMA controller handles the actual data moves by stealing memory cycles from the CPU for its own memory accesses whenever notified by the peripheral that a data item is waiting to be accepted or delivered.

At the end of the commissioned series of transfers, the DMA controller typically informs the CPU by way of an interrupt. As part of the pertaining service routine, the CPU does then inspect some status register in the DMA controller to find out whether the transfer has been successfully completed or prematurely aborted.

Table A.15   The basic I/O transfer protocols compared.

	Input/Output Data Transfer Scheme		
	Polling	Interrupt-driven	DMA
Hardware overhead	close to none, PIA port bit	small, interrupt mechanism	moderate, DMA controller
CPU burden	high as CPU has to idle in a loop	moderate, once per data item	minimal, once per data series
Response time	unpredictable, depends on loop	a few instruction cycles unless masked	almost immediate, a few clock cycles
Transfer rate	moderate	fair	high

As an improvement to polling, it is possible to combine most of its simplicity with the efficiency of interrupt-driven I/O by having a **timer** periodically trigger the interrupt line. In the interrupt service routine, the CPU then polls one peripheral device after the other and branches to a specific service routine for each that has a request pending.

A.8 SUMMARY

- Digital hardware deals with bit patterns rather than with numbers. These patterns assume a meaning as numbers only when interpreted according to the specific number representation scheme that the designers had in mind.

- As zero-latency loops may give rise to unpredictable behavior, it is best to avoid them.

- Be prepared to observe unexpected transient pulses at the output of any combinational logic unless you have proof to the contrary.

- While all datasheets and textbooks mention setup time, hold time, and propagation delay, it is not possible to understand how a simple shift register works without introducing the concept of contamination delay.

- Any bistable is either an
  - Edge-triggered flip-flop, a
  - Level-sensitive latch, or an
  - Unclocked bistable.

  It is most important to keep those apart technically and linguistically.

- Before starting up any kind of electronic design automation tool, it is important to
  - Know what its optimization criteria and limitations are,
  - Find out whether it does indeed apply to the problem at hand,
  - Develop an understanding of the available options and controls such as to
  - Set all options and control knobs to suitable values.

- Three basic protocols are available for organizing the transfer of data between a peripheral device and a microcomputer:
  - Periodically polling the peripheral for service requests in a program loop.
  - Having a special signal from the peripheral interrupt regular program execution.
  - By bypassing the CPU with the aid of an extra direct memory access controller.

FINITE STATE MACHINES

B

This chapter is divided into two major sections. Section B.1 reviews the classes of finite state machines used in electronics design and their equivalence relationships. Although this material is strongly related to automata theory — or actually part of it — no attempt is made to cover the theory since there are excellent and comprehensive textbooks on the subject. Rather, emphasis is on a number of mathematical facts relevant to hardware design that are not normally found in such references. Section B.2 then looks at finite state machines more from an implementation point of view, yet without committing to any specific technology.

B.1 ABSTRACT AUTOMATA

Automata theory is a mathematical discipline concerned with fundamental issues of discrete computation such as formal languages and grammars, parsing, decidability and computability. The underlying formal models are crude abstractions that essentially simplify computing equipment to transducers that, while changing from state to state, convert a given input string into some output string. Most issues relevant to digital design such as hardware architecture, computer arithmetics, parasitic states, state encoding, transient effects, delays, synchronization, etc. are neglected, which raises the question

"Why study the abstract subject of automata theory in the context of electronics design?"

The motivation is threefold:

Functional specification. Describing what a digital system has to do is not always easy. Automata theory often helps to specify the relationship between a circuit's inputs and outputs in a more formal way, especially for control- and protocol-oriented tasks.

Modeling and verification. When viewed from outside, the behavior of an entire system can be modeled as a single finite state machine. While usually not a very efficient approach to constructing a circuit, this abstraction proves useful for verifying a system's behavior by simulation and testing.

Synthesis. Almost any practical system is composed of a number of cooperating subsystems, each of which can in turn be modeled as finite state machine. At a certain level of detail, any synchronous circuit is patterned after a specific type of automaton.

"What do we mean by finite state machine then?"

Definition B.1. A deterministic automaton is a system, which at discrete moments of time $t = 0, T, 2T, 3T, \ldots, kT, (k+1)T, \ldots$ satisfies the following conditions:

1. The input to the system can be chosen from a set of possible stimuli I.
2. The system subject to an input i can be in just one of a set of possible states S and does output one out of a set of possible responses O.
3. The state of the system and the input to it uniquely define the state which the system is going to assume in the next such moment of time.

Throughout this text, we will stay within the framework of **discrete** and **deterministic** automata so defined. We will further limit our discussion to **finite state machines** (FSM) where each of the three sets I, S, and O is restricted to a finite number of elements. Please note that, in general, I, S, and O do not have the same cardinality, although they may happen to do so.

B.1.1 MEALY MACHINE

There is more than one way to express a behavior that conforms with the above. A first approach uses two equations

$$o(k) = g(i(k), s(k)) \tag{B.1}$$

$$s(k+1) = f(i(k), s(k)) \tag{B.2}$$

where $i \in I$, $s \in S$, and $o \in O$. g is termed **output function** and f **transition function** or next state function. In addition, at $k = 0$ the automaton is assumed to be in a special state $s_0 \in S$ which is called **start state**.

Equations (B.1) and (B.2) together form a Mealy model. More precisely, we speak of a Mealy automaton if the present output depends on the present input. Many real-world examples are of more simple nature which gets reflected by the absence of one or more terms from (B.1) and (B.2). See section B.1.5 for a detailed classification scheme.

Instead of stating transition function f and output function g as equations, any finite state machine can be completely described by listing next state and output values for any combination of present state and input. Such a list is termed a **state table**, see fig.B.1 for an example.

The well-known **state graph**, aka state transition diagram, is nothing else than a pictorial representation of a state table. Each state is represented by a vertex and each transition by a directed edge. A short arrow identifies the start state. Where an input symbol causes a state to persist, i.e. to fall back on itself, a loop is drawn.

So far, we have been concerned with FSM behavior exclusively, but we should also ask for a hardware structure that behaves accordingly. The result from expressing output function and transition function with a data dependency graph (DDG), is shown in fig.B.2 for a Mealy machine. As will be shown in section B.2.5, this straightforward solution is not necessarily also the most efficient one, however.

Example

$I = \{a,b\}$, $S = \{p,q,r,t\}$, and $O = \{0,1\}$.

$i(k)$	a	b	a	b
$s(k)$	\multicolumn{2}{c}{$o(k)$}	\multicolumn{2}{c}{$s(k+1)$}
p	1	0	r	t
q	0	1	p	q
r	1	0	q	r
t	0	1	r	p

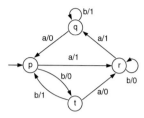

FIGURE B.1

Mealy machine with four states. State table (left) and state graph (right).

□

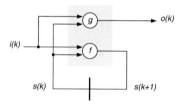

FIGURE B.2

DDG of a Mealy automaton.

B.1.2 MOORE MACHINE

The Moore model differs from the Mealy model in that the present output depends on the present state exclusively; there is no input literal $i(k)$ in output function g. As a consequence, the output is allowed to change only as a result of a state transition.[1]

$$o(k) = g(s(k)) \tag{B.3}$$

$$s(k+1) = f(i(k), s(k)) \tag{B.4}$$

As shown below, the more restrictive formulation for the output function (B.3) gets reflected by omissions in the network structure, state table, and state graph of Moore automata.

---

[1] Because state transitions are restricted to discrete moments of time kT, the Moore model is sometimes said to have a synchronous output. This interpretation commonly found in texts on automata theory abstracts from propagation delay. In practice, a Moore output will settle to its final value a couple of gate delays after the active clock edge and remain constant until the next active clock edge.

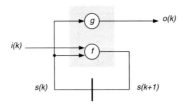

FIGURE B.3

DDG of a Moore automaton.

Example

$I = \{a,b\}$, $S = \{u,v,w,x,y,z\}$, and $O = \{0,1\}$.

$i(k)$		a	b
$s(k)$	$o(k)$	$s(k+1)$	
u	1	z	w
v	0	z	w
w	0	x	u
x	0	y	x
y	1	v	y
z	1	y	x

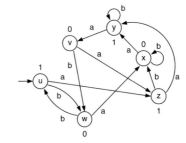

FIGURE B.4

Moore machine with six states. State table (left) and state graph (right).

☐

Both Mealy and Moore automata are pervasive in electronics circuits, primarily in controllers that govern sequential data processing and data exchange operations.

B.1.3 MEDVEDEV MACHINE

Some particular situations in electronic hardware design prohibit logic operations in the output function.[2] Mathematicians use the name Medvedev to designate a subclass of Moore machines where the output function g has degenerated to the identity function.[3]

$$o(k) = s(k) \tag{B.5}$$
$$s(k+1) = f(i(k), s(k))$$

---

[2] E.g. because of the inertial and transient effects associated with the pertaining combinational networks, see section B.2.4. Also, the direct observability and controllability of Medvedev outputs via scan path test structures is welcome in many controller applications.

[3] Medvedev automata are also known as "finite acceptors" or as "automata without output" because studying state transitions alone is sufficient for many problems from automata and formal language theory.

In the context of digital circuits where state and output symbols are encoded as binary vectors, it makes sense to slightly relax this definition. We will speak of a Medvedev automaton even when state bits are dropped from the output or when state bits are duplicated in the output. State and output vectors are thus allowed to differ in cardinality as long as this is just a matter of wiring and, hence, involves no logic gates. Most counters are practical examples of Medvedev automata, see fig.B.5 for a hardware structure.

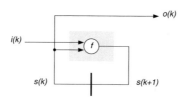

DDG of a Medvedev automaton.

B.1.4 RELATIONSHIPS BETWEEN FINITE STATE MACHINE MODELS

Two questions of both theoretical and practical importance are

"Under what conditions is it possible to replace a finite state machine by another one?" and "How do the various classes of automata relate to each another?"

Since finite state machines can be viewed as transducers that convert some input string into some output string, the most natural way of defining **functional equivalence** is as follows.

Definition B.2. Two automata are considered equivalent if they always yield identical strings of output symbols for any identical strings of input symbols.

It is important to understand that the state graphs of equivalent automata need not be isomorphic. This is because the above definition refers to input and output quantities only. Put in other words, automata are abstracted to black boxes in the context of equivalence. Figs.B.8 and B.9, for instance, show a pair of Mealy machines that are equivalent but not isomorphic.

Equivalence of Mealy and Moore machines in the context of automata theory

While the above definition of equivalence works fine when comparing Mealy automata with Mealy automata or Moore automata with Moore automata, there is a technical problem when comparing machines across the two classes. A Mealy machine can respond to an input change at any time whereas the response of a Moore machine is necessarily deferred to after the next active clock edge. Put differently, Moore outputs have latency 1 and Mealy outputs latency 0. No matter what the first input $i(0)$ looks like, the first symbol in the output string from a Moore machine is $o(0) = g(s_0)$, only later can the input affect the output. This implies that Moore automata take $n + 1$ computation periods to process a total of n consecutive input symbols while Mealy automata require only n periods. The number of output symbols released from the two models differ accordingly.

As a workaround, the requirement for equivalence is relaxed as follows in the context of automata theory because that theory is primarily concerned with the mapping between symbol strings.

Definition B.3. A Mealy and a Moore automaton are considered equivalent if they always yield identical strings of output symbols for any identicals string of input symbols, when the first output symbol — which is associated with the start state — is deleted from the output string of the Moore automaton.

This somewhat academic understanding gives rise to a well-known result

Theorem B.1. *For any Mealy automaton there exists an equivalent Moore automaton in the broad sense of definition B.3, and vice versa.*

At first sight this may seem surprising because the output function of the Mealy model is more general than that of the Moore model. We will sketch a constructive proof, i.e. two algorithms for converting a Moore machine into an equivalent Mealy machine, and vice versa. More details can be obtained from [230], an excellent textbook on abstract automata and formal languages.

Converting a Moore machine into an equivalent Mealy model is very easy. For every vertex of the state graph delete the output symbol associated with the vertex and attach it to all edges that enter that vertex. Clearly, the procedure will leave the number of states unchanged.

Coming up with a Moore model for a Mealy machine cannot simply follow the inverse procedure because any attempt to assign a vertex the output symbol associated with its incoming edges must lead to a conflict unless all incoming edges agree in their outputs. Thus, whenever a conflict arises, the vertex is split into as many copies as there are distinct output symbols attached to its incoming edges. All output symbols can so be transferred from the incoming edges to their respective copy of the vertex. Each copy keeps the full set of outgoing edges attached to the original vertex. The process is repeated for the successor nodes until all vertices have been visited. Please note, that the number of states may — and often will — increase dramatically when going from a Mealy to a Moore machine.

Incidentally, we conclude from the two conversion procedures that (a) for any Moore automaton there exists an equivalent Mealy automaton with a smaller or equal number of states, and (b) the above Moore-to-Mealy conversion algorithm does not, in general, lead to a solution with the minimum possible number of states.

Example
In the broad sense of automata theory, the two state machines described in figs.B.1 and B.4 are actually equivalent. Their respective output strings are opposed in table B.1 for two input strings chosen at random.

Table B.1 Output strings of equivalent Mealy and Moore automata compared.

k	0	1	2	3	4	7	8
automaton				output			
input	a	b	b				
Mealy	1	0	0				
Moore	[1]	1	0	0			
input	b	b	a	a	b	a	
Mealy	0	1	1	1	1	0	
Moore	[1]	0	1	1	1	1	0

□

Equivalence of Mealy and Moore machines in the context of hardware design

From an engineering point of view, the extra output symbol $o(0)$ and the timewise offset of all subsequent symbols cannot be abstracted from. After all, few applications will tolerate replacing a finite state machine by another one the output of which lags or leads by a full clock cycle. Unless latency is indeed uncritical for the application at hand, equivalence must, therefore, be understood in the stricter sense of its original definition which implies that

Theorem B.2. *A Mealy automaton and a Moore automaton can never be equivalent in the more narrow sense of definition B.2.*

Equivalence of Moore and Medvedev machines

Theorem B.3. *For any Moore automaton there exists an equivalent Medvedev automaton in the more narrow sense of definition B.2, and vice versa.*

The proof is by showing how to build an equivalent Medvedev automaton from a Moore automaton.[4] The problem with designing a Medvedev automaton is that $|O| = |S|$ while, in general, this is not the case for Moore automata where $|O| \leq |S|$. To get around this difficulty, we allow the state symbol to be composed of two parts, namely a left part, which we will also use as output symbol, and a right part, that will remain hidden from the outside world. In any case must their concatenation yield a unique state symbol. Note that this trick does not introduce actual logic operations into the output function and is, therefore, consistent with our definition of Medvedev automata. However, it breaks the constraint that forced $O = S$.

The conversion algorithm works as follows. Consider the state graph of the Moore automaton. Assign every state its output symbol as left part of its state symbol. If all states are uniquely labeled, the right parts remain empty and conversion is completed since the automaton had already followed the Medvedev model in the first place. Otherwise, assign each state a right part such as to make it unique. The easiest way to do so is simply to copy the state symbols from the initial Moore automaton. When

---

[4] The inverse transform is trivial since any Medvedev automaton is a Moore automaton by definition.

fed with the same stimuli, the resulting automaton will always output the same responses as the Moore automaton.

Perhaps a more intuitive conversion procedure is given in fig.B.6.

The number of states of a Medvedev automaton so obtained is always equal to that of the original Moore model. It takes more bits to encode the wider Medvedev state symbols, though.

Example

The Medvedev automaton shown in fig.B.7 has been obtained from the Moore automaton of fig.B.4 by way of the constructive algorithm stated above.

$I = \{a,b\}$, $S = \{0c,0d,0e,1c,1d,1e\}$, and $O = \{0,1\}$.

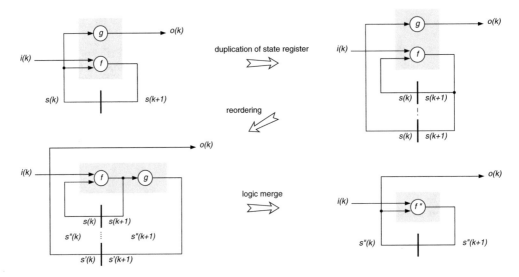

Turning a Moore machine into an equivalent Medvedev machine.

$i(k)$		a	b
$s(k)$	$o(k)$	$s(k+1)$	
1c	1	1e	0d
0c	0	1e	0d
0d	0	0e	1c
0e	0	1d	0e
1d	1	0c	1d
1e	1	1d	0e

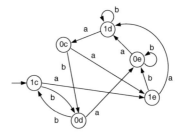

Medvedev machine with six states. State table (left) and state graph (right).

□

B.1.5 TAXONOMY OF FINITE STATE MACHINES

The table below, which is patterned after a Karnaugh map, classifies deterministic automata according to what actually determines their next state and their output.

Several subclasses, the fields of which are marked with special characters in table B.2, make no sense from a technical point of view. Here are the reasons why:

† Unobservable automata are useless in engineering applications.
‡ There is no point in controlling a state that exerts no influence on the output.
§ Having the output depend on a fixed and thus effectively inexistent state makes no sense.

Other subclasses have been given a name because they find widespread applications as digital function blocks. They are identified by capital letters as listed below.

Table B.2   Taxonomy of finite state machines.

				output function g depends on			
				—	state	input and state	input
transition function f depends on		—		†	§	§	C
			state	† ‡	A		‡
	input	and	state	† ‡	O	Y	‡
	depends on	input		† ‡	D		‡
				Moore model		Mealy model	

	Subclass	Example
Y	Full Mealy automaton[5]	Controller
O	Full Moore automaton[5]	Controller, cell of cellular automaton
A	Autonomous automaton	Clock generator, pseudo random number generator
D	Delay automaton	Pipeline stage (combinational logic plus register)
C	Combinational logic	Full adder, unpipelined multiplier

---

[5] The word "full" is meant to imply that no term has been dropped from the general equations (B.1) (B.2) and (B.3) (B.4) respectively.

Consider the state table of some finite state machine and assume that two states have exactly the same entries in their present output fields and also in their next state fields. As an example, this applies to states "7" and "10" in fig.B.8. From a graph point of view, this means their outgoing edges are labelled in exactly the same way and point to exactly the same vertices. It is intuitively clear that any two such states must appear to be the same when nothing but the machine's inputs and outputs are observed.

Definition B.4. Two states of a finite state machine are considered indistinguishable if the machine can be placed in either of the two and responds with identical strings of output symbols to any string of input symbols.

Indistinguishable states are also known as equivalent states and as **redundant states**. Merging them has no effect on a machine's behavior. The new automaton so obtained will necessarily be equivalent to the original one, but simpler to implement. State reduction is the process of collapsing redundant states until no equivalent state machine with a smaller number of states exists. Collapsing all states that have identical state table entries does not suffice, however, as two states can have distinct next state fields and still be perfectly indistinguishable.

Theorem B.4. *Two states of a finite state machine are indistinguishable iff they have (a) identical outputs and (b) go to indistinguishable successor states for any possible input symbol.*

The difficulty with applying this theorem to state reduction directly lies in its recursiveness. A more practical approach is the **implication chart algorithm** due to Paull and Unger and nicely described in [231], for instance. Luckily, there is no need for designers do that manually as automatic state reduction is part of HDL synthesis. We thus refrain from presenting algorithmic details and are content to show an FSM before and after state reduction.

Example

The state graph depicted in fig.B.8 has been chosen for demonstration purposes with no particular application in mind.

$i(k)$	00	01	10	11	00	01	10	11
$s(k)$		$o(k)$				$s(k+1)$		
1	1	1	1	1	1	1	1	2
2	0	0	0	0	2	8	8	3
3	0	0	0	0	3	5	5	4
4	0	0	0	0	4	5	5	2
5	0	0	0	0	6	5	5	9
6	0	0	0	0	7	11	11	3
7	1	0	0	1	1	1	1	1
8	0	0	0	0	9	8	8	6
9	0	0	0	0	10	11	11	4
10	1	0	0	1	1	1	1	1
11	0	0	0	0	1	1	1	1

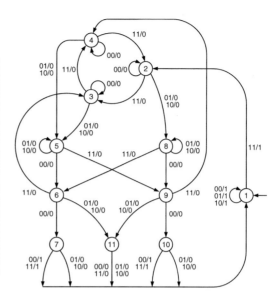

FIGURE B.8

Original state machine. State table (left) and state graph (right).

Systematic state reduction not only confirms (7,10) as indistinguishable states, but also reveals equivalences between (2,3,4), (5,8), and (6,9). While the new state machine shown in fig.B.9 preserves the input-to-output relationship, the number of states has dropped from 11 to 6.

$i(k)$	00	01	10	11	00	01	10	11
$s(k)$		$o(k)$				$s(k+1)$		
1	1	1	1	1	1	1	1	2
2	0	0	0	0	2	5	5	2
5	0	0	0	0	6	5	5	6
6	0	0	0	0	7	11	11	2
7	1	0	0	1	1	1	1	1
11	0	0	0	0	1	1	1	1

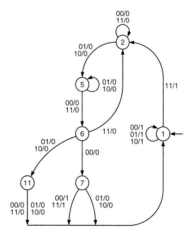

FIGURE B.9

Reduced state machine. State table (left) and state graph (right).

□

B.2 PRACTICAL ASPECTS AND IMPLEMENTATION ISSUES

How to turn finite state machines into electronic hardware is discussed in the main text. Yet, several practical issues can be understood from a mathematical background alone.

B.2.1 PARASITIC STATES AND SYMBOLS

Input symbols, states, and output symbols must ultimately be encoded as binary vectors. As it takes $w_x \geq \lceil \log_2 |X| \rceil$ bits to uniquely encode the $|X|$ elements of a set X, the code vector may assume is 2^{w_x} distinct values. This implies that $2^{w_x} - |X| \geq 0$ code values exist that do not correspond to any element $x \in X$. Such unused values that result as a side effect from binary coding are termed **parasitic** or residual.

As a consequence, a finite state machine implemented with two-valued electronics will exhibit parasitic input symbols unless $2^{w_i} = |I|$, parasitic states unless $2^{w_s} = |S|$, and parasitic output symbols unless $2^{w_o} = |O|$. Being careful engineers, we ask ourselves

"What happens if, by accident, a finite state machine is presented with some parasitic input symbol or falls into some parasitic state?"[6]

From a mathematical perspective, neither transition function f nor output function g are defined. In practice, the circuit logic will generate some outputs in a deterministic but unspecified way. Thus, while the designer's intention is to build a state graph of $|S|$ vertices each of which has an out-degree $|I|$, the actual result is a supergraph of 2^{w_s} vertices with out-degree 2^{w_i} that includes the original graph as a subgraph. Figure B.10 illustrates this by way of a Medvedev automaton that implements a controllable up/down counting function modulo 5.

Depending on the exact characteristics of the supergraph, a physical automaton may react in various ways in response to a parasitic input symbol: The automaton may show no discernible reaction, may produce just one mistaken output symbol, may move to some regular state via a transition unplanned for, or may fall into a parasitic state. In the latter case, two different outcomes are possible: The state machine may either return to the regular subgraph after a number of clocks, or may get trapped in a dead state or in a circular path forever, a dramatic situation known as **lockup** condition. Fig.B.10a shows all shades of how a physical circuit can fail in response to a parasitic input.

Observation B.1. *In the presence of irregular conditions, parasitic states and symbols left undealt with hold the dangers of serious circuit malfunctioning and of permanent lockup.*

"What can the digital designer do about parasitic finite state machine behavior?"

---

[6] Such unforeseen situations may occur as a consequence from interference, transmission errors, switching noise, poor synchronization, ionizing radiation, hot plug-in, or temporary sagging of power.

Broadly speaking, **fault tolerance** and **graceful degradation** are the goals of any engineering activity. They imply that a system confronted with irregular input data or some other form of disturbance shall

- absorb it with as little impact on its internal functioning as possible,
- continue to produce the most meaningful and/or least offensive output, and
- confine the consequences of any temporary failure to the shortest possible time span.

Example

$I = \{h,u,d\}$, $S = \{0,1,2,3,4\}$, and $O = \{0,1,2,3,4\}$.

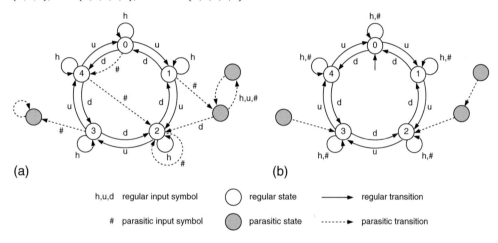

(a)   (b)

h,u,d regular input symbol	◯ regular state	⟶ regular transition
# parasitic input symbol	⬤ parasitic state	┈▶ parasitic transition

FIGURE B.10

State graph of a modulo 5 up/down counter with parasitic states and input symbols left loose (a) compared to a safer version (b). Note that (b) leaves room for further improvement.

☐

In the context of finite state machines, five measures must contribute towards these goals.

1. Collapse all parasitic input symbols to carefully selected regular ones.
2. Make sure that all parasitic states reconverge to the original subgraph. The standard practice is to explicitly indicate a regular successor state for each parasitic state before logic synthesis is undertaken.
3. Assign inoffensive output symbols to all parasitic states.
4. Provide some means for forcing the automaton into start state s_0 from any other state by adding an extra reset mechanism.

Finite state machines that adhere to these guidelines are sometimes qualified as **fail safe**. Note that, from a mathematical point of view, measures 1. and 2. extend the domain of f and g to include all parasitic values of i and s, whereas measure 4. depends on an ancillary mechanism that is independent of f and g. In the occurrence of the above example, a safer version of the modulo 5 up/down counter is shown in fig.B.10b. Albeit at a somewhat different level of abstraction, the measure below is as important as the ones mentioned before.

5. Notify the next higher level in the system hierarchy, e.g. by way of an error signal or message, whenever a parasitic state or input symbol has been detected. This avoids any innocent interpretation of corrupted FSM output and makes it possible for the superordinate system levels to decide on corrective action.

B.2.2 MEALY-, MOORE-, MEDVEDEV-TYPE, AND COMBINATIONAL OUTPUT BITS

The fact that output symbols get encoded as binary vectors gives rise to another subtlety. Consider an automaton where a subset of the output bits depends on the present state exclusively, that is, where some of the bits do not get affected by the present input. In analogy to the classification scheme for automata, such outputs are termed **Moore-type outputs** with "decoded Moore-type outputs" and "unconditional outputs" being synonyms.

Clearly, the state machine as a whole remains a Mealy automaton as long as there exist other output bits — called **Mealy-type outputs**, aka "conditional outputs" — that actually are a function of the present input as well.

Similarly, a Mealy or a Moore machine may or may not include **Medvedev-type outputs**, aka "undecoded Moore-type outputs" and "direct outputs". As the name suggests, such output lines are nothing else than bits tapped from the state register either in direct or in complemented form. Their switching occurs essentially aligned to the clock.

Last but not least, an FSM may or may not feature **combinational outputs**, i.e. bits that depend on the present input exclusively. Combinational and Mealy-type outputs are sometimes subsumed as **through paths**; only Mealy machines can sport them.

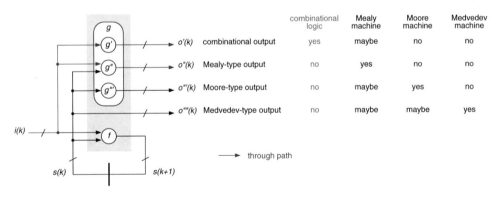

		combinational logic	Mealy machine	Moore machine	Medvedev machine
$o'(k)$	combinational output	yes	maybe	no	no
$o''(k)$	Mealy-type output	no	yes	no	no
$o'''(k)$	Moore-type output	no	maybe	yes	no
$o''''(k)$	Medvedev-type output	no	maybe	maybe	yes

——→ through path

FIGURE B.11

Finite state machine with output bits broken into four subsets. Each subset depends on $i(k)$ and $s(k)$ in a different way and is labeled accordingly.

Being knowledgeable about output types has been found to be useful not only during circuit design, but also during logic simulation, timing analysis, and prototype testing. An example is to follow soon.

B.2.3 THROUGH PATHS AND LOGIC INSTABILITY

The presence of a through path in a state machine holds a serious danger. Any external circuitry that uses the FSM's present output to determine the FSM's present input may give rise to logic contradictions. Uncontrolled oscillations may then develop because Mealy and combinational outputs instantly respond to new input. What makes the problem particularly treacherous is that contradictions and oscillations may actually occur for a limited subset of states and input values exclusively, while the design behaves in a totally inconspicuous way in all other situations.

Example

Consider the control loop below, where `act`, `ini`, and `dcr` are integer variables and . . . stands for some unspecified data manipulations.

```
. . .
act := ini
repeat
   . . .
   act := act - dcr
until act < 0
. . .
```

A possible hardware structure is depicted in fig.B.12a. A finite state machine interprets the carry/borrow bit from the ALU and controls ALU operation and data transport paths. Fig.B.12b shows the tiny portion of a Mealy-type state graph relevant to the above loop computations. The intention is to subtract `dcr` from `act` until the ALU produces a borrow. Everything works fine as long as `act≥dcr`. When this relation ceases to hold, however, the circuit enters an oscillatory regime caused by the mutual and contradictory dependency of `op2` and `borrow` in a feedback loop that encompasses the ALU, the controller, and the leftmost multiplexer. This unstable condition is at all made possible by the attempt to carry out subtraction and decision making in a single clock cycle which concept differs from how a microprocessor would evaluate the above piece of code.

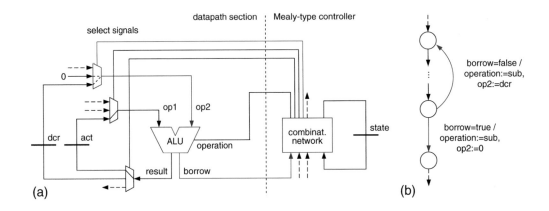

Instability in a Mealy-type controller subject to combinational feedback. Block diagram (a), portion of controller state graph (b).

☐

Observation B.2. *Instability may or may not develop in Mealy machines if the surrounding logic provides immediate feedback from the output to the input of the same machine. The existence of a zero-latency feedback path from one or more output bits of a through path to one or more input bits of the same through path is a necessary precondition for this to happen.*

Basically there are four options for staying clear of instability in automata:

- ○ Select an automaton with non-zero latency, see table B.5.
- ○ Include latency by making feedback paths start from Moore or Medvedev bits exclusively.
- ○ Add a latency register to the surrounding circuitry.
- ○ Formally prove that no logic instability exists in spite of the zero-latency loop.

B.2.4 SWITCHING HAZARDS

From our discussion of transient effects in digital circuitry, we know that almost any combinational network has the potential of developing brief unwanted pulses known as hazards. As the output function g of an FSM is no exception, both Mealy and Moore automata must be suspected to generate hazards unless one has proof to the contrary. Medvedev machines, in contrast, cannot give rise to hazards because they lack a combinational network between state register and output.

Signals from any automaton can be made hazard-free by adding extra flip-flops at the output to align their switching to the clock if need be. The term **registered outputs** is often used to discern those bits that pass through such a **resynchronization register** from the normal ones that are taken from the output logic directly.

Practically speaking, when hazard-free outputs are to be combined with minimum latency, one can either add a resynchronization register to a Mealy automaton or build a Medvedev machine by encoding its state such that the output bits can be tapped from the state register directly, the conversion procedure of fig.B.6 should help. Either option also eliminates the risk of instability, see table B.5 for an overview.

B.2.5 HARDWARE COSTS

The costs of an FSM are given by the w_s bistables that maintain its state, the logic gates for computing transition function f and output function g, plus the necessary wiring. Note that f and g share part of the logic gates in typical hardware implementations. Although the costs of these resources are not the same in a full-custom IC as in field-programmable logic (FPL), a number of observations can be made.

Concurrency, hierarchy and modularity are key to efficiency

Classical state graphs tend to explode in size even when systems of very moderate complexity are being described.[7] The reasons are as follows.

- State graphs are flat with no levels of abstraction, they lack any notion of hierarchy.
- State graphs lack modularity, they do not distinguish between mechanisms.
- State graphs cannot model concurrent activities other than by a single global state.

More importantly, combinatorial explosion is also a problem for electronic hardware if a circuit is organized as one finite state machine. Beyond a certain complexity, it is much more efficient to partition the desired functionality into a bunch of smaller cooperating automata, a pattern which is sometimes also referred to as **linked state machines** (LSM). This not only permits one to compose state and complex behavior from many simple models, but also makes it possible to use specialized and highly efficient subcircuits for implementing subfunctions such as counters, lookup tables (LUT), en/decoders, etc.

Observation B.3. *While it is always possible to model a clocked sequential system as one Mealy-type automaton, a cluster of cooperating smaller automata (including counters, shift registers, etc.) is a much more efficient model for designing and verifying digital circuits.*

Example

A synchronous circuit is to generate a binary pseudo random sequence of length 15 at a rate of 1/4 of its clock. Conceptually, what we need is an autonomous automaton with 60 states. There are two basic options. The entire functionality is either packed into a single FSM, or it is decomposed into a divide-by-4 counter plus a 4 bit linear feedback shift register (LFSR) working under control of that counter. Table B.3 shows the hardware costs. The structured approach saves 4/5 of the area when compared to a flat machine the states of which are randomly encoded.

---

[7] A more succinct visual formalism are the **statecharts** proposed by David Harel [232] [233]. In a nutshell
statecharts = state graphs + hierarchy + concurrency + interprocess communication
Statecharts help a lot to expose orthogonalities in behavioral models. CAE tools for editing and simulating statecharts are commercially available. Most of them are capable of generating program code for microprocessors, some of them also generate HDL code for further processing by synthesis tools. Yet, competition seems to disallow companies to give credit to Harel for proposing the statechart formalism that all their tools have in common.

Table B.3   Hardware costs of a pseudo random sequence generator organized in various ways.

structure	state assignment	flip-flops	std cells	nets	size [GE]	area [Mλ^2]
counter & LFSR	native	6	9	13	60	0.15
flat Moore FSM	adjacent	6	26	33	90	0.23
,, ,,	random (typ.)	6	102	107	261	0.76
,, ,,	one hot	60	69	130	488	1.16

□

State reduction

State reduction, see section B.1.6, always has the benefit of eliminating unnecessary clutter from state tables and state graphs. In most cases, state reduction also pays off in terms of circuit complexity and performance because it introduces new don't care conditions, gives more room for finding a better state encoding, and, thereby, leads to a more economical solution.

Example
Table B.4 juxtaposes two standard cell implementations of the FSM specified earlier in figs.B.8 (unreduced) and B.9 (reduced). The relevant cost factor is the total area occupied after routing. Although the example is an artificial one, the figures indicate that the benefit from state reduction is mainly due to the simplification of the combinational network and not so much a matter of doing away with a flip-flop or two.

Table B.4   Impact of state reduction on hardware costs for the finite state machine of fig.B.8.

number of states	state assignment	flip-flops	std cells	nets	size [GE]	area [Mλ^2]
11	adjacent	4	35	38	94	0.26
6	adjacent	3	14	20	44	0.12

□

Observation B.4. *It normally pays to eliminate redundant states from a state graph or state table prior to translating it into hardware.*

This need not always be true, however, since using less bits for storing a machine's state can sometimes increase the number of terms and literals in transition and/or output functions, inflate combinational circuitry, and offset the savings obtained from using fewer bistables.

State encoding

State encoding, aka state assignment, is the process of deciding on how the various states are going to be mapped onto vectors of binary digits. An obvious requirement is that each state is assigned a unique bit vector. The number of bits required for uniquely encoding $|S|$ states is

$$w_s \geq \lceil \log_2 |S| \rceil \tag{B.6}$$

where $\lceil x \rceil$ denotes the least integer not smaller than x. We speak of **minimum bit encoding** when (B.6) is made to hold with equality because the number of bistables is then minimal.

From a purely <u>functional</u> point of view, state encoding is immaterial because any unique state assignment necessarily leads to a correct circuit that is equivalent to those resulting from all other mappings.[8]

From an <u>efficiency</u> point of view, some state encodings will yield smaller and faster circuits than others. Energy dissipation and testability are also likely to differ. As the implications from the subsequent steps in the design process — logic synthesis, placement, and routing — are difficult to anticipate, one might be tempted to complete the design and to evaluate different schemes on the grounds of the final result. The number n_s of truly distinct state assignments, i.e. those that cannot be derived from others simply by permuting and/or complementing state bits has been known since the late 1950s [234] and is given in (B.7). Numeric evaluation quickly tells that, even for small automata, it is not computationally feasible to find the best state assignment using an enumerative approach.

$$n_s = \frac{(2^{w_s} - 1)!}{(2^{w_s} - |S|)! w_s!} \tag{B.7}$$

From an <u>engineering</u> point of view, we are not so much interested in the absolute optimum but in finding a good state assignment that results in a near-minimal hardware solution with reasonable effort. A variety of heuristic approaches have been devised, all of which attempt to carry out state encoding such as to minimize the number of literals in the FSM's logic equations in some way or another. The techniques differ in whether they target two-level or multi-level logic equations and in how they resolve conflicts between contradicting requirements.

Adjacent state assignment is the name for a class of heuristics that assumes two-level logic in sum-of-products form. The idea behind all such heuristics essentially is to lower the number of product terms and the number of literals by assigning states that have similar entries in the state table codes that differ in a single bit. Put differently, they make similar states adjacent in the Karnaugh map. Somewhat surprisingly, this approach has been found to have beneficial effects on multi-level logic implementations too and is, therefore, quite popular.

As the name suggests, **one-hot state encoding** uses a binary vector of length $|S|$ and assigns each state a code where a single digit is logic 1 and all others are 0.[9] When compared to minimum-bit state encoding, one-hot encoding typically results in many more bistables but, at the same time, also in a

---

[8] Designers typically leave state assignment to a synthesis tool and often ignore the actual binary codes selected. However, testing and debugging will involve gaining access to state registers, e.g. by way of a scan path or by physical probing. Interpreting those binary codes then requires knowledge of the state encoding chosen.

[9] Please make sure you understand that one-hot state encoding is not in contradiction to state reduction.

less combinational logic. Whether this pays off or not depends on the application, it certainly did not in the example of table B.3. In FPL devices with limited routing resources it sometimes is the only way to implement substantial finite state machines. On the negative side, one-hot encoding brings about a huge number of parasitic states, namely $2^{|S|} - |S|$. Making all of them reconverge to the main subgraph as described in section B.2.1 often proves unwieldy.

B.3 SUMMARY

- The key characteristics of the most common automata are collected below. Observe that responding to a new input within the on-going computation period (latency 0) and unconditional stability (no through path) are mutually exclusive.

Table B.5   **Six types of automata and their key characteristics.**

resynchronization register	class of automaton		
	Mealy	Moore	Medvedev
no	latency 0 possibly unstable hazards likely	latency 1 stable hazards likely	latency 1 stable hazard-free
yes	latency 1 stable hazard-free	latency 2 stable hazard-free	latency 2 stable hazard-free

- Designing with finite state machines involves the following steps:
  1. Partitioning the desired functionality into a cluster of cooperating automata whenever this is advantageous from an economy or modularity point of view.
  2. Selecting the appropriate type of automaton and the types for its outputs.
  3. Detailed specification and verification of statechart, state graph, or state table.
  4. State reduction.
  5. Deciding on how to safely handle parasitic inputs and states, if any.
  6. State assignment.
  7. Minimization of combinational functions.
  8. Designing the circuit logic from the components or library cells available.

Today's electronic design automation packages routinely cover steps 4 and 6 through 8. Also available are software tools for the editing of state graphs and for visualizing inputs, state transitions, and outputs at a behavioral level.

SYMBOLS AND CONSTANTS

C.1 ABBREVIATIONS

aka	also known as
ckt	circuit
iff	if and only if
wrt	with respect to
SysVer	SystemVerilog

For technical and scientific acronyms please check the Index.

C.2 MATHEMATICAL SYMBOLS

Quantity	Unit	Explanation
α	1	MOSFET velocity saturation index
α	1	defect clustering factor
α_k	1	node activity
β	1	BJT current gain
β	A/V^2	MOSFET gain factor
$\beta_{\square}$	A/V^2	process gain factor
Γ	1	cycles per data item
δ	1	duty cycle
ϵ_0	As/Vm	permittivity of vacuum, aka electric constant
ϵ_r	1	relative permittivity, once known as dielectric constant
$\theta_{a,c,j}$	K or °C	temperature (of ambient air, case, and junction respectively)
Θ	s^{-1}	data throughput
λ	1	MOSFET channel length modulation factor
Λ	m	pitch of virtual layout grid
μ	m^2/Vs	carrier mobility
μ_0	Vs/Am	permeability of vacuum, aka magnetic constant
μ_r	1	relative permeability
ϱ	kg/m^3	density
ρ	Ωm	resistivity
σ	S/m	conductivity
σ_k	1	crossover energy quota
Φ	J or eV	work function
Ψ	W/Hz $=$ J	dissipated power per switching rate
ω	1	overdrive factor
A	m^2	area or, as a generalization, circuit size in gate equivalents [GE]
AT	m^{2}s	size-time product, alternatively in [GEs]
B	1	base in a positional number system
c	USD	cost, occasionally in [EUR] or [CHF]
c	m/s	speed of light in a medium
c_0	m/s	speed of light in vacuum
c_{ox}	F/m^2	gate capacitance per area
C	F	capacitance
d	1	iterative decomposition factor
d	m	diameter
D	m^{-2}	defect density
E	m	enclosure of one layout structure around another
E	J	energy, occasionally in [eV]
E_{ch}	J	energy dissipated for charging and discharging
E_{cr}	J	energy dissipated due to crossover currents
E_{lk}	J	energy dissipated due to leakage

E_{rr}	J	energy dissipated by resistive loads
$E_{C,V}$	J or eV	electron energy (at lower conduction and upper valence band edge)
E_F	J or eV	Fermi level
E_G	J or eV	bandgap
E_{vac}	J or eV	free electron energy ("vacuum level")
E	V/m	electric field strength
EOT	m	equivalent oxide thickness
f_{clk}	Hz	clock frequency
f_{cp}	Hz	computation rate
f_d	Hz	edge rate
$f_{to\,ff}$	Hz	flip-flop toggling rate
F	m	minimum half pitch of lines with staggered contacts
F	V/m	electric field strength
G	1	total gate count
G, g	S	conductance
h	m	geometric height, thickness
H	m	minimum half pitch
H_{fin}	m	fin height
i	1	index indicating bit (or digit) position within a data word
i	1	input symbol for a finite state machine
I	1	set of input symbols for a finite state machine
I, i	A	current
I_d	A	MOSFET drain current
I_f	A	forward current
I_{oup}	A	output current
I_r	A	reverse current
I_s	A	MOSFET source current
J	A/m^2	current density
k	J/K or eV/K	Boltzmann constant
$K_{P,\theta,V}$	1	derating factor (for process, temperature and voltage respectively)
K_1	s	metastability parameter of a bistable
K_2	Hz	metastability parameter of a bistable
l	m	geometric length
L	1	latency in computation periods
L	H	inductance
L	m	length of MOSFET gate
L_{eff}	m	effective MOSFET gate length
L_{drawn}	m	drawn MOSFET gate length
m	1	MOSFET body effect coefficient
$m_{Chem.}$	g/mole	molar mass
M	m	minimum feature size
$n_{m,f}$	1	number of chips (manufactured and functional respectively)
n	1	index of refraction

n	m^{-3}	electron concentration
$N_{A,D}$	m^{-3}	doping concentration (of acceptors and donors respectively)
N_{Avo}	mole^{-1}	Avogadro's number
o	1	output symbol from a finite state machine
O	1	set of output symbols from a finite state machine
p	1	pipelining factor, loop unfolding factor
p	m^{-3}	hole concentration
P	1	number of lithographic patterning steps
P	m	minimum line pitch
P	W	power
q	1	replication factor
q_e	C	elementary charge
Q	C	charge
r_{cap}	$\text{s/F} = \Omega$	load factor
r_{sl}	V/s	slew rate
R	Ω	electrical resistance
$R_{\square}$	$\Omega/\square = \Omega$	sheet resistance
R_{θ}	K/W	thermal resistance
s	1	time sharing factor
s	1	state of a finite state machine
s_0	1	start state of a finite state machine
s_{ra}	1	slope sensitivity factor
S	1	set of states of a finite state machine
S	m	spacing between two layout structures
S	$\text{V/decade} = \text{V}$	MOSFET subthreshold slope
t_{al}	s	allowed time to recover from metastability
t_{cd}	s	contamination delay
t_{di}	s	clock distribution delay
t_{fa}	s	fall time
t_{hi}	s	high time
t_{ho}	s	hold time
t_{id}	s	insertion delay
t_{it}	s	intrinsic cell delay
t_{jt}	s	clock jitter
t_{lo}	s	low time
t_{lp}	s	longest path delay
t_{mr}	s	metastability resolution time
t_{MOSFET}	s	intrinsic MOSFET delay
t_{MTBE}	s	mean time between errors
t_{ox}	m	gate dielectric thickness
t_{pd}	s	propagation delay
t_{pu}	s	pulse width
t_{ra}	s	ramp width

t_{ri}	s	rise time
t_{sk}	s	clock skew
t_{sl}	s	slack
t_{sp}	s	shortest path delay
t_{su}	s	setup time
T	s	time per data item, period
T_{clk}	s	clock period
T_{cp}	s	computation period
U, u	V	voltage[1]
U_θ	V	thermal voltage
U_{bi}	V	junction built-in voltage
U_{bs}	V	MOSFET body-to-source voltage
U_{dd}	V	supply voltage
U_{ds}	V	MOSFET drain-to-source voltage
U_f	V	forward voltage
U_{gs}	V	MOSFET gate-to-source voltage
U_{ih}	V	input high voltage
U_{il}	V	input low voltage
U_{inp}	V	input voltage
U_{inv}	V	inverter threshold voltage
U_{nm}	V	noise margin
U_{oh}	V	output high voltage
U_{ol}	V	output low voltage
U_{oup}	V	output voltage
U_{pn}	V	junction anode-to-cathode voltage
U_r	V	reverse voltage
U_{th}	V	MOSFET threshold voltage
U_{trip}	V	trip voltage
v	1	voltage amplification
V	m^3	volume
$w_{i,o,s}$	1	word width (for input, output, and state respectively)
w	m	geometric width
W	m	width of MOSFET gate or other layout structure
W_{fin}	m	fin width
X	m	extension of one layout structure beyond another
y_f	1	fabrication yield
$\#_{items}$	1	number of items

---

[1] American writers usually write V for the quantity voltage and V for the unit volt. In accordance with recommendations by the International Electrotechnical Commission (IEC) and the Système International d'Unités (SI), we use U as quantity symbol for voltage and V as unit symbol for volt to clearly distinguish the two. Also, V collides with volume. A renowned US publication that adheres to this practice much as our text does is [235].

C.3 PHYSICAL AND MATERIAL CONSTANTS

Table C.2   Selected physical constants.

Avogadro's number N_{Avo}	$6.022 \cdot 10^{23}$ /mole
Boltzmann constant k	$1.381 \cdot 10^{-23}$ J/K = $86.17 \cdot 10^{-6}$ eV/K
Planck constant $\hbar = \frac{h}{2\pi}$	$0.1055 \cdot 10^{-33}$ Js = $0.6582 \cdot 10^{-15}$ eVs
Absolute zero	0 K = -273.15 °C
Elementary or electron charge q_e	$0.1602 \cdot 10^{-18}$ C
Permittivity of vacuum ϵ_0	$8.854 \cdot 10^{-12}$ As/Vm (= F/m)
Permeability of vacuum μ_0	$4\pi \cdot 10^{-7}$ Vs/Am = $1.257 \cdot 10^{-6}$ Vs/Am (= H/m)
Speed of light in vacuum c_0	$299.8 \cdot 10^6$ m/s
Thermal voltage $U_\theta = \frac{k\theta_j}{q_e}$	25.9 mV @ 300 K junction temperature

Please note: The properties of thin films may considerably differ from those of bulk materials, and the same holds true when comparing narrow ribbons or small dots to large sheets. Permittivity is not a constant either and tends to diminish with frequency, although not necessarily in a monotonic fashion.

A note on carbon allotropes

What makes **carbon** so special is that it comes in many allotropic variations [245]. **Diamond** forms a tetrahedral crystal lattice where carbon atoms sit in the corners and are held together by covalent bonds exactly as in monocrystalline silicon. This spatial arrangement with four strong bonds oriented around each nucleus renders diamond extremely hard and durable. The large bandgap of about 5.5 eV makes it an electrical insulator and optically transparent.

Table C.3 Key properties of selected materials (mostly after [239] [240] and Wikipedia augmented by [241] for GaN, [242] for InGaAs, [243] for InSb, and [244] for SiC 4H). The data on graphene are incomplete and must be considered unreliable as the material has no history as an industrial semiconductor so far. A nice entry point for further information is http://en.wikipedia.org/wiki/List_of_semiconductor_materials

Material	Semimetal	Semiconductors				
	C	InSb	InAs	Ge	In$_{.53}$Ga$_{.47}$As	Si
crystallographic variety [a]	G	Z	Z	D	Z	D
Bandgap E_G at 300 K [eV]	≈ 0	0.17	0.354	0.661	0.74...0.75	1.12
Relative permittivity ϵ_r at 0 Hz		16.8	15.15	16.2	13.9	11.7
Approx. breakdown field [$\frac{kV}{mm}$]			40	10	200	30
Electron mobility μ_e at 300 K [$\frac{cm^2}{Vs}$]	[b]	77 000	40 000	3900	8450	1400
Hole mobility μ_h at 300 K [$\frac{cm^2}{Vs}$]		850	500	1900	300	450
Saturated electron velocity [$10^6\frac{cm}{s}$]						10
Molar mass m [$\frac{g}{mole}$]	12.01	236.58	189.74	72.64	168.54	28.09
Density ϱ [$\frac{g}{cm^3}$]		5.77	5.68	5.323	5.504	2.329
Lattice spacing [nm]		0.6479	0.60583	0.56461	0.58687[c]	0.54309
Melting point [°C]		527	942	937	≈ 1100	1412
Thermal conductivity [$\frac{W}{cm K}$]	≈ 50[d]	0.18	0.27	0.58	0.05	1.3

(continued)	Semiconductors		Wide-gap semiconductors				Insulator
Material	InP	GaAs	SiC		GaN	C	SiO$_2$
crystallographic variety	Z	Z	6H	4H	W	D	amorph.
Bandgap E_G at 300 K [eV]	1.344	1.424	3.03	3.26	3.49	≈ 5.5	8...9
Rel. permittivity ϵ_r at 0 Hz	12.5	12.9	9.66	9.7	9.0	5.7	≈ 3.9
Approx. breakdown field [$\frac{kV}{mm}$]	50	40	≈ 300	220	300	1000	500...1000
Electron mob. μ_e at 300 K [$\frac{cm^2}{Vs}$]	4600	8500	[e]	700	< 2000	2200	
Hole mobility μ_h [$\frac{cm^2}{Vs}$]	200	400				1800	
Satur. electron velocity [$10^6\frac{cm}{s}$]	10	13	20	20	13	27	
Molar mass m [$\frac{g}{mole}$]	145.79	144.63	40.10	40.10	83.73	12.01	60.08
Density ϱ [$\frac{g}{cm^3}$]	4.787	5.320	3.21	3.21	6.1	3.515	2.27
Lattice spacing [nm]	0.58687	0.56525	0.308	0.307		0.35668	
Melting point [°C]	1060	1240	2830	2830	600	4030	≈ 1700
Thermal conductivity [$\frac{W}{cm K}$]	0.68	0.55	4.9	4.56	[f]	20	0.014

[a] D = diamond, Z = zincblende, W = wurtzite, G = graphene (planar monocrystalline sheet).

[b] Carrier mobilities in 2D sheets are extremely dependent on conditions and vary between 200 000 cm^2/ Vs for large-area single-layer graphene sheets with no charged impurities and microscopic ripples [236] and less than 200 cm^2/Vs for 10 nm wide nanoribbons where strong electron-phonon scattering occurs [237].

[c] Matches lattice of InP.

[d] From [238].

[e] Highly anisotropic.

[f] Numerical indications in the literature vary from 0.66 W/cm K in [235] to 2.5 W/cm K in www.kymatech.comt (retrieved 18. Feb 2011)

Graphene is the name given to a single layer of carbon atoms where each nucleus sits in the corner of a hexagon and is covalently bonded to three neighbors such as to form a planar lattice reminiscent of chicken wire [246] [247]. **Graphite** is a three-dimensional structure where many such sheets are held together by much weaker Van der Waal forces which explains why graphite appears soft and slick. Bandgap is almost zero which means that valence and conduction bands barely touch. As a consequence, graphite has a mediocre electrical conductivity and is sometimes termed a semimetal or metalloid.

Carbon nanotubes (CNT) can be thought of as graphene sheets that have rolled up. Note that the rolling up occurs at specific and discrete angles (just compare with forming tubes from a gift wrap paper that carries some periodic pattern). CNTs exist as multi-walled and as single-walled hollow cylinders, straight or twisted. Diameter is a little over 1 nm for a single-walled nanotube, and up to 50 nm for multi-walled nanotubes. Depending on diameter and rolling angle, CNTs can have band gaps as low as zero (as metal), as high as that of silicon, and almost anywhere in between. **Fullerenes** are spherical macromolecules that resemble soccer balls. Carbon nanotubes and fullerenes share an extraordinary strength and stability.

The lampblack deposits that form when organic fuels are burned with a lack of sufficient oxygen largely consist of **amorphous carbon** with no long-range pattern of atomic positions.

Table C.4   Selected conductor materials (except for Al 0.5% Cu, TiN and ITO, data are from [235] and refer to pure bulk material).

Material	Resistivity ρ at 300 K $[10^{-9}\ \Omega m]$	Thermal cond. at 300 K $[\frac{W}{cm\,K}]$	Melting temp. $[°C]$	
Cu	17.25	4.01	1085	used for interconnect lines
Al 0.5% Cu	≈ 30			”
W	54.4	1.74	3422	used in contact/via plugs
Al	27.33	2.37	660	used in metal gates
TiN[a]	300...700	0.291	2930	”
Ag	16.3	4.29	962	given for comparison
Au	22.7	3.17	1064	”
Mo	55.2	1.38	2623	”
Fe	99.8	0.802	1538	”
Ta	135	0.575	3017	”
Ti	390[b]	0.219	1668	”
ITO[c]	≈ 1000		1800...2200	” , used in LCDs

[a] A crystalline ceramic, data according to Wikipedia (retrieved 7. Oct 2010).
[b] At 273 K; $420 \cdot 10^{-9}$ Ωm at 293 K according to Wikipedia (retrieved 7. Oct 2010).
[c] Indium tin oxide is a transparent conductor that typically contains 90% In_2O_3 and 10% SnO_2 by weight; resistivity after Thin Solid Films 411:1(1-5); 2002; melting point after Wikipedia (retrieved 7. Jan 2011).

Table C.5 Selected interconnect dielectrics (bulk materials after [235] unless otherwise stated).

Material	Relative permittivity ϵ_r	Dielectric strength E_{max} [kV/mm]	
Alumina Al_2O_3	9.3...11.5	13.4	used for packages
Ceramic substrates	7...8		,,
Epoxy resins	≈ 4.2		,, and PCBs
Epoxy laminate FR4	≈ 4	≈ 38	used for PCBs
Silicate glass SiO_2	1.8^a...3.9	470...670	inorganic ILD
Fluorinated silicate glass (FSG) SiOF	3.0...3.7		,,
Hydrogen silsesquioxane (HSQ)	3.0...2.7	> 400	,,
Carbon-doped oxide (CDO) SiOC	$\approx 2.4^b$...3.3		
Organosilicate glass (SiCOH)	$\approx 1.8^c$...2.9		
Polyimides	3.0...3.6		organic ILD
Parylene	2.6		,,
Benzocyclobutane	2.6		,,
Polyarylene ether	2.3		,,
Polytetrafluoroethylene (Teflon ^{TM})	2.1	$87...173^d$	,,
Dry air	1.00	≈ 1	given for comparison

[a] At 75% porosity.

[b] Nanoporous.

[c] Idem.

[d] Just 60 kV/mm according to Wikipedia (retrieved 26. Nov 2009).

Table C.6 Selected gate dielectrics (numbers from [248] [249] [250] [251] [252] [253] [254]).

Material	Relative permittivity ϵ_r	
silicon dioxide SiO_2	3.9	traditional
nitrided silicon oxide aka oxynitride SiO_xN_y	≈ 5.1	
silicon nitride Si_3N_4	7.5	
aluminum oxide Al_2O_3	8...11.5	
hafnium silicon oxynitride HfSiON	≈ 9...11	
hafnium (IV) silicate $HfSiO_4$	≈ 11	
hafnium silicate $(HfO_2)_x(SiO_2)_{1-x}$	≈ 12	for $x = 0.6$...0.7
hafnium aluminate HfAlO	≈ 15	
hafnium aluminum oxynitride HfAlON	≈ 18	
strontium hafnium oxide $SrHfO_3$	≈ 19	IBM
hafnium oxide HfO_2	≈ 21...25	Intel HKMG @ 45 nm
zirconium oxide ZrO_2	22...28	Intel "TeraHertz"
lanthanum aluminate $LaAlO_3$	25.1	
tantalum pentoxide Ta_2O_5	27	
titanium dioxide (rutile) TiO_2	> 25	
barium strontium nitrate $Ba\,|Sr(NO_3)_2$	> 25	
sodium beta-alumina (SBA) $NaAl_{11}O_{17}$	170...30	@ 50 Hz and 1 MHz resp.
strontium titanate ^{a} $SrTiO_3$	≈ 200	ceramic capacitors, DRAMs

[a] Strontium titanate belongs to the family of perovskite ceramics and is given here for reference.

Bibliography

[1] Bill McClean. 2001 IC Industry at the Crossroads. *Semiconductor International*, 24(1), January 2001.

[2] Charles E. Stroud. *A Designer's Guide to Built-in Self-Test*. Springer, 2002.

[3] Niraj K. Jha and Sandeep Gupta. *Testing of Digital Systems*. Cambridge University Press, 2003.

[4] Michael L. Bushnell and Vishwani Agrawal. *Essentials of Electronic Testing*. Kluwer Academic Publishers, 2000.

[5] Kai-hui Chang, Igor L. Markov, and Valeria Bertacco. Automating Postsilicon Debugging and Repair. *IEEE Computer*, 41(7):47–54, July 2008.

[6] Alberto Sangiovanni-Vincentelli. Corsi e Ricorsi: The EDA Story. *IEEE Solid-State Circuits Magazine*, 2(3):6–25, Summer 2010.

[7] Alberto Sangiovanni-Vincentelli. The Tides of EDA. *IEEE Design & Test of Computers*, 20(6):59–75, November/December 2003.

[8] Dmitri B. Strukov and Konstantin K. Likharev. Reconfigurable Nano-Crossbar Architectures. In Rainer Waser, editor, *Nanoelectronics and Information Technology*. Wiley, 3rd edition, 2012. http://www.ece.ucsb.edu/ strukov/papers/2012/Was2012.pdf.

[9] Ian Kuon and Jonathan Rose. Measuring the Gap Between FPGAs and ASICs. *IEEE Transactions on Computer-Aided Design*, 26(2):203–215, February 2007.

[10] Man-Ho Ho et al. Architecture and Design Flow for a Highly Efficient Structured ASIC. *IEEE Transactions on Very Large Scale Integration (VLSI) Systems*, 21(3):424–433, March 2013.

[11] Anonymous. Expect a Breakthrough Advantage in Next-Generation FPGAs, June 2013. http://www.altera.com/literature/wp/wp-01199-next-generation-FPGAs.pdf, retrieved 29.1.2014.

[12] Nick Mehta. Xilinx UltraScale Architecture for High-Performance, Smarter Systems, December 2013. http://www.xilinx.com/support/documentation/white_papers/wp434-ultrascale-smarter-systems.pdf, retrieved 29.1.2014.

[13] Team of authors. FPGA Central. http://www.fpgacentral.com, retrieved 17.2.2014.

[14] Markus Wannemacher. Die aufzu Halbleiterhersteller-Ecke. http://www.aufzu.de/semi/halbleit.html.

[15] Clive Maxfield, editor. *FPGAs World Class Designs*. Newnes, Burlington MA, 2009.

[16] Ian Grout. *Digital Systems Design with FPGAs and CPLDs*. Newnes, Burlington MA, 2008.

[17] Clive Maxfield. *The Design Warrior's Guide to FPGAs*. Newnes, Burlington MA, 2004.

[18] Bob Zeidman. *Designing with FPGAs and CPLDs*. CMP Books, Lawrence KS, 2002.

[19] Donald G. Bailey, editor. *Design for Embedded Image Processing on FPGAs*. John Wiley & Sons, Singapore, 2011.

[20] A. Curiger and H. Bonnenberg and R. Zimmermann and N. Felber and H. Kaeslin and W. Fichtner. VINCI: VLSI Implementation of the new Secret-Key Block Cipher IDEA. In *Proceedings of the IEEE 1993 Custom Integrated Circuits Conference*, pages 15.5.1–4, San Diego CA, 1993. IEEE.

[21] Nirmal R. Saxena et al. Dependable Computing and Online Testing in Adaptive and Configurable Systems. *IEEE Design & Test of Computers*, 17(1):29–41, January/March 2000.

[22] Ingrid Verbauwhede, Patrick Schaumont, and Henry Kuo. Design and Performance Testing of a 2.29Gb/s Rijndael Processor. *IEEE Journal on Solid State Circuits*, 38(3):569–571, March 2003. (encryption only, max. throughput is 2.29Gbit/s with 256bit blocks).

[23] Victoria Goode. Virtual Components for the Division Operation. Master thesis, Integrated Systems Laboratory, ETH Zurich, 2008.

[24] Zaher Baidas, Andrew D. Brown, and Alan Christopher Williams. Floating-Point Behavioral Synthesis. *IEEE Transactions on Computer-Aided Design*, 20(7):828–839, July 2001.

[25] Dimitris Bariamis, Dimitris Maroulis, and Dimitris K. Iakovidis. Adaptable, Fast Area-Efficient Architecture for Logarithm Approximation with Arbitrary Accuracy on FPGA. *Journal of Signal Processing Systems for Signal, Image, and Video Technology*, 58(3):301–310, March 2010.

[26] Clay S. Turner. A Fast Binary Logarithm Algorithm. *IEEE Signal Processing Magazine*, 27(5):124,140, September 2010.

[27] Ingrid Verbauwhede and Alireza Hodjat. High-Throughput Programmable Cryptoprocessor. *IEEE Micro*, 24(3):34–45, May/June 2004.

[28] Tilman Glökler, Andreas Hoffmann, and Heinrich Meyr. Methodical Low-Power ASIP Design Space Exploration. *Journal of VLSI Signal Processing Systems*, 33(3):229–246, March 2003.

[29] Sven Woop, Joerg Schmittler, and Philipp Slusallek. RPU: A Programmable Ray Processing Unit for Realtime Ray Tracing. In *Proceedings of the ACM SIGGRAPH conference*, pages 434–444, Los Angeles, July/August 2005. ACM.

[30] Russell Tessier and Wayne Burleson. Reconfigurable Computing for Digital Signal Processing: A Survey. *Journal of VLSI Signal Processing Systems*, 28(1/2):7–27, May/June 2001.

[31] Neil Jacobson. *The in-system configuration handbook: a designer's guide to ISC*. Kluwer Academic Publishers, Hingham MA, 2004.

[32] John Villasenor and Brad Hutchings. The Flexibility of Configurable Computing. *IEEE Signal Processing*, 15(5):67–84, September 1998.

[33] P.H.W. Leong et al. Pilchard - A Reconfigurable Computing Platform with Memory Slot Interface. In *Proceedings of the IEEE Symposium on Field-Programmable Custom Computing Machines (FCCM)*, Rohnert Park, CA, 2001. IEEE.

[34] Sami Khawam et al. The Reconfigurable Instruction Cell Array. *IEEE Transactions on Very Large Scale Integration (VLSI) Systems*, 16(1):75–85, January 2008.

[35] John L. Hennessy and David A. Patterson. *Computer Architecture, a Quantitative Approach*. Morgan Kaufmann Publishers, San Mateo CA, fourth edition, 2007.

[36] Mohamed Rafiquzzaman. *Fundamentals of Digital Logic and Microcomputer Design*. John Wiley & Sons, Hoboken NJ, 2005.

[37] Sam Naffziger. Microprocessor of the Future: Commodity or Engine of Growth? *IEEE Solid State Circuits Magazine*, 1(1):76–82, 2009.

[38] Yale Patt. Requirements, Bottlenecks and Good Fortune: Agents for Microprocessor Evolution. *Proceedings of the IEEE*, 89(11):1553–1559, November 2001.

[39] Shekhar Borkar and Andrew A. Chien. The Future of Microprocessors. *Communications of the ACM*, 54(5):67–77, 2011.

[40] Doug Burger et al. Scaling to the End of Silicon with EDGE Architectures. *IEEE Computer*, 37(7):44–54, July 2004.

[41] Doug Burger and James R. Goodman. Billion-Transistor Architectures: There and Back Again. *IEEE Computer*, 37(3):22–28, March 2004.

[42] Randy Goldberg and Lance Riek. *A Practical Handbook of Speech Coders*. CRC Press, Boca Raton, 2000.

[43] John Stephen Walther. The Story of Unified CORDIC. *Journal of VLSI Signal Processing*, 25(2):107–112, June 2000. (part of special issue on CORDIC).

[44] Ray Andraka. A survey of CORDIC algorithms for FPGA based computers. In *Proceedings of the 1998 ACM/SIGDA Sixth International Symposium on Field-Programmable Gate Arrays*, Monterey CA, February 1998. DOI: 10.1145/275107.275139.

[45] Yu Hen Hu. CORDIC-Based VLSI Architectures for Digital Signal Processing. *IEEE Signal Processing Magazine*, 9(3):16–35, July 1992.

[46] A. Burg, M. Borgmann, M. Wenk, M. Zellweger, W. Fichtner, and H. Bölcskei. VLSI Implementation of MIMO Detection using the Sphere Decoder Algorithm. *IEEE Journal of Solid-State Circuits*, 40(7):1566–1577, 2005.

[47] Jay R. Southard. MacPitts: An Approach to Silicon Compilation. *IEEE Computer*, 16(12):74–82, December 1983.

[48] Keshab K. Parhi. *VLSI Digital Signal Processing Systems*. John Wiley & Sons, New York, 1999.

[49] Boaz Porat. From Academe to Industry (or from Writing Papers to Making Chips): Experiences and Conclusions. *IEEE Signal Processing Magazine*, 20(4):8–11, July 2003.

[50] Johannes Widmer. Soft-Output Viterbi Equalizer supporting RX Diversity. Master thesis plus private communication, Integrated Systems Laboratory, ETH Zurich, 2012.

[51] X. Lai, J. L. Massey, and S. Murphy. Markov Ciphers and Differential Cryptanalysis. In *Advances in Cryptology – EUROCRYPT '91*, pages 8–13, Berlin, 1991. Springer.

[52] H. Bonnenberg and A. Curiger and N. Felber and H. Kaeslin and X. Lai. VLSI Implementation of a New Block Cipher. In *Proceedings of the International Conference on Computer Design*, Cambridge MA, October 1991. IEEE.

[53] Robert G. Swartz. Ultra-High Speed Multiplexer/Demultiplexer Architectures. *International Journal of High Speed Electronics*, 1(1):73–99, 1990.

[54] Jarmo Takala and Konsta Punkka. Scalable FFT Processors and Pipelined Butterfly Units. *Journal of VLSI Signal Processing Systems*, 43(2/3):113–123, June 2006.

[55] K. Babionitakis et al. Fully Systolic FFT Architecture for Giga-sample Applications. *Journal of Signal Processing Systems for Signal, Image, and Video Technology*, 58(3):281–299, March 2010.

[56] Keshab K. Parhi. Approaches to Low-Power Implementations of DSP Systems. *IEEE Transactions on Circuits and Systems I: Fundamental Theory and Applications*, 48(10):1214–1224, October 2001.

[57] Doris Keitel-Schulz and Norbert Wehn. Embedded DRAM Development: Technology, Physical Design, and Application Issues. *IEEE Design & Test of Computers*, 18(3):7–15, May/June 2001.

[58] Sreedhar Natarajan, Shine Chung, Lluis Paris, and Ali Keshavarzi. Searching for the Dream Embedded Memory. *IEEE Solid State Circuits Magazine*, 3(1):34–44, 2009.

[59] Kiyoo Itoh. *VLSI Memory Chip Design*. Springer, New York, 2001.

[60] Sung-Mo Kang and Yusuf Leblebici. *CMOS Digital Integrated Circuits*. McGraw Hill, Boston, 2003.

[61] John E. Ayers. *Digital Integrated Circuits, Analysis and Design*. CRC Press, Boca Raton FL, 2004.

[62] Kiat-Seng Yeo and Kaushik Roy. *Low-Voltage, Low-Power VLSI Subsystems*. McGraw-Hill, New York, 2005.

[63] Roberto Bez, Emilio Camerlenghi, Alberto Modelli, and Angelo Visconti. Introduction to Flash Memory. *Proceedings of the IEEE*, 91(4):489–502, April 2003.

[64] Takashi Kobayashi, Hideaki Kurata, and Katustaka Kimura. Trends in High-Density Flash Memory Technology. *IEICE Transactions*, E87-C(10):1656–1663, October 2004.

[65] Marco Sanvido, Frank Chu, Anand Kulkarni, and Robert Selinger. NAND Flash Memory and its Role in Storage Architectures. *Proceedings of the IEEE*, 96(11):1864–1874, November 2008.

[66] Rino Micheloni et al. Non-Volatile Memories for Removable Media. *Proceedings of the IEEE*, 97(1):148–160, January 2003.

[67] Narendra Shenoy. Retiming: Theory and Practice. *Integration, the VLSI journal*, 22(1-2):1–21, August 1997.

[68] Sachin Sapatnekar. *Timing*. Kluwer Academic Publishers, Boston, MA, 2004.

[69] Charles E. Leiserson and James B. Saxe. Retiming Synchronous Circuitry. *Algorithmica*, 6(1):5–35, 1991.

[70] Hervé J. Touati and Robert K. Brayton. Computing the Initial State of Retimed Circuits. *IEEE Transactions on Computer-Aided Design*, 12(1):157–162, January 1993.

[71] Nikolay Petkov. *Systolic Parallel Processing*. North-Holland, Amsterdam, 1993.

[72] Charles E. Leiserson and James B. Saxe. Optimizing Synchronous Systems. *Journal of VLSI and Computer Systems*, 1(1):41–67, 1983.

[73] Katsuhiko Hayashi, Kaushal K. Dhar, Kazunori Sugahara, and Kotaro Hirano. Design of High-Speed Digital Filters Suitable for Multi-DSP Implementation. *IEEE Transactions on Circuits and Systems*, 33(2):202–217, February 1986.

[74] Peter M. Kogge. *The Architecture of Pipelined Computers*. McGraw-Hill, New York, 1981.

[75] Keshab K. Parhi. Finite word effects in pipelined recursive filters. *IEEE Transactions on Acoustics, Speech and Signal Processing*, 39(6):1451–1454, June 1991.

[76] Mehdi Hatamian and Keshab K. Parhi. A 85-MHz Fourth-Order Programmable IIR Digital Filter Chip. *IEEE Journal of Solid-State Circuits*, 27(2):175–183, February 1992.

[77] Horng-Dar Lin and David G. Messerschmitt. Finite State Machine has Unlimited Concurrency. *IEEE Transactions on Circuits and Systems*, 38(5):465–475, May 1991.

[78] Jun Ma, Keshab K. Parhi, and Ed F. Deprettere. A Unified Algebraic Transformation Approach for Parallel Recursive and Adaptive Filtering and SVD Algorithms. *IEEE Transactions on Signal Processing*, 49(2):424–437, February 2001.

[79] R. Zimmermann and A. Curiger and H. Bonnenberg and H. Kaeslin and N. Felber and W. Fichtner. A 177 Mb/s VLSI Implementation of the International Data Encryption Standard. *IEEE Journal of Solid-State Circuits*, 29(3):303–307, March 1994.

[80] M. Wenk, L., Bruderer, C. Studer, and A. Burg. Area- and Throughput-Optimized VLSI Architecture of Sphere Decoding. In *Proc. of 18th IEEE Int. Conf. on VLSI and Systems-on-Chip*, September 2010.

[81] Gilles Privat and Alain D. Wittmann. Pipelined recursive filter architectures for subband image coding. *Integration, the VLSI journal*, 14(3):361–379, February 1993.

[82] Peter B. Denyer and David Renshaw. *VLSI Signal Processing: A Bit-Serial Approach*. Addison-Wesley Publishing Company, Wokingham, England, 1985.

[83] Stewart G. Smith and Peter B. Denyer. *Serial-Data Computation*. Kluwer Academic Publishers, Boston, 1988.

[84] Stanley A. White. Applications of Distributed Arithmetic to Digital Signal Processing: A Tutorial Review. *IEEE Acoustics, Speech, and Signal Processing Magazine*, 6(3):4–19, July 1989.

[85] Distributed Arithmetic Laplacian Filter. XCELL No. 20, 1996. Xilinx Inc.

[86] Les Mintzer. FIR Filters with Field-Programmable Gate Arrays. *Journal of VLSI Signal Processing*, 6(2):119–127, August 1993.

[87] Kyung-Saeng Kim and Kwyro Lee. Low-Power and Area-Efficient FIR Filter Implementation Suitable for Multiple Taps. *IEEE Transactions on Very Large Scale Integration (VLSI) Systems*, 11(1):150–153, February 2003.

[88] G. Fettweis, L. Thiele, and H. Meyr. Algorithm transformations for unlimited parallelism. In *Proc. of the International Symposium on Circuits and Systems*. IEEE, New Orleans, 1990.

[89] Bernard Carré. *Graphs and Networks*. Clarendon Press, Oxford, 1979.

[90] Bernard Sklar. How I Learned to Love the Trellis. *IEEE Signal Processing Magazine*, 20(3):87–102, May 2003.

[91] Irfan Habib, Özgün Paker, and Sergei Sawitzki. Design Space Exploration of Hard-Decision Viterbi Decoding: Algorithm and VLSI Implementation. *IEEE Transactions on Very Large Scale Integration (VLSI) Systems*, 18(5):689–696, May 2010.

[92] Marc Biver and Hubert Kaeslin and Carlo Tommasini. In-Place Updating of Path Metrics in Viterbi Decoders. *IEEE Journal of Solid-State Circuits*, 24(4):1158–1160, August 1989.

[93] Alan Allan et al. 2001 Technology Roadmap for Semiconductors. *Computer*, 35(1):42–53, January 2002.

[94] Mark L. Chang and Scott Hauck. Précis: A Usercentric Word-Length Optimization Tool. *IEEE Design & Test of Computers*, 22(4):349–361, July/August 2005.

[95] Frank K. Gürkaynak. *GALS System Design: Side-Channel-Attack-Secure Cryptographic Accelerators*. PhD thesis, ETH, Zürich, 2006.

[96] Peter J. Ashenden and Jim Lewis. *VHDL-2008, Just the New Stuff*. Morgan Kaufmann Publishers, San Francisco, 2008.

[97] Stuart Sutherland, Don Mills, and Chris Spear. (More) Standard Gotchas, subtleties in the Verilog and SystemVerilog standards that every engineer should know! In *Synopsys User Group Conference Proc. (SNUG 2006/7)*, Boston/San Jose, 2006/7. http://www.sutherland-hdl.com/papers/2006-SNUG-Boston_standard_gotchas_paper.pdf and http://www.sutherland-hdl.com/papers/2007-SNUG-SanJose_gotcha_again_paper.pdf.

[98] IEEE, New York. *IEEE Standard for SystemVerilog - Unified Hardware Design, Specification, and Verification Language*, 2013. (IEEE Standard 1800-2012).

[99] David F. Bacon, Rodric Rabbah, and Sunil Shukla. FPGA Programming for the Masses. *Communications of the ACM*, 56(4):56–63, April 2013.

[100] Mark Zwolinski. *Digital System Design with SystemVerilog*. Addison-Wesley, Upper Saddle River, NJ, 2010.

[101] Stuart Sutherland, Simon Davidmann, and Peter Flake. *SystemVerilog for Design*. Springer, New York, 2nd edition, 2006.

[102] Clifford E. Cummings and Arturo Salz. SystemVerilog Event Regions, Race Avoidance & Guidelines. In *Synopsys User Group Conference Proc. (SNUG 2006)*, pages 1–42, Boston, 2006. http://www.sunburst-design.com/papers/.

[103] Chris Spear. *SystemVerilog for Verification*. Springer, New York, 2nd edition, 2010.

[104] Dirkjan Jongeneel and Ralph H.J.W. Otten. Technology Mapping for Area and Speed. *Integration, the VLSI Journal*, 29(1):45–66, March 2000.

[105] OpAr, Notes and Exercises of Arithmetics (with Java applets). http://users-tima.imag.fr/cis/guyot/Cours/Oparithm/english/Op_Ar2.htm, retrieved 28.2.14.

[106] Stephen Bailey. Comparison of VHDL, Verilog and SystemVerilog, 2006. http://boydtechinc.com/btf/archive/att-1977/01-LanguageWhitePaper.pdf.

[107] Clifford E. Cummings. SystemVerilog - Is This The Merging of Verilog & VHDL? In *Synopsys User Group Conference Proc. (SNUG 2003)*, pages 1–22, Boston, 2003. http://www.sunburst-design.com/papers/.

[108] IEEE, New York. *IEEE Standard VHDL Language Reference Manual*, 2009. (IEEE Standard 1076-2008).

[109] IEEE, New York. *IEEE Standard VHDL Language Reference Manual*, 2002. (IEEE Standard 1076-2002).

[110] Reto Zimmermann. VHDL AMS Syntax (IEEE Standard 1076.1-1999). http://dz.ee.ethz.ch/support/ic/vhdl/vhdlams_syntax.html.

[111] Reto Zimmermann. VHDL Syntax (IEEE Standard 1076-1993). http://dz.ee.ethz.ch/support/ic/vhdl/vhdl93_syntax.html.

[112] Jayaram Bhasker. *A Guide to VHDL Syntax*. Prentice Hall, Englewood Cliffs NJ, 1995.

[113] Volnei A. Pedroni. *Circuit Design and Simulation with VHDL*. MIT Press, Cambridge MA, 2010.

[114] Jürgen Reichardt and Bernd Schwarz. *VHDL-Synthese, Entwurf digitaler Schaltungen und Systeme*. Oldenbourg Wissenschaftsverlag, München, 5th edition, 2009.

[115] Peter J. Ashenden. *The Designer's Guide to VHDL*. Morgan Kaufmann Publishers, San Francisco, 3rd edition, 2008.

[116] Pong P. Chu. *RTL Hardware Design Using VHDL, Coding for Efficiency, Portability and Scalability*. Wiley-Interscience, Hoboken NJ, 2006.

[117] Paul Molitor and Jörg Ritter. *VHDL, eine Einführung*. Pearson Studium, München, 2004.

[118] Peter J. Ashenden, Gregory D. Peterson, and Darrell A. Teegarden. *The System Designer's Guide to VHDL-AMS*. Morgan Kaufmann, San Francisco, 2003.

[119] Sudhakar Yalamanchili. *Introductory VHDL: From Simulation to Synthesis*. Prentice Hall, Upper Saddle River NJ, 2001.

[120] Ulrich Heinkel et al. *The VHDL Reference*. John Wiley & Sons, Chichester, GB, 2000.

[121] Mark Zwolinski. *Digital System Design with VHDL*. Prentice Hall, Harlow, GB, 2000.

[122] K.C. Chang. *Digital Systems Design with VHDL and Synthesis, An Integrated Approach*. IEEE Computer Society Press, Los Alamitos CA, 1999.

[123] K.C. Chang. *Digital Design and Modeling with VHDL and Synthesis*. IEEE Computer Society Press, Los Alamitos CA, 1997.

[124] Zainalabedin Navabi. *VHDL Analysis and Modeling of Digital Systems*. McGraw-Hill, New York, 2nd edition, 1998.

[125] Jayaram Bhasker. *A VHDL Synthesis Primer*. Star Galaxy Publishing, Allentown PA, 1996.

[126] Janick Bergeron. *Writing Testbenches, functional verification of HDL Models*. Kluwer Academic Publishers, Boston, 2000.

[127] Klaus Lagemann. The Hamburg VHDL Archive. http://tams-www.informatik.uni-hamburg.de/vhdl, retrieved 27.2.2014.

[128] Ray Salemi. (FPGA Simulation) A SystemVerilog Primer for VHDL Engineers. Webpage, 2009. http://www.fpgasimulation.com.

[129] Janick Bergeron. *Writing Testbenches using SystemVerilog*. Springer, New York, 2006.

[130] Aleksandar Milenković. (System)Verilog Tutorial. http://www.ece.uah.edu/ milenka/npage/-data/cpe527/Lecture_Notes/verilog_synth.pdf, retrieved 27.2.14.

[131] Doulos. SystemVerilog Tutorials. http://www.doulos.com/knowhow/sysverilog/tutorial/.

[132] Peter J. Ashenden and Philip A. Wilsey. Protected Shared Variables in VHDL: IEEE Standard 1076a. *IEEE Design & Test of Computers*, 16(4):74–83, October/November/December 1999.

[133] Alain Vachoux. Analog and Mixed-Signal Extensions to VHDL. *Analog Integrated Circuits and Signal Processing*, 16(2):97–112, June 1998. (IEEE Standard 1076.1).

[134] François Pêcheux, Christophe Lallement, and Alain Vachoux. VHDL-AMS and Verilog-AMS as Alternative Hardware Description Languages for Efficient Modeling of Multidiscipline Systems. *IEEE Transactions on Computer-Aided Design*, 24(2):204–225, February 2005.

[135] Michael Keating and Pierre Bricaud. *Reuse Methodology Manual for Systems-on-a-Chip Designs*. Kluwer Academic Publishers, Boston, 1998.

[136] Peter Sinander. *VHDL Modelling Guidelines*. European Spage Agency (ESA estec), 1994. http://www.esa.int/TEC/Microelectronics/SEMS7EV681F_0.html.

[137] NXP. *CoReUse and Qcore*. e-books and specialized EDA tools. IPextreme, 2008. http://www.ip-extreme.com/coreuse.shtml, retrieved 27.1.14.

[138] Michael C. McFarland. Formal Verification of Sequential Hardware: A Tutorial. *IEEE Transactions on Computer-Aided Design*, 12(5):633–654, May 1993.

[139] Niklaus Wirth. *Compilerbau*. B. G. Teubner, Stuttgart, 1977.

[140] Sameh Asaad et al. A cycle-accurate, cycle-reproducible multi-FPGA system for accelerating multi-core processor simulation. In *Proceedings of the ACM/SIGDA international symposium on Field Programmable Gate Arrays (FPGA '12)*, 2012. DOI: 10.1145/2145694.2145720.

[141] J. L. Lions et al. Ariane 5 Flight 501 Failure. Report by the Inquiry Board, European Space Agency, 1996.

[142] Dick Price. Pentium FDIV flaw - lessons learned. *IEEE Micro*, 15(2):86–88, April 1995.

[143] David Goldberg. Computer Arithmetic. In David A. Patterson and John L. Hennessy, editors, *Computer Architecture, a Quantitative Approach*. Morgan Kaufmann Publishers, San Mateo CA, second edition, 1996.

[144] H. P. Sharangpani and M. L. Barton. Statistical Analysis of Floating Point Flaw in the Pentium Processor. Technical bulletin, Intel Corporation, 1994.

[145] Jim Lewis. Open Source VHDL Verification Methodology. http://osvvm.org/ and also http://www.doulos.com/knowhow/vhdl_designers_guide/OSVVM/.

[146] Hubert Kaeslin. *Digital Integrated Circuit Design, from VLSI Architectures to CMOS Fabrication*. Cambridge University Press, Cambridge UK, 2008.

[147] Brian Bailey et al. *TLM-Driven Design and Verification Methodology*. Cadence Design Systems Inc., San Jose CA, 2010.

[148] Harry Foster. Response checkers, monitors, and assertions. In Dhiraj K. Pradhan and Ian G. Harris, editors, *Practical Design Verification*. Cambridge University Press, Cambridge UK, 2009. assertion based verification.

[149] Gregg D. Lahti. Test Benches: The Dark Side of IP Reuse. In Synopsys User Group, editor, *SNUG 2000*. Synopsys, San Jose, 2000.

[150] Accellera. Universal Verification Methodology. http://www.accellera.org/community/uvm/ and also http://www.doulos.com/knowhow/sysverilog/uvm/tutorial_0/.

[151] Maher N. Mneimneh and Karem A. Sakallah. Principles of Sequential Equivalence Checking. *IEEE Design & Test of Computers*, 22(3):248–257, May/June 2005.

[152] E. Clarke, O. Grumberg, and D. Peled. *Model Checking*. MIT Press, Cambridge MA, 2000.

[153] Christoph Kern and Mark R. Greenstreet. Formal verification in hardware design: A survey. *ACM Transaction on Design Automation of Electronic Systems*, 4(2):123–193, April 1999.

[154] Farn Wang. Formal verification of timed systems: A survey and perspective. *Proceedings of the IEEE*, 92(8):1283–1305, August 2004.

[155] Steven M. Nowick and Montek Singh. High-Performance Asynchronous Pipelines: An Overview. *IEEE Design & Test of Computers*, pages 8–22, September/October 2011.

[156] Alain J. Martin and Mika Nyström. Asynchronous Techniques for System-on-Chip Design. *Proceedings of the IEEE*, 94(6):1089–1120, June 2006.

[157] Jens Sparsø and Steve Furber. *Principles of Asynchronous Circuit Design, A Systems Perspective*. Kluwer Academic Publishers, Boston, MA, 2001.

[158] Scott Hauck. Asynchronous Design Methodologies: An Overview. *Proceedings of the IEEE*, 83(1):69–93, January 1995.

[159] Peter Alfke. Just Say NO to Asynchronous Design. In *User Guide and Tutorials*. Xilinx Inc., San Jose, CA, 1991.

[160] Steve Knapp. KISS those asynchronous-logic problems good-bye. *Personal Engineering and Instrumentation News*, pages 53–55, November 1997. http://www.fpga-site.com/kiss.html, retrieved 27.2.14.

[161] Kees van Berkel et al. Asynchronous Does Not Imply Low Power, But ... In Anantha Chandrakasan and Robert Brodersen, editors, *Low-Power CMOS Design*, pages 227–232. IEEE Press, Piscataway NJ, 1998.

[162] Alex Kondratyev and Kelvin Lwin. Design of Asynchronous Circuits Using Synchronous CAD Tools. *IEEE Design & Test of Computers*, 19(4):107–117, July/August 2002. (incl. a note by Alain J. Martin on practical asynchronous circuits).

[163] Kenneth S. Stevens, Pankaj Golani, and Peter A. Beerel. Energy and Performance Models for Synchronous and Asynchronous Communication. *IEEE Transactions on Very Large Scale Integration (VLSI) Systems*, 19(3):369–382, March 2011.

[164] Paul Teehan, Mark Greenstreet, and Guy Lemieux. A Survey and Taxonomy of GALS Design Styles. *IEEE Design & Test of Computers*, 24(5):418–428, September/October 2007.

[165] Frank Gürkaynak, Stephan Oetiker, Hubert Kaeslin, Norbert Felber, and Wolfgang Fichtner. GALS at ETH Zurich: Success or Failure? In *Proceedings of the 12th IEEE International Symposium on Asynchronous Circuits and Systems (ASYNC 2006)*, Grenoble, March 2006.

[166] Kenneth Y. Yun and Ryan P. Donohue. Plausible Clocking: A First Step Toward Heterogeneous Systems. In *Proceedings of the ICCD-96*, pages 118–123, 1996.

[167] Fenghao Mu and Christer Svensson. Self-tested self-synchronization circuit for mesochronous clocking. *IEEE Transactions on Circuits and Systems II: Analog and Digital Signal Processing*, 48(2):129–140, February 2001.

[168] Ingemar Söderquist. Globally Updated Mesochronous Design Style. *IEEE Journal of Solid-State Circuits*, 38(7):1242–1249, July 1993.

[169] Miloš Krstić, Eckhard Grass, Pascal Vivet, and Frank Gürkaynak. Globally Asynchronous, Locally Synchronous Circuits: Overview and Outlook. *IEEE Design & Test of Computers*, 24(5):430–441, September/October 2007.

[170] Andreas Burg, Frank K. Gürkaynak, Hubert Kaeslin and Wolfgang Fichtner. Variable Delay Ripple Carry Adder with Carry Chain Interrupt Detection. In *Proc. of the IEEE International Symposium on Circuits and Systems*, Bangkok, May 2003. IEEE.

[171] James E. Buchanan. *BiCMOS/CMOS Systems Design*. McGraw-Hill, New York, 1991.

[172] Kai Hwang. *Computer Arithmetics*. John Wiley & Sons, New York, 1979.

[173] Reto Zimmermann and Rod Whitby. Emacs VHDL Mode Home Page. http://www.iis.ee.ethz.ch/~zimmi/emacs/vhdl-mode.html, retrieved 27.2.14.

[174] Jose Luis Neves and Eby G. Friedman. Buffered Clock Tree Synthesis with Non-Zero Clock Skew Scheduling for Increased Tolerance to Process Parameter Variations. *Journal of VLSI Signal Processing*, 16(2/3):149–161, June/July 1997.

[175] Xun Liu, Marios C. Papaefthymiou, and Eby G. Friedman. Retiming and Clock Scheduling for Digital Circuit Optimization. *IEEE Transactions on Computer-Aided Design*, 21(2):184–203, February 2002.

[176] Jeng-Liang Tsai et al. Yield-Driven False-Path-Aware Clock Skew Scheduling. *IEEE Design & Test of Computers*, 22(3):214–222, May/June 2005.

[177] Tim Horel and Gary Lauterbach. UltraSparc-III. *IEEE Micro*, 19(3):73–85, May/June 1999.

[178] Antonio G. M. Strollo, Ettore Napoli, and Carlo Cimino. Analysis of Power Dissipation in Double Edge-Triggered Flip-Flops. *IEEE Transactions on Very Large Scale Integration (VLSI) Systems*, 8(5):624–629, October 2000.

[179] Shi-Zheng Eric Lin, Chieh Changfan, Yu-Chin Hsu, and Fur-Shing Tsai. Optimal Time Borrowing Analysis and Timing Budgeting Optimization for Latch-Based Design. *ACM Transactions on Design Automation of Electronic Systems*, 7(1):217–230, January 2002.

[180] James J. Engel et al. Design methodology for IBM ASIC products. *IBM Journal of Research and Development*, 40(4):387–406, July 1996.

[181] J. D. Warnock. The Circuit and Physical Design of the POWER4 Microprocessor. *IBM Journal of Research and Development*, 46(1), 2002. DOI: 10.1147/rd.461.0027.

[182] Derek C. Wong, Giovanni De Micheli, Michael J. Flynn, and Robert E. Huston. A Bipolar Population Counter Using Wave Pipelining to Achieve $2.5\times$ Normal Clock Frequency. *IEEE Journal of Solid-State Circuits*, 27(5):745–753, May 1992.

[183] Wayne P. Burleson, Maciej J. Ciesielski, Fabian Klass, and Wentai Liu. Wave Pipelining: A Tutorial and Research Survey. *IEEE Transactions on Very Large Scale Integration (VLSI) Systems*, 6(3):464–474, September 1998.

[184] Kenneth D. Wagner. Clock System Design. *IEEE Design & Test of Computers*, pages 9–27, October 1988.

[185] William J. Dally and John W. Poulton. *Digital Systems Engineering*. Cambridge University Press, Cambridge UK, 2008.

[186] Daniel Dobberpuhl. A 200MHz 64Bit Dual Issue CMOS Microprocessor. In *Proceedings of the International Solid-State Circuits Conference*, San Francisco, February 1992. IEEE. (DEC 21064 alpha chip).

[187] Eric S. Fetzer. Using Adaptive Circuits to Mitigate Process Variations in Microprocessor Design. *IEEE Design & Test of Computers*, 23(6):438–451, November/December 2006. (Intel itanium 2 Montecito).

[188] Phillip J. Restle et al. A Clock Distribution Network for Microprocessors. *IEEE Journal of Solid-State Circuits*, 36(5):792–799, May 2001.

[189] Ana Sonia Leon et al. A Power-Efficient High-Throughput 32-Thread SPARC Processor. *IEEE Journal of Solid-State Circuits*, 42(1):7–16, January 2007. (Sun Niagara).

[190] D. Pham et al. The Design and Implementation of a First-Generation CELL Processor. In *International Solid-State Circuits Conference Digest of Technical Papers*, 2005.

[191] Krishnaswamy Nagaraj et al. Architectures and Circuit Techniques for Multi-Purpose Digital Phase Lock Loops. *IEEE Transactions of Circuits and Systems I: Regular Papers*, 60(3):517–528, March 2013.

[192] John Rogers, Calvin Plett, and Foster Dai. *Integrated Circuit Design for High-Speed Frequency Synthesis*. Artech House, Norwood MA, 2006.

[193] R. Jacob Baker. *CMOS Circuit Design, Layout and Simulation*. IEEE Press, Piscataway NJ, 2005.

[194] Chih-Kong Ken Yang. Delay-Locked Loops - An Overview. In Behzad B. Razavi, editor, *Phase-Locking in High-Performance Systems*, pages 13–22. IEEE Press, Piscataway NJ, 2003.

[195] Richard C. Walker. Designing Bang-Bang PLLs for Clock and Data Recovery in Serial Data Transmission Systems. In Behzad B. Razavi, editor, *Phase-Locking in High-Performance Systems*, pages 34–55. IEEE Press, Piscataway NJ, 2003.

[196] Norman Dodel and Heinrich Klar. 10Gb/s Bang-Bang Clock and Data Recovery (CDR) for optical transmission systems. *Advances in Radio Science*, 3:293–297, 2005.

[197] Marco Zanuso et al. Noise Analysis and Minimization in Bang-Bang Digital PLLs. *IEEE Transactions on Circuits and Systems II: Express Briefs*, 56(11):835–839, November 2009.

[198] Lin Zhang et al. A Low-Power Clock Distribution Scheme for High-Performance Microprocessors. *IEEE Transactions on Very Large Scale Integration (VLSI) Systems*, 16(9):1251–1256, September 2008.

[199] Bruno W. Garlepp et al. A Portable Digital DLL for High-Speed CMOS Circuits. *IEEE Journal of Solid-State Circuits*, 34(5):632–644, May 1999.

[200] Simon Tam et al. Clock Generation and Distribution for the First IA-64 Microprocessor. *IEEE Journal of Solid-State Circuits*, 35(11):1545–1552, November 1990.

[201] Nasser Kurd et al. Next Generation Intel Core Micro-Architecture (Nehalem) Clocking. *IEEE Journal of Solid-State Circuits*, 44(4):1121–1129, April 2009.

[202] Masafumi Nogawa and Yusuke Ohtomo. A Data-Transition Look-Ahead DFF Circuit for Statistical Reduction in Power Consumption. *IEEE Journal of Solid-State Circuits*, 33(5):702–706, May 1998.

[203] Ad M. G. Peeters. *Single-Rail Handshake Circuits*. PhD thesis, Eindhoven University of Technology, Eindhoven, 1996.

[204] Clifford E. Cummings. Simulation and Synthesis Techniques for Asynchronous FIFO Design. In *Synopsys User Group Conference Proc. (SNUG 2002)*, pages 1–23, San Jose, 2002. http://www.sunburst-design.com/papers/CummingsSNUG2002SJ_FIFO1.pdf.

[205] Ran Ginosar. Metastability and Synchronizers: A Tutorial. *IEEE Design & Test of Computers*, pages 23–35, September/October 2011. In fig.11b right, add a third flip-flop in front of the two that synchronize the Ack signal entering the Sender.

[206] Clifford E. Cummings and Peter Alfke. Simulation and Synthesis Techniques for Asynchronous FIFO Design with Asynchronous Pointer Comparisons. In *Synopsys User Group Conference Proc. (SNUG 2002)*, pages 1–18, San Jose, 2002. http://www.sunburst-design.com/papers/CummingsSNUG2002SJ_FIFO2.pdf.

[207] Andrea Pfister. *Metastability in Digital Circuits with Emphasis on CMOS Technology*. PhD thesis, ETH Zurich, Zurich, 1989.

[208] Jens U. Horstmann, Hans W. Eichel, and Robert L. Coates. Metastability Behavior of CMOS ASIC Flip-Flops in Theory and Test. *IEEE Journal on Solid State Circuits*, 24(1):146–157, February 1989.

[209] Peter Alfke. *Metastability Delay and Mean Time Between Failure in Virtex-II Pro FFs*. Xilinx, San Jose CA, 2002.

[210] Takahiko Kozaki et al. A 156-Mb/s Interface CMOS LSI for ATM Switching Systems. *IEICE Transactions on Communications*, E76-B(6):684–693, June 1993.

[211] Ran Ginosar. Fourteen Ways to Fool Your Synchronizer. In *Proceedings Ninth IEEE International Symposium on Asynchronous Circuits and Systems (ASYNC 2003)*, pages 89–96, Vancouver, May 2003.

[212] J. Juan-Chico, M.J. Bellido, A.J. Acosta, M. Valencia, and J.L. Huertas. Analysis of Metastable Operation in a CMOS Dynamic D-Latch. *Analog Integrated Circuits and Signal Processing*, 14:143 – 157, 1997.

[213] Jun Zhou, David J. Kinniment, Charles E. Dike, Gordon Russell, and Alexandre Yakovlev. On-Chip Measurement of Deep Metastability in Synchronizers. *IEEE Journal on Solid State Circuits*, 43(2):550–557, February 2008.

[214] Suk-Jin Kim, Jeong-Gun Lee, and Kiseon Kim. A Parallel Flop Synchronizer and the Handshake Interface for Bridging Asynchronous Domains. *IEICE Transactions on Fundamental of Electronics, Communications, and Computer Science*, E87-A(12):3166–3173, December 2004.

[215] Lindsay Kleeman and Antonio Cantoni. Metastable Behavior in Digital Systems. *IEEE Design & Test of Computers*, 4(6):4–19, December 1987.

[216] David J. Kinniment, Alexandre Bystrov, and Alex V. Yakovlev. Synchronization Circuit Performance. *IEEE Journal on Solid State Circuits*, 37(2):202–209, February 2002.

[217] Thaddeus J. Gabara, Gregory J. Cyr, and Charles E. Stroud. Metastability of CMOS Master/Slave Flip-Flops. *IEEE Transactions of Circuits and Systems II: Analog and Digital Signal Processing*, 39(10):734–740, October 1992.

[218] Saurabh Verma and Ashima S. Dabare. Understanding Clock Domain Crossing Issues, retrieved 21.9.2012. http://www.eetimes.com/design/eda-design/4018520/Understanding-Clock-Domain-Crossing-Issues.

[219] Zvi Kohavi and Niraj K. Jha. *Switching and Finite Automata Theory*. Cambridge University Press, Cambridge UK, 3rd edition, 2009.

[220] Jaakko T. Astola and Radomir S. Stankovic. *Fundamentals of Switching Theory and Logic Design*. Springer, Dordrecht, 2006.

[221] Edward J. McCluskey. *Logic Design Principles*. Prentice-Hall, Englewood Cliffs, NJ, 1986.

[222] Anil K. Maini. *Digital Electronics, Principles, Devices and Applications*. John Wiley & Sons, Chichester, 2007.

[223] John F. Wakerly. *Digital Design, Principles and Practice*. Pearson Prentice Hall, Upper Saddle River NJ, 2006.

[224] Mike Tooley. *Electronic Circuits, Fundamentals and Applications*. Newnes, Oxford UK, 2006.

[225] Randy H. Katz and Gaetano Borriello. *Contemporary Logic Design*. Pearson Prentice Hall, Upper Saddle River NJ, 2nd edition, 2005.

[226] Thomas L. Floyd. *Digital Fundamentals*. Prentice-Hall International Inc., London, 10th edition, 2009.

[227] Robert K. Brayton, Gary D. Hachtel, Curtis T. McMullen, and Alberto L. Sangiovanni-Vincentelli. *Logic Minimization Algorithms for VLSI Synthesis*. Kluwer Academic Publishers, Boston, 1984.

[228] Maitham Shams, Jo C. Ebergren, and Mohamed I. Elmasry. Modeling and Comparing CMOS Implementations of the C-Element. *IEEE Transactions on Very Large Scale Integration (VLSI) Systems*, 6(4):563–567, December 1998.

[229] Teresa H. Meng. *Synchronization Design for Digital Systems*. Kluwer Academic Publishers, Boston, MA, 1991.

[230] Daniel I. A. Cohen. *Introduction to Computer Theory*. John Wiley & Sons, New York, 1986.

[231] Douglas Lewin and David Protheroe. *Design of Logic Systems*. Chapman & Hall, London, 1992.

[232] David Harel. *Algorithmics*. Addison-Wesley, Reading MA, 3rd edition, 2004.

[233] David Harel. Statecharts: a Visual Formalism for Complex Systems. *Science of Computer Programming*, 8(3):231–274, June 1987.

[234] Edward J. McCluskey and Stephen H. Unger. A note on the number of internal variable assignments for sequential switching circuits. *IRE Transactions on Electronic Computers*, EC-8:439–440, December 1959.

[235] Davis R. Lide, editor. *CRC Handbook of Chemistry and Physics*. CRC Press, Boca Raton FL, 90th edition, 2009.

[236] S.V. Morozov et al. Giant Intrinsic Carrier Mobilities in Graphene and its Bilayer. *Physical Review Letters*, 100(1), 2008.

[237] B. Radisavljevic, A. Radenovic, J. Brivio, V. Giacometti, and A. Kis. Single-layer MoS_2 transistors. Nature Nanotechnology Online, January 2011. DOI: 10.1038/NNANO.2010.279.

[238] Alexander A. Balandin et al. Superior Thermal Conductivity of Single-Layer Graphene. *Nano Letters*, 8(3):902–907, February 2008.

[239] M. Levinshtein, S. Rumyantsev, and M. Shur, editors. *Handbook Series on Semiconductor Parameters, Vol.1 Si, Ge, C (Diamond), GaAs, GaP, GaSb, InAs, InP, InSb*. World Scientific Publishing, Singapore, 1996.

[240] M. Levinshtein, S. Rumyantsev, and M. Shur, editors. *Handbook Series on Semiconductor Parameters, Vol.2 Ternary and Quaternary III-V Compounds*. World Scientific Publishing, Singapore, 1999.

[241] Lester F. Eastman and Umesh K. Mishra. The Toughest Transistor Yet. *IEEE Spectrum*, 39(5):28–33, May 2002.

[242] Akio Sasaki, Yujiro Imamura, and Toshinori Takagi. Electron mobility and energy gap of $In_{0.53}Ga_{0.47}As$ on InP substrate. *Journal of Applied Physics*, 47(12), December 1976. http://dx.doi.org/10.1063/1.322570.

[243] Marc Heyns and Wilman Tsai. Ultimate Scaling of CMOS Logic Devices with Ge and III-V Materials. *MRS Bulletin*, 34(7):485–488, July 2009.

[244] Nagarajan Sridhar. Power Electronics in Automotive Applications, November 2013. http://www.ti.com/lit/wp/slyy052/slyy052.pdf, retrieved 18.2.2014.

[245] Andreas Hirsch. The era of carbon allotropes. *Nature Materials*, 9(11):868–869, November 2010.

[246] A.K. Geim and K.S. Novoselov. The rise of graphene. *Nature materials*, 6(3):183–191, March 2007.

[247] Andre K. Geim and Philip Kim. Carbon Wonderland. *Scientific American*, 298(4):68–75, April 2008.

[248] P.K. Vasudev and P.M. Zeitzoff. Si-ULSI with a Scaled-Down Future. *IEEE Circuits & Devices*, 14(2):19–29, March 1998. (special issue on NTRS '97).

[249] Yee-Chia Yeo, Tsu-Jae King, and Chenming Hu. MOSFET Gate Leakage Modeling and Selection Guide for Alternative Gate Dielectrics Based on Leakage Considerations. *IEEE Transactions on Electron Devices*, 50(4):1027–1035, April 2003. oxynitride, gate oxide leakage.

[250] J. Zhu, Z. G. Liu, and Y. R. Li. HfAlON films fabricated by pulsed laser ablation for high-k gate dielectric applications. *Materials Letters*, 59(7):821–825, March 2005. HfAlON, gate oxide.

[251] Christophe Rossel et al. Field-effect transistors with $SrHfO_3$ as gate oxide. *Applied Physics Letters*, 89(5):053506, May 2006. Hf, gate oxide.

[252] I.Z. Mitrovica et al. Electrical and structural properties of hafnium silicate thin films. *Microelectronics Reliability*, 47(4-5):645–648, April-May 2007. Hf, gate oxide.

[253] Bhola N. Pal, Bal Mukund Dhar, Kevin C. See, and Howard E. Katz. Solution-deposited sodium beta-alumina gate dielectrics for low-voltage and transparent field-effect transistors. *Nature Materials*, 8(11):898–903, November 2009. Intro by Hagen Klauk, p853.

[254] Rahul Suri et al. Energy-band alignment of Al_2O_3 and HfAlO gate dielectrics deposited by atomic layer deposition on 4HSiC. *Applied Physics Letters*, 96(4), 2010.

Index

Note: Page numbers in *italics* refer to figures and tables.

565

Printed in the United States
By Bookmasters